CATIA® Reference Guide

Paul Carman and Paul Tigwell

CATIA® Reference Guide
By Paul Carman and Paul Tigwell

Published by:
OnWord Press
2530 Camino Entrada
Santa Fe, NM 87505-4835 USA

Carol Leyba, Publisher
David Talbott, Acquisitions Editor
Barbara Kohl, Associate Editor
Daril Bentley, Senior Editor
Jean Cooksey, Development Editor
Cynthia Welch, Production Manager
Liz Bennie, Director of Marketing
Lauri Hogan, Marketing Services Manager
Lynne Egensteiner, Cover designer, Illustrator

Second Edition, 1998

SAN 694-0269

10 9 8 7 6 5 4 3 2 1

Printed in the United States of America

Library of Congress Cataloging-in-Publication Data

Carman, Paul, 1954-

CATIA reference guide / Paul Carman and Paul Tigwell.

p. cm.

Includes index.

ISBN 1-56690-155-3

1. Computer-aided engineering--Software. 2. CATIA (Computer file)

I. Tigwell, Paul, 1953- . II. Title.

TA345.C367 1998

670' .285'5369—dc21

98-9653

CIP

Trademarks

Warning and Disclaimer

About the Authors

Paul Carman, a designer with 20,000 hours of experience working with CATIA, obtained a higher national certificate in mechanical engineering from Reading College of Technology, and is a member of the Institute of Incorporated Mechanical Engineers. He joined the Joint European Torus (JET) nuclear fusion experiment at Culham (near Oxford, England) in 1976, and has been providing CATIA training since January 1987. At present, Paul serves as co-coordinator of CATIA training and problem solving at JET.

After obtaining a higher national certificate in mechanical engineering from Reading College Technology, **Paul Tigwell** worked in the drawing offices of various companies before joining JET in 1979. Paul has over 20,000 hours of experience working with CATIA, and has served as co-coordinator of CATIA training and problem solving at JET since 1995.

Acknowledgments

We wish to express our appreciation to the JET Project for choosing CATIA and thus giving us the opportunity to work with and provide training in CATIA; to Dr. Michael Pick at JET for the introduction to OnWord Press; to Henri Duquenoy (head of the JET Design Office) for his cooperation; to Mark Claxton, Dave Robson, and particularly Krishan Purahoo for system support; and to our good friend Tim Potter for our original CATIA training.

The Authors

Next, I would like to thank my family and friends for their support and encouragement during the writing of *INSIDE CATIA*, as well as my partner in this venture, Paul Tigwell ("Tiggers"), who helped to make it a pleasure. Finally, I wish to dedicate this book to my wife, Jacqueline ("Ginge").

Paul Carman

I would like to thank my wife, Sandra, and my boys, Dean and Alex, for their patience and encouragement during the writing of this book. Thanks must also go to my family and friends for their encouragement and also to the "other" Paul for being the ideal writing partner.

Paul Tigwell

Contents

Introduction xi
Command List xv
ANALYSIS in DRAW Mode 1
- NUMERIC 1
- RELATIVE 2
- LOGICAL 4

ANALYSIS in 2D SPACE Mode 5
- NUMERIC 5
- RELATIVE 7
- LOGICAL 7
- CURVE 8

ANALYSIS in 3D SPACE Mode 9
- NUMERIC 9
- RELATIVE 11
- LOGICAL 11
- CURVE 12
- INERTIA 13

AUXVIEW in DRAW Mode 14
- CREATE 14
- CHANGE 16
- MODIFY 16
- DELETE 20
- RENAME 20
- TRANSFER 21
- DTAILING 21

AUXVIEW2 22
- AUXVIEW2 Menu Structure 24
- USE 25
- DEFAULT 43
- UPD ALL 48

AXIS in DRAW Mode 49
- CREATE 49
- CHANGE 49
- SWAP 50
- INVERT 50
- FIXED 50
- UNFIXED 50
- RENAME 51

AXIS in 2D SPACE Mode 52
- CREATE 52
- CHANGE 53
- SWAP 53
- INVERT 53
- FIXED 54
- UNFIXED 54
- ANALYZE 54

AXIS in 3D SPACE Mode 56
- CREATE 56
- CHANGE 57
- SWAP 58
- INVERT 58
- FIXED 58
- UNFIXED 58
- ANALYZE 59

COMBIVU in DRAW Mode 60
- LINES 60
- SECTION 61
- COMBINE 61

CURVE2 in DRAW Mode 63
- CIRCLE 63
- ELLIPSE 66
- CONIC 67
- SPLINE 69
- PARALLEL 70
- APPROXIM 71
- CONNECT 71
- INVERT 72

CURVE2 in 2D SPACE Mode 73
- CIRCLE 73
- ELLIPSE 77
- CONIC 78
- PTS CST 79
- PARALLEL 80
- APPROXIM 81
- CONNECT 81
- TGT CONT 83
- CVT CONT 83
- INVERT 84

CURVE2 in 3D SPACE Mode 85
PARALLEL 85
CONNECT 86
PTS CST 88
APPROXIM 88
TGT CONT 89
CVT CONT 89
DEPTH 90
CIRCLE 91
HELIX 91
SPINE 92
INVERT 92
DETAIL in DRAW Mode 93
DITTO 94
COPY 96
MODIFY 96
EXPLODE 98
CREATE 98
DELETE 98
MANAGE 99
CHANGE 101
TRANSFER 102
DETAIL in 3D SPACE Mode 103
DITTO 103
COPY 104
MODIFY 105
EXPLODE 106
CREATE 107
DELETE 107
MANAGE 108
CHANGE 110
TRANSFER 110
DIMENS2 in DRAW Mode 112
DIMENS2 Menu Structure 113
CREATE 117
MODIFY 127
ADD 134
DELETE 134
MANAGE 134
DRAFT in DRAW Mode 137
CHANGE 137
CREATE 137
DELETE 138
RENAME 138
TRANSFER 138
COPY 138
DRW>SPC in SPACE Mode 140
CREATE 140
DRW STD 141
ANNOTAT 142
CUSTOM 148
PATTERN 153
ERASE in DRAW and SPACE Modes . . 154
ERASE 154
PACK 154
NO SHOW/SHOW 155
NO PICK/PICK 155
FILE in DRAW and SPACE Modes 156
FILE 156
READ 157
WRITE 157
COPY 157
DELETE 158
MOVE 158
RENAME 159
CREATE 159
CALL SV 159
COMMENT 160
KEYBOARD 162
EXIT 164
GRAPHIC in DRAW Mode 165
MODIFY 166
VERIFY 168
ANALYZE 169
GRAPHIC in SPACE Mode 170
MOD GEN 171
MOD SPEC 173
MOD VISU 173
VERIFY 175
ANALYZE 176
GROUP in DRAW Mode 178
GRP_1, GRP_2, GRP_3 178
GROUP in SPACE Mode 182
GRP_1, GRP_2, GRP_3 182
IDENTIFY in DRAW and SPACE Modes 186
RENAME 186
UPDATE 187
RENUMBER 187

IMAGE in DRAW and SPACE Modes . 189
WINDOW 189
SCREEN 193
LOCAL TR 196
IUA in DRAW and SPACE Modes 197
FILE 197
EXECUTE 198
MODIFY 199
Using IUA Commands Transparently 199
KEEP in DRAW and SPACE Modes . . . 202
SELECT 202
KEEP 204
MERGE 205
LAYER in DRAW and SPACE Modes . 206
FILTER 206
LAYER 211
IDENTIFY 212
LIBRARY in DRAW and SPACE Modes 213
FILE 213
FAMILY 214
READ 214
WRITE 214
DELETE 215
MODIFY 215
UPDATE 215
LIMIT1 in DRAW Mode 217
RELIMIT 217
CORNER 218
MACHINE 219
BREAK 222
CONCATEN 222
EXTRAPOL 223
LIMIT1 in 2D SPACE Mode 224
RELIMIT 224
CORNER 225
MACHINE 226
BREAK 229
CONCATEN 230
EXTRAPOL 230
APPR CRV 230
LIMIT1 in 3D SPACE Mode 231
RELIMIT 231
CORNER 232
MACHINE 233
BREAK 235
CONCATEN 236
EXTRAPOL 236
APPR CRV 237
LINE in DRAW Mode 238
Vicinity Selection 238
Rubber Banding 239
Line Limitation 239
PT-PT 241
PARALLEL 243
HORIZONT or VERTICAL 244
NORMAL 246
MEDIAN 247
BISECT 248
ANGLE 248
COMPON 249
TANGENT 250
MEAN 252
MODIFY 252
GRID 254
LINE in 2D SPACE Mode 258
Vicinity Selection 258
Rubber Banding 259
Horizontal and Vertical 259
Line Limitation 259
PT-PT 261
PARALLEL 263
HORIZONT or VERTICAL 265
NORMAL 266
MEDIAN 267
BISECT 268
ANGLE 269
COMPON 270
TANGENT 271
MEAN 272
MODIFY 273
EDGE 274
LINE in 3D SPACE Mode 275
Vicinity Selection 275
Rubber Banding 276
Line Limitation 276
PT-PT 278
PARALLEL 279
NORMAL 280

INTERSEC ... 282
PROJECT ... 282
ANGLE ... 283
COMPON ... 284
EDGE ... 285
TANGENT ... 285
MEAN ... 286
POL EDGE ... 286
MODIFY ... 286
MARK UP in DRAW Mode ... 288
AXIS ... 288
ARROW ... 289
MERGE in DRAW and SPACE Modes ... 292
SELECT ... 292
MERGE ... 293
MODELS in DRAW and SPACE Modes 294
MANAGE ... 294
MODIFY ... 295
COPY ... 296
BREAKOUT ... 297
WORKAREA ... 297
PARAM3D in SPACE Mode ... 300
PROFILE ... 300
PRIMITIV ... 302
LINK ... 303
UNLINK ... 303
MODIFY ... 303
PARAMETER ... 304
TRANSFOR ... 305
DIMENSION ... 305
ANALYZE ... 308
DELETE ... 308
ADVANCED ... 309
PATTERN in DRAW Mode ... 311
AUTO ... 313
SELECT ... 314
REPLACE ... 315
VISUALTN ... 315
ANALYZE ... 317
UPDATE ... 317
PLANE in SPACE Mode ... 318
THROUGH ... 318
EQUATION ... 319
PARALLEL ... 319
NORMAL ... 320
ANGLE ... 321
ORIENTN ... 321
MEAN ... 321
PRL WINW ... 322
EDGES ... 322
SPACES ... 323
LIMITS ... 324
PLOT in DRAW and SPACE Modes ... 326
QUICK ... 327
FILE ... 332
DATABASE ... 338
POINT in DRAW Mode ... 339
VICINITY SELECTION ... 339
PROJ INT ... 339
PROJECT ... 341
COORD ... 342
LIMITS ... 344
SPACES ... 345
TANGENT ... 346
GRID ... 347
POINT in 2D SPACE Mode ... 351
VICINITY SELECTION ... 351
PROJ INT ... 351
PROJECT ... 352
COORD ... 354
LIMITS ... 356
SPACES ... 357
TANGENT ... 359
POINT in 3D SPACE Mode ... 360
VICINITY SELECTION ... 360
PROJ INT ... 360
PROJECT ... 361
COORD ... 362
LIMITS ... 363
SPACES ... 365
TANGENT ... 368
SETS in DRAW and SPACE Modes ... 369
CHANGE ... 369
CREATE ... 369
DELETE ... 369
TRANSFER ... 370
COPY ... 370
LINK ... 370

SHAPE in DRAW Mode 371
CREATE 371
OFFSET..................... 373
MODIFY 374
OPERATE 375
CLOSE 376
SOLIDE in SPACE Mode 377
CREATE..................... 378
OPERATN 389
ANALYZE 396
MODIFY 398
EXTRACT 405
UPDATE 407
PART EDITOR 407
SOLIDM in SPACE Mode 410
CREATE 411
OPERATN 417
ANALYZE 419
MODIFY 422
EXTRACT 429
VISU STD 429
UPDATE 430
RESTORE 430
SPC->DR2 in DRAW Mode 431
CUT 431
PROJECT 432
SPC->DRW in DRAW Mode 434
CUT 434
PROJECT 434
STANDARD in DRAW
and SPACE Modes 436
SPACE ELT 436
DRAW ELT 439
COLOR 439
MODEL 440
LINETYPE 442
SYMBOL in DRAW Mode 443
SYMBOL 443
COPY 446
MODIFY 446
EXPLODE 448
DEFINE 448
DELETE 449
MANAGE 449
EXTRACT 452
VANISH 452
TEXT in DRAW Mode 453
T.NODE 453
STANDARD 454
TEXT in SPACE Mode 457
CREATE 457
MODIFY 458
ERASE 458
MOVE 459
EDIT 459
SHOW 460
TEXTD2 in DRAW Mode 461
CREATE 461
MODIFY 476
ADD 479
DELETE 479
MANAGE 480
TRANSFOR in DRAW Mode 482
CREATE 482
APPLY 486
ERASE 487
ANALYZE 487
FREEHAND 488
UNDO 489
TRANSFOR in 2D SPACE Mode 490
CREATE 490
APPLY 494
STORE...................... 496
MANAGE 496
TRANSFOR in 3D SPACE Mode 498
CREATE 498
APPLY 504
STORE 505
MANAGE 506
UTILITY 508
CatUtil Icon 508
Utilities 509
Appendix A 511
Element Identifiers Used in CATIA . . 511
Appendix B 514
Keywords Used for Multiple Selection 514
Combining Multiple Selection
Keywords 516

Appendix C 518
CATIA General Commands 518
Appendix D 521
CATIA Pull-down Menus 521
File 521
Select 522
View 523
Filter 524
Options 525
Tools 525
Window 527
Help 527
Appendix E 529
Engineering Symbols 529
Appendix F 530
Fixed Menu and Dialog Area 530
Appendix G 534
Display and Manipulation Window . . 534
STD Option Menu 534
2D Option Menu 536
3D Option Menu 538
COL Option Menu 540
Index 543

Introduction

CATIA is produced by Dassault Systemes (Paris, France), and is considered to be among the most powerful CAD systems in use today. CATIA, a highly interactive 3D CAD/CAM system, is an acronym for computer aided three-dimensional integrated application. The software provides two-dimensional drafting and three-dimensional modeling facilities together with several design analysis tools and comprehensive manufacturing facilities.

The authors have used CATIA for ten years. In their opinion, the software is without rival. The JET Project at Culham, England, is testimony to CATIA's versatility. JET is the most successful nuclear fusion experimental machine in the world; the present configuration would not have been possible without CATIA. CATIA is also used extensively by some of the world's leading aerospace and automotive manufacturers.

Any CAD system is only as good as the person who uses it. With this in mind, the authors have attempted to provide reference works thus far unavailable. The *CATIA Reference Guide* can be used as a companion to *INSIDE CATIA* (OnWord Press, 1998), or as a standalone reference addition to CATIA users' libraries. All functions and options used or mentioned in *INSIDE CATIA* are covered in this reference.

Book Organization

An alphabetical listing of first, second, and third level options covered in this reference is located after the Contents section. Selected fourth level options are included as well. The options list includes the function and command sequence as appropriate.

A total of 61 functions are organized alphabetically. Options at all levels of the command hierarchy are discussed in each of the 61 parts.

The seven appendices provide information on element identifiers, keywords and multiple selection, general commands, pull-down menus, engineering symbols, fixed menu and dialog area, and the Display and Manipulation window.

Finally, the topical index is focused on answers to conceptual questions. For instance, assume that you do not know the name of the command(s) that would permit you to restore display attributes of SPACE elements visible in DRAW views. In this instance, you could consult the index by searching for "restore display attributes of SPACE elements visible in DRAW views." The index listing references the relevant function and provides the page number.

Typography Conventions and Function Section Presentation

Commands that can be selected at the first level of the CATIA command hierarchy are called *functions*. Functions covered in the reference are organized alphabetically, from ANALYSIS in DRAW mode to UTILITY. Commands selected at the second to fifth levels of the command hierarchy are called *options*, and occasionally suboptions.

Discussion of options within function sections is organized according to the command hierarchy. In other words, the option (and its respective suboptions) listed first after selecting the function is discussed first, the second is discussed next, and so on. For example, as seen in the tree diagram for the ANALYSIS in DRAW mode function, the options are listed in the following order: NUMERIC, RELATIVE, and LOGICAL. Subsequently, the discussion begins with NUMERIC and its suboptions, and proceeds through LOGICAL and its suboptions.

Every function section includes the information listed below.

- Cross-reference to *INSIDE CATIA*.
- Concise definition and usage
- Tree diagram depicting command hierarchy
- Prompts by option
- Actions required to achieve specific objectives by each option (and suboption)
- Illustrations of windows and results as appropriate
- Notes, tips, and warnings

Typical sample portions of a function section appear below.

Every section opens with the function name at the top of the page, followed by an INSIDE CATIA cross-reference, a brief description of the function, and a tree diagram of the option hierarchy.

PATTERN in DRAW Mode

Chapter 7

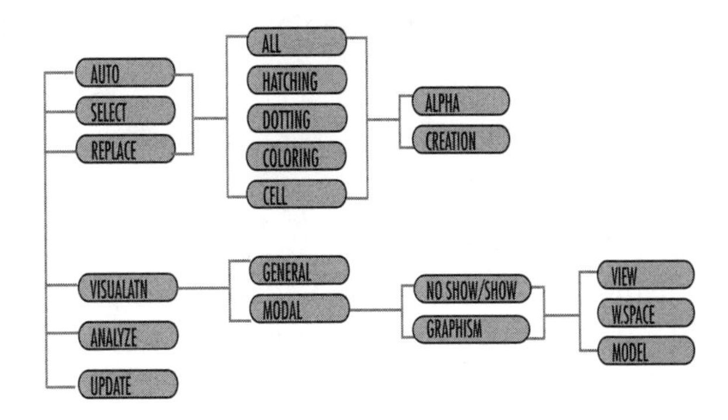

Option menu for PATTERN in DRAW mode.

As indicated above, explanatory material and illustrations may appear at this juncture and throughout the discussion of options. First level options (AUTO, SELECT, REPLACE, VISUALATN, ANALYZE, and UPDATE in this case) are introduced in the discussion in a large bold type underlined with a gray bar. A brief description of the first level option follows, along with the selection sequence.

Typically, the user can select from more than one option after choosing a first level option. Such choices are indicated in the Select line. If there are no second level options, the appropriate prompt appears below the Select line. Next, discussion of second level options is introduced by describing the action pertaining to the suboption in a solid italicized typeface, following by Select and Prompt lines.

VISUALTN

Visualize patterns and modify pattern visualization.

Select: PATTERN > VISUALTN | GENERAL or MODAL

Define general pattern visualization

Select: PATTERN > VISUALTN | GENERAL

Prompt: YES : NO SHOW PATTERN

Prompt: YES : SHOW PATTERN

Click on the YES button to visualize or hide all patterns already created. In the NO SHOW mode, no patterns will be visualized. In the SHOW mode, patterns will be visible or invisible according to how the selective display is defined by a shape. (See VISUALTN | MODAL | NO SHOW/SHOW.)

Of course, many CATIA second level options provide the opportunity to select still more options, and so on. Discussion of third and fourth level options is introduced by descriptions of the pertinent action or objective, and appears in a typeface specific to the option level. The Select sequence and Prompt lines are included. Appearing below is a sample of a third level option heading and associated discussion for the PATTERN function.

Transfer patterns into show or no show mode

Select: PATTERN > VISUALTN | MODAL | NO_SHOW or SHOW | VIEW or W.SPACE or MODEL

Prompt: SEL SHAP // YES : SWAP

- To transfer patterns into the NO SHOW mode (from SHOW mode), select pattern(s) to be transferred.
- To transfer patterns into the SHOW mode (from NO SHOW mode), click on the YES button to display patterns in NO SHOW mode and then select the pattern(s) to be transferred.
- If all patterns are to be transferred, select the ALL PATTERNS command button.

The book also includes notes, tips, and warnings throughout.

✓ **NOTE:** *Information on features, exceptions, and caveats that is not straightforward, immediately obvious, or intuitive appears in notes.*

➾ **TIP:** *Tips are the fruit of experience, and are aimed at saving you time and stress.*

✗ **WARNING:** *Warnings are intended to help you avoid committing yourself to unexpected or undesirable results.*

Command List

3 PTS, CURVE2 > CONIC in 2D SPACE mode, 78
3 PTS, CURVE2 > CONIC in DRAW mode, 67
5 PTS, CURVE2 > CONIC in 2D SPACE mode, 79
5 PTS, CURVE2 > CONIC in DRAW mode, 69
ABSOLUTE, ANALYSIS > INERTIA | COMBINE in 3D SPACE mode, 13
ABSOLUTE, ANALYSIS > NUMERIC | COMPUTE in 2D SPACE mode, 5
ABSOLUTE, ANALYSIS > NUMERIC | COMPUTE in 3D SPACE mode, 9
ABSOLUTE, ANALYSIS > NUMERIC | CREATE in 2D SPACE mode, 6
ADD NODE, TEXTD2 > MODIFY | LEADER | PATH in DRAW mode, 478
ADD PT, CURVE2 > PTS CST in 2D SPACE mode, 79
ADD PT, CURVE2 > PTS CST in 3D SPACE mode, 88
ADD, AUXVIEW2 > USE | BACK_PLN, 37
ADD, AUXVIEW2 > USE | BREAKOUT, 39
ADD, AUXVIEW2 > USE | CLIP, 38
ADD, AUXVIEW2 > USE | SECTION, 40
ADD, DIMENS > MODIFY | TEXT in DRAW mode, 132
ADD, DIMENS2 in DRAW mode, 134
ADD, MODELS > WORKAREA | MODIFY in DRAW and SPACE modes, 298
ADD, SHAPE > MODIFY in DRAW mode, 374
ADD, TEXTD2 > MODIFY | LEADER in DRAW mode, 477
ADD, TEXTD2 in DRAW mode, 479
ADD_CALLOUT, AUXVIEW2 > USE | CLIP, 38
ADD_CALLOUT, AUXVIEW2 > USE | PLANE, 34
ADD_CALLOUT, AUXVIEW2 > USE | SECTION, 40
ADVANCED, PARAM3D in SPACE mode, 309
ALIGNMENT, SOLIDM > ANALYZE | POSITN in SPACE mode, 422
ALL, PARAM3D > DELETE in SPACE mode, 309
ALL, PATTERN > AUTO in DRAW mode, 313
ALL, PATTERN > SELECT in DRAW mode, 314
ANALYSIS, AUXVIEW2 > USE | DIMENS, 43
ANALYZE, AXIS in 2D SPACE mode, 54
ANALYZE, AXIS in 3D SPACE mode, 59
ANALYZE, DETAIL > MANAGE in 3D SPACE mode, 108
ANALYZE, DETAIL > MANAGE in DRAW mode, 99
ANALYZE, GRAPHIC in DRAW mode, 169
ANALYZE, GRAPHIC in SPACE mode, 176
ANALYZE, LAYER > FILTER in DRAW and SPACE modes, 210
ANALYZE, LAYER > LAYER in DRAW and SPACE modes, 212
ANALYZE, MODELS > MANAGE in DRAW and SPACE modes, 294
ANALYZE, PARAM3D in SPACE mode, 308
ANALYZE, PATTERN in DRAW mode, 317
ANALYZE, SOLIDE in SPACE mode, 396
ANALYZE, SOLIDM > in SPACE mode, 419
ANALYZE, SYMBOL > MANAGE in DRAW mode, 449
ANALYZE, TRANSFOR > MANAGE in 2D SPACE mode, 496
ANALYZE, TRANSFOR > MANAGE in 3D SPACE mode, 506
ANALYZE, TRANSFOR in DRAW mode, 487
ANGLE, DIMENS > CREATE in DRAW mode, 121
ANGLE, LIMIT1 > MACHINE | CHAMFER in 2D SPACE mode, 227
ANGLE, LIMIT1 > MACHINE | CHAMFER in 3D SPACE mode, 233
ANGLE, LIMIT1 > MACHINE | CHAMFER in DRAW mode, 219
ANGLE, LINE in 2D SPACE mode, 269
ANGLE, LINE in 3D SPACE mode, 283
ANGLE, LINE in DRAW mode, 248
ANGLE, PLANE > EDGES in SPACE mode, 323
ANGLE, PLANE in SPACE mode, 321
ANGLES, TRANSFOR > CREATE in 3D SPACE mode. 503
ANNOTAT, DRW STD, 142
APPLICTN, KEEP > SELECT in DRAW and SPACE modes, 202

APPLY, LAYER > FILTER in DRAW and SPACE modes, 206
APPLY, MODELS > WORKAREA in DRAW and SPACE modes, 297
APPLY, TRANSFOR in 2D SPACE mode, 494
APPLY, TRANSFOR in 3D SPACE mode, 504
APPLY, TRANSFOR in DRAW mode, 486
APPR CRV, LIMIT1 in 2D SPACE mode, 230
APPR CRV, LIMIT1 in 3D SPACE mode, 237
APPROXIM, CURVE2 in 2D SPACE mode, 81
APPROXIM, CURVE2 in 3D SPACE mode, 88
APPROXIM, CURVE2 in DRAW mode, 71
APPROXIM, SOLIDE > EXTRACT in SPACE mode, 406
APPROXIM, SOLIDM > EXTRACT | CUT in SPACE mode, 429
APPROXIM, SPC->DR2 > CUT in DRAW mode, 431
APPROXIM, SPC->DR2 > PROJECT in DRAW mode, 432
ARROW, MARK UP in DRAW mode, 289
ASSEMBLE, SOLIDM > OPERATN in SPACE mode, 419
ASSIGN, PARAM3D > PARAMETER in SPACE mode, 304
ATTRIBUT, GROUP > GRP_1, 2, or 3 | EXCLUDE | CURRENT in DRAW mode, 180
ATTRIBUT, GROUP > GRP_1, 2, or 3 | EXCLUDE | W.SPACE in DRAW mode, 180
ATTRIBUT, GROUP > GRP_1, 2, or 3 | INCLUDE | CURRENT in DRAW mode, 180
ATTRIBUT, GROUP > GRP_1, 2, or 3 | INCLUDE | W.SPACE in DRAW mode, 180
ATTRIBUT, GROUP > GRP_1, 2, or 3 | EXCLUDE | CURRENT in SPACE mode, 185
ATTRIBUT, GROUP > GRP_1, 2, or 3 | EXCLUDE | W.SPACE in SPACE mode, 185
ATTRIBUT, GROUP > GRP_1, 2, or 3 | INCLUDE | CURRENT in SPACE mode, 185
ATTRIBUT, GROUP > GRP_1, 2, or 3 | INCLUDE | W.SPACE in SPACE mode, 185
AUTO, PATTERN in DRAW mode, 313
AUXILIARY VIEW, AUXVIEW2 > USE | VIEW | NEW, 28
AXIS, CURVE2 > ELLIPSE in 2D SPACE mode, 77
AXIS, CURVE2 > ELLIPSE in DRAW mode, 67
AXIS, MARK UP in DRAW mode, 288
BACK_PLN, AUXVIEW2 > USE, 37
BALLOON, TEXTD2 > CREATE in DRAW mode, 474
BISECT, LINE in 2D SPACE mode, 268
BISECT, LINE in DRAW mode, 248
BLNK LIN, DIMENS > MANAGE | VISUALTN in DRAW mode, 136
BLNK LIN, TEXTD2 > MANAGE | VISUALATN in DRAW mode, 481
BOOLE, LAYER > FILTER | APPLY in DRAW and SPACE modes, 207
BREAK, LIMIT1 in 2D SPACE mode, 229
BREAK, LIMIT1 in 3D SPACE mode, 235
BREAK, LIMIT1 in DRAW mode, 222
BREAKOUT, AUXVIEW2 > USE, 39
BREAKOUT, MODELS in DRAW and SPACE modes, 297
CALL SV, FILE in DRAW and SPACE modes, 159
CANONIC, SOLIDE > CREATE in SPACE mode, 378
CANONIC, SOLIDM > CREATE in SPACE mode, 411
CATALOG, SOLIDE > CREATE | FEATURE in SPACE mode, 388
CELL, PATTERN > AUTO in DRAW mode, 313
CELL, PATTERN > SELECT in DRAW mode, 314
CENTER, AUXVIEW > MODIFY | VIEW in DRAW mode, 19
CENTER, CURVE2 > CIRCLE | MULT-TGT in DRAW mode, 66
CENTER, CURVE2 > CIRCLE | MULT-TGT in SPACE mode, 76
CENTER, CURVE2 > ELLIPSE in 2D SPACE mode, 77
CENTER, CURVE2 > ELLIPSE in DRAW mode, 66
CENTER, IMAGE > WINDOW | MODIFY | PLANE in DRAW and SPACE modes, 190
CHAIN, POINT > COORD in 2D SPACE mode, 355
CHAIN, POINT > COORD in 3D SPACE mode, 363
CHAIN, POINT > COORD in DRAW mode, 343

CHAMFER, DIMENS2 > CREATE in DRAW mode, 126
CHAMFER, LIMIT1 > MACHINE in 2D SPACE mode, 226
CHAMFER, LIMIT1 > MACHINE in 3D SPACE mode, 233
CHAMFER, LIMIT1 > MACHINE in DRAW mode, 219
CHAMFER, SOLIDE > OPERATN in SPACE mode, 394
CHANGE, AUXVIEW in DRAW mode, 16
CHANGE, AXIS in 2D SPACE mode, 53
CHANGE, AXIS in 3D SPACE mode, 57
CHANGE, AXIS in DRAW mode, 49
CHANGE, DETAIL in 3D SPACE mode, 110
CHANGE, DETAIL in DRAW mode, 101
CHANGE, DIMENS > MODIFY | SYMBOL in DRAW mode, 131
CHANGE, DIMENS > MODIFY | TOL in DRAW mode, 131
CHANGE, DRAFT in DRAW mode, 137
CHANGE, SETS in DRAW and SPACE modes, 369
CHILDREN, ANALYSIS > LOGICAL in 2D SPACE mode, 5
CHILDREN, ANALYSIS > LOGICAL in 3D SPACE mode, 12
CHILDREN, PARAM3D > ANALYZE | RELATION in SPACE mode, 308
CHOOSE, GRAPHIC > MOD GEN in SPACE mode, 171
CHOOSE, GRAPHIC > MODIFY in DRAW mode, 166
CHOOSE, GRAPHIC > MOD SPEC in SPACE mode, 173
CHOOSE, GRAPHIC > MOD VISU in SPACE mode, 173
CHOOSE, GRAPHIC > VERIFY | CHOOSE in SPACE mode, 175
CHOOSE, GRAPHIC > VERIFY in DRAW mode, 168
CIR CNT, DIMENS > CREATE | DIAMETER in DRAW mode, 122
CIR CNT, DIMENS > CREATE | RADIUS in DRAW mode, 124
CIR TGT, DIMENS > CREATE | RADIUS in DRAW mode, 125
CIR TGT, DIMENS > CREATE | DIAMETER in DRAW mode, 123
CIRCLE, CURVE2 in 2D SPACE mode, 73
CIRCLE, CURVE2 in 3D SPACE mode, 91
CIRCLE, CURVE2 in DRAW mode, 63
CLIP, AUXVIEW2 > USE, 37
CLOSE, LIMIT1 > RELIMIT in 2D SPACE mode, 225
CLOSE, LIMIT1 > RELIMIT in 3D SPACE mode, 232
CLOSE, LIMIT1 > RELIMIT in DRAW mode, 218
CLOSE, SHAPE > CREATE | POLYGON in DRAW mode, 371
CLOSE, SHAPE in DRAW mode, 376
CLOSE, SOLIDE > CREATE | COMPLEX in SPACE mode, 416
COLOR, SOLIDE > MODIFY | DRESS UP in SPACE mode, 405
COLOR, SOLIDM > MODIFY | DRESS UP in SPACE mode, 428
COLOR, STANDARD in DRAW and SPACE modes, 439
COLORING, PATTERN > AUTO in DRAW mode, 313
COLORING, PATTERN > SELECT in DRAW mode, 314
COMBINE, ANALYSIS > INERTIA in 3D SPACE mode, 13
COMBINE, COMBIVU in DRAW mode, 61
COMBINE, TRANSFOR > CREATE in DRAW mode, 486
COMBINE, TRANSFOR > MANAGE in 2D SPACE mode, 497
COMBINE, TRANSFOR > MANAGE in 3D SPACE mode, 506
COMMENT, FILE in DRAW and SPACE modes, 160
COMP CRV, LIMIT1 > CONCATEN in 2D SPACE mode, 230
COMP CRV, LIMIT1 > CONCATEN in 3D SPACE mode, 236
COMPACT, DETAIL > COPY | LIBRARY in 3D SPACE mode, 104
COMPACT, DETAIL > COPY | MODEL in 3D SPACE mode, 104
COMPACT, DETAIL > COPY | LIBRARY in DRAW mode, 96
COMPACT, DETAIL > COPY | MODEL in DRAW mode, 94
COMPACT, DETAIL > DITTO | LIBRARY in 3D SPACE mode, 104
COMPACT, DETAIL > DITTO | LIBRARY in DRAW mode, 96
COMPACT, DETAIL > DITTO | MODEL in DRAW mode, 94

COMPACT, SYMBOL > SYMBOL | LIBRARY in DRAW mode, 445
COMPLETE, CURVE2 > ELLIPSE | AXIS in 2D SPACE mode, 77
COMPLETE, CURVE2 > ELLIPSE | AXIS in DRAW mode, 67
COMPLETE, CURVE2 > ELLIPSE | CENTER in 2D SPACE mode, 77
COMPLETE, CURVE2 > ELLIPSE | CENTER in DRAW mode, 66
COMPLEX, SOLIDE > CREATE in SPACE mode, 387
COMPLEX, SOLIDM > CREATE in SPACE mode, 415
COMPON, LINE in 2D SPACE mode, 270
COMPON, LINE in 3D SPACE mode, 284
COMPON, LINE in DRAW mode, 249
COMPUTE, ANALYSIS > INERTIA in 3D SPACE mode, 13
COMPUTE, ANALYSIS > NUMERIC in 2D SPACE mode, 15
COMPUTE, ANALYSIS > NUMERIC in 3D SPACE mode, 9
COMPUTE, CURVE2 > PTS CST in 2D SPACE mode, 79
COMPUTE, CURVE2 > PTS CST in 3D SPACE mode, 88
COMPUTE, PARAM3D > ADVANCED in SPACE mode, 310
COMPUTE, SOLIDE > ANALYZE | POSITN | INTERFER in SPACE mode, 397
COMPUTE, SOLIDM > ANALYZE | POSITN | INTERFER in SPACE mode, 421
CONCATEN, LIMIT1 in 2D SPACE mode, 230
CONCATEN, LIMIT1 in 3D SPACE mode, 236
CONCATEN, LIMIT1 in DRAW mode, 222
CONE, DIMENS > CREATE | DIAMETER in DRAW mode, 123
CONE, SOLIDM > CREATE | CANONIC in SPACE mode, 413
CONG, SOLIDE > CREATE | CANONIC in SPACE mode, 386
CONIC, CURVE2 in 2D SPACE mode, 78
CONIC, CURVE2 in DRAW mode, 67
CONIC, IMAGE > WINDOW | DEFINE in DRAW and SPACE modes, 192
CONNECT, CURVE2 in 2D SPACE mode, 81
CONNECT, CURVE2 in 3D SPACE mode, 86
CONNECT, CURVE2 in DRAW mode, 71
CONSTRN, PARAM3D > ANALYZE in SPACE mode, 308
CONSTRN, PARAM3D > PROFILE in SPACE mode, 301
CONTOUR, SOLIDE > MODIFY | GEOMETRY in SPACE mode, 400
CONTOUR, SOLIDM > MODIFY | GEOMETRY in SPACE mode, 425
COORD, POINT in 2D SPACE mode, 354
COORD, POINT in 3D SPACE mode, 362
COORD, POINT in DRAW mode, 342
COPY VIEW, AUXVIEW2 > USE | VIEW | NEW, 29
COPY, DETAIL in 3D SPACE mode, 104
COPY, DETAIL in DRAW mode, 96
COPY, DRAFT in DRAW mode, 138
COPY, FILE in DRAW and SPACE modes, 157
COPY, MODELS > WORKAREA | CREATE in DRAW and SPACE modes, 298
COPY, MODELS in DRAW and SPACE modes, 296
COPY, PLOT > FILE | SHEET in DRAW and SPACE modes, 333
COPY, PLOT > FILE | WINDOW in DRAW and SPACE modes, 336
COPY, SETS in DRAW and SPACE modes, 370
COPY, SYMBOL in DRAW mode, 446
CORNER, LIMIT1 in 2D SPACE mode, 225
CORNER, LIMIT1 in 3D SPACE mode, 232
CORNER, LIMIT1 in DRAW mode, 218
CREATE, ANALYSIS > NUMERIC in 2D SPACE mode, 6
CREATE, ANALYSIS > NUMERIC in 3D SPACE mode, 10
CREATE, ANALYSIS > RELATIVE | REPEAT in 2D SPACE mode, 7
CREATE, ANALYSIS > RELATIVE | SINGLE in 2D SPACE mode, 7
CREATE, AUXVIEW in DRAW mode, 14
CREATE, AXIS in 2D SPACE mode, 52
CREATE, AXIS in 3D SPACE mode, 56
CREATE, AXIS in DRAW mode, 49
CREATE, CURVE2 > CIRCLE in 3D SPACE mode, 91
CREATE, CURVE2 > CONNECT | TYPE 2 in 2D SPACE mode, 81
CREATE, CURVE2 > CONNECT | TYPE 2 in 3D SPACE mode, 86
CREATE, CURVE2 > CONNECT | TYPE 3 in 2D SPACE mode, 82
CREATE, CURVE2 > CONNECT | TYPE 3 in 3D SPACE mode, 87
CREATE, DETAIL in DRAW mode, 98

CREATE, DETAIL in 3D SPACE mode, 107
CREATE, DIMENS2 in DRAW mode, 117
CREATE, DRAFT in DRAW mode, 137
CREATE, DRW STD > ANNOTAT | DESCRIPT, 145
CREATE, DRW STD > ANNOTAT | STANDARD, 152
CREATE, DRW>SPC in SPACE mode, 140
CREATE, DRW STD > PATTERN, 153
CREATE, FILE in DRAW and SPACE modes, 159
CREATE, LAYER > FILTER in DRAW and SPACE modes, 208
CREATE, MODELS > WORKAREA in DRAW and SPACE modes, 297
CREATE, PARAM3D > PARAMETER in SPACE mode, 304
CREATE, PARAM3D > PROFILE in SPACE mode, 302
CREATE, PLOT > FILE | SHEET in DRAW and SPACE modes, 332
CREATE, PLOT > FILE | WINDOW in DRAW and SPACE modes, 335
CREATE, PLOT > QUICK in DRAW and SPACE modes, 327
CREATE, POINT > GRID | UNLIMITED in DRAW mode, 349
CREATE, SETS in DRAW and SPACE modes, 369
CREATE, SHAPE in DRAW mode, 371
CREATE, SOLIDE in SPACE mode, 378
CREATE, SOLIDM > in SPACE mode, 411
CREATE, TEXT in SPACE mode, 457
CREATE, TEXTD2 in DRAW mode, 461
CREATE, TRANSFOR in 2D SPACE mode, 490
CREATE, TRANSFOR in 3D SPACE mode, 498
CREATE, TRANSFOR in DRAW mode, 483
CRTL OFF, IUA > EXECUTE in DRAW and SPACE modes, 198
CRTL ON, IUA > EXECUTE in DRAW and SPACE modes, 198
CST, POINT > LIMITS in 2D SPACE mode, 357
CST, POINT > LIMITS in 3D SPACE mode, 365
CST, POINT > SPACES in 2D SPACE mode, 358
CST, POINT > SPACES in 3D SPACE mode, 367
CUBOID, SOLIDE > CREATE | CANONIC in SPACE mode, 378
CUBOID, SOLIDM > CREATE | CANONIC in SPACE mode, 411
CUR.FILT, LAYER > FILTER | MODIFY in DRAW and SPACE modes, 209
CUR SET, IDENTIFY > RENAME | DISPLAY in DRAW and SPACE modes, 186
CUR SET, IDENTIFY > RENAME | LAYER in DRAW and SPACE modes, 187
CUR SET, IDENTIFY > RENAME | LIST in DRAW and SPACE modes, 186
CUR SET, IDENTIFY > RENAME | TYPE in DRAW and SPACE modes, 187
CUR_TRAP, GROUP > GRP_1, 2 or 3 | EXCLUDE in SPACE mode, 182
CUR_TRAP, GROUP > GRP_1, 2 or 3 | INCLUDE in SPACE mode, 182
CURRENT, DETAIL > EXPLODE in 3D SPACE mode, 106
CURRENT, DETAIL > EXPLODE in DRAW mode, 98
CURRENT, DETAIL > MANAGE | LAYER in DRAW mode, 100
CURRENT, DETAIL > MANAGE | REPLACE in DRAW mode, 100
CURRENT, ERASE > ERASE in DRAW and SPACE modes, 154
CURRENT, ERASE > NO PICK/PICK in DRAW and SPACE modes, 155
CURRENT, ERASE > NO SHOW/SHOW in DRAW and SPACE modes, 155
CURRENT, ERASE > PACK in DRAW and SPACE modes, 154
CURRENT, GROUP > GRP_1, 2 or 3 | EXCLUDE in DRAW mode, 180
CURRENT, GROUP > GRP_1, 2 or 3 | EXCLUDE in SPACE mode, 184
CURRENT, GROUP > GRP_1, 2 or 3 | INCLUDE in DRAW mode, 180
CURRENT, GROUP > GRP_1, 2 or 3 | INCLUDE in SPACE mode, 184
CURRENT, LAYER > FILTER | RESET in DRAW and SPACE modes, 210
CURRENT, SYMBOL > MANAGE | LAYER in DRAW mode, 451
CURRENT, SYMBOL > MANAGE | VERIFY in DRAW mode, 450
CURVAT, CURVE2 > SPLINE > IMPOSE in DRAW mode, 70
CURVAT, MARK UP > AXIS in DRAW mode, 289

CURVE, ANALYSIS in 2D SPACE mode, 8
CURVE, ANALYSIS in 3D SPACE mode, 12
CURVE, LIMIT1 > CONCATEN in 2D SPACE mode, 230
CURVE, LIMIT1 > CONCATEN in 3D SPACE mode, 236
CURVES, LIMIT1 > MACHINE | SHT JOG in 2D SPACE mode, 229
CURVES, LIMIT1 > MACHINE | SHT JOG in DRAW mode, 221
CUSTOM, DRW STD, 148
CUT, SOLIDE > EXTRACT in SPACE mode, 405
CUT, SOLIDM > EXTRACT in SPACE mode, 429
CUT, SPC->DR2 in DRAW mode, 431
CUT, SPC->DRW in DRAW mode, 434
CVT CONT, CURVE2 in 2D SPACE mode, 83
CVT CONT, CURVE2 in 3D SPACE mode, 89
CYLIND, IMAGE > WINDOW | DEFINE in DRAW and SPACE modes, 192
CYLINDER, DIMENS > CREATE | DIAMETER in DRAW mode, 120
CYLINDER, DIMENS > CREATE | RADIUS in DRAW mode, 125
CYLINDER, SOLIDE > CREATE | CANONIC in SPACE mode, 380
CYLINDER, SOLIDM > CREATE | CANONIC in SPACE mode, 411
DAT FEAT, TEXTD2 > CREATE in DRAW mode, 472
DAT TARG, TEXTD2 > CREATE in DRAW mode, 473
DATABASE, PLOT in DRAW and SPACE modes, 338
DBL RAD, LIMIT1 > MACHINE | MACH JOG in 2D SPACE mode, 228
DBL RAD, LIMIT1 > MACHINE | MACH JOG in 3D SPACE mode, 235
DBL RAD, LIMIT1 > MACHINE | MACH JOG in DRAW mode, 221
DEFAULT, AUXVIEW2, 43
DEFINE, IMAGE > WINDOW in DRAW and SPACE modes, 192
DEFINE, IMAGE > SCREEN in DRAW and SPACE modes, 194
DEFINE, SYMBOL in DRAW mode, 448
DEL NODE, TEXTD2 > MODIFY | LEADER | PATH in DRAW mode, 478
DEL PT, CURVE2 > PTS CST in 2D SPACE mode, 79
DEL PT, CURVE2 > PTS CST in 3D SPACE mode, 88
DEL, AUXVIEW2 > USE | BACK_PLN, 37
DEL, AUXVIEW2 > USE | BREAKOUT, 40
DEL, AUXVIEW2 > USE | SECTION, 40
DEL, AUXVIEW2 > USE | TEXT, 35
DEL, AUXVIEW2 > USE | VIEW, 29
DEL_CALLOUT, AUXVIEW2 > USE | CLIP, 38
DEL_CALLOUT, AUXVIEW2 > USE | PLANE, 34
DEL_CALLOUT, AUXVIEW2 > USE | SECTION, 41
DELETE, AUXVIEW in DRAW mode, 20
DELETE, DETAIL in DRAW mode, 98
DELETE, DETAIL in 3D SPACE mode, 107
DELETE, DIMENS > MODIFY | TEXT in DRAW mode, 132
DELETE, DIMENS > MODIFY | TOL in DRAW mode, 132
DELETE, DIMENS2 in DRAW mode, 134
DELETE, DRAFT in DRAW mode, 138
DELETE, FILE > KEYBOARD in DRAW and SPACE modes, 164
DELETE, FILE in DRAW and SPACE modes, 158
DELETE, IMAGE > SCREEN in DRAW and SPACE modes, 196
DELETE, IMAGE > WINDOW in DRAW and SPACE modes, 193
DELETE, LAYER > FILTER in DRAW and SPACE modes, 209
DELETE, LIBRARY in DRAW and SPACE modes, 215
DELETE, MODELS > WORKAREA | CREATE in DRAW and SPACE modes, 298
DELETE, PARAM3D > DIMENSION in SPACE mode, 307
DELETE, PARAM3D > PARAMETER in SPACE mode, 305
DELETE, PARAM3D in SPACE mode, 308
DELETE, PLOT > FILE | WINDOW in DRAW and SPACE modes, 336
DELETE, PLOT > FILE in DRAW and SPACE modes, 333
DELETE, PLOT > QUICK in DRAW and SPACE modes, 329
DELETE, POINT > GRID | UNLIMITED in DRAW mode, 349
DELETE, SETS in DRAW and SPACE modes, 369

DELETE, SOLIDE > MODIFY | OPERATN in SPACE mode, 402
DELETE, SOLIDM > MODIFY | OPERATN in SPACE mode, 426
DELETE, SYMBOL in DRAW mode, 449
DELETE, TEXTD2 > MODIFY | LEADER in DRAW mode, 447
DELETE, TEXTD2 in DRAW mode, 479
DEPTH, CURVE2 in 3D SPACE mode, 90
DESCRIPT, DRW STD > ANNOTAT, 142
DETAIL VIEW, AUXVIEW2 > USE | VIEW | NEW, 28
DETAIL, DETAIL > MANAGE | ANALYZE in DRAW mode, 99
DETAIL, KEEP > SELECT | GEOMETRY in DRAW and SPACE modes, 203
DETAIL, LIBRARY > FAMILY in DRAW and SPACE modes, 214
DETAIL, LIBRARY > MODIFY in DRAW and SPACE modes, 215
DETAIL, LIBRARY > READ in DRAW and SPACE modes, 214
DETAIL > LIBRARY > UPDATE in DRAW and SPACE modes, 216
DETAIL, LIBRARY > WRITE in DRAW and SPACE modes, 214
DETAIL, MERGE > SELECT in DRAW and SPACE modes, 293
DIAMETER, CURVE2 > CIRCLE in 2D SPACE mode, 74
DIAMETER, CURVE2 > CIRCLE in DRAW mode, 4
DIAMETER, DIMENS > CREATE in DRAW mode, 121
DIM LINE, DIMENS > MODIFY in DRAW mode, 127
DIMENS, AUXVIEW2 > USE, 42
DIMENSION, PARAM3D in SPACE mode, 305
DIRECT, DETAIL > DELETE | UNUSED in DRAW mode, 98
DIRECT, FILE > COPY in DRAW and SPACE modes, 157
DIRECT, FILE > DELETE in DRAW and SPACE modes, 158
DIRECT, FILE > MOVE in DRAW and SPACE modes, 158
DIRECT, FILE > RENAME in DRAW and SPACE modes, 159
DIRECT, LAYER > FILTER | CREATE in DRAW and SPACE modes, 208
DIRECT, LAYER > FILTER | APPLY in DRAW and SPACE modes, 206
DIRECT, PLOT > FILE | DELETE in DRAW and SPACE modes, 329
DISCRETN, GRAPHIC > MOD VISU | CHOOSE in SPACE mode, 174
DISPLAY, GRAPHIC > MOD VISU | CHOOSE in SPACE mode, 174
DISPLAY, IDENTIFY > RENAME in DRAW and SPACE modes, 186
DISTANCE, DIMENS > CREATE in DRAW mode, 117
DISTANCE, DRW STD > CUSTOM | DESCRIPT, 148
DITTO, DETAIL in 3D SPACE mode, 103
DITTO, DETAIL in DRAW mode, 94
DITTO, LAYER > FILTER | APPLY | DIRECT in DRAW and SPACE modes, 207
DITTO, LAYER > FILTER | MODIFY | CUR FILT in DRAW and SPACE modes, 209
DOTTING, PATTERN > AUTO in DRAW mode, 313
DOTTING, PATTERN > SELECT in DRAW mode, 314
DOWNSTRM, PARAM3D > ANALYZE | RELATION in SPACE mode, 308
DRAFT, SOLIDE > OPERATN in SPACE mode, 392
DRAW ELT, STANDARD in DRAW and SPACE modes, 439
DRESS UP, SOLIDE > MODIFY in SPACE mode, 404
DRESS UP, SOLIDM > MODIFY in SPACE mode, 428
DRESS, AUXVIEW2 > USE | VIEW, 30
DROP, AUXVIEW2 > USE | VIEW, 32
DROP, DETAIL > MANAGE in 3D SPACE mode, 109
DROP, DETAIL > MANAGE in DRAW mode, 101
DROP, MODELS > WORKAREA | MODIFY in DRAW and SPACE modes, 298
DROP, SYMBOL > MANAGE in DRAW mode, 451
DRW ELEM, GROUP > GRP_1, 2 or 3 | EXCLUDE | CURRENT in DRAW mode, 181
DRW ELEM, GROUP > GRP_1, 2 or 3 | INCLUDE | CURRENT in DRAW mode, 181
DTAILING, AUXVIEW in DRAW mode, 21
DUPLICAT, AUXVIEW > CREATE in DRAW mode, 15

DUPLICAT, AUXVIEW > CREATE in DRAW mode, 15
DUPLICAT, CURVE2 > CIRCLE | MODIFY in 2D SPACE mode, 76
DUPLICAT, CURVE2 > CIRCLE | MODIFY in DRAW mode, 66
DUPLICAT, CURVE2 > CIRCLE in 3D SPACE mode, 91
DUPLICAT, DETAIL > TRANSFER in 3D SPACE mode, 110
DUPLICAT, DETAIL > TRANSFER in DRAW mode, 102
DUPLICAT, LINE > MODIFY in DRAW mode, 252
DUPLICAT, LINE > MODIFY in 3D SPACE mode, 286
DUPLICAT, SOLIDE > MODIFY | OPERATN in SPACE mode, 403
DUPLICAT, SOLIDM > MODIFY | OPERATN in SPACE mode, 426
DUPLICAT, SYMBOL > DEFINE in DRAW mode, 448
DUPLICAT, SYMBOL > MODIFY | FLIPX in DRAW mode, 447
DUPLICAT, SYMBOL > MODIFY | FLIPY in DRAW mode, 447
DUPLICAT, SYMBOL > MODIFY | ROTATE in DRAW mode, 447
DUPLICAT, SYMBOL > MODIFY | SCALE in DRAW mode, 447
DUPLICAT, SYMBOL > MODIFY | SCALE in DRAW mode, 447
DUPLICAT, SYMBOL > MODIFY | SYMMETRY in DRAW mode, 447
DUPLICAT, SYMBOL > MODIFY | TRANSLATE in DRAW mode, 446
DUPLICAT, TRANSFOR > APPLY in 2D SPACE mode, 494
DUPLICAT, TRANSFOR > APPLY in 3D SPACE mode, 504
DUPLICAT, LINE > MODIFY in 2D SPACE mode, 273
DUPLICAT, LINE > MODIFY in DRAW mode, 252
DUPLICAT, MODELS > MODIFY in DRAW and SPACE modes, 295
EDGE, DIMENS > CREATE | DIAMETER in DRAW mode, 124
EDGE, DIMENS > CREATE | RADIUS in DRAW mode, 126
EDGE, LINE in 2D SPACE mode, 274
EDGE, SOLIDE > OPERATN | FILLET in SPACE mode, 392
EDGE, SOLIDM > MODIFY | VISUALATN in SPACE mode, 428
EDGECRV, SOLIDE > EXTRACT in SPACE mode, 406
EDGES, PLANE in SPACE mode, 322
EDGESKIN, SOLIDE > EXTRACT in SPACE mode, 406
EDIT, DIMENS > MODIFY | TEXT in DRAW mode, 132
EDIT, PARAM3D > PARAMETER in SPACE mode, 304
EDIT, TEXT in SPACE mode, 459
EDIT, TEXTD2 > MODIFY | TEXT in DRAW mode, 476
ELEMENT, ANALYSIS > LOGICAL in DRAW mode, 4
ELEMENT, GROUP > GRP_1, 2 or 3 | EXCLUDE | CURRENT in DRAW mode, 180
ELEMENT, GROUP > GRP_1, 2 or 3 | EXCLUDE | W.SPACE in DRAW mode, 180
ELEMENT, GROUP > GRP_1, 2 or 3 | INCLUDE | CURRENT in DRAW mode, 180
ELEMENT, GROUP > GRP_1, 2 or 3 | INCLUDE | W.SPACE in DRAW mode, 180
ELEMENT, GROUP > GRP_1, 2 or 3 | EXCLUDE | CURRENT in DRAW mode, 184
ELEMENT, GROUP > GRP_1, 2 or 3 | EXCLUDE | W.SPACE in DRAW mode, 184
ELEMENT, GROUP > GRP_1, 2 or 3 | INCLUDE | CURRENT in DRAW mode, 184
ELEMENT, GROUP > GRP_1, 2 or 3 | INCLUDE | W.SPACE in DRAW mode, 184
ELEMENT, IDENTIFY > RENAME | DISPLAY in DRAW and SPACE modes, 186
ELEMENT, IDENTIFY > RENAME | LAYER in DRAW and SPACE modes, 187
ELEMENT, IDENTIFY > RENAME | LIST in DRAW and SPACE modes, 186
ELEMENT, IDENTIFY > RENAME | TYPE in DRAW and SPACE modes, 187

ELEMENT, KEEP > SELECT | APPLICTN in DRAW and SPACE modes, 204
ELEMENT, KEEP > SELECT | GEOMETRY in DRAW and SPACE modes, 203
ELEMENT, MODELS > COPY in DRAW and SPACE modes, 296
ELEMENT, SOLIDE > ANALYZE | POSITN | INTERFER in SPACE mode, 397
ELEMENT, SOLIDM > ANALYZE | POSITN | INTERFER in SPACE mode, 421
ELEMENT, SOLIDM > MODIFY | VISUALATN in SPACE mode, 428
ELEMENT, TRANSFOR > APPLY | DUPLICAT in 2D SPACE mode, 505
ELEMENT, TRANSFOR > APPLY | DUPLICAT in 3D SPACE mode, 494
ELEMENT, TRANSFOR > APPLY | REPLACE in 2D SPACE mode, 504
ELLIPSE, CURVE2 in 2D SPACE mode, 77
ELLIPSE, CURVE2 in DRAW mode, 66
EMPTY, MODELS > WORKAREA | APPLY in DRAW and SPACE modes, 297
EMPTY, MODELS > WORKAREA | CREATE in DRAW and SPACE modes, 319
EQUATION, PLANE in SPACE mode, 130
ERASE, DIMENS > MODIFY | EXT LINE in DRAW mode, 154
ERASE, ERASE in DRAW and SPACE modes, 154
ERASE, SHAPE > MODIFY in DRAW mode, 374
ERASE, TEXT in SPACE mode, 459
ERASE, TRANSFOR > MANAGE in 2D SPACE mode, 496
ERASE, TRANSFOR > MANAGE in 3D SPACE mode, 506
ERASE, TRANSFOR in DRAW mode, 487
EULER, TRANSFOR > CREATE in 3D SPACE mode, 504
EXACT, SPC->DR2 > CUT in DRAW mode, 431
EXACT, SPC->DR2 > PROJECT in DRAW mode, 432
EXCLUDE, GROUP > GRP_1 or 2 or 3 in DRAW mode, 179
EXCLUDE, GROUP > GRP_1 or 2 or 3 in SPACE mode, 182
EXECUTE, IUA in DRAW and SPACE modes, 198
EXIT, FILE in DRAW and SPACE modes, 164
EXPLODE, DETAIL in 3D SPACE mode, 106
EXPLODE, DETAIL in DRAW mode, 98
EXPLODE, SYMBOL in DRAW mode, 448
EXT LINE, DIMENS > MODIFY in DRAW mode, 130
EXTRACT, SOLIDE in SPACE mode, 405
EXTRACT, SOLIDM in SPACE mode, 429
EXTRACT, SYMBOL in DRAW mode, 452
EXTRAPOL, LIMIT1 in 2D SPACE mode, 230
EXTRAPOL, LIMIT1 in 3D SPACE mode, 236
EXTRAPOL, LIMIT1 in DRAW mode, 223
FACE FACE, SOLIDE > OPERATN | FILLET in SPACE mode, 393
FACE, SOLIDE > EXTRACT in SPACE mode, 407
FAKE DIM, DIMENS > MODIFY | VALUE in DRAW mode, 128
FAMILY, ANALYSIS > LOGICAL in 2D SPACE mode, 8
FAMILY, ANALYSIS > LOGICAL in 3D SPACE mode, 12
FAMILY, GROUP > GRP_1, 2 or 3 | EXCLUDE | CURRENT in DRAW mode, 180
FAMILY, GROUP >GRP_1, 2 or 3 | EXCLUDE | W.SPACE in DRAW mode, 180
FAMILY, GROUP > GRP_1, 2 or 3 | INCLUDE | CURRENT in DRAW mode, 180
FAMILY, GROUP > GRP_1, 2 or 3 | INCLUDE | W.SPACE in DRAW mode, 180
FAMILY, GROUP > GRP_1, 2 or 3 | EXCLUDE | CURRENT in SPACE mode, 184
FAMILY, GROUP >GRP_1, 2 or 3 | EXCLUDE | W.SPACE in SPACE mode, 184
FAMILY, GROUP > GRP_1, 2 or 3 | INCLUDE | CURRENT in SPACE mode, 184
FAMILY, GROUP > GRP_1, 2 or 3 | INCLUDE | W.SPACE in SPACE mode, 184
FAMILY, LIBRARY in DRAW and SPACE modes, 214
FAMILY, MODELS > COPY in DRAW and SPACE modes, 296
FAMILY, TRANSFOR > APPLY | DUPLICAT in 2D SPACE mode, 495
FAMILY, TRANSFOR > APPLY | DUPLICAT in 3D SPACE mode, 505

FAMILY, TRANSFOR > APPLY | REPLACE in 2D SPACE mode, 495
FAMILY, TRANSFOR > APPLY | REPLACE in 3D SPACE mode, 504
FEATURE, SOLIDE > CREATE in SPACE mode, 388
FILE, FILE in DRAW and SPACE modes, 156
FILE, IUA in DRAW and SPACE modes, 197
FILE, LIBRARY in DRAW and SPACE modes, 213
FILE, PLOT in DRAW and SPACE modes, 332
FILLET, SOLIDE > OPERATN in SPACE mode, 392
FILTER, AUXVIEW2 > USE | VIEW, 31
FILTER, LAYER in DRAW and SPACE modes, 206
FIT LEAD, TEXTD2 > CREATE | TEXT in DRAW mode, 470
FITTED, TEXTD2 > CREATE | TEXT in DRAW mode, 469
FIXED, AXIS in 2D SPACE mode, 54
FIXED, AXIS in 3D SPACE mode, 58
FIXED, AXIS in DRAW mode, 50
FLIPX, DETAIL > MODIFY in DRAW mode, 97
FLIPX, SYMBOL > MODIFY in DRAW mode, 447
FLIPY, DETAIL > MODIFY in DRAW mode, 97
FLIPY, SYMBOL > MODIFY in DRAW mode, 447
FRAME, AUXVIEW > MODIFY in DRAW mode, 16
FREE TGT, CURVE2 > PTS CST in 2D SPACE mode, 79
FREE TGT, CURVE2 > PTS CST in 3D SPACE mode, 88
FREE, CURVE2 > SPLINE in DRAW mode, 70
FREEHAND, TRANSFOR in DRAW mode, 488
FSUR, SOLIDE > MODIFY | DRESS UP | RENAME in SPACE mode, 404
FSUR, SOLIDM > MODIFY | DRESS UP | RENAME in SPACE mode, 428
FULL IN, GROUP > EXCLUDE | CUR_TRAP or WSP_TRAP in SPACE mode, 183
FULL IN, GROUP > EXCLUDE | TRAP in DRAW mode, 179
FULL IN, GROUP > INCLUDE | CUR_TRAP or WSP_TRAP in SPACE mode, 183
FULL IN, GROUP > INCLUDE | TRAP in DRAW mode, 179
FULL OUT, GROUP > EXCLUDE | CUR_TRAP or WSP_TRAP in SPACE mode, 184
FULL OUT, GROUP > EXCLUDE | TRAP in DRAW mode, 179
FULL OUT, GROUP > INCLUDE | CUR_TRAP or WSP_TRAP in SPACE mode, 184
FULL OUT, GROUP > INCLUDE | TRAP in DRAW mode, 179
FUNNEL, DIMENS > MODIFY | EXT LINE in DRAW mode, 130
GAPS, AUXVIEW2 > USE | PLANE, 34
GAPS, AUXVIEW2 > USE | SECTION, 41
GDIM, PARAM3D > DIMENSION | POSITN in SPACE mode, 307
GENERAL, CURVE2 > DEPTH in 3D SPACE mode, 90
GENERAL, LAYER > FILTER | APPLY | DIRECT in DRAW and SPACE modes, 207
GENERAL, LAYER > FILTER | MODIFY | CUR.FILT in DRAW and SPACE modes, 209
GENERAL, PATTERN > VISUALATN in DRAW mode, 316
GENERAL, STANDARD > SPAC ELT in DRAW and SPACE modes, 436
GEOM TOL, TEXTD2 > CREATE in DRAW mode, 470
GEOM, MERGE > SELECT | SET in DRAW and SPACE modes, 293
GEOMETRY, KEEP > SELECT in DRAW and SPACE modes, 202
GEOMETRY, SOLIDE > MODIFY in SPACE mode, 398
GEOMETRY, SOLIDM > MODIFY in SPACE mode, 422
GRAPH, AUXVIEW2 > USE | CLIP, 38
GRAPH, AUXVIEW2 > USE | PLANE, 35
GRAPH, AUXVIEW2 > USE | SECTION, 42
GRAPHIC, DIMENS > MODIFY | TEXT in DRAW mode, 133
GRAPHIC, DIMENS > MODIFY | VALUE in DRAW mode, 129
GRAPHIC, DIMENS > MODIFY in DRAW mode, 133

GRAPHIC, TEXTD2 > MODIFY | TEXT in DRAW mode, 476
GRAPHISM, PATTERN > VISUALATN | MODAL in DRAW mode, 316
GRID, DETAIL > COPY | LIBRARY | COMPACT in DRAW mode, 96
GRID, DETAIL > COPY | LIBRARY | STANDARD in DRAW mode, 96
GRID, DETAIL > DITTO | LIBRARY | COMPACT in DRAW mode, 96
GRID, DETAIL > DITTO | MODEL | COMPACT in DRAW mode, 96
GRID, DETAIL > DITTO | MODEL | STANDARD in DRAW mode, 95
GRID, DETAIL > DITTO | LIBRARY | STANDARD in DRAW mode, 96
GRID, DETAIL > DITTO | MODEL | STANDARD in DRAW mode, 95
GRID, LINE in DRAW mode, 254
GRID, POINT in DRAW mode, 347
GRID, SYMBOL > COPY | LIBRARY | COMPACT in DRAW mode, 446
GRID, SYMBOL > COPY | LIBRARY | STANDARD in DRAW mode, 446
GRID, SYMBOL > COPY | MODEL | COMPACT in DRAW mode, 446
GRID, SYMBOL > COPY | MODEL | STANDARD in DRAW mode, 445
GRID, SYMBOL > SYMBOL | MODEL | COMPACT in DRAW mode, 445
GRID, SYMBOL > SYMBOL | MODEL | STANDARD in DRAW mode, 445
GROUP, GROUP > GRP_1, 2 or 3 | EXCLUDE | CURRENT in DRAW mode, 180
GROUP, GROUP > GRP_1, 2 or 3 | EXCLUDE | W.SPACE in DRAW mode, 180
GROUP, GROUP > GRP_1, 2 or 3 | INCLUDE | CURRENT in DRAW mode, 180
GROUP, GROUP > GRP_1, 2 or 3 | INCLUDE | W.SPACE in DRAW mode, 180
GROUP, GROUP > GRP_1, 2 or 3 | EXCLUDE | CURRENT in SPACE mode, 184
GROUP, GROUP > GRP_1, 2 or 3 | EXCLUDE | W.SPACE in SPACE mode, 184
GROUP, GROUP > GRP_1, 2 or 3 | INCLUDE | CURRENT in SPACE mode, 184
GROUP, GROUP > GRP_1, 2 or 3 | INCLUDE | W.SPACE in SPACE mode, 184
GRP_1, 2 or 3, GROUP in DRAW mode, 178
GRP_1, 2 or 3, GROUP in SPACE mode, 182
HATCHING, PATTERN > AUTO in DRAW mode, 313
HATCHING, PATTERN > SELECT in DRAW mode, 314
HELIX, CURVE2 in 3D SPACE mode, 91
HLR, GRAPHIC > MOD VISU | CHOOSE in SPACE mode, 174
HLR, STANDARD > SPAC ELT | SPEC ELT in DRAW and SPACE modes, 438
HORIZONT, LINE > PT-PT | SEGMENT in 2D SPACE mode, 262
HORIZONT, LINE > PT-PT | SEGMENT in DRAW mode, 245
HORIZONT, LINE > PT-PT | UNLIM in 2D SPACE mode, 262
HORIZONT, LINE in 2D SPACE mode, 264
HORIZONT, LINE in DRAW mode, 244
HORIZONT, MARK UP > ARROW in DRAW mode, 290
HORIZONT, POINT > PROJECT in 2D SPACE mode, 353
HORIZONT, POINT > PROJECT in DRAW mode, 341
HORIZONT, TEXT > T.NODE | CREATE in DRAW mode, 453
HOR VERT, POINT > PROJECT in DRAW mode, 341
HOR VERT, POINT> PROJECT in SPACE mode, 353
IDENTIFY, LAYER in DRAW and SPACE modes, 212
IMP TGT, CURVE2 > PTS CST in 2D SPACE mode, 79
IMP TGT, CURVE2 > PTS CST in 3D SPACE mode, 88
IMPORT, SOLIDE > CREATE in SPACE mode, 389
IMPOSE, CURVE2 > SPLINE in DRAW mode, 69
INACTIVE, SOLIDE > OPERATN in SPACE mode, 395
INCLUDE, GROUP > GRP_1, 2 or 3 in DRAW mode, 178
INERTIA, ANALYSIS in 3D SPACE mode, 13

INERTIA, SOLIDE > ANALYZE | SELF in SPACE mode, 396
INERTIA, SOLIDM > ANALYZE | SELF in SPACE mode, 420
INSERT, SOLIDE > MODIFY | OPERATN in SPACE mode, 402
INSERT, SOLIDM > MODIFY | OPERATN in SPACE mode, 425
INTERFER, SOLIDE > ANALYZE | POSITN in SPACE mode, 397
INTERFER, SOLIDM > ANALYZE | POSITN in SPACE mode, 421
INTERRUP, DIMENS > MODIFY | EXT LINE in DRAW mode, 130
INTERSEC, LINE in 3D SPACE mode, 282
INTERSEC, SHAPE > OPERATE in DRAW mode, 375
INTERSEC, SOLIDE > OPERATN in SPACE mode, 390
INTERSEC, SOLIDM > MODIFY | OPERATN | INSERT in SPACE mode, 417
INTERSEC, SOLIDM > OPERATN in SPACE mode, 417
INVERT, AUXVIEW2 > USE | PLANE, 34
INVERT, AXIS in 2D SPACE mode, 35
INVERT, AXIS in 3D SPACE mode, 58
INVERT, AXIS in DRAW mode, 50
INVERT, CURVE2 in 2D SPACE mode, 84
INVERT, CURVE2 in 3D SPACE mode, 92
INVERT, CURVE2 in DRAW mode, 72
INVERT, DIMENS > MODIFY | SYMBOL in DRAW mode, 131
INVERT, TRANSFOR > CREATE in DRAW mode, 486
INVERT, TRANSFOR > MANAGE in 2D SPACE mode, 496
INVERT, TRANSFOR > MANAGE in 3D SPACE mode, 506
ISOLATE, DIMENS > MANAGE in DRAW mode, 135
ISOLATE, PARAM3D > PARAMETER in SPACE mode, 304
ISOLATE, TEXTD2 > MANAGE in DRAW mode, 480
ISOMETRIC VIEW, AUXVIEW2 > USE | VIEW | NEW, 28
KEEP, KEEP in DRAW and SPACE modes, 204
KEYBOARD, FILE in DRAW and SPACE modes, 162
LAYER, DETAIL > MANAGE in 3D SPACE mode, 109
LAYER, DETAIL > MANAGE in DRAW mode, 100
LAYER, GROUP > GRP_1, 2 or 3 | EXCLUDE | CURRENT in DRAW mode, 180
LAYER, GROUP > GRP_1, 2 or 3 | EXCLUDE | W.SPACE in DRAW mode, 180
LAYER, GROUP > GRP_1, 2 or 3 | INCLUDE | CURRENT in DRAW mode, 180
LAYER, GROUP > GRP_1, 2 or 3 | INCLUDE | W.SPACE in DRAW mode, 180
LAYER, GROUP > GRP_1, 2 or 3 | EXCLUDE | CURRENT in SPACE mode, 184
LAYER, GROUP > GRP_1, 2 or 3 | EXCLUDE | W.SPACE in SPACE mode, 184
LAYER, GROUP > GRP_1, 2 or 3 | INCLUDE | CURRENT in SPACE mode, 184
LAYER, GROUP > GRP_1, 2 or 3 | INCLUDE | W.SPACE in SPACE mode, 184
LAYER, IDENTIFY > RENAME in DRAW and SPACE modes, 187
LAYER, LAYER in DRAW and SPACE modes, 211
LAYER, STANDARD > COLOR | MODIFY in DRAW and SPACE modes, 439
LAYER, SYMBOL > MANAGE in DRAW mode, 451
LAYOUT, PLOT > FILE in DRAW and SPACE modes, 337
LAYOUT, PLOT > QUICK in DRAW and SPACE modes, 329
LEADER, TEXTD2 > CREATE | TEXT in DRAW mode, 467
LEADER, TEXTD2 > MODIFY in DRAW mode, 477
LENGTH, DIMENS > CREATE in DRAW mode, 120
LENGTH, DIMENS > MODIFY | EXT LINE in DRAW mode, 130
LENGTH, LIMIT1 > MACHINE | CHAMFER in 2D SPACE mode, 227
LENGTH, LIMIT1 > MACHINE | CHAMFER in 3D SPACE mode, 234

LENGTH, LIMIT1 > MACHINE | CHAMFER in DRAW mode, 219
LIMITED, POINT > GRID in DRAW mode, 347
LIMITS, PLANE in SPACE mode, 324
LIMITS, POINT in 2D SPACE mode, 356
LIMITS, POINT in 3D SPACE mode, 363
LIMITS, POINT in DRAW mode, 344
LINE, TRANSFOR > CREATE | AFFINITY in 3D SPACE mode, 502
LINE, TRANSFOR > CREATE | SYMMETRY in 2D SPACE mode, 491
LINE, TRANSFOR > CREATE | SYMMETRY in 3D SPACE mode, 499
LINE, TRANSFOR > CREATE | SYMMETRY in DRAW mode, 483
LINEAR, MARK UP > AXIS in DRAW mode, 288
LINES, COMBIVU in DRAW mode, 60
LINETYPE, GRAPHIC > MODIFY in DRAW mode, 166
LINETYPE, STANDARD in DRAW and SPACE modes, 442
LINK, PARAM3D in SPACE mode, 303
LINK, SETS in DRAW and SPACE modes, 370
LIST, IDENTIFY > RENAME in DRAW and SPACE modes, 186
LIST, KEEP > SELECT | APPLICTN in DRAW and SPACE modes, 204
LIST, MODELS > WORKAREA in DRAW and SPACE modes, 299
LIST, PLOT > FILE | WINDOW in DRAW and SPACE modes, 337
LIST, PLOT > QUICK in DRAW and SPACE modes, 329
LNG PT, LIMIT1 > MACHINE | SHT JOG | PARAL LN in DRAW mode, 221
LNG TGT, LIMIT1 > MACHINE | SHT JOG | PARAL LN in 2D SPACE mode, 228
LNG TGT, LIMIT1 > MACHINE | SHT JOG | PARAL LN in DRAW mode, 221
LOAD, IUA > FILE in DRAW and SPACE modes, 198
LOCAL TR, IMAGE in DRAW and SPACE modes, 196
LOCAL, PARAM3D > TRANSFOR in SPACE mode, 305
LOCATE, PLOT > FILE | LAYOUT in DRAW and SPACE modes, 337
LOCATE, PLOT > QUICK | LAYOUT in DRAW and SPACE modes, 330
LOCATION, DIMENS > MODIFY | VALUE in DRAW mode, 128
LOCATION, TEXTD2 > MODIFY in DRAW mode, 478
LOCK, AUXVIEW2 > USE | VIEW, 32
LOCK, DRW STD > ANNOTAT | DESCRIPT, 145
LOCK, DRW STD > PATTERN, 153
LOGICAL, ANALYSIS in 2D SPACE mode, 7
LOGICAL, ANALYSIS in 3D SPACE mode, 11
LOGICAL, ANALYSIS in DRAW mode, 4
MACH JOG, LIMIT1 > MACHINE in 2D SPACE mode, 227
MACH JOG, LIMIT1 > MACHINE in 3D SPACE mode, 234
MACH JOG, LIMIT1 > MACHINE in DRAW mode, 220
MACHINE, LIMIT1 in 2D SPACE mode, 226
MACHINE, LIMIT1 in 3D SPACE mode, 233
MACHINE, LIMIT1 in DRAW mode, 219
MACRO, SOLIDE > CREATE in SPACE mode, 389
MANAGE, DETAIL in 3D SPACE mode, 108
MANAGE, DETAIL in DRAW mode, 99
MANAGE, DIMENS2 in DRAW mode, 134
MANAGE, MODELS in DRAW and SPACE modes, 294
MANAGE, PLOT > QUICK | PLOTTING in DRAW and SPACE modes, 332
MANAGE, SYMBOL in DRAW mode, 449
MANAGE, TEXTD2 in DRAW mode, 480
MANAGE, TRANSFOR in 2D SPACE mode, 496
MANAGE, TRANSFOR in 3D SPACE mode, 506
MEAN, LINE in 2D SPACE mode, 272
MEAN, LINE in 3D SPACE mode, 286
MEAN, LINE in DRAW mode, 252
MEAN, PLANE in SPACE mode, 322
MEDIAN, LINE in 2D SPACE mode, 267
MEDIAN, LINE in DRAW mode, 247
MERGE, KEEP in DRAW and SPACE modes, 205
MERGE, MERGE in DRAW and SPACE modes, 293
MERGE, PLOT > FILE | SHEET in DRAW and SPACE modes, 334
MOD GEN, GRAPHIC in SPACE mode, 171
MOD SPEC, GRAPHIC in SPACE mode, 173
MOD VISU, GRAPHIC in SPACE mode, 173

MODAL, PATTERN > VISUALATN in DRAW mode, 316
MODEL, DETAIL > DITTO in 3D SPACE mode, 104
MODEL, DETAIL > DITTO in DRAW mode, 94
MODEL, DETAIL > EXPLODE in 3D SPACE mode, 106
MODEL, DETAIL > EXPLODE in DRAW mode, 98
MODEL, IUA > FILE in DRAW and SPACE modes, 198
MODEL, STANDARD in DRAW and SPACE modes, 440
MODEL, SYMBOL > COPY in DRAW mode, 446
MODEL, SYMBOL > SYMBOL in DRAW mode, 441
MODIFY, AUXVIEW in DRAW mode, 16
MODIFY, AUXVIEW2 > USE | BACK_PLN, 37
MODIFY, AUXVIEW2 > USE | FRAME, 36
MODIFY, AUXVIEW2 > USE | PLANE, 34
MODIFY, AUXVIEW2 > USE | SECTION, 41
MODIFY, AUXVIEW2 > USE | TEXT, 35
MODIFY, CURVE2 > CIRCLE in 2D SPACE mode, 76
MODIFY, CURVE2 > CIRCLE in 3D SPACE mode, 91
MODIFY, CURVE2 > CIRCLE in DRAW mode, 66
MODIFY, DETAIL in 3D SPACE mode, 105
MODIFY, DETAIL in DRAW mode, 96
MODIFY, DIMENS2 in DRAW mode, 127
MODIFY, GRAPHIC in DRAW mode, 166
MODIFY, IMAGE > SCREEN in DRAW and SPACE modes, 193
MODIFY, IMAGE > WINDOW in DRAW and SPACE modes, 189
MODIFY, IUA in DRAW and SPACE modes, 199
MODIFY, LAYER > FILTER in DRAW and SPACE modes, 209
MODIFY, LIBRARY in DRAW and SPACE modes, 215
MODIFY, LINE in 2D SPACE mode, 273
MODIFY, LINE in 3D SPACE mode, 286
MODIFY, LINE in DRAW mode, 252
MODIFY, MODELS > WORKAREA in DRAW and SPACE modes, 298
MODIFY, MODELS in DRAW and SPACE modes, 295
MODIFY, PARAM3D in SPACE mode, 303
MODIFY, PLOT > FILE | WINDOW in DRAW and SPACE modes, 335
MODIFY, PLOT > QUICK in DRAW and SPACE modes, 327
MODIFY, SHAPE in DRAW mode, 374
MODIFY, SOLIDE in SPACE mode, 398
MODIFY, SOLIDM > in SPACE mode, 422
MODIFY, STANDARD > COLOR in DRAW and SPACE modes, 439
MODIFY, SYMBOL in DRAW mode, 446
MODIFY, TEXT in SPACE mode, 458
MODIFY, TEXTD2 in DRAW mode, 476
MONO-TGT, CURVE2 > CIRCLE in 2D SPACE mode, 75
MONO-TGT, CURVE2 > CIRCLE in DRAW mode, 65
MOV NODE, TEXTD2 > MODIFY | LEADER | PATH in DRAW mode, 477
MOVE, AUXVIEW > MODIFY | FRAME in DRAW mode, 16
MOVE, AUXVIEW2 > USE | DIMENS, 42
MOVE, AUXVIEW2 > USE | VIEW, 29
MOVE, CURVE2 > CONNECT | TYPE 2 in 2D SPACE mode, 82
MOVE, CURVE2 > CONNECT | TYPE 2 in 3D SPACE mode, 87
MOVE, CURVE2 > CONNECT | TYPE 3 in 3D SPACE mode, 88
MOVE, CURVE2 > SPLINE in DRAW mode, 70
MOVE, FILE in DRAW and SPACE modes, 158
MOVE, MODELS > MODIFY | DUPLICAT in DRAW and SPACE modes, 296
MOVE, MODELS > MODIFY | REPLACE in DRAW and SPACE modes, 296
MOVE, PLOT > FILE | SHEET in DRAW and SPACE modes, 333
MOVE, SHAPE > MODIFY in DRAW mode, 374
MOVE, SOLIDE > MODIFY | GEOMETRY in SPACE mode, 398
MOVE, SOLIDM > MODIFY | GEOMETRY in SPACE mode, 422
MOVE, TEXT in SPACE mode, 459
MOVE, TRANSFOR > CREATE in 2D SPACE mode, 493
MOVE, TRANSFOR > CREATE in 3D SPACE mod, 503

MOVE, TRANSFOR > CREATE in DRAW mode, 485
MOVE_CALLOUT, AUXVIEW2 > USE | PLANE, 34
MULTI, DETAIL > COPY | LIBRARY | COMPACT in DRAW mode, 96
MULTI, DETAIL > COPY | LIBRARY | STANDARD in DRAW mode, 96
MULTI, DETAIL > DITTO | LIBRARY | COMPACT in DRAW mode, 96
MULTI, DETAIL > DITTO | LIBRARY | STANDARD in DRAW mode, 96
MULTI, SYMBOL > COPY | MODEL | COMPLEX in DRAW mode, 444
MULT-TGT, CURVE2 > CIRCLE in DRAW mode, 65
MULT-TGT, CURVE2 > CIRCLE in 2D SPACE mode, 76
NEW BGD, AUXVIEW > CREATE in DRAW mode, 14
NEW BGD, AUXVIEW > CREATE in DRAW mode, 14
NEW, AUXVIEW2 > USE | VIEW, 26
NO PICK, SETS > CHANGE in DRAW and SPACE modes, 369
NO PICK or PICK, ERASE in DRAW and SPACE modes, 155
NO SHOW, AUXVIEW > MODIFY | FRAME in DRAW mode, 18
NO SHOW, AUXVIEW > MODIFY | FRAME in DRAW mode, 18
NO SHOW, AUXVIEW2 > USE | FRAME, 36
NO SHOW, DIMENS > MANAGE | VISUALTN in DRAW mode, 136
NO SHOW, PARAM3D > DIMENSION in SPACE mode, 306
NO_SHOW or SHOW, PATTERN > VISUALTN | MODAL in DRAW mode, 316
NO SHOW/SHOW, ERASE in DRAW and SPACE modes, 155
NO SHOW, TEXTD2 > MANAGE | VISUALTN in DRAW mode, 481
NO TRIM, LIMIT1 > CORNER in 2D SPACE mode, 226
NO TRIM, LIMIT1 > CORNER in 3D SPACE mode, 233
NO TRIM, LIMIT1 > MACHINE | CHAMFER | ANGLE in 2D SPACE mode, 227
NO TRIM, LIMIT1 > MACHINE | CHAMFER | ANGLE in 3D SPACE mode, 233
NO TRIM, LIMIT1 > MACHINE | CHAMFER | LENGTH in 2D SPACE mode, 227
NO TRIM, LIMIT1 > MACHINE | MACH JOG | DBL RAD in 2D SPACE mode, 228
NO TRIM, LIMIT1 > MACHINE | MACH JOG | SINGL RAD in 2D SPACE mode, 228
NO TRIM, LIMIT1 > CORNER in 2D SPACE mode, 226
NO TRIM, LIMIT1 > CORNER in DRAW mode, 219
NO TRIM, LIMIT1 > MACHINE | CHAMFER | LENGTH in 2D SPACE mode, 234
NO TRIM, LIMIT1 > MACHINE | MACH JOG | DBL RAD in DRAW mode, 221
NO TRIM, LIMIT1 > MACHINE | MACH JOG | DBL RAD in 3D SPACE mode, 234
NO TRIM, LIMIT1 > MACHINE | MACH JOG | SINGL RAD in DRAW mode, 220
NO TRIM, LIMIT1 > MACHINE | MACH JOG | SINGL RAD in 3D SPACE mode, 234
NO TRIM, SHAPE > OPERATE | INTERSEC in DRAW mode, 375
NO TRIM, SHAPE > OPERATE | SUBRACT in DRAW mode, 375
NORMAL, LINE in 2D SPACE mode, 266
NORMAL, LINE in 3D SPACE mode, 280
NORMAL, LINE in DRAW mode, 246
NORMAL, PLANE > EDGES in SPACE mode, 323
NORMAL, PLANE in SPACE mode, 320
NORMAL, POINT > GRID | LIMITED in DRAW mode, 347
NORMAL, TEXT > T.NODE | CREATE in DRAW mode, 453
NORMAL, TRANSFOR > CREATE | AFFINITY | PLANE in 3D SPACE mode, 501
NORMAL, TRANSFOR > CREATE | AFFINITY in 2D SPACE mode, 492
NORMAL, TRANSFOR > CREATE | AFFINITY in DRAW mode, 484
NORMAL, TRANSFOR > CREATE | SYMMETRY | LINE in 2D SPACE mode, 491
NORMAL, TRANSFOR > CREATE | SYMMETRY | LINE in 3D SPACE mode, 500
NORMAL, TRANSFOR > CREATE | SYMMETRY | LINE in DRAW mode, 483
NORMAL, TRANSFOR > CREATE | SYMMETRY | PLANE in 3D SPACE mode, 499
NUM DISP, DIMENS > MODIFY | VALUE in DRAW mode, 129
NUMERIC, ANALYSIS in 2D SPACE mode, 5

NUMERIC, ANALYSIS in 3D SPACE mode, 9
NUMERIC, ANALYSIS in DRAW mode, 1
NUMERIC, SOLIDE > ANALYZE | SELF in SPACE mode, 396
NUMERIC, SOLIDM > ANALYZE | SELF in SPACE mode, 420
OBLIQUE, TRANSFOR > CREATE | AFFINITY | PLANE in 3D SPACE mode, 501
OBLIQUE, TRANSFOR > CREATE | AFFINITY in 2D SPACE mode, 493
OBLIQUE, TRANSFOR > CREATE | AFFINITY in DRAW mode, 485
OBLIQUE, TRANSFOR > CREATE | SYMMETRY | LINE in 2D SPACE mode, 491
OBLIQUE, TRANSFOR > CREATE | SYMMETRY | LINE in DRAW mode, 483
OBLIQUE, TRANSFOR > CREATE | SYMMETRY | PLANE in 3D SPACE mode, 500
OFFSET, CURVE2 > PARALLEL in 2D SPACE mode, 80
OFFSET, CURVE2 > PARALLEL in DRAW mode, 70
OFFSET, SHAPE in DRAW mode, 373
OFFSET, SOLIDE > CREATE | COMPLEX in SPACE mode, 388
OFFSET, SOLIDM > CREATE | COMPLEX in SPACE mode, 415
OPEN, SHAPE > CREATE | POLYGON in DRAW mode, 373
OPERATE, SHAPE in DRAW mode, 375
OPERATN, SOLIDE > MODIFY in SPACE mode, 402
OPERATN, SOLIDE in SPACE mode, 398
OPERATN, SOLIDM > in SPACE mode, 417
OPERATN, SOLIDM > MODIFY in SPACE mode, 425
ORIENTAT, TEXTD2 > MODIFY in DRAW mode, 478
ORIENTN, PLANE in SPACE mode, 321
ORIG.DEF, SYMBOL > DELETE | USED in DRAW mode, 449
ORTHOGNL, POINT > PROJECT in 2D SPACE mode, 353
ORTHOGNL, POINT > PROJECT in 3D SPACE mode, 361
ORTHOGNL, POINT > PROJECT in DRAW mode, 341
OTHER, LAYER > FILTER | MODIFY | CUR FILT in DRAW and SPACE modes, 210
PACK, ERASE in DRAW and SPACE modes, 154
PANEL, IUA > FILE in DRAW and SPACE modes, 197
PARAL LN, LIMIT1 > MACHINE | SHT JOG in 2D SPACE mode, 228
PARAL LN, LIMIT1 > MACHINE | SHT JOG in DRAW mode, 221
PARALLEL, CURVE2 in 2D SPACE mode, 80
PARALLEL, CURVE2 in 3D SPACE mode, 85
PARALLEL, CURVE2 in DRAW mode, 70
PARALLEL, LINE in 2D SPACE mode, 263
PARALLEL, LINE in 3D SPACE mode, 279
PARALLEL, LINE in DRAW mode, 243
PARALLEL, PLANE > EDGES in SPACE mode, 323
PARALLEL, PLANE in SPACE mode, 319
PARALLEL, TEXT > T.NODE | CREATE in DRAW mode, 453
PARAMETER, PARAM3D in SPACE mode, 304
PARENT, PARAM3D > ANALYZE | RELATION in SPACE mode, 308
PARENTS, ANALYSIS > LOGICAL in 2D SPACE mode, 8
PARENTS, ANALYSIS > LOGICAL in 3D SPACE mode, 12
PARM TXT, PLOT > FILE | LAYOUT in DRAW and SPACE modes, 337
PARM TXT, PLOT > QUICK | LAYOUT in DRAW and SPACE modes, 331
PARM, AUXVIEW2 > USE | VIEW, 33
PARM, SOLIDE > ANALYZE | SELF in SPACE mode, 396
PARM, SOLIDE > MODIFY | GEOMETRY in SPACE mode, 401
PARM, SOLIDM > ANALYZE | SELF in SPACE mode, 420
PARM, SOLIDM > MODIFY | GEOMETRY in SPACE mode, 424
PART IN, GROUP > GRP_1, 2 or 3 | EXCLUDE | CUR_TRAP or WSP_TRAP in SPACE mode, 183
PART IN, GROUP > GRP_1, 2 or 3 | INCLUDE | CUR_TRAP or WSP_TRAP in SPACE mode, 183
PART OUT, GROUP > GRP_1, 2 or 3 | EXCLUDE | CUR_TRAP or WSP_TRAP in SPACE mode, 183

PART OUT, GROUP > GRP_1, 2 or 3 | INCLUDE | CUR_TRAP or WSP_TRAP in SPACE mode, 183
PART IN, GROUP > GRP_1, 2 or 3 | EXCLUDE | TRAP in DRAW mode, 179
PART IN, GROUP > GRP_1, 2 or 3 | INCLUDE | TRAP in DRAW mode, 179
PART OUT, GROUP > GRP_1, 2 or 3 | EXCLUDE | TRAP in DRAW mode, 179
PART OUT, GROUP > GRP_1, 2 or 3 | INCLUDE | TRAP in DRAW mode, 179
PART-ARC, CURVE2 > CIRCLE in 2D SPACE mode, 75
PART-ARC, CURVE2 > CIRCLE in DRAW mode, 64
PARTIAL, CURVE2 > ELLIPSE | AXIS in 2D SPACE mode, 77
PARTIAL, CURVE2 > ELLIPSE | AXIS in DRAW mode, 67
PARTIAL, CURVE2 > ELLIPSE | CENTER in 2D SPACE mode, 77
PARTIAL, CURVE2 > ELLIPSE | CENTER in DRAW mode, 67
PATH, TEXTD2 > MODIFY | LEADER in DRAW mode, 477
PATTERN, DRW STD, 153
PATTERN, PATTERN in DRAW mode, 313
PICK or NO PICK, LAYER > LAYER | VERIFY in DRAW and SPACE modes, 211
PICK or NO PICK, SETS > CHANGE in DRAW and SPACE modes, 369
PIPE, SOLIDE > CREATE | CANONIC in SPACE mode, 387
PIPE, SOLIDM > CREATE | CANONIC in SPACE mode, 414
PLANE, GROUP > GRP_1, 2 or 3 | INCLUDE or EXCLUDE | W.SPACE in SPACE mode, 185
PLANE, IMAGE > WINDOW | MODIFY in DRAW and SPACE modes, 189
PLANE, TRANSFOR > CREATE | SYMMETRY in 3D SPACE mode, 499
PLOTTING, PLOT > FILE in DRAW and SPACE modes, 338
PLOTTING, PLOT > QUICK in DRAW and SPACE modes, 331
POINT, CURVE2 > SPLINE > IMPOSE in DRAW mode, 69
POINT, TRANSFOR > CREATE | SYMMETRY in 2D SPACE mode, 492
POINT, TRANSFOR > CREATE | SYMMETRY in 3D SPACE mode, 500
POINT, TRANSFOR > CREATE | SYMMETRY in DRAW mode, 484
POL EDGE, LINE in 3D SPACE mode, 286
POLAR, POINT > COORD | CHAIN in DRAW mode, 343
POLAR, POINT > COORD | CHAIN in 2D SPACE mode, 355
POLAR, POINT > COORD | REPEAT in DRAW mode, 243
POLAR, POINT > COORD | REPEAT in 2D SPACE mode, 355
POLAR, POINT > COORD | SINGLE in DRAW mode, 343
POLAR, POINT > COORD | SINGLE in 2D SPACE mode, 355
POLYGON, SHAPE > CREATE in DRAW mode, 371
POSITN, PARAM3D > DIMENSION in SPACE mode, 307
POSITN, SOLIDE > ANALYZE in SPACE mode, 397
POSITN, SOLIDM > ANALYZE in SPACE mode, 421
PREVIEW, PLOT > FILE in DRAW and SPACE modes, 338
PREVIEW, PLOT > QUICK | PLOTTING in DRAW and SPACE modes, 332
PRIMARY VIEW, AUXVIEW2 > USE | VIEW | NEW, 26
PRIMITIV, PARAM3D in SPACE mode, 302
PRIMITIV, PARAM3D > DELETE in SPACE mode, 309
PRIMSKIN, SOLIDE > EXTRACT in SPACE mode, 406
PRIMSOL, SOLIDE > EXTRACT in SPACE mode, 406
PRIMVOL, SOLIDE > EXTRACT in SPACE mode, 406
PRIMVOL, SOLIDM > EXTRACT in SPACE mode, 429
PRINCIPAL VIEW, AUXVIEW2 > USE | VIEW | NEW, 26
PRINT, PLOT > FILE | PLOTTING in DRAW and SPACE modes, 338
PRINT, PLOT > PLOTTING | QUICK in DRAW and SPACE modes, 331
PRISM, SOLIDE > CREATE | CANONIC in SPACE mode, 379
PRISM, SOLIDM > CREATE | CANONIC in SPACE mode, 412
PRL WINW, PLANE in SPACE mode, 322

PROC, IUA > FILE in DRAW and SPACE modes, 197
PROFILE, PARAM3D > DELETE in SPACE mode, 309
PROFILE, PARAM3D in SPACE mode, 300
PROFILE, PLOT > QUICK | PLOTTING in DRAW and SPACE modes, 332
PROJ INT, POINT in 2D SPACE mode, 351
PROJ INT, POINT in 3D SPACE mode, 360
PROJ INT, POINT in DRAW mode, 339
PROJECT, LINE in 3D SPACE mode, 282
PROJECT, POINT in 2D SPACE mode, 352
PROJECT, POINT in 3D SPACE mode, 361
PROJECT, POINT in DRAW mode, 341
PROJECT, SOLIDE > CREATE | COMPLEX in SPACE mode, 388
PROJECT, SOLIDE > EXTRACT in SPACE mode, 405
PROJECT, SOLIDM > COMPLEX in SPACE mode, 416
PROJECT, SOLIDM > EXTRACT in SPACE mode, 429
PROJECT, SPC->DR2 in DRAW mode, 432
PROJECT, SPC->DRW in DRAW mode, 434
PT/LN, PLANE > EDGES in SPACE mode, 322
PT-PT, LINE in 2D SPACE mode, 261
PT-PT, LINE in 3D SPACE mode, 278
PT-PT, LINE in DRAW mode, 241
PTS CST, CURVE2 in 2D SPACE mode, 79
PTS CST, CURVE2 in 3D SPACE mode, 88
PTS, POINT > LIMITS in 2D SPACE mode, 358
PTS, POINT > LIMITS in 3D SPACE mode, 363
PTS, POINT > SPACES in 2D SPACE mode, 356
PTS, POINT > SPACES in 3D SPACE mode, 365
PYRAMID, SOLIDE > CREATE | CANONIC in SPACE mode, 387
PYRAMID, SOLIDM > CREATE | CANONIC in SPACE mode, 414
QUICK, PLOT in DRAW and SPACE modes, 327
RADIUS, CURVE2 > CIRCLE in 2D SPACE mode, 74
RADIUS, CURVE2 > CIRCLE in DRAW mode, 64
RADIUS, DIMENS > CREATE in DRAW mode, 124
RE_CUT, AUXVIEW2 > USE | VIEW | FILTER, 32
RE_USE, AUXVIEW2 > USE | VIEW | FILTER, 31
READ, FILE in DRAW and SPACE modes, 157
READ, LIBRARY in DRAW and SPACE modes, 214
READ, PLOT > FILE | SHEET in DRAW and SPACE modes, 333
RECALL, IMAGE > SCREEN in DRAW and SPACE modes, 195
RECALL, IMAGE > WINDOW in DRAW and SPACE modes, 193
RECTANG, POINT > COORD | CHAIN in 2D SPACE mode, 355
RECTANG, POINT > COORD | CHAIN in 3D SPACE mode, 362
RECTANG, POINT > COORD | CHAIN in DRAW mode, 343
RECTANG, POINT > COORD | REPEAT in 2D SPACE mode, 355
RECTANG, POINT > COORD | REPEAT in 3D SPACE mode, 363
RECTANG, POINT > COORD | REPEAT in DRAW mode, 343
RECTANG, POINT > COORD | SINGLE in 2D SPACE mode, 355
RECTANG, POINT > COORD | SINGLE in 3D SPACE mode, 362
RECTANG, POINT > COORD | SINGLE in DRAW mode, 343
RECTANGL, SHAPE > CREATE | POLYGON in DRAW mode, 372
REFRAME, IMAGE > WINDOW | MODIFY | PLANE in DRAW and SPACE modes, 191
RELATION, PARAM3D > ADVANCED in SPACE mode, 309
RELATION, PARAM3D > ANALYZE in SPACE mode, 308
RELATION, PARAM3D > DELETE in SPACE mode, 309
RELATIVE, ANALYSIS > INERTIA | COMBINE in 3D SPACE mode, 13
RELATIVE, ANALYSIS > INERTIA | COMPUTE in 3D SPACE mode, 13
RELATIVE, ANALYSIS > NUMERIC | COMPUTE in 2D SPACE mode, 5
RELATIVE, ANALYSIS > NUMERIC | COMPUTE in 3D SPACE mode, 9
RELATIVE, ANALYSIS in 2D SPACE mode, 7
RELATIVE, ANALYSIS in 3D SPACE mode, 11
RELATIVE, ANALYSIS in DRAW mode, 3
RELATIVE, SOLIDE > ANALYZE | POSITN in SPACE mode, 397

RELATIVE, SOLIDM > ANALYZE | POSITN in SPACE mode, 421
RELATN, SOLIDE > ANALYZE | POSITN | INTERFER in SPACE mode, 397
RELATN, SOLIDM > ANALYZE | POSITN | INTERFER in SPACE mode, 421
RELIMIT, LIMIT1 in 2D SPACE mode, 224
RELIMIT, LIMIT1 in 3D SPACE mode, 231
RELIMIT, LIMIT1 in DRAW mode, 217
RENAME, AUXVIEW in DRAW mode, 20
RENAME, AXIS in DRAW mode, 51
RENAME, DETAIL > MANAGE in 3D SPACE mode, 109
RENAME, DETAIL > MANAGE in DRAW mode, 101
RENAME, DRAFT in DRAW mode, 138
RENAME, DRW STD > ANNOTAT | DESCRIPT, 146
RENAME, DRW STD > PATTERN, 153
RENAME, FILE in DRAW and SPACE modes, 159
RENAME, IDENTIFY in DRAW and SPACE modes, 186
RENAME, IMAGE > SCREEN in DRAW mode, 196
RENAME, IMAGE > WINDOW in DRAW and SPACE modes, 193
RENAME, LAYER > FILTER in DRAW and SPACE modes, 209
RENAME, PLOT > FILE | SHEET in DRAW and SPACE modes, 334
RENAME, SOLIDE > MODIFY | DRESS UP in SPACE mode, 404
RENAME, SOLIDM > MODIFY | DRESS UP in SPACE mode, 428
RENAME, SYMBOL > MANAGE in DRAW mode, 451
RENUMBER, IDENTIFY in DRAW and SPACE modes, 187
REPEAT, ANALYSIS > RELATIVE in 2D SPACE mode, 7
REPEAT, ANALYSIS > RELATIVE in 3D SPACE mode, 11
REPEAT, POINT > COORD in 2D SPACE mode, 355
REPEAT, POINT > COORD in 3D SPACE mode, 363
REPEAT, POINT > COORD in DRAW mode, 343
REPEAT, POINT > PROJ INT in 2D SPACE mode, 352
REPEAT, POINT > PROJ INT in 3D SPACE mode, 361
REPEAT, POINT > PROJ INT in DRAW mode, 339
REPLACE, CURVE2 > CIRCLE | MODIFY in 2D SPACE mode, 76
REPLACE, CURVE2 > CIRCLE | MODIFY in DRAW mode, 66
REPLACE, CURVE2 > CIRCLE in 3D SPACE mode, 91
REPLACE, DETAIL > MANAGE in 3D SPACE mode, 108
REPLACE, DETAIL > MANAGE in DRAW mode, 100
REPLACE, DETAIL > TRANSFER in 3D SPACE mode, 110
REPLACE, DETAIL > TRANSFER in DRAW mode, 102
REPLACE, LINE > MODIFY in 2D SPACE mode, 273
REPLACE, LINE > MODIFY in 3D SPACE mode, 286
REPLACE, LINE > MODIFY in DRAW mode, 252
REPLACE, MODELS > MODIFY in DRAW and SPACE modes, 295
REPLACE, PATTERN in DRAW mode, 315
REPLACE, SOLIDE > MODIFY | OPERATN in SPACE mode, 402
REPLACE, SOLIDM > MODIFY | OPERATN in SPACE mode, 426
REPLACE, SYMBOL > DEFINE in DRAW mode, 448
REPLACE, SYMBOL > MANAGE in DRAW mode, 450
REPLACE, SYMBOL > MODIFY | FLIPX in DRAW mode, 447
REPLACE, SYMBOL > MODIFY | FLIPY in DRAW mode, 447
REPLACE, SYMBOL > MODIFY | ROTATE in DRAW mode, 447
REPLACE, SYMBOL > MODIFY | SCALE in DRAW mode, 447
REPLACE, SYMBOL > MODIFY | SYMMETRY in DRAW mode, 447
REPLACE, SYMBOL > MODIFY | TRANSLATE in DRAW mode, 446
REPLACE, TRANSFOR > APPLY in 2D SPACE mode, 494
REPLACE, TRANSFOR > APPLY in 3D SPACE mode, 504

RESET, GRAPHIC > MOD VISU in SPACE mode, 175
RESET, GROUP > GRP_1, 2 or 3 in DRAW mode, 181
RESET, GROUP > GRP_1, 2 or 3 in SPACE mode, 185
RESET, LAYER > FILTER in DRAW and SPACE modes, 210
RESTORE, DIMENS > MODIFY | EXT LINE in DRAW mode, 131
RESTORE, PARAM3D > DIMENSION in SPACE mode, 306
RESTORE, SOLIDM > in SPACE mode, 430
RESTORE, STANDARD > COLOR | TABLE in DRAW and SPACE modes, 440
RETRIEVE, FILE > KEYBOARD in DRAW and SPACE modes, 163
REVOLUTN, SOLIDE > CREATE | CANONIC in SPACE mode, 381
REVOLUTN, SOLIDM > CREATE | CANONIC in SPACE mode, 412
ROTATE, AUXVIEW > MODIFY | VIEW in DRAW mode, 19
ROTATE, AUXVIEW2 > USE | VIEW, 33
ROTATE, CURVE2 > DEPTH in 3D SPACE mode, 90
ROTATE, DETAIL > MODIFY in 3D SPACE mode, 105
ROTATE, DETAIL > MODIFY in DRAW mode, 97
ROTATE, MODELS > MODIFY | DUPLICAT in DRAW and SPACE modes, 295
ROTATE, MODELS > MODIFY | REPLACE in DRAW and SPACE modes, 295
ROTATE, TRANSFOR > CREATE in 2D SPACE mode, 491
ROTATE, TRANSFOR > CREATE in 3D SPACE mode, 499
ROTATE, TRANSFOR > CREATE in DRAW mode, 483
ROTATE, TRANSFOR > FREEHAND in DRAW mode, 488
ROUGHNESS, TEXTD2 > CREATE in DRAW mode, 475
SAME BGD, AUXVIEW > CREATE in DRAW mode, 15
SAME BGD, AUXVIEW > CREATE in DRAW mode, 15
SAME, GRAPHIC > MODIFY in DRAW mode, 167
SAME, GRAPHIC > MOD GEN in SPACE mode, 172
SAME, GRAPHIC > MOD SPEC in SPACE mode, 173
SAME, GRAPHIC > MOD VISU in SPACE mode, 174
SAME, GRAPHIC > VERIFY in DRAW mode, 169
SAME, GRAPHIC > VERIFY in SPACE mode, 176
SAME, SETS > COPY in DRAW and SPACE modes, 370
SAME, SPC->DRW > PROJECT in DRAW mode, 434
SAVE, STANDARD > COLOR | TABLE in DRAW and SPACE modes, 440
SC/FRAME, AUXVIEW > MODIFY | VIEW in DRAW mode, 18
SC/FRAME, IMAGE > WINDOW | MODIFY | PLANE in DRAW and SPACE modes, 190
SCALE FRAME, AUXVIEW2 > USE | FRAME, 36
SCALE, AUXVIEW > MODIFY | VIEW in DRAW mode, 18
SCALE, AUXVIEW2 > USE | VIEW, 33
SCALE, DETAIL > MODIFY in 3D SPACE mode, 105
SCALE, DETAIL > MODIFY in DRAW mode, 97
SCALE, SOLIDE > MODIFY | GEOMETRY in SPACE mode, 400
SCREEN, IMAGE in DRAW and SPACE modes, 193
SECTION CUT, AUXVIEW2 > USE | VIEW | NEW, 28
SECTION VIEW, AUXVIEW2 > USE | VIEW | NEW, 27
SECTION, AUXVIEW2 > USE, 40
SECTION, COMBIVU in DRAW mode, 61
SEGMENT, ANGLE | XY-XZ in 3D SPACE mode, 283
SEGMENT, ANGLE | XY-YZ in 3D SPACE mode, 283
SEGMENT, ANGLE | XZ-YZ in 3D SPACE mode, 283
SEGMENT, LINE > ANGLE in 2D SPACE mode, 269
SEGMENT, LINE > ANGLE in DRAW mode, 248
SEGMENT, LINE > BISECT in 2D SPACE mode, 268

SEGMENT, LINE > BISECT in DRAW mode, 248
SEGMENT, LINE > COMPON in 2D SPACE mode, 270
SEGMENT, LINE > COMPON in 3D SPACE mode, 284
SEGMENT, LINE > COMPON in DRAW mode, 249
SEGMENT, LINE > HORIZONT in 2D SPACE mode, 265
SEGMENT, LINE > HORIZONT in DRAW mode, 245
SEGMENT, LINE > MEDIAN in 2D SPACE mode, 268
SEGMENT, LINE > MEDIAN in DRAW mode, 247
SEGMENT, LINE > NORMAL in 2D SPACE mode, 266
SEGMENT, LINE > NORMAL in DRAW mode, 246
SEGMENT, LINE > PARALLEL in 2D SPACE mode, 263
SEGMENT, LINE > PARALLEL in 3D SPACE mode, 279
SEGMENT, LINE > PARALLEL in DRAW mode, 243
SEGMENT, LINE > POL EDGE in 3D SPACE mode, 286
SEGMENT, LINE > PT-PT in 2D SPACE mode, 262
SEGMENT, LINE > PT-PT in 3D SPACE mode, 278
SEGMENT, LINE > PT-PT in DRAW mode, 241
SEGMENT, LINE > TANGENT in 2D SPACE mode, 271
SEGMENT, LINE > TANGENT in 3D SPACE mode, 285
SEGMENT, LINE > TANGENT in DRAW mode, 251
SEGMENT, LINE > VERTICAL in DRAW mode, 245
SELECT, KEEP in DRAW and SPACE modes, 202
SELECT, MERGE in DRAW and SPACE modes, 292
SELECT, PATTERN in DRAW mode, 314
SELF, SOLIDE > ANALYZE in SPACE mode, 396
SELF, SOLIDM > ANALYZE in SPACE mode, 420
SEPARATE, IMAGE > SCREEN | MODIFY in DRAW and SPACE modes, 194
SET, GROUP > GRP_1, 2 or 3 | EXCLUDE | W.SPACE in DRAW mode, 181
SET, GROUP > GRP_1, 2 or 3 | EXCLUDE | W.SPACE in SPACE mode, 185
SET, GROUP > GRP_1, 2 or 3 INCLUDE | W.SPACE in DRAW mode, 181
SET, GROUP > GRP_1, 2 or 3 INCLUDE | W.SPACE in SPACE mode, 185
SET, IDENTIFY > RENAME | DISPLAY in DRAW and SPACE modes, 186
SET, IDENTIFY > RENAME | LIST in DRAW and SPACE modes, 186
SET, KEEP > SELECT | APPLICTN in DRAW and SPACE modes, 204
SET, KEEP > SELECT | GEOMETRY in DRAW and SPACE modes, 203
SET, MERGE > SELECT in DRAW and SPACE modes, 293
SET, MODELS > COPY in DRAW and SPACE modes, 297
SET, PARAM3D > TRANSFOR in SPACE mode, 305
SET, STANDARD > COLOR | MODIFY in DRAW and SPACE modes, 439
SET, TRANSFOR > APPLY | DUPLICAT in 2D SPACE mode, 495
SET, TRANSFOR > APPLY | DUPLICAT in 3D SPACE mode, 505
SET, TRANSFOR > APPLY | REPLACE in 2D SPACE mode, 494
SET, TRANSFOR > APPLY | REPLACE in 3D SPACE mode, 504
SEWING, SOLIDE > OPERATN in SPACE mode, 395
SEWING, SOLIDM > MODIFY | OPERATN | INSERT in SPACE mode, 425
SHEET, PLOT > FILE in DRAW and SPACE modes, 332
SHELL, SOLIDE > OPERATN in SPACE mode, 394
SHOW BOX, DIMENS > MANAGE | VISUALTN in DRAW mode, 135
SHOW BOX, TEXTD2 > MANAGE | VISUALATN in DRAW mode, 480
SHOW TXT, TEXTD2 > MANAGE | VISUALATN in DRAW mode, 480
SHOW, AUXVIEW > MODIFY | FRAME in DRAW mode, 18
SHOW, AUXVIEW2 > USE | FRAME, 36

SHOW, PARAM3D > DIMENSION in SPACE mode, 306
SHOW, TEXT in SPACE mode, 460
SHT JOG, LIMIT1 > MACHINE in 2D SPACE mode, 228
SHT JOG, LIMIT1 > MACHINE in 3D SPACE mode, 235
SHT JOG, LIMIT1 > MACHINE in DRAW mode, 221
SIMILTRY, TRANSFOR > CREATE in DRAW mode, 485
SIMPLE, TEXTD2 > CREATE | TEXT in DRAW mode, 465
SIMULTRY, TRANSFOR > CREATE in 2D SPACE mode, 493
SINGL RAD, LIMIT1 > MACHINE | MACH JOG in 2D SPACE mode, 227
SINGL RAD, LIMIT1 > MACHINE | MACH JOG in 3D SPACE mode, 234
SINGL RAD, LIMIT1 > MACHINE | MACH JOG in DRAW mode, 220
SINGLE, ANALYSIS > RELATIVE in 2D SPACE mode, 7
SINGLE, ANALYSIS > RELATIVE in 3D SPACE mode, 11
SINGLE, DETAIL > COPY | LIBRARY | COMPACT in DRAW mode, 96
SINGLE, DETAIL > COPY | LIBRARY | STANDARD in DRAW mode, 96
SINGLE, DETAIL > DITTO | LIBRARY | COMPACT in DRAW mode, 96
SINGLE, DETAIL > DITTO | LIBRARY | STANDARD in DRAW mode, 96
SINGLE, POINT > COORD in 2D SPACE mode, 355
SINGLE, POINT > COORD in 3D SPACE mode, 362
SINGLE, POINT > COORD in DRAW mode, 343
SINGLE, POINT > PROJ INT in 2D SPACE mode, 351
SINGLE, POINT > PROJ INT in 3D SPACE mode, 360
SINGLE, POINT > PROJ INT in DRAW mode, 339
SINGLE, SYMBOL > COPY | LIBRARY | COMPACT in DRAW mode, 444
SINGLE, SYMBOL > COPY | LIBRARY | STANDARD in DRAW mode, 444
SIZE, AUXVIEW > MODIFY | FRAME in DRAW mode, 17
SLANT, DIMENS > MODIFY | EXT LINE in DRAW mode, 130
SLOPE, DIMENS2 > CREATE in DRAW mode, 127
SOL TYPE, SOLIDE > MODIFY in SPACE mode, 401
SORT OUT, SOLIDE > OPERATN in SPACE mode, 395
SORT OUT, SOLIDM > OPERATN in SPACE mode, 419
SPAC ELT, STANDARD in DRAW and SPACE modes, 436
SPACE, AUXVIEW > MODIFY | VIEW in DRAW mode, 20
SPACE, IMAGE > WINDOW | MODIFY in DRAW and SPACE modes, 191
SPACES, DETAIL > COPY | LIBRARY | COMPACT in DRAW mode, 96
SPACES, DETAIL > COPY | LIBRARY | STANDARD in DRAW mode, 96
SPACES, DETAIL > DITTO | LIBRARY | COMPACT in DRAW mode, 96
SPACES, DETAIL > DITTO | MODEL | COMPACT in DRAW mode, 95
SPACES, DETAIL > DITTO | LIBRARY | STANDARD in DRAW mode, 96
SPACES, DETAIL, > DITTO | MODEL | STANDARD in DRAW mode, 95
SPACES, PLANE in SPACE mode, 323
SPACES, POINT in 2D SPACE mode, 357
SPACES, POINT in 3D SPACE mode, 365
SPACES, POINT in DRAW mode, 345
SPACES, SYMBOL > COPY | LIBRARY | COMPLEX in DRAW mode, 444
SPC ELEM, GROUP > GRP_1, 2 or 3 | INCLUDE or EXCLUDE | CURRENT or W.SPACE in SPACE mode, 185
SPEC ELT, STANDARD > SPAC ELT in DRAW and SPACE modes
SPECIFIC, MERGE > SELECT | SET in DRAW and SPACE modes, 293
SPHERE, SOLIDE > CREATE | CANONIC in SPACE mode, 386
SPHERE, SOLIDM > CREATE | CANONIC in SPACE mode, 413
SPHERIC, POINT > COORD | CHAIN in 3D SPACE mode, 363
SPHERIC, POINT > COORD | REPEAT in 3D SPACE mode, 363
SPHERIC, POINT > COORD | SINGLE in 3D SPACE mode, 362

SPINE, CURVE2 in 3D SPACE mode, 92
SPLINE, CURVE2 in DRAW mode, 69
SPLIT, SHAPE > OPERATE in DRAW mode, 376
SPLIT, SOLIDE > OPERATN in SPACE mode, 394
SPLIT, SOLIDM > OPERATN in SPACE mode, 418
SQUARE, SHAPE > CREATE | POLYGON in DRAW mode, 372
STANDARD, ANALYSIS > LOGICAL in 2D SPACE mode, 7
STANDARD, ANALYSIS > LOGICAL in 3D SPACE mode, 12
STANDARD, CURVE2 > PARALLEL in 2D SPACE mode, 80
STANDARD, CURVE2 > PARALLEL in DRAW mode, 70
STANDARD, DETAIL > COPY | LIBRARY in 3D SPACE mode, 104
STANDARD, DETAIL > COPY | MODEL in 3D SPACE mode, 104
STANDARD, DETAIL > COPY | LIBRARY in DRAW mode, 96
STANDARD, DETAIL > COPY | MODEL in DRAW mode, 96
STANDARD, DETAIL > DITTO | LIBRARY in DRAW mode, 96
STANDARD, DETAIL > DITTO | MODEL in DRAW mode, 94
STANDARD, DRW STD > ANNOTAT, 152
STANDARD, GRAPHIC > MODIFY in DRAW mode, 167
STANDARD, GRAPHIC > MOD SPEC in SPACE mode, 173
STANDARD, GRAPHIC > VERIFY in DRAW mode, 169
STANDARD, GRAPHIC > MOD GEN in SPACE mode, 173
STANDARD, GRAPHIC > VERIFY in SPACE mode, 176
STANDARD, LINE > PT-PT | SEGMENT in 2D SPACE model, 262
STANDARD, LINE > PT-PT | UNLIM in 2D SPACE mode, 262
STANDARD, SETS > COPY in DRAW and SPACE modes, 370
STANDARD, SPC->DRW > PROJECT in DRAW mode, 380
STANDARD, SYMBOL > COPY | MODEL in DRAW mode, 445
STANDARD, SYMBOL > SYMBOL | LIBRARY in DRAW mode, 444
STANDARD, SYMBOL > SYMBOL | MODEL in DRAW mode, 443
STANDARD, TEXT in DRAW mode, 454
STARTUP, DIMENS > MANAGE in DRAW mode, 136
STARTUP, TEXTD2 > MANAGE in DRAW mode, 481
STORE, FILE > KEYBOARD in DRAW and SPACE modes, 164
STORE, IMAGE > SCREEN in DRAW and SPACE modes, 195
STORE, IMAGE > WINDOW in DRAW and SPACE modes, 193
STORE, TRANSFOR in 2D SPACE mode, 496
STORE, TRANSFOR in 3D SPACE mode, 505
STRING, GROUP > GRP_1, 2 or 3 | EXCLUDE | CURRENT in DRAW mode, 180
STRING, GROUP > GRP_1, 2 or 3 | EXCLUDE | W.SPACE in DRAW mode, 180
STRING, GROUP > GRP_1, 2 or 3 | INCLUDE | CURRENT in DRAW mode, 180
STRING, GROUP > GRP_1, 2 or 3 | INCLUDE | W.SPACE in DRAW mode, 180
STRING, GROUP > GRP_1, 2 or 3 | EXCLUDE | CURRENT in SPACE mode, 184
STRING, GROUP > GRP_1, 2 or 3 | EXCLUDE | W.SPACE in SPACE mode, 184
STRING, GROUP > GRP_1, 2 or 3 | INCLUDE | CURRENT in SPACE mode, 184
STRING, GROUP > GRP_1, 2 or 3 | INCLUDE | W.SPACE in SPACE mode, 184
STRING, IDENTIFY > RENAME in DRAW and SPACE modes, 187
SUBTRACT, SHAPE > OPERATE in DRAW mode, 375
SUBTRACT, SOLIDE > OPERATN in SPACE mode, 391
SUBTRACT, SOLIDM > OPERATN in SPACE mode, 418
SURFACE, SOLIDM > CREATE in SPACE mode, 416

SWAP, AXIS in 2D SPACE mode, 53
SWAP, AXIS in 3D SPACE mode, 58
SWAP, AXIS in DRAW mode, 50
SWEEP, SOLIDE > CREATE | CANONIC in SPACE mode, 381
SYMBOL, DIMENS > MODIFY in DRAW mode, 131
SYMBOL, DRW STD > ANNOTAT | DESCRIPT, 146
SYMBOL, LIBRARY > FAMILY in DRAW and SPACE modes, 214
SYMBOL, LIBRARY > MODIFY in DRAW and SPACE modes, 215
SYMBOL, LIBRARY > READ in DRAW and SPACE modes, 214
SYMBOL, LIBRARY > UPDATE in DRAW and SPACE modes, 216
SYMBOL, LIBRARY > WRITE in DRAW and SPACE modes, 214
SYMBOL, SYMBOL in DRAW mode, 443
SYMBOL, TEXTD2 > MODIFY in DRAW mode, 478
SYMBOLS, SYMBOL > DELETE | USED in DRAW mode, 449
SYMMETRY, DETAIL > MODIFY in 3D SPACE mode, 106
SYMMETRY, DETAIL > MODIFY in DRAW mode, 97
SYMMETRY, MODELS > MODIFY | DUPLICAT in DRAW and SPACE modes, 296
SYMMETRY, MODELS > MODIFY | REPLACE in DRAW and SPACE modes, 296
SYMMETRY, SYMBOL > MODIFY in DRAW mode, 447
SYMMETRY, TRANSFOR > CREATE in 2D SPACE mode, 491
SYMMETRY, TRANSFOR > CREATE in 3D SPACE mode, 499
SYMMETRY, TRANSFOR > CREATE in DRAW mode, 483
T.NODE, TEXT in DRAW mode, 453
TABLE, STANDARD > COLOR in DRAW and SPACE modes, 440
TANGENT, CURVE2 > SPLINE > IMPOSE in DRAW mode, 69
TANGENT, LINE in 2D SPACE mode, 271
TANGENT, LINE in 3D SPACE mode, 285
TANGENT, LINE in DRAW mode, 250
TANGENT, POINT in 2D SPACE mode, 359
TANGENT, POINT in 3D SPACE mode, 368
TANGENT, POINT in DRAW mode, 346
TEXT, DIMENS > MODIFY in DRAW mode, 132
TEXT, TEXTD2 > CREATE in DRAW mode, 465
TEXT, TEXTD2 > MODIFY in DRAW mode, 467
TGT CONT, CURVE2 in 2D SPACE mode, 83
TGT CONT, CURVE2 in 3D SPACE mode, 89
THICK, SOLIDE > OPERATN in SPACE mode, 391
THREAD, MARK UP > AXIS in DRAW mode, 289
THREE-PT, CURVE2 > CIRCLE in 2D SPACE mode, 74
THREE-PT, CURVE2 > CIRCLE in DRAW mode, 64
THROUGH, PLANE in SPACE mode, 318
TOL, DIMENS > MODIFY in DRAW mode, 131
TORUS, SOLIDE > CREATE | CANONIC in SPACE mode, 386
TORUS, SOLIDM > CREATE | CANONIC in SPACE mode, 414
TRANSFER, AUXVIEW in DRAW mode, 21
TRANSFER, DETAIL in 3D SPACE mode, 110
TRANSFER, DETAIL in DRAW mode, 102
TRANSFER, DRAFT in DRAW mode, 138
TRANSFER, LAYER > LAYER in DRAFT and SPACE modes, 211
TRANSFER, SETS in DRAW and SPACE modes, 370
TRANSFOR, PARAM3D in SPACE mode, 305
TRANSLAT, AUXVIEW > MODIFY | VIEW in DRAW mode, 19
TRANSLAT, CURVE2 > DEPTH in 3D SPACE mode, 90
TRANSLAT, DETAIL > MODIFY in DRAW mode, 96
TRANSLAT, DETAIL > MODIFY in 3D SPACE mode, 105
TRANSLAT, IMAGE > SCREEN | MODIFY in DRAW and SPACE modes, 194
TRANSLAT, TRANSFOR > CREATE in 2D SPACE mode, 490
TRANSLAT, TRANSFOR > CREATE in 3D SPACE mode, 498
TRANSLAT, TRANSFOR > CREATE in DRAW mode, 483

TRANSLAT, TRANSFOR > FREEHAND in DRAW mode, 488
TRANSLATE, DETAIL > MODIFY in DRAW mode, 96
TRANSLATE, IMAGE > WINDOW | MODIFY | PLANE in DRAW and SPACE modes, 190
TRANSLATE, MODELS > MODIFY | DUPLICAT in DRAW and SPACE modes, 295
TRANSLATE, MODELS > MODIFY | REPLACE in DRAW and SPACE modes, 295
TRANSLATE, SYMBOL > MODIFY in DRAW mode, 446
TRANSPAR, GRAPHIC > MOD VISU | CHOOSE in SPACE mode, 174
TRAP, GROUP > GRP_1, 2 or 3 | EXCLUDE in DRAW mode, 178
TRAP, GROUP > GRP_1, 2 or 3 | INCLUDE in DRAW mode, 178
TRI TGT, SOLIDE > OPERATN | FILLET in SPACE mode, 393
TRIM ALL, LIMIT1 > CORNER in 2D SPACE mode, 225
TRIM ALL, LIMIT1 > CORNER in 3D SPACE mode, 232
TRIM ALL, LIMIT1 > CORNER in DRAW mode, 218
TRIM ALL, LIMIT1 > MACHINE | CHAMFER | ANGLE in 2D SPACE mode, 227
TRIM ALL, LIMIT1 > MACHINE | CHAMFER | ANGLE in 3D SPACE mode, 233
TRIM ALL, LIMIT1 > MACHINE | CHAMFER | ANGLE in DRAW mode, 219
TRIM ALL, LIMIT1 > MACHINE | CHAMFER | LENGTH in 2D SPACE mode, 227
TRIM ALL, LIMIT1 > MACHINE | CHAMFER | LENGTH in 3D SPACE mode, 233
TRIM ALL, LIMIT1 > MACHINE | CHAMFER | LENGTH in DRAW mode, 219
TRIM ALL, LIMIT1 > RELIMIT in 2D SPACE mode, 224
TRIM ALL, LIMIT1 > RELIMIT in 3D SPACE mode, 231
TRIM ALL, LIMIT1 > RELIMIT in DRAW mode, 217
TRIM ALL, SHAPE > OPERATE | INTERSEC in DRAW mode, 375
TRIM ALL, SHAPE > OPERATE | SUBTRACT in DRAW mode, 375
TRIM EL1, LIMIT1 > CORNER in 2D SPACE mode, 226
TRIM EL1, LIMIT1 > CORNER in 3D SPACE mode, 233
TRIM EL1, LIMIT1 > RELIMIT in 2D SPACE mode, 225
TRIM EL1, LIMIT1 > RELIMIT in 3D SPACE mode, 232
TRIM EL1, LIMIT1 > RELIMIT in DRAW mode, 218
TRIM EL1, LIMIT1 > CORNER in DRAW mode, 218
TRIM EL1, LIMIT1 > MACHINE | CHAMFER | ANGLE in 2D SPACE mode, 233
TRIM EL1, LIMIT1 > MACHINE | CHAMFER | LENGTH in 2D SPACE mode, 234
TRIM EL1, SHAPE > OPERATE | INTERSEC in DRAW mode, 375
TRIM EL1, SHAPE > OPERATE | SUBTRACT in DRAW mode, 375
TRIM EL2, LIMIT1 > MACHINE | MACH JOG | DBL RAD in 2D SPACE mode, 228
TRIM EL2, LIMIT1 > MACHINE | MACH JOG | DBL RAD in 3D SPACE mode, 235
TRIM EL2, LIMIT1 > MACHINE | MACH JOG | SINGL RAD in 2D SPACE mode, 227
TRIM EL2, LIMIT1 > MACHINE | MACH JOG | SINGL RAD in 3D SPACE mode, 234
TRIM EL2, LIMIT1 > MACHINE | MACH JOG | DBL RAD in DRAW mode, 221
TRIM EL2, LIMIT1 > MACHINE | MACH JOG | SINGL RAD in DRAW mode, 220
TRIM EL2, SHAPE > OPERATE | INTERSEC in DRAW mode, 375
TRIM EL2, SHAPE > OPERATE | SUBRACT in DRAW mode, 375
TRUE DIM, DIMENS > MODIFY | VALUE in DRAW mode, 129
TYPE, GROUP > GRP_1, 2 or 3 | EXCLUDE | CURRENT in DRAW mode, 180
TYPE, GROUP > GRP_1, 2 or 3 | EXCLUDE | W.SPACE in DRAW mode, 180
TYPE, GROUP > GRP_1, 2 or 3 | INCLUDE | CURRENT in DRAW mode, 180
TYPE, GROUP > GRP_1, 2 or 3 | INCLUDE | W.SPACE in DRAW mode, 180
TYPE, GROUP > GRP_1, 2 or 3 | EXCLUDE | CURRENT in SPACE mode, 184
TYPE, GROUP > GRP_1, 2 or 3 | EXCLUDE | W.SPACE in SPACE mode, 184

TYPE, GROUP > GRP_1, 2 or 3 | INCLUDE | CURRENT in SPACE mode, 184
TYPE, GROUP > GRP_1, 2 or 3 | INCLUDE | W.SPACE in SPACE mode, 184
TYPE, IDENTIFY > RENAME in DRAW and SPACE modes, 187
TYPE, STANDARD > COLOR | MODIFY in DRAW and SPACE modes, 439
UN_CUT, AUXVIEW2 > USE | VIEW | FILTER, 31
UN_USE, AUXVIEW2 > USE | VIEW | FILTER, 31
UNDO, TRANSFOR in DRAW mode, 489
UNFIXED, AXIS in 2D SPACE mode, 54
UNFIXED, AXIS in 3D SPACE mode, 58
UNFIXED, AXIS in DRAW mode, 50
UNION, SHAPE > OPERATE in DRAW mode, 375
UNION, SOLIDE > OPERATN in SPACE mode, 390
UNION, SOLIDM > OPERATN in SPACE mode, 417
UNLIM, ANGLE | XY-XZ in 3D SPACE mode, 284
UNLIM, ANGLE | XY-YZ in 3D SPACE mode, 284
UNLIM, ANGLE | XZ-YZ in 3D SPACE mode, 284
UNLIM, LINE > ANGLE in 2D SPACE mode, 269
UNLIM, LINE > ANGLE in DRAW mode, 249
UNLIM, LINE > BISECT in 2D SPACE mode, 268
UNLIM, LINE > BISECT in DRAW mode, 248
UNLIM, LINE > COMPON in 2D SPACE mode, 270
UNLIM, LINE > COMPON in 3D SPACE mode, 284
UNLIM, LINE > COMPON in DRAW mode, 250
UNLIM, LINE > HORIZONT in 2D SPACE mode, 265
UNLIM, LINE > HORIZONT in DRAW mode, 245
UNLIM, LINE > MEDIAN in 2D SPACE mode, 268
UNLIM, LINE > MEDIAN in DRAW mode, 247
UNLIM, LINE > NORMAL in 2D SPACE mode, 267
UNLIM, LINE > NORMAL in 3D SPACE mode, 281
UNLIM, LINE > NORMAL in DRAW mode, 247
UNLIM, LINE > PARALLEL in 2D SPACE mode, 264
UNLIM, LINE > PARALLEL in 3D SPACE mode, 279
UNLIM, LINE > PARALLEL in DRAW mode, 244
UNLIM, LINE > POL EDGE in 3D SPACE mode, 286
UNLIM, LINE > PROJECT in 3D SPACE mode, 282
UNLIM, LINE > PT-PT in 2D SPACE mode, 262
UNLIM, LINE > PT-PT in 3D SPACE mode, 278
UNLIM, LINE > PT-PT in DRAW mode, 241
UNLIM, LINE > TANGENT in 2D SPACE mode, 272
UNLIM, LINE > TANGENT in 3D SPACE mode, 285
UNLIM, LINE > TANGENT in DRAW mode, 251
UNLIM, LINE > VERTICAL in 2D SPACE mode, 265
UNLIM, LINE > VERTICAL in DRAW mode, 245
UNLIMITED, POINT > GRID in DRAW mode, 349
UNLINK, PARAM3D in SPACE mode, 303
UNLOCK, DRW STD > ANNOTAT | DESCRIPT, 146
UNLOCK, DRW STD > PATTERN, 153
UNSPEC, CURVE2 > CIRCLE | MULT-TGT in 2D SPACE mode, 76
UNSPEC, CURVE2 > CIRCLE | MULT-TGT in DRAW mode, 65
UNSPEC, IMAGE > WINDOW | DEFINE in DRAW and SPACE modes, 192
UNSPEC, MARK UP > ARROW in DRAW mode, 290
UNSPEC, POINT > GRID | LIMITED in DRAW mode, 348
UNSPEC, POINT > PROJECT in 2D SPACE mode, 354
UNSPEC, POINT > PROJECT in 3D SPACE mode, 362
UNSPEC, POINT > PROJECT in DRAW mode, 342
UNSPEC, SHAPE > CREATE in DRAW mode, 373

UNSPEC, TRANSFOR > CREATE in 2D SPACE mode, 494
UNSPEC, TRANSFOR > CREATE in 3D SPACE mode, 503
UNUSED, DETAIL > DELETE in 3D SPACE mode, 107
UNUSED, DETAIL > DELETE in DRAW mode, 98
UNUSED, SYMBOL > DELETE in DRAW mode, 449
UPD ALL, AUXVIEW2, 48
UPDATE, AUXVIEW2 > USE | BACK_PLN, 37
UPDATE, AUXVIEW2 > USE | BREAKOUT, 39
UPDATE, AUXVIEW2 > USE | CLIP, 37
UPDATE, AUXVIEW2 > USE | DIMENS, 42
UPDATE, AUXVIEW2 > USE | FRAME, 36
UPDATE, AUXVIEW2 > USE | PLANE, 33
UPDATE, AUXVIEW2 > USE | SECTION, 40
UPDATE, AUXVIEW2 > USE | TEXT, 35
UPDATE, AUXVIEW2 > USE | VIEW, 26
UPDATE, DETAIL > MANAGE in 3D SPACE mode, 108
UPDATE, DETAIL > MANAGE in DRAW mode, 99
UPDATE, DIMENS > MODIFY | TEXT in DRAW mode, 132
UPDATE, IDENTIFY in DRAW and SPACE modes, 187
UPDATE, LIBRARY in DRAW and SPACE modes, 215
UPDATE, MODELS > MANAGE in DRAW and SPACE modes, 294
UPDATE, PATTERN in DRAW mode, 317
UPDATE, SOLIDE in SPACE mode, 407
UPDATE, SOLIDM > in SPACE mode, 430
UPDATE, SYMBOL > MANAGE in DRAW mode, 450
UPDATE, TEXTD2 > MODIFY | TEXT in DRAW mode, 477
UPGRADE, DRW STD > ANNOTAT, 150
USE, AUXVIEW2, 43
USED DETAIL > DELETE in 3D SPACE mode, 107
USED, DETAIL > DELETE in DRAW mode, 99
USED, SYMBOL > DELETE in DRAW mode
VALUE, DIMENS > MODIFY in DRAW mode, 128
VALUE, PARAM3D > DIMENSION | POSITN in SPACE mode, 307
VALUE, PARAM3D > DIMENSION in SPACE mode, 307
VANISH, SYMBOL in DRAW mode, 452
VERIFY, DETAIL > MANAGE in 3D SPACE mode, 109
VERIFY, DETAIL > MANAGE in DRAW mode, 100
VERIFY, DIMENS > MANAGE in DRAW mode, 135
VERIFY, GRAPHIC in DRAW mode, 168
VERIFY, GRAPHIC in SPACE mode, 175
VERIFY, LAYER in DRAW and SPACE modes, 211
VERIFY, SYMBOL > MANAGE in DRAW mode, 450
VERT HOR, POINT > PROJECT in 2D SPACE mode, 354
VERT HOR, POINT > PROJECT in DRAW mode, 342
VERTICAL, LINE in 2D SPACE mode, 265
VERTICAL, LINE in DRAW mode, 244
VERTICAL, MARK UP > ARROW in DRAW mode, 290
VERTICAL, POINT > PROJECT in 2D SPACE mode, 353
VERTICAL, POINT > PROJECT in DRAW mode, 341
VERTICAL, TEXT > T.NODE | CREATE in DRAW mode, 453
VIEW, AUXVIEW > MODIFY in DRAW mode, 18
VIEW, AUXVIEW2 > USE, 25
VIEW, GROUP > GRP_1, 2 or 3 | EXCLUDE | W.SPACE in DRAW mode, 181
VIEW, GROUP > GRP_1, 2 or 3 | INCLUDE | W.SPACE in DRAW mode, 181
VIEW, LAYER > FILTER | APPLY | DIRECT in DRAW and SPACE modes, 207
VIEW, STANDARD > COLOR | MODIFY in DRAW and SPACE modes, 440
VISU STD, SOLIDM > in SPACE mode, 429
VISUALATN, DETAIL > DELETE | UNUSED in DRAW mode, 99
VISUALATN, FILE > COPY in DRAW and SPACE modes, 157
VISUALATN, FILE > DELETE in DRAW and SPACE modes, 158
VISUALATN, FILE > MOVE in DRAW and SPACE modes, 158

VISUALATN, FILE > RENAME in DRAW and SPACE modes, 159
VISUALATN, PLOT > QUICK | DELETE in DRAW and SPACE modes, 329
VISUALATN, POINT > GRID | UNLIMITED in DRAW mode, 349
VISUALATN, SOLIDM > MODIFY in SPACE mode, 427
VISUALATN, TEXTD2 > MANAGE in DRAW mode
VISUALTN, DIMENS > MANAGE in DRAW mode, 135
VISUALTN, DRW STD > ANNOTAT | DESCRIPT, 142
VISUALTN, PATTERN in DRAW mode, 316
VOL, SOLIDM > CREATE | COMPLEX in SPACE mode, 415
VOLUME, SOLIDE > COMPLEX in SPACE mode, 388
VOLUME, SOLIDE > EXTRACT in SPACE mode, 405
W.SPACE, DETAIL > EXPLODE in 3D SPACE mode, 106
W.SPACE, DETAIL > EXPLODE in DRAW mode, 98
W.SPACE, DETAIL > MANAGE in DRAW mode, 100
W.SPACE, DETAIL > MANAGE | LAYER in DRAW mode, 100
W.SPACE, DETAIL > MANAGE | REPLACE in DRAW mode, 100
W.SPACE, ERASE > ERASE in DRAW and SPACE modes, 154
W.SPACE, ERASE > NO PICK/PICK in DRAW and SPACE modes, 155
W.SPACE, ERASE > NO SHOW/SHOW in DRAW and SPACE modes, 155
W.SPACE, ERASE > PACK in DRAW and SPACE modes, 154
W.SPACE, LAYER > FILTER | RESET in DRAW and SPACE modes, 210
W.SPACE, SYMBOL > MANAGE | LAYER in DRAW mode, 451
W.SPACE, SYMBOL > MANAGE | VERIFY in DRAW mode, 450
WINDOW, IMAGE in DRAW and SPACE modes, 189
WINDOW, PLOT > FILE in DRAW and SPACE modes, 334
WORKAREA, MODELS in DRAW and SPACE modes, 297
WRITE, FILE in DRAW and SPACE modes, 157
WRITE, LIBRARY in DRAW and SPACE modes, 214
WRTFILE, FILE > COPY in DRAW and SPACE modes, 158
WRTFILE, FILE > MOVE in DRAW and SPACE modes, 158
WSP_TRAP, GROUP > GRP_1, 2 or 3 | EXCLUDE in SPACE mode, 182
WSP_TRAP, GROUP > GRP_1, 2 or 3 | INCLUDE in SPACE mode, 182
X-AXIS, AXIS > CREATE in 2D SPACE mode, 52
X-AXIS, AXIS > CREATE in 3D SPACE mode, 56
XY-XZ, LINE > ANGLE in 3D SPACE mode, 283
XY-YZ, LINE > ANGLE in 3D SPACE mode, 283
XZ, YZ, LINE > ANGLE in 3D SPACE mode, 283
Y-AXIS, AXIS > CREATE in 2D SPACE mode, 52
Y-AXIS, AXIS > CREATE in 3D SPACE mode, 56
Z-AXIS, AXIS > CREATE in 3D SPACE mode, 56
ZM/FRAME, IMAGE > SCREEN | MODIFY in DRAW and SPACE modes, 193
ZOOM, IMAGE > WINDOW | MODIFY | PLANE in DRAW and SPACE modes, 190
ZOOM, IMAGE > SCREEN | MODIFY in DRAW and SPACE modes, 194

ANALYSIS in DRAW Mode

Chapter 4

The ANALYSIS function in DRAW mode is used to analyze the numerical and relative values of and the logijcal links between DRAW elements.

Option menu for ANALYSIS in DRAW mode.

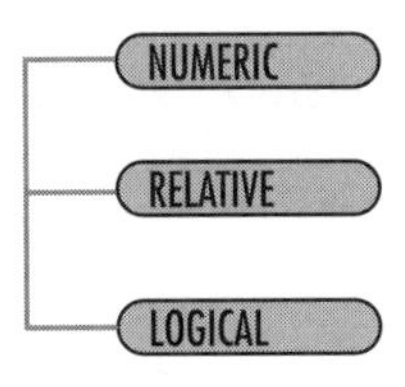

NUMERIC

Analyze numerical characteristics of DRAW elements.

Select: ANALYSIS > NUMERIC

Prompt: SEL: ELEM

Element analyzed	Additional prompt	Information displayed
PTD	None	• Coordinates of point relative to current axis system.
LND	SEL PT / / SEL ELEM / / NO:END	• Coordinates of end points. • Angles relative to current axis. • Angles relative to two-axis reference system. • Length of line. • Equation of line.
CIRD	SEL PT / / SEL ELEM / / NO:END	• Radius of circle. • Coordinates of center point of circle relative to current axis. • Length of line. • Angular limits of circle arc.
ELLD	SEL PT / / SEL ELEM / / NO:END	• Coordinates of center of ellipse relative to current axis system. • Coordinates of element foci relative to current axis system. • Length of minor and major axes. • Orientation of major axis relative to current axis system. • Parametric limitations of ellipse arc.

Element analyzed	Additional prompt	Information displayed
PARD	SEL PT / / SEL ELEM / / NO:END	• Coordinates of parabola focus relative to current axis system. • Parametric limitations. • Axis orientation relative to current axis system. • Equation of direction line relative to current axis system.
HYPD	SEL PT / / SEL ELEM / / NO:END	• Coordinates of hyperbola center relative to current axis system. • Coordinates of focus relative to current axis system. • Length of major and minor axes. • Parametric limitations. • Angle between two asymptotes. • Axis orientation relative to current axis system.
SPLD	SEL PT / / SEL ELEM / / NO:END*	• Length of spline. • Number of elementary arcs. • Arc degree, in cases of single arc in spline.
GDPT or GDLN	None	• Base point coordinates of current axis system. • Pitch of grid. • Angles of both grid directions relative to current axis system.
SHAP	None	Message: BLANKING MAY MODIFY ANALYS. RESULTS. NOTE: Blanking occurs when hatching (a shape) is automatically omitted around text and dimensions. • Total perimeter of shape (i.e., sum of internal and external perimeters). • Coordinates of center of gravity relative to current axis. • Area of shape. • Main moments of shape inertia. • Angles of main axes of inertia relative to current axis system.
AXSD	None	• Coordinates of point of origin relative to absolute axis system. • Components of direction vectors relative to absolute axis system. • Angles between each axis relative to a horizontal line. • Coordinates of center of associated SPACE axis system. • Two vectors defining SPACE plane. • SPACE plane equation.

RELATIVE

Analyze relative position of two DRAW elements.

Select: ANALYSIS > RELATIVE

First prompt: SEL ELEM / / SEL VU

Selection	Second prompt	Information displayed
Two PTD elements	SEL ELEM	• Coordinates of two points relative to current axis. • Components of vector connecting points. • Components of unit vector connecting points. • Length of segment line connecting points.
PTD, then LND	SEL ELEM	• Coordinates of point relative to axis system. • Coordinates of projection point on line relative to current axis system. • Tangent to line at projection point relative to current axis. • Point to line distance. NOTE: If an element selected is a segment line, the KEY 0: LINE SUPPORT message displays. Type 0 and press <Enter>; the segment will be interpreted as an infinite line.
PTD, then CIRD	SEL ELEM	• Coordinates of point. • Radius of circle. If selected point is not center of circle: • Coordinates of circle. • Coordinates of either one or two projection points on circle. • Tangents at projection points on circle. • Distance(s) between projection point(s) and selected point.
PTD, then CRVD	SEL ELEM	• Coordinates of point relative to current axis system. • Coordinates of nearest projection point on current axis system. • Tangent at projection point on curve in current axis system. • Point to curve distance.
Two LND elements	SEL ELEM	• If lines are parallel, distance between two lines. • If lines intersect, coordinates of point relative to current axis system, components of two tangent vectors relative to current axis system, angle between two vectors.
LND, then CIRD	SEL ELEM	• If there is no intersection, distance between two elements (distance between circle center and line minus circle radius). • If there is one intersection: - Coordinates of intersection point relative to current axis. - Components of tangent vector on line. - Components of tangent vector on curve. - Angle between two tangent vectors. • If there are several intersections, the intersection points.

Selection	Second prompt	Information displayed
LND, then CRVD	SEL ELEM	• If line is unlimited and there is no intersection, minimum distance between line and points on curve where tangent is parallel to line. • If there is one intersection, coordinates of point intersection relative to current axis system, components of tangent vector on line, and angle between two tangent vectors. • If there are several intersections, the intersections.
CRVD, then CIRD Two CIRD elements	SEL ELEM	• If two CIRD elements are selected and there is no intersection, the distance between the centers of the two circles. • If there is one intersection, the coordinates of the point of intersection relative to the current axis system, the components of the vector tangent to the second curve, the components of the vector tangent to the first curve, and the angle between the two tangent vectors. • If there are several intersections, the intersection points.

LOGICAL

Analyze logical links of selected elements.

Select: ANALYSIS > LOGICAL

Prompt: SEL: ELEM

Select an element. ISOLATED ELEMENT displays if selected element has no logical links with other elements. ELEMENT IN HIGHLIGHTED FAMILY displays if element is linked to other elements, such as dimensions or notes.

ANALYSIS in 2D SPACE Mode

Chapters 9, 10, 11

Analyze numerical characteristics and relative positions or variations in curvature and gradient of elements.

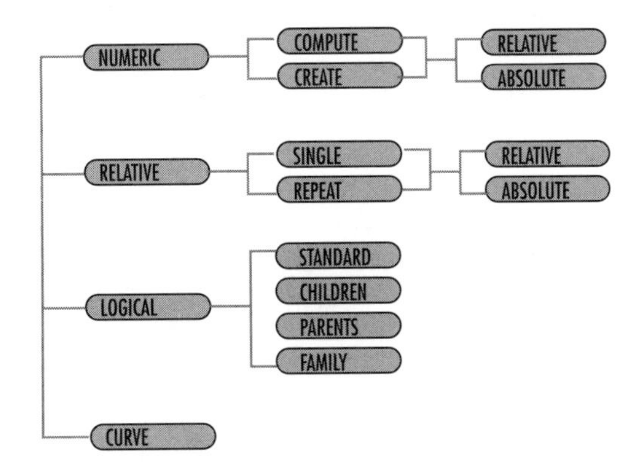

Option menu for ANALYSIS in 2D SPACE mode.

NUMERIC

Analyze numerical characteristics of and create temporary 2D SPACE elements.

Select: ANALYSIS > NUMERIC | COMPUTE | RELATIVE or ABSOLUTE

Prompt: SEL: ELEM

Analyzing numerical characteristics

Element selected	Additional prompt	Information displayed
PT	None	• Coordinates of point.
LN	None	• Coordinates of line end points. • Components of a line's unit direction vector. • Line direction angles with respect to current axis system. • Length of line.

Element selected	Additional prompt	Information displayed
CRV	SEL PT1 / / SEL: ELEM	• Length of curve. • Number of arcs along curve and degree of curve. If curve has several arcs, click on YES button to obtain degree of each arc, or click on NO to proceed to next interaction. If analysis of logic edges associated with curve is required, click on YES. • Identifier of representation support (plane or surface), or Cartesian equation of implicit planar support and number of arcs of edge representation. Click on YES button as many times as necessary to analyze the following edge representations.
CIR, ELL, PAR, HYP (e.g., conics)	SEL PT1 / / SEL: ELEM	• Equation of plane containing conic. • Circle: center, coordinates of center, radius, and angular limitations. • Hyperbola or ellipse: coordinates of center, coordinates of focus(es), length of first semi-axis, length of second semi-axis, eccentricity, parametric limitations, direction angles of major axis, and center and/or focus(es). • Parabola: coordinates of focus and parametric limitations, direction angles of axis, and focus.
CCV	SEL: ELEM / / YES: ARC DEG. / / NO: CONT	• Total length of composite curve. • Number of curve elements per type (LN, CRV, conic) with respective reference number. NOTE: A reference number is allocated to each element of the composite curve and displays it at the midpoint of the element. Select the curve element to be analyzed or key in the number of the curve element to be analyzed, and the characteristics of the selected element type.
AXS	None	• Coordinates of three-axis system origin. • Components of three vectors. • Euler angles system analysis.
FAC	KEY DENSITY / / YES. STD.	• Key in face density or click on YES button to accept current density. • Area, mass, and coordinates of center of gravity. • Three main axes of inertia. • Three moments of inertia. • Area of the external and internal domains if face has several domains.

Create temporary elements in model

Select: ANALYSIS > NUMERIC | CREATE

Prompt: SEL TEMPORARY ELEMENT

To create a single element, select the element and then click on the YES button. Clicking on YES without selecting an element creates all temporary elements in the model. This applies to all ANALYSIS options described above.

RELATIVE

Define relative position of two elements and make temporary elements permanent.

Select: ANALYSIS > RELATIVE | SINGLE or REPEAT | RELATIVE or ABSOLUTE

Define relative position of two elements

Select: ANALYSIS > RELATIVE | SINGLE or REPEAT

First prompt: SEL ELEM

Second prompt: SEL ELEM

Elements analyzed (in any paired combination) include PT, LN, CRV, CIR, ELL, PAR, HYP, and CCV.

SINGLE enables the two current elements to be used once. REPEAT means that the first (base) element can be used repeatedly. For each element or pair, the window displays the smallest distance, deviation, greatest deviation, perpendicular projection, and greatest perpendicular projection.

Make temporary elements permanent

Select: ANALYSIS > RELATIVE | SINGLE or REPEAT | CREATE

Prompt: SEL TEMPORARY ELEM / / YES: ALL

LOGICAL

Analyze logical characteristics of 2D SPACE element(s)

Select: ANALYSIS > LOGICAL | STANDARD or CHILDREN or PARENTS or FAMILY

Define logical characteristics of element

Select: ANALYSIS > LOGICAL | STANDARD

Element selected	Prompt	Information displayed
CRV or LN	SEL ELEM	• Number of edge representations. • Element identifiers.
FAC	SEL ELEM	• Element identifier of plane on which base lies. • Number of domains for domain edge.

Highlight identifiers to create selected element by type

Select: ANALYSIS > LOGICAL | CHILDREN

Prompt: SEL ELEM

Highlight identifiers to create selected element by type

Select: ANALYSIS > LOGICAL | PARENTS

Prompt: SEL ELEM

Highlight identifiers pertaining to family of selected element by type

Select: ANALYSIS > LOGICAL | FAMILY

Prompt: SEL ELEM

CURVE

Analyze variations in curve tangents and curvature. Also used to smooth curves such as splines to improve their shape, this option is used primarily in the creation and modification of surfaces. The latter topic is beyond the scope of this book.

Select: ANALYSIS > CURVE

Prompt: SEL CCV/PIP/NET

ANALYSIS in 3D SPACE Mode

INSIDE CATIA

Chapter 12

The ANALYSIS function in 3D SPACE mode is used to analyze numerical characteristics, relative position, logical relationships, inertial characteristics, or variations in curvature of elements.

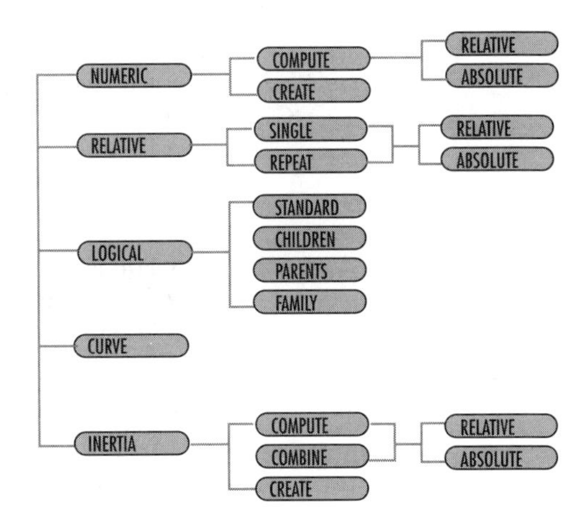

Option menu for ANALYSIS in 3D SPACE mode.

NUMERIC

Analyze numerical characteristics of 3D SPACE elements.

Select: ANALYSIS > NUMERIC | COMPUTE or CREATE

Analyze elements

Select: ANALYSIS > NUMERIC | COMPUTE | RELATIVE or ABSOLUTE

Prompt: SEL: ELEM

The following table depicts information displayed in the alphanumeric window by element types indicated using the specified prompt.

Element selected	Prompt	Window display
PT	SEL: ELEM	• Point coordinates.

Element selected	Prompt	Window display
LN	SEL: ELEM	• Coordinates of end points. • Components of line vector. • Line direction angles with respect to current axis system. • Line length.
NBRC	SEL: ELEM	• Curve length. • Number of arcs along nurbs curve and degree of curve.
CRV	SEL ELEM // YES: ARC DEG. // NO: CONT	• Curve length. • Number of arcs along curve. • Degree of arc in instances of single arc. Click on YES button to obtain degree of each arc. Click on NO to proceed to next interaction.
CIR, ELL, PAR, HYP	SEL PT! // SEL ELEM	• Equation of plane containing conic. • Center coordinates. • Radius and angular limitation of circle. • Center and focus coordinates. • Length of semi and second semi-axis. • Eccentricity and parametric limitations. • Direction of major axis and center and/or foci for hyperbola or ellipse. • Focus and focus coordinates for parabola.
CCV	SEL ELEM	• Total length of composite curve. • Number of curve elements per type.
PLN	SEL ELEM	• Equation of plane AX + BY + CZ = D. A, B, and C are components of unit vector normal to plane. Absolute value of D is distance from plane to origin.
NRBS	SEL ELEM	• Number of patches and maximum degee of surface along u and v.
SUR or NET	YES: AREA & CTR GVTY // SEL/IND PT	• Number of patches and maximum degree of surface. Additional characteristics displayed are beyond scope of this book.
AXS	SEL ELEM	• Coordinates of axis origin. • Components of three vectors. • Various angle characteristics.
DIT	Depends on element type within ditto	• Coordinates of ditto positioning point. • Components of three vectors of ditto three-axis system. • Ditto scale.

Create temporary elements in model

Select: ANALYSIS > NUMERIC | CREATE

Prompt: SEL TEMPORARY ELEMENT // YES: ALL

In the course of using any of the ANALYSIS options described above, click on the YES button to create all temporary elements in the model. To create one or more temporary elements, select the desired temporary element. If the element is a line, key in line length or click on the YES button to create infinite lines.

RELATIVE

Analyze relative positions of two 3D SPACE elements.

Select: ANALYSIS > RELATIVE | SINGLE or REPEAT | RELATIVE or ABSOLUTE

Prompt: SEL ELEM

Elements analyzed: PT, LN, NRBC, CRV, CIR, ELL, PAR, HYP, CCV, PLN, NRBS, SUR, FAC, SKI NET

The characteristics displayed for each element follow.

- Smallest distance between selected elements
- Deviation or largest distance from all points on the first element to all points on the second element
- Greatest deviation
- Perpendicular projection
- Greatest perpendicular projection

LOGICAL

Analyze logical characteristics of 3D SPACE elements.

Select: ANALYSIS > LOGICAL | STANDARD or CHILDREN or PARENTS or FAMILY

Prompt: SEL ELEM

The following table depicts the information displayed in the alphanumeric window for indicated element types using the specified prompt.

Element selected	Prompt	Window display
CRV or LN	SEL ELEM	Edge representations and curve identifiers of defining curves.
FAC	SEL ELEM	Element on which bottom of face lies, number of domains, and domain edges.
SUR or PLN	SEL ELEM	Identifiers of elements lying on surface.
VOL	SEL ELEM	Number of domains and faces limiting these domains.

Analyze logical characteristics of selected element

Select: ANALYSIS > LOGICAL | STANDARD

Prompt: SEL ELEM

Highlight identifiers logically linked to selected element

Select: ANALYSIS > LOGICAL | CHILDREN

Prompt: SEL ELEM

Highlight identifiers to be logically linked to selected element to be created

Select: ANALYSIS > LOGICAL | PARENTS

Prompt: SEL ELEM

Highlight identifiers pertaining to family of selected element

Select: ANALYSIS > LOGICAL | FAMILY

Prompt: SEL ELEM

CURVE

Analyze variations in curve tangents and curvature. While also used to smooth curves such as splines to improve their shape, this function is used primarily in the creation and modification of surfaces. (Creation and modification of surfaces is beyond the scope of this reference.)

Select: ANALYSIS | CURVE

Prompt: SEL CRV

INERTIA

Analyze inertial characteristics of 3D SPACE elements.

Select: ANALYSIS | INERTIA | COMPUTE or COMBINE | RELATIVE or ABSOLUTE

Analyze inertial characteristics of single 3D SPACE element

Select: ANALYSIS > INERTIA | COMPUTE | RELATIVE or ABSOLUTE

Prompt: MSELW FAC/VOL/POL/SOL/STR/PIP

Select one of the elements listed in the above prompt, and either define the density or click on the YES button to select the standard density. The window displays the following information.

- Surface or volume
- Mass
- Wetted surface
- Coordinates of center of gravity
- Three main axes of inertia
- Associated three main moments of inertia

Click on YES to store results if combined analysis is required later.

Analyze inertial characteristics of group of elements (e.g., solid assembly)

Select: ANALYSIS > INERTIA | COMBINE | RELATIVE or ABSOLUTE

Prompt: MSELW FAC/VOL/POL/SOL

Successively select elements of the same type whose respective analyses have been stored. As each element is selected, it is combined with previously selected elements. Inertial characteristics as described under the INERTIA entry are displayed in the alphanumeric window.

AUXVIEW in DRAW Mode

Chapters 6, 17

The AUXVIEW function is used to create and modify DRAW mode views. An AUXVIEW is an orthographic view, but unlike views created on a drawing board, CATIA creates views that are linked to each other to enable geometry from one view to be used to generate geometry in another view.

AUXVIEWs can be created using third angle projection (U.S. view convention) or first angle projection (European view convention).

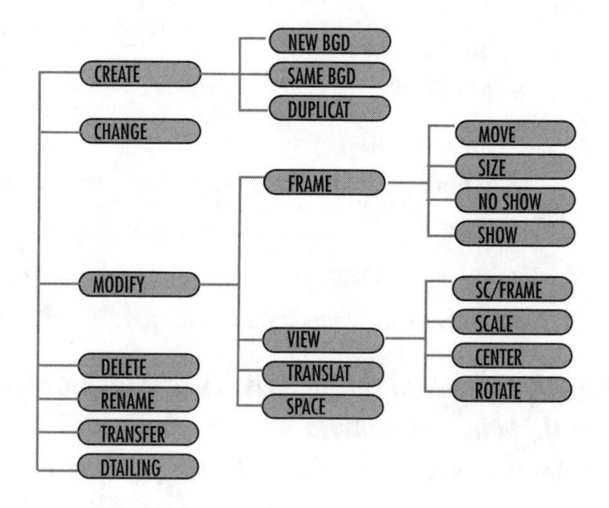

Option menu for AUXVIEW in DRAW mode.

CREATE

Create new view.

Select: AUXVIEW > CREATE | NEW BGD or SAME BGD or DUPLICAT

Create view with new background

Select: AUXVIEW > CREATE | NEW BGD

Prompt: SEL/IND SPC WINW // SEL LN/PLN

If all views in a model are to be linked, be sure to click on the YES button when creating new views.

Use one of the following alternatives for creating a new view while using either DRAW or SPACE elements to define the view.

Create view from existing DRAW elements

1. Select an LND (line) element from the defining view.
2. Click on the YES button to create the new view with the same origin as the defining view. (This alternative will usually be preferred to the following option.) Alternatively, select a PTD (point) element. The origin of the new view will be the projection of the point onto the selected line.
3. If necessary, key in *1* for third angle projection (U.S. convention) or *2* for first angle projection (European convention). This step is necessary only if you wish to change the view convention to that which was selected the last time AUXVIEW was used.
4. Indicate a position for the new view (recall the convention you are using).
5. Key in the view identifier, and if necessary, key in the view scale value.

Create view from existing SPACE elements

1. To define the view using SPACE elements, you have the following alternatives: (a) Select a PLN (plane) element. (b) Select an axis of the SPACE axis system. The plane is defined by the other two axes of the axis system. For instance, if the X axis is selected, the defined plane will be XY. (c) Key in the identifier for the SPACE mode window (e.g., *XYZ* will define the view the same as the SPACE image). (d) Select any element from the SPACE mode window. The created view will be the same definition as the SPACE mode window.
2. Indicate a position for the new view (recall the convention you are using).
3. Key in the view identifier, and if necessary, key in the view scale value.

Create new view with same background as defining view

Select: AUXVIEW > CREATE | SAME BGD

Prompt: SEL VIEW // YES:LIST

The two views will be linked in such a way that changes to one view will also take effect in the other linked view, although view filters can be added that apply only to the selected view.

1. Select an existing view, or click on the YES button to use the current view.
2. Indicate a position for the new view (recall the convention you are using).
3. Key in the view identifier, and if necessary, key in the view scale value.

Create duplicate of defining view

Select: AUXVIEW > CREATE | DUPLICAT

Prompt: SEL VIEW // YES:CUR VIEW

The new view is a duplicate of the defining view, including its background plane and all elements in that view. Changes to the new view will not affect the defining view as is the case with views created using SAME BGD.

1. Select the view to be copied, or click on the YES button to copy the current view.
2. Indicate a position for the new view (recall the convention you are using).
3. Key the view identifier, and if necessary, key in the view scale value.

CHANGE

Change current view.

Select: AUXVIEW > CHANGE

Prompt: SEL VIEW // YES:CUR VIEW

The current view will be shown dimmed and will be nonselectable. Select an element in the view you wish to make current, or click on the YES button to display the list of available views (the current view will not be listed) and select the view you wish to make current.

MODIFY

Modify view characteristics, such as create and modify a view clipping frame, change the scale or orientation of a view, move a view, or enable the SPACE elements of a view to be shown.

Select: AUXVIEW > MODIFY | FRAME or VIEW or TRANSLAT or SPACE

Move, size, or hide view clipping frame

Select: AUXVIEW > MODIFY | FRAME | MOVE or SIZE or NO SHOW or SHOW

Move position of view clipping frame

Select: AUXVIEW > MODIFY | FRAME | MOVE

Prompt: SEL VIEW SEL FRAME ELEM // SEL PT

Alternatives for moving the clipping frame follow.

✓ ***NOTE:*** *The clipping frame is always shown with dashed lines.*

- **Use frame elements and a point.** If the view to be modified is not the current view, select the required view. Select an element of the clipping frame in the working view. Select a PTD (point) element. The frame will move so that the frame element lies on the PTD element selected.

- **Use frame elements and a line**. If the view to be modified is not the current view, select the desired view. Select an element of the clipping frame in the working view. Select a LND (line) element. The frame will move so that the frame element lies on the selected LND element.
- **Use working view frame elements and frame elements of another view**. If the view to be modified is not the current view, select the required view. Select an element of the clipping frame in the working view. Select an element of the clipping frame of another view. The frame will move so that the frame elements coincide.
- **Define frame center point**. If the view to be modified is not the current view, select the required view. Select a PTD (point) element in the current view and the frame center will move to that point.

Modify size of clipping frame

Select: AUXVIEW > MODIFY | FRAME | SIZE

Prompt: SEL VIEW // SEL PT (if the current view lacks a clipping frame)

Prompt: SEL VIEW // YES:INFINITE FRAME KEY SCL // SEL PT // SEL FRAME ELEM (if the current view has a clipping frame)

Change frame scale

If the view to be modified is not the current view, select the desired view. Use one of the following methods. Both are possible only if the frame is not infinite. (a) Key in a number; the dimensions of the frame will be multiplied by this number, but its center will not move. (b) Select a PTD (point) element in the working view. Key in a scale ratio. The relative position of the selected element will remain unchanged but the frame dimensions will be modified by the keyed ratio.

Modify frame dimensions

1. If the view to be modified is not the current view, select the desired view.
2. Use one of the following methods. (a) Select two PTD (point) elements pertaining to the working view. The points define the diagonal of the clipping frame. (b) Indicate two positions in the working view. The points define the diagonal of the clipping frame. (c) Select a PTD in the working view. Key in the dimensions of the clipping frame (length, width). The selected PTD will become the center of the frame with respect to keyed dimensions. (d) Select an element of the working view frame. Key in the dimensions of the clipping frame (length, width). The position of the selected frame element remains unchanged, but the frame size changes to the keyed dimen-

sions. (e) Click on the YES button to create an infinite frame. Obviously this will work only if the frame has previously been changed.

Line up frame elements with other elements

If the view to be modified is not the current view, select the required view. Select one side of the frame of the working view. Select a PTD (point) element and the selected frame element will move to this point.

Change frame element visibility

Select: AUXVIEW > MODIFY | FRAME | NO SHOW or SHOW

Prompt: SEL VIEW YES:NO SHOW CUR VIEW FRAME

Prompt: SEL VIEW YES:SHOW CUR VIEW FRAME

✓ **NOTE:** *When using this option the display of the frame is switched from SHOW to NO SHOW or from NO SHOW to SHOW.*

Select the view whose frame display you wish to change, or click on the YES button to change the frame display of the current view. Results of this option take effect outside of the AUXVIEW function.

✓ **NOTE:** *Display of the frame remains visible while working with any AUXVIEW option.*

Modify and mainpulate view

Select: AUXVIEW > MODIFY | VIEW | SC/FRAME or SCALE or CENTER or ROTATE or TRANSLAT or SPACE

Modify scale and center of view with clipping frame

Select: AUXVIEW > MODIFY | VIEW | SC/FRAME

Prompt: SEL 1ST PT // SEL VIEW

If the view to be modified is not the current view, select the required view. Select two PTD (point) elements. The two points define the diagonal of a rectangle. The view will then be scaled and its center moved so the rectangle is inscribed in the clipping frame.

Modify view scale

Select: AUXVIEW > MODIFY | VIEW | SCALE

Prompt: KEY SCL // SEL VIEW

If the view to be modified is not the current view, select the required view. Key in the required view scale. The current scale will be displayed in the message area.

Translate view center without translating respective clipping frame

Select: AUXVIEW > MODIFY | VIEW | CENTER

Prompt: SEL VIEW SEL 1ST ELEM // KEY DX,DY

If the view to be modified is not the current view, select the required view. Alternatives follow.

- Select two PTD (point) or LND (line) elements. The view will move to position the first selected element onto the second selected element.
- Key in the DX,DY translation components for the required translation of the view center.
- Select a PTD (point) element and click on the YES button to translate the view center to the selected point.
- Select a LND (line) element and key in the displacement value for the view center to move normal to the selected line.

Rotate view

Select: AUXVIEW > MODIFY | VIEW | ROTATE

Prompt: KEY ANG // SEL VIEW SEL 1ST LN

If the view to be modified is not the current view, select the required view. Key in the required rotation angle in decimal degrees. An alternative method is to select two LND (line) elements that form the angle for the required rotation. If necessary, click on the YES button to rotate the view through 180°.

Translate view and clipping frame

Select: AUXVIEW > MODIFY | VIEW | TRANSLAT

Prompt: SEL VIEW SEL 1ST ELEM // KEY DX,DY

If the view to be modified is not the current view, select the required view. Alternatives follow.

- Select two PTD (point) or LND (line) elements; the view will move to position the first selected element onto the second selected element.
- Key in the DX,DY translation components for the required translation of the view.
- Select a PTD (point) element and click on the YES button to translate the view center to the selected point.
- Select an LND (line) element and key in the displacement value for the view center to move normal to the selected line.

Modify transparency of view regarding SPACE elements

Select: AUXVIEW > MODIFY | VIEW | SPACE

Prompt: SEL VIEW YES:SHOW 3D CUR VIEW

Prompt: SEL VIEW YES:NO SHOW CUR VIEW

✓ ***NOTE 1:*** *When creating an AUXVIEW, the SPACE elements (if they exist) are always available to be viewed in the created AUXVIEW.*

✓ ***NOTE 2:*** *When using this option the display of the SPACE elements is switched from SHOW to NO SHOW or from NO SHOW to SHOW.*

Select the view whose SPACE elements display you wish to change, or click on the YES button to change the SPACE elements display of the current view.

DELETE

Delete views.

Select: AUXVIEW > DELETE

Prompt: SEL VIEW // YES:LIST

Select an element in the view that you wish to delete; the selected view will be highlighted.

As an alternative, you can click on the YES button to display the list of views. Select the required view from the list; the selected view will be highlighted. Click on the YES button to confirm view deletion.

RENAME

Change view identifier.

Select: AUXVIEW > RENAME

Prompt: SEL VIEW // YES:LIST

Select an element in the view that you wish to rename; the selected view will change to dimmed mode.

As an alternative, click on the YES button to display the list of views. Select the required view from the list, or key in the view identifier. Key in the new identifier for the selected view.

TRANSFER

Transfer elements from one view to another.

Select: AUXVIEW > TRANSFER

Prompt: SEL NEW VIEW

Select the view into which you wish to transfer elements; the selected view is changed to dimmed mode. Select from other views the elements that you wish to transfer. Multiselection options are available. Click on the YES button to accept the transfer. If the element selected is part of a family, the complete family will be transferred. Click on the YES button to end the transfer operation.

✓ **NOTE:** *The most common reason for transferring elements is when text has been placed in the wrong view. Such "misplacement" occurs through indicating text position when you are not in the view that you thought you were in.*

DTAILING

Change the visualization of dress up elements in a view (i.e., patterns, texts, dimensions, and mark-up arrows).

Select: AUXVIEW > DTAILING

Prompt: SEL VIEW YES:NO DRESSING IN CUR VIEW

Prompt: SEL VIEW YES:DRESSING IN CUR VIEW

✓ **NOTE:** *When using this option the visualization of dress up elements is switched from showing to hidden or from hidden to showing.*

Select the view whose visualization you wish to change, or click on the YES button to change the visualization of the current view.

✓ **NOTE:** *The AUXVIEW function is available for use within a detail workspace but SCALE is the only available option.*

AUXVIEW2

This function is used to create and modify orthographic views. Standard and advanced menu options are available. The advanced menu is accessed by selecting the Advanced User Interface icon from the DEFAULT option.

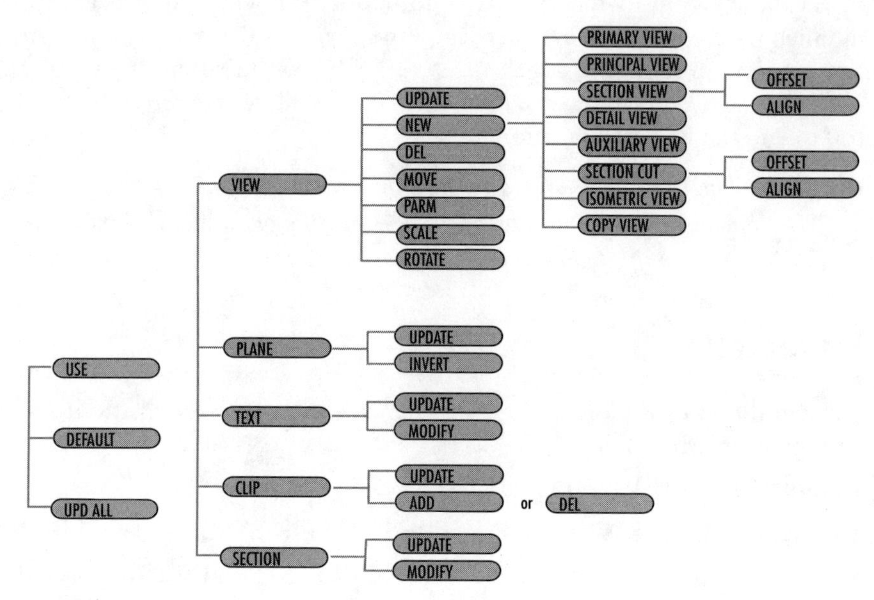

AUXVIEW2 standard menu options.

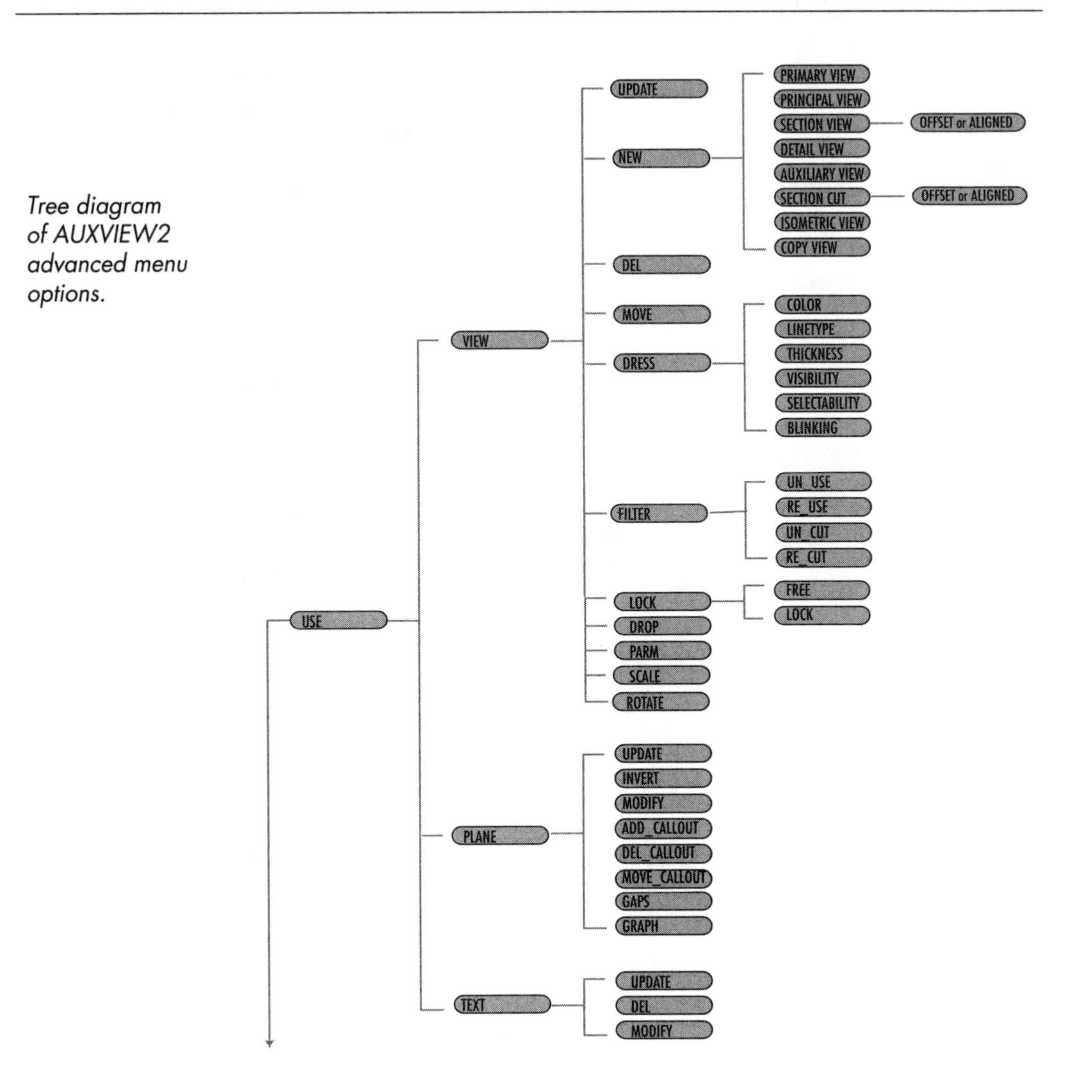

Tree diagram of AUXVIEW2 advanced menu options.

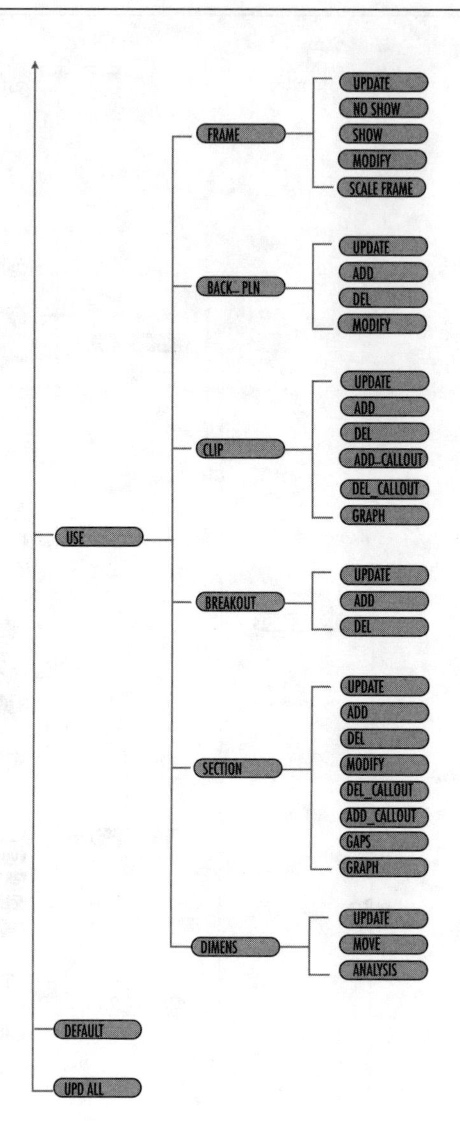

AUXVIEW2 Menu Structure

The AUXVIEW2 menu structure differs significantly from most other functions in CATIA in that the options are selected from two additional windows—Objects and Actions—rather than the menu at the right of the main window.

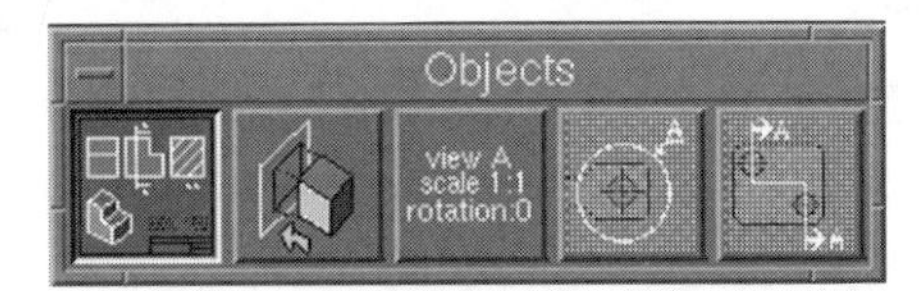

Objects window, standard mode.

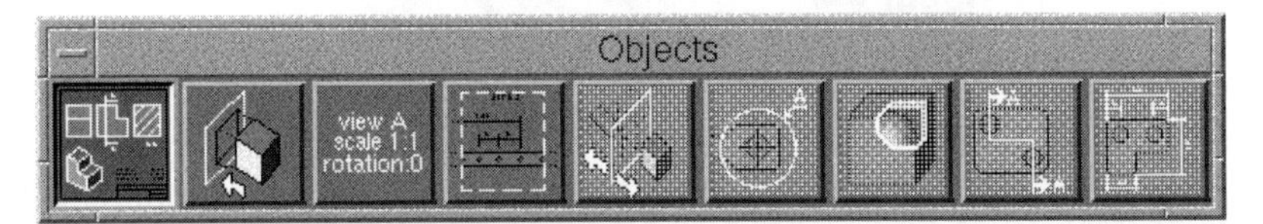

Objects window, advanced mode.

Action window.

CATIA's side menu and the Action window feature word options. However, the Objects window displays options as icons. Options for both AUXVIEW2 windows are explained below. The explanation shown for each Object window icon appears in the message area when the cursor is passed over the icon.

View icon manages view entity.

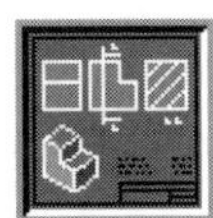

Plane icon defines new current view projection.

Text icon manages text associated with a view.

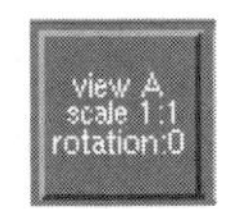

Frame icon manages a view frame.

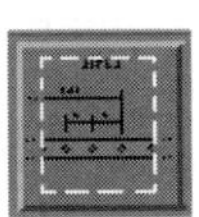

Back_pln icon performs a back clipping.

Clip icon manages view clipping.

Breakout icon manages a simple (or unspec) view.

Section icon manages view graphical and

Dimens icon manages dimensions.

USE

Create and modify orthographic views. From the USE option, you can select numerous suboptions. The following options are found in the Advanced user interface.

Select: AUXVIEW2 > USE | VIEW or PLANE or TEXT or FRAME or BACK PLN or CLIP or BREAKOUT or SECTION or DIMENS

View object

Select: AUXVIEW2 > USE | VIEW | UPDATE or NEW or DEL or MOVE or DRESS or FILTER or LOCK or DROP or PARM or SCALE or ROTATE

Update individual view

Select: AUXVIEW2 > USE | VIEW | UPDATE

Prompt: YES: confirm update

This option would be used, for example, if the solids from which the view was created have been modified. Click on the YES button to update the current view.

✓ **NOTE:** *The view to be updated must be the current view. You can change the current view using the VU button on the fixed menu.*

Define and create view types

Select: AUXVIEW2 > USE | VIEW | NEW | PRIMARY VIEW or PRINCIPAL VIEW or SECTION VIEW or DETAIL VIEW or AUXILIARY VIEW or SECTION CUT or ISOMETRIC VIEW or COPY VIEW

Define primary view

Select: AUXVIEW2 > USE | VIEW | NEW | PRIMARY VIEW

Prompt: Sel origin // Sel lnd

A primary view is defined by a plane in 3D SPACE. It is typically the first view to be created, and the view from which additional orthographic views are generated. To create a primary view, you must work with a split screen containing a SPACE window and a DRAW window.

1. To define the primary view, from the SPACE window select two lines that lie on the same plane, or a point and then a line. The center of the primary view is where the lines intersect (or the point you select).
2. To position the new view click anywhere in the DRAW window using the second mouse button.

✓ **NOTE:** *Once the primary view is created, the menu options change to the USE | VIEW | MOVE option, which allows you to move the view to a new position if necessary.*

Define principal view

A principal view is defined by orthographic projections.

Select: AUXVIEW2 > USE | VIEW | NEW | PRINCIPAL VIEW

Prompt: Sel view // Ind view definition

1. Two diagonal highlighted lines that intersect at the center of the current view display. If the current view is where you wish to create a principal view, indicate in one of the four quadrants. For example, if you indicate in the right quadrant you produce a side view of the primary view. The view displayed depends on whether you define first (European convention) or third (U.S. convention) angle projection in the DEFAULT window.
2. If the current view is incorrect for creating the principal view, select the required defining view and then proceed with step 1.

Define section view

A section view is an orthographic sectional view that displays geometry beyond the cutting plane.

Select: AUXVIEW2 > USE | VIEW | NEW | SECTION VIEW

Prompt: Sel view // Sel elem // Ind ptd

Navigate to and select the appropriate view, and then proceed with one of the following alternatives.

- In order to define the cutting plane, select a line. If the direction of the cutting plane is incorrect it can be inverted by selecting one of the vector arrows. Next, click on the YES button to end the selection. To position the section view, indicate a position where the view is required. This position can be moved using AUXVIEW2 > USE | VIEW | MOVE.
- If two points are selected, the cutting plane will be defined between the selected points. If the direction of the cutting plane is incorrect it can be inverted by selecting one of the vector arrows. Next, click on the YES button to end the selection. To position the section view, indicate a position where the view is required. This position can be moved using AUXVIEW2 > USE | VIEW | MOVE.
- If a staggered section is required, select two lines, three points, or any combination of points and lines to define the staggered section. Next, use one of the methods described above to complete the process.

When you select more than one line or point, the CUT OPTION window appears. In the window you can define whether the staggered section is parallel (OFFSET) to or angled (ALIGNED) with respect to the current axis.

✓ **NOTE 1:** *Hatching applied to the section view is inserted at the same angle. To alter the angle or pitch of the hatching, select PATTERN > VISUALTN | MODAL | GRAPHISM. If you use this option, additional updates using AUXVIEW2 update options will leave the modified hatchings unchanged.*

✓ **NOTE 2:** *The steps required to create a section view and cut are identical.*

Define detail view

A detail view is enlarged to clarify information.

Select: AUXVIEW2 > USE | VIEW | NEW | DETAIL VIEW

Prompt: SEL PTD // Sel view // SEL SURFACE

1. Navigate to and select the appropriate view.
2. Select or indicate two points to define the profile of the detail view. Click on the YES button to end profile selection.
3. Indicate a position for the detail view.

Define auxiliary view

An auxiliary view is defined by any line not necessarily normal or parallel to the view axis.

Select: AUXVIEW2 > USE | VIEW | NEW | AUXILIARY VIEW

Prompt: Sel lnd // Sel view

1. Navigate to and select the appropriate view.
2. Select a line to define the plane on which the view will be created. The auxiliary view is normal to the line selected.
3. Indicate a position for the view. The created view follows first or third angle projection rules.

Define section cut

A section cut is an orthographic sectional view that displays only the geometry on the cutting plane. Refer to steps under "Define section view" above.

Select: AUXVIEW2 > USE | VIEW | NEW | SECTION CUT

Prompt: Sel view // Sel elem // Ind ptd

Define isometric view

To create an isometric view, use a split screen containing a SPACE window and a DRAW window.

Select: AUXVIEW2 > USE | VIEW | NEW | ISOMETRIC VIEW

Prompt: New plane: Sel pln   Sel SPACE WINDOW

1. To define the orientation of the isometric view, use one of the methods listed below.
   - Select a plane.
   - If the SPACE window is in the correct orientation, select a SPACE element which defines a parallel window plane.

- Key in XYZ coordinates to define an isometric view according to the axis system of the SPACE model.

2. To position the isometric view, indicate a position where the view is required. This position can be moved using AUXVIEW2 > USE | VIEW | MOVE.

✓ **NOTE:** *If the created view is the inverse of what you require, invert the view using AUXVIEW2 > USE | PLANE | INVERT, and then rotate the view by -90°. Alternatively, modify the creation plane using PLANE > ORIENTN and recreate the view.*

Define copy view

A copy view duplicates an existing view.

Select: AUXVIEW2 > USE | VIEW | NEW | COPY VIEW

Prompt: Location: Sel ptd // Sel view

1. Navigate to and select the appropriate view.
2. Indicate a position where the duplicated view is required. This position can be moved using AUXVIEW2 > USE | VIEW | MOVE.

Delete view

Select: AUXVIEW2 > USE | VIEW | DEL

Prompt: YES: Delete selected view(s)

When this option is selected, a window appears that lists all views available for deletion. Select view(s) from the screen or the displayed list; the views you select are highlighted. Click on the YES button to complete the process.

✓ **NOTE:** *To select the current view, click on the Current View icon in the window list.*

Translate view to new position

Select: AUXVIEW2 > USE | VIEW | MOVE

Prompt: Location: Sel ptd // Sel 1st elem   Sel view // Key DH,DV

This option translates a view to a new position and, if desired, aligns it with other views. Once you select the view to be altered, several methods exist for moving views, depending on whether you wish to align the views or arbitrarily position them.

- If the current view is correct and alignment is not an issue, simply indicate a position and the view moves. Its position is determined by its axis, which is placed on the position you indicate.
- To align the current view with another view, select a line or point in the current view, the view to be moved, and a line or point in another view. The current

view is moved to align the two lines or points with each other, if the orthographic projection of the views is correct.

- Key in the translation coordinates in the form *DH,DV* (e.g., *50,0* to move the view 50mm in the horizontal direction and 0mm in the vertical direction).

Change graphical representation of elements

With the DRESS option, you can change the graphical representation of elements created in AUXVIEW2. Using this option instead of the GRAPHICS function allows you to change all elements associated with a solid with a single selection instead of selecting each element in turn. In addition, AUXVIEW2 remembers the selections you make if the view is subsequently updated, while the GRAPHIC function does not. DRESS options are summarized in the next table.

Option	Result
COLOR	Change color of one or more elements.
LINETYPE	Change line type of one or more elements.
THICKNESS	Change thickness of one or more elements.
VISIBILITY	Control visibility (SHOW/NO SHOW) of one or more elements.
SELECTABILITY	Control selection (PICK/NO PICK) of one or more elements.
BLINKING	Highlight one or more elements.

Select: AUXVIEW2 > USE | VIEW | DRESS

Prompt: SEL ELEMENT

Once you select the DRESS option, you must determine which graphical option to change. While the steps below are focused on changing color, the process for changing other options is the same.

✓ **NOTE:** *The DRESS option changes only the graphical properties of elements in the current view.*

1. Select COLOR and key in the number corresponding to the new color (e.g., 3 equals green).
2. Select a DRAW element of the solid you wish to change. All elements associated with the solid are changed.
3. Select a SPACE element of the solid you wish to change. All DRAW elements in the AUXVIEW2 associated with the solid are changed.

✓ **NOTE:** *If the elements are part of a ditto, you cannot change their graphical representation using the DRESS option.*

Filter views

Select: AUXVIEW > USE | VIEW | FILTER | UN_USE or RE_USE or UN_CUT or RE_CUT

Eliminate individual solid from selected view

Select: AUXVIEW2 > USE | VIEW | FILTER | UN_USE

Prompt: MSL SPC ELEMENT   Sel view // SEL DRAW ELEMENT

Navigate to and select the appropriate view. Choose one of the following alternatives: (a) Select a DRAW element associated with the solid you wish to eliminate. All elements associated with the solid are deleted from the current view. (b) Select a SPACE element of the solid. All DRAW elements in the AUXVIEW2 mode associated with the solid are deleted from the current view.

✓ **NOTE:** *If the solid is part of a ditto, you must use the second alternative. All solids in the selected ditto are eliminated.*

Restore solid following UN_USE

Select: AUXVIEW2 > USE | VIEW | FILTER | RE_USE

Prompt: MSL SPC ELEMENT   Sel view // SEL DRAW ELEMENT

To reuse a solid after UN_USE is invoked, take the following steps.

1. Navigate to and select the appropriate view.
2. Select a SPACE element of the solid; selectable solids are highlighted. All DRAW elements in the AUXVIEW2 mode associated with the solid are added to the current view.

Uncut individual solid in section view or cut

Select: AUXVIEW2 > USE | VIEW | FILTER | UN_CUT

Prompt: MSL SPC ELEMENT   Sel view // SEL DRAW ELEMENT

The UN_CUT option is used after a section view of an assembly is created and certain components (e.g., center shafts) are unnecessary.

Navigate to and select the appropriate view. Choose one of the following alternatives: (a) Select a DRAW element associated with the solid to uncut. The section is undone and the solid appears uncut. (b) Select a SPACE element of the solid. All DRAW elements in the AUXVIEW2 mode associated with the solid appear uncut.

✓ **NOTE:** *If the solid is part of a ditto, the second alternative must be followed. All solids in the selected ditto are uncut.*

Cut solid in section view or cut following UN_CUT

Select: AUXVIEW2 > USE | VIEW | FILTER | RE_CUT

Prompt: MSL SPC ELEMENT Sel view // SEL DRAW ELEMENT

To resection a solid after you have invoked UN_CUT, take the following steps.

1. Navigate to and select the appropriate view.
2. Choose one of the following alternatives (a) Select a DRAW element associated with the solid you wish to recut. The section is redone and the solid appears sectioned. (b) Select a SPACE element of the solid. The solid is recut.

✓ **NOTE:** *If the solid is part of a ditto, the second alternative in step 2 must be followed. All solids in the selected ditto are recut.*

Lock (free) views for modification only under AUXVIEW2

Elements contained in a view locked in AUXVIEW2 cannot be deleted using ERASE or changed graphically using GRAPHIC.

Select: AUXVIEW2 > USE | VIEW | LOCK | LOCK

Prompt: Sel view

Select any element in the desired view; the view is locked.

Select: AUXVIEW2 > USE | VIEW | LOCK | FREE

Prompt: Sel view

Select any element in the desired (locked) view; the view is freed (unlocked).

Isolate views to prevent revision

Select: AUXVIEW2 > USE | VIEW | DROP

Prompt: Sel view // YES : CONFIRM DROP

This option isolates views to prevent them from being subsequently revised using UPDATE or another AUXVIEW2 option.

✗ **WARNING:** *Once a view has been dropped, the operation cannot be reversed unless you delete and recreate the view.*

1. Navigate to and select the appropriate view.
2. Select any element.
3. Click on the YES button; the view is dropped.

Change parameters in DEFAULT panel

Select: AUXVIEW2 > USE | VIEW | PARM

Prompt: Sel view

This option changes certain parameters set in the DEFAULT panel before the view was created.

1. Navigate to and select the appropriate view.
2. Select any element in the proper view. A window appears which allows you to change the following parameters. (See DEFAULT option for description of parameters.)
    - Hidden lines, projection mode, intensity, texture.
    - Pattern, graphical attributes, scratch, or layer for old elements, view update.
    - Scale, rotation.
3. After completing changes of parameteres in the window, select AUXVIEW2 > USE | UPDATE to update the view and apply the changes.

Change view scale

Select: AUXVIEW2 > USE | VIEW | SCALE

Prompt: Scaling ratio: Sel ptd1   Sel view // Key scale

Navigate to and select the appropriate view, and then key in a new scale. Alternatively, you can indicate or select two points to define a scaling ratio.

Rotate view

Select: AUXVIEW2 > USE | VIEW | ROTATE

Prompt: Rotation1: Sel lnd   Sel view // Key angle

Navigate to and select the appropriate view, and then key in a new view angle. Alternatively, select lines to define the rotation angle.

Manipulate view (plane)

Select: AUXVIEW2 > USE | PLANE | UPDATE or INVERT or MODIFY or ADD_CALLOUT or DEL_CALLOUT or MOVE_CALLOUT or GAPS or GRAPH

Update individual view

Select: AUXVIEW2 > USE | PLANE | UPDATE

Prompt: YES: confirm update

Update an individual view (e.g., solids from which the view was created were modified). Click on the YES button to update the current view.

Invert view

Select: AUXVIEW2 > USE | PLANE | INVERT

This option is typically used when the view was created facing the wrong direction. Once you select this option, the current view is automatically inverted.

Modify defining plane in view

Select: AUXVIEW2 > USE | PLANE | MODIFY

Prompt: Sel: SPACE WINDOW Sel view // NEW_PLN : SEL PLN

Upon selecting the defining view, select a new plane from the SPACE window.

Add callout to view

Select: AUXVIEW2 > USE | PLANE | ADD_CALLOUT

Prompt: Sel view

Once you select the defining view, the callout (i.e., view lines and arrows) is added.

Delete callout in view

Select: AUXVIEW2 > USE | PLANE | DEL_CALLOUT

Prompt: Sel: view // YES: delete callout

Navigate to and select the view to be modified, and then click on the YES button.

Move callout in current view

Select: AUXVIEW2 > USE | PLANE | MOVE_CALLOUT

Prompt: first point : Sel ptd // Sel view

Navigate to and select the desired view and indicate or select two points to define the new callout position.

Insert gaps in callout lines

Select: AUXVIEW2 > USE | PLANE | GAPS

Prompt: Element to be gapped: sel lnd Sel view// YES: reset all gaps

To insert gaps in callout lines (in a section or view), navigate to the desired view, select the callout line to be modified, and indicate two points to define the required gap.

To remove existing gaps, click on the YES button to clear the gaps.

Change graphical representation of section definition

Select: AUXVIEW2 > USE | PLANE | GRAPH

Prompt: Sel: callout txtn // Sel: view

Navigate to the proper view. A callout parameter window appears, from which you can modify the following parameters. (See the DEFAULT option entry for description of parameters.)

- Callout presentation
- Arrow type
- Anchor point
- Length
- Identity

Changing any of the above parameters automatically changes the callout attributes. If the identity is changed, all associated identities are also changed.

Manipulate text in view

Select: AUXVIEW2 > USE | TEXT | UPDATE or DEL or MODIFY

Update view

Select: AUXVIEW2 > USE | TEXT | UPDATE

Prompt: YES: confirm update

Update an individual view (e.g., solids from which the view was created were modified). Click on the YES button up update the current view.

Delete text in view

Select: AUXVIEW2 > USE | TEXT | DEL

To delete all AUXVIEW2 text from an individual view, simply make the above selection. The text is automatically deleted from the current view.

Modify text in view

Select: AUXVIEW2 > USE | TEXT | MODIFY

Prompt: Text position: Sel ptd // sel view

If the section or view identifications are changed, associated text in other views is also changed.

1. Navigate to and select the desired view.

2. A parameter window appears, from which AUXVIEW2 text can be edited, added, or deleted. Select text or indicate a new position to move text.

Frame manipulation

Select: AUXVIEW2 > USE | FRAME | UPDATE or NO SHOW or SHOW or MODIFY or SCALE FRAME

Update individual view

Select: AUXVIEW2 > USE | FRAME | UPDATE

Prompt: YES: confirm update

Update an individual view (e.g., solids from which the view was created were modified). Click on the YES button to update the current view.

Hide or make frames visible

Select: AUXVIEW2 > USE | FRAME | NO SHOW or SHOW

Prompt: Sel: view

1. Navigate to the view in which frame visibility will be changed.
2. Select the SHOW or NO SHOW icon from the Active window; the selection is applied to all frames in the current view.

Modify or create view frames

Select: AUXVIEW2 > USE | FRAME | MODIFY

Prompt: SEL PTD // Sel: frame lnd   Sel: view // YES: infinite

Add frame to view

1. Select the view to which a frame will be added.
2. Indicate two diagonal corner positions to define the frame.

Modify existing view frame

1. Select the view in which the frame will be modified.
2. Indicate two diagonal corner positions to define the new frame.

Change frame scale without altering size

Select: AUXVIEW2 > USE | FRAME | SCALE FRAME

Prompt: First point: Sel ptd

1. Select the view in which frame scale will be modified.

2. Indicate two diagonal points inside the frame; the two points define the scale ratio of the view within the frame.

Manipulate back planes in views

Select: AUXVIEW2 > USE | BACK_PLN | UPDATE or ADD or DEL or MODIFY

Update view

Select: AUXVIEW2 > USE | BACK_PLN | UPDATE

Prompt: YES: confirm update

Update an individual view (e.g., solids from which the view was created were modified). Click on the YES button to update the current view.

Add back plane to view

Select: AUXVIEW2 > USE | BACK_PLN | ADD

Prompt: Sel: view // NEW_PLN: SEL: PLN

This option adds a back plane to an existing view, beyond which no geometry is visible. Navigate to the appropriate view, and then select a new defining plane from the SPACE window. The view is automatically updated.

Delete back plane in view

Select: AUXVIEW2 > USE | BACK_PLN | DEL

Upon making the above selection, all back planes applied to the current view are automatically deleted.

Modify back plane in view

Select: AUXVIEW2 > USE | BACK_PLN | MODIFY

Prompt: Sel: view // NEW_PLN: SEL PLN

Navigate to and select the desired view, and then select a new defining back plane from the SPACE window. The view is automatically updated.

Manipulate clipping frame in view

Select: AUXVIEW2 > USE | CLIP | UPDATE or ADD or DEL or ADD_CALLOUT or DEL_CALLOUT or GRAPH

Update view

Select: AUXVIEW2 > USE | CLIP | UPDATE

Prompt: YES: confirm update

Update an individual view (e.g., solids from which the view was created were modified). Click on the YES button to update the current view.

Add clipping frame

Select: AUXVIEW2 > USE | CLIP | ADD

Prompt: SEL: PTD // Sel view

This option adds a clipping frame to a view in order to create an irregularly shaped detail view.

1. Navigate to and select the desired view.
2. Select or indicate a series of points to define the clipping frame profile.
3. Click on the YES button to end the profile selection; the view is automatically updated.

Delete clipping frame

Select: AUXVIEW2 > USE | CLIP | DEL

Prompt: Sel: view // YES: del clipping

Navigate to the appropriate view. Click on the YES button to delete the clipping frame; the view is automatically updated.

Add callout to view containing clipping frame

Select: AUXVIEW2 > USE | CLIP | ADD_CALLOUT

Prompt: Sel: the defining view

1. Navigate to the appropriate view and verify that it contains a clipping.
2. Select the view from which the clipped view can be defined. A callout circle with identification is added to the view.

Delete callout in view containing clipping frame

Select: AUXVIEW2 > USE | CLIP | DEL_CALLOUT

Prompt: Sel: view // YES: delete callout

This option deletes a callout added using the previous option. Select the view containing a callout, and click on the YES button to delete the callout.

Change graphical representation of clipping callout

Select: AUXVIEW2 > USE | CLIP | GRAPH

Prompt: Text position: Sel: ptd // Sel: view

1. Navigate to and select the appropriate view.
2. A callout parameter window appears from which you can modify the following parameters. (See the DEFAULT option entry for a description of each parameter.)
   - Callout presentation
   - Arrow type
   - Length
   - Identity
3. Indicate or select a point to define a new position for the text identity.

Modification of any parameter automatically changes the callout attributes. If the identity is changed, all associated identities are also changed.

Break out manipulation in view

Select: AUXVIEW2 > USE | BREAKOUT | UPDATE or ADD or DEL

Update view

Select: AUXVIEW2 > USE | BREAKOUT | UPDATE

Prompt: YES: confirm update

Update an individual view (e.g., solids from which the view was created were modified). Click on the YES button to update the current view.

Add breakout

Select: AUXVIEW2 > USE | BREAKOUT | ADD

Prompt: Sel: solid // SEL PTD // Sel view

This option creates a part section within an existing view or section (e.g., part of a cover plate could be removed to show the internal components of an assembly). Navigate to and select the appropriate view, and then use one of the following alternatives.

- Select a solid from the SPACE window; the solid is removed from the current view.
- Indicate a series of points in the current view to define the profile of the part section. Click on the YES button to end the profile selection. Select two planes from the SPACE window to define the depth of the part section. All solids are removed from the area defined by the profile and planes.

✓ **NOTE:** *Additional part sections can be performed within the first part section.*

Delete breakout

This option deletes part sections created using the ADD option.

Delete breakout created by selecting a solid from the SPACE window

Select: AUXVIEW2 > USE | BREAKOUT | DEL

Prompt: Sel: lnd // Sel: solid

When this option is selected, solids removed from the current view using BREAKOUT are highlighted in the SPACE window. Select the solid to be included in the current view; the view is automatically updated.

Delete breakout defined by profile and planes

Select: AUXVIEW2 > USE | BREAKOUT | DEL

Prompt: Sel: lnd // Sel: solid

When this option is selected, breakout profiles in the current view are highlighted. Upon selecting a line in the profile, the breakout part section is deleted; the view is automatically updated.

Manipulate section in view

Select: AUXVIEW2 > USE | SECTION | UPDATE or ADD or DEL or MODIFY or DEL_CALLOUT or ADD_CALLOUT or GAPS or GRAPH

Update view

Select: AUXVIEW2 > USE | SECTION | UPDATE

Prompt: YES: confirm update

Update an individual view (e.g., solids from which the view was created were modified). Navigate to and select the appropriate view, and then click on the YES button to update the current view.

Change view to section view

Select: AUXVIEW2 > USE | SECTION | ADD

Prompt: Sel: the defining view

Navigate to and select the appropriate view to be used for defining the section in the current view. A CUT OPTION window appears, which allows you to create a SECTION VIEW or SECTION CUT. Use the same steps to generate a SECTION VIEW or SECTION CUT. (See AUXVIEW2 > USE | VIEW | NEW for more information.)

Delete section

Select: AUXVIEW2 > USE | SECTION | DEL

Prompt: Sel: view // YES: delete section

This option deletes a section created using the ADD option. Navigate to and select the appropriate view, and then click on the YES button to delete the section. The view is automatically updated.

Modify section

Select: AUXVIEW2 > USE | SECTION | MODIFY

Prompt: MODIFY: SEL: PTD // Sel: view

This option modifies the position of the section definition arrows (thereby modifying the section).

1. Navigate to and select the appropriate view.
2. Select the section callout and indicate new positions for the section arrows.
3. Update the view using AUXVIEW2 > USE | SECTION | UPDATE, and then click on the YES button to confirm the update.

Delete section callout

Select: AUXVIEW2 > USE | SECTION | DEL_CALLOUT

Prompt: Sel: view // YES: delete callout

This option deletes the callout (section lines and identification) from a view. Navigate to and select the appropriate view, and then click on the YES button to confirm deletion.

Restore deleted section callout

Select: AUXVIEW2 > USE | SECTION | ADD_CALLOUT

Prompt: Sel: the defining view

This option restores callouts deleted using the DEL_CALLOUT option. Navigate to and select the view to which the callout should be added.

Insert gaps in callout lines

Select: AUXVIEW2 > USE | SECTION | GAPS

Prompt: Element to be gapped: sel lnd   Sel: view // YES: reset all gaps

This option inserts gaps into callout lines (in a section or view).

1. Navigate to and select the appropriate view.
2. Select the callout line to be modified.
3. Indicate two points to define the required gap.
4. If existing gaps should be removed, click on the YES button to clear the gaps.

Change graphical representation of section definition

Select: AUXVIEW2 > USE | SECTION | GRAPH

Prompt: Sel: callout txtn // Sel: view

1. Navigate to and select the appropriate view.
2. A CALLOUT PARAMETER window appears, from which you can modify the following parameters. (See the DEFAULT option entry for a description of each parameter.)
   - Callout presentation
   - Arrow type
   - Anchor point
   - Length
   - Identity

Changing any parameter automatically changes the callout attributes. If the identity is changed, all associated identities are also changed.

Move and analyze dimensions created automatically from parameterized solid

To automatically create dimensions from parameterized solids, select the dimension option in the default panel.

Select: AUXVIEW2 > USE | DIMENS | UPDATE or MOVE or ANALYSIS

Update view

Select: AUXVIEW2 > USE | DIMENS | UPDATE

Prompt: YES: confirm update

Update an individual view (e.g., solids from which the view was created were modified). Navigate to and select the appropriate view, and then click on the YES button to update the current view.

Move dimensions between views

Select: AUXVIEW2 > USE | DIMENS | MOVE

Prompt: Annd to be moved: msel annd   Sel receiving view

This option allows automatically created dimensions to be moved from one view to another.

1. Select the dimension to be moved.
2. Select the view to receive the dimension.
3. Click on the YES button to end movement and select another dimension.

✓ ***NOTE:*** *Dimensions created in AUXVIEW2 can be modified using DIMENS2 > MODIFY.*

Analyze dimensions

Select: AUXVIEW2 > USE | DIMENS | ANALYSIS

This option analyzes automatically created dimensions. When this option is selected, the AUTO DIM ANALYSIS window appears on the screen and displays the following information.

- RELATIONS
- GENERATED DIMS: Lists the number of automatically created dimensions.
- FALSE DIMS: Lists the number of false dimensions created automatically. (False dimensions are not projected orthogonally onto the view plane, and therefore have values different from parameterized solid values.)

DEFAULT

Although the DEFAULT option is not the first option on the menu, it is often the first option that must be selected because it defines the appearance of views created in AUXVIEW2. Changing DEFAULT settings will change subsequently created views but will not affect existing views. To change the parameters of existing views, refer to the USE option.

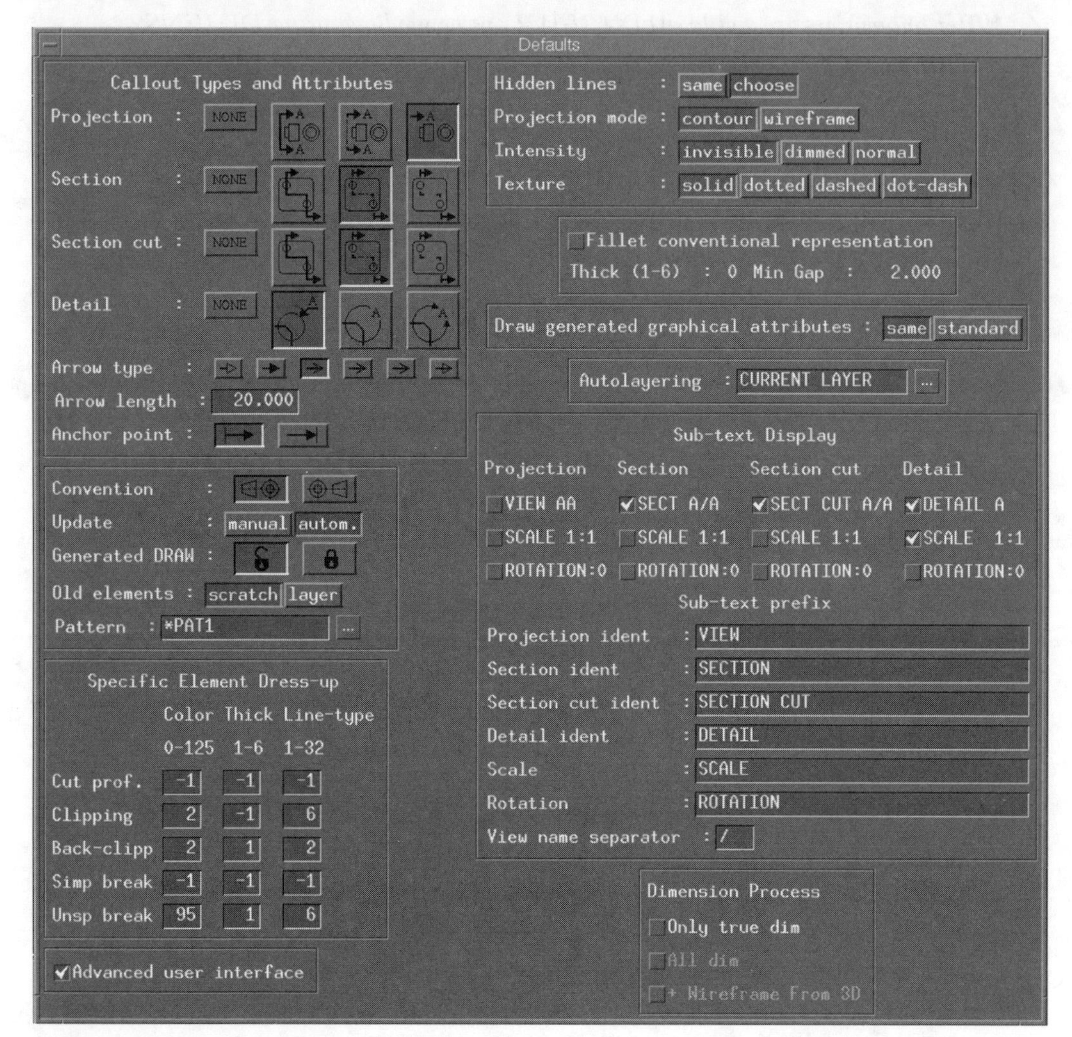

DEFAULT option in AUXVIEW2.

DEFAULT panel contents are described below.

Callout types and attributes

These options define how section lines, projection lines, and arrows are displayed

Projection

Defines how projection view callout lines are shown.

- NONE: No callout is shown.

- Callout lines are solid with arrows at either end.
- Callout lines are shown dot-dashed with arrows at either end.
- An arrow without callout lines is shown.

Section

Defines how section view callout lines are shown.

- NONE: No callout is shown.
- Callout lines are solid with arrows at either end.
- Callout lines are shown dot-dashed with arrows at either end.
- No callout lines are shown but an arrow is shown at either end of the section definition line.

Section cutl

Defines how section cut view callout lines are shown.

- NONE: No callout is shown.
- Callout lines are solid with arrows at either end.
- Callout lines are shown dot-dashed with arrows at either end.
- No callout lines are shown but an arrow is shown at either end of the section definition line.

Detail

Defines how detail view callout lines are displayed.

- NONE: No callout is shown.
- A solid callout circle with a text arrow is shown.
- A callout circle is shown dot-dashed with text on the circle.
- A callout circle is shown dot-dashed with text on the circle and arrow heads on either end.

Arrow types

Defines how the arrow heads will appear. Six choices are available.

✓ **NOTE:** *Arrowhead size is defined in the MARK UP function under the ARROW option.*

Arrow length

Defines the length of arrow leader lines. Key in a number such that the arrow length is greater than arrow head length.

Anchor point

Defines the attachment point for the arrow leader.

1. The arrow leader is attached at the end of the leader. (ANSI)
2. The arrow is attached at the point of the arrow. (ISO)

Element appearance

The hidden lines, projection mode, intensity, and texture options define how the created elements appear in the view (e.g., intensity and texture allow the user to show hidden lines in various line types).

Hidden lines

Defines whether the current choice of solid hidden line in SPACE mode is used for the creation of the new view.

- SAME. SPACE definition of a solid hidden line.
- CHOOSE. Define hidden lines.
- CHOICE. Define any hidden lines used.

Projection mode

Defines how solids are projected into new view.

- CONTOUR. Only the contour of a solid is projected.
- WIREFRAME. Wireframe representation of the solid is projected into the new view.

Intensity

Defines how hidden lines are displayed when the CHOOSE option is selected in HIDDEN LINES.

- INVISIBLE. No hidden lines are shown.
- DIMMED. Hidden lines are shown dimmed (i.e., with a line thickness of 1).
- NORMAL. Hidden lines are shown in the current line thickness.

Texture

Defines how hidden lines are graphically displayed.

- SOLID. Line type 1.
- DOTTED. Line type 2.
- DASHED. Line type 3.
- DOT-DASHED. Line type 4.

Fillet conventional representation

When this option is switched on, the display for fillets created in OPERATE | FILLET is modified.

Thick

Defines the thickness of the line used for the fillet representation. Keying in a number between 1 and 6 defines the line thickness used; the numbers relate to line thickness as defined in the STANDARD function.

Min gap

Defines the gap at the end of the line used for the fillet representation.

Draw generated graphical attributes

This option allows the attributes from the SPACE elements or the DRAW standard to be used for the creation of new views.

Same

Uses the same graphical attributes applied to SPACE elements being projected into the view.

Standard

Uses the graphical attributes set in DRAW STANDARD options.

Additional DEFAULT options

The DEFAULT option also includes the options listed below.

- **Sub-text display.** Defines the text applied to views and sections and the wording used for identification. Sub-text display options can be toggled on or off, and the sub-text prefix choices can be entered in the appropriate box.
- **Dimensional process.** Controls automatic dimensioning from parameterized solid models.
- **Advanced user interface.** Controls the options available in the OBJECTS window. With this option on or off, available choices are shown in menu tree illustrations.

- **Specific element dress-up.** Defines the graphical representation of lines used for section and clipping lines.
- **Convention.** Defines first (European) or third (U.S.) angle projection.
- **Update.** Creates a view that shows only SPACE (Manual) or DRAW (Automatic) elements.
- **Generated DRAW.** Locks view upon creation.

✓ **NOTE:** *Locked views can be modified only with the use of AUXVIEW2 options.*

- **Old elements.** Automatically erases old elements after a view is updated (scratch) or places old elements on a chosen layer (layer).
- **Pattern.** Defines the pattern (or hatching) applied to a section. Selecting the arrow at the right of the Pattern text box displays all available hatching patterns.

UPD ALL

Select: UPD ALL

Prompt: YES: CONFIRM UPDATE

Update all views with a single command. Compare this option with USE | UPDATE, which update views one at a time. To update all views, click on the YES button.

✓ **NOTE 1:** *If the design involves several large views, the UPD ALL process can be time-consuming.*

✓ **NOTE 2:** *AUXVIEW2 can be used in multimodel and single model environments. When updating a view created in a multimodel environment, verify that all required models are in place before you select this option; otherwise, the information in the updated view may be lost.*

AXIS in DRAW Mode

Chapter 6

The AXIS function is used to create new axis systems and change the current axis system.

Option menu under AXIS in DRAW mode.

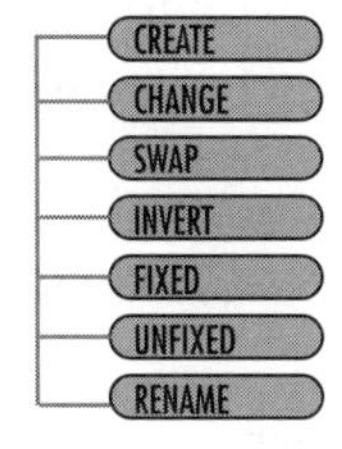

CREATE

Create new axis system in current view.

Select: AXIS > CREATE

Prompt: SEL PT/LN

1. Define the axis origin by selecting a PTD (point) element or selecting two intersecting LND (line) elements. (The origin will be the point where the lines intersect.)
2. Define the H axis direction by first selecting an LND (line) element. The axis system will be created with the H axis parallel to the element you selected. Alternatively, key in an angle value in degrees. The axis system will be created with the H axis at the defined angle from the current axis system.

✓ **NOTE:** *When a new axis system is created it automatically becomes the current axis system. Non-current axis systems are distinguished by dashed lines.*

TIP: *The new axis system can be reversed by clicking on the YES button.*

CHANGE

Change current axis system.

Select: AXIS > CHANGE

Prompt: SEL AXS

Select any other available axis system.

SWAP

Swap individual axes of an axis system

Select: AXIS > SWAP

Prompt: SEL AXS

Select the axis system you wish to swap. The orientation of the axes will be reversed (i.e., the V axis will become the H axis, and vice versa).

INVERT

Invert axes of axis system.

Select: AXIS > INVERT

Prompt: SEL AXSD

Axes pointing to the right will point to the left (and vice versa), and axes pointing downward will point upward (and vice versa).

FIXED

Fix unfixed axis.

Select: AXIS > FIXED

Prompt: SEL AXS

Select an axis that you wish to fix. (Note that you can only select *unfixed* axes.)

✓ ***NOTE:*** *New axis systems are by default fixed. In brief, they will not move when a transformation is applied to the current view. When an axis is unfixed (see below), it will move with the transformation.*

UNFIXED

Unfix axis.

Select: AXIS > UNFIXED

Prompt: SEL AXS

Select an axis in order to move the axis when a transformation is applied. (Note that only *fixed* axes are selectable.)

RENAME

Change identity of axes in axis system.

Select: AXIS > RENAME

Prompt: SEL AXS

Select the axis system to rename, and then key in the new identities separated by a comma (e.g., *H,V*).

AXIS in 2D SPACE Mode

Chapter 13

The AXIS function is used to create and modify axis systems and change the current axis system. The only difference between 2D SPACE and 3D SPACE function menus is the CREATE option. In 2D SPACE you can only use the X or Y axes to create a new axis system.

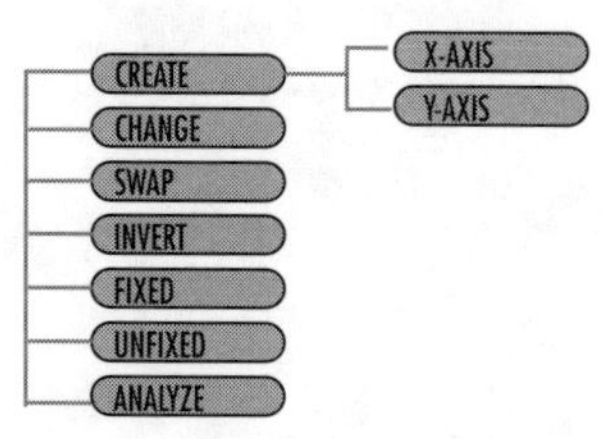

Option menu for AXIS in 2D SPACE mode.

The four displays of axis systems in SPACE mode are listed in the next table.

Type of axis system	Display mode
Current direct	Solid lines
Noncurrent direct	Dashed lines
Current reversed	Dotted lines
Noncurrent reversed	Dot-dashed lines

CREATE

Create new axis system.

Select: AXIS > CREATE | X-AXIS or Y-AXIS

Prompt: SEL PT/LN

When creating new axis systems you have the choice of creating an X or Y axis system (i.e., making it the privileged axis). The following descriptions assume that the X axis is privileged. Axis systems can be created in two ways.

- Define the origin by selecting a PT (point) element. An axis system displays with the origin positioned on the point and axes parallel to those of the current axis system.

- Select a LN (line) element; an axis system displays with the origin as the projection of the current axis system origin point to the selected line. The X axis is colinear with this line.

✓ **NOTE 1:** *In either case the Z axis will be perpendicular to the current 2D plane and will appear dimmed.*

✓ **NOTE 2:** *When a new axis system is created, it automatically becomes the current axis system. Noncurrent axis systems are distinguished by dashed lines.*

If necessary, you can rotate the axis system about its origin point using one of the following methods: (a) Key in an angle value; the axis system rotates about the Z axis. (b) Select an LN (line) element; the X axis becomes colinear with this line.

Once you have defined the system's origin point and orientation, click on the YES button to accept and make current the displayed axis system.

✓ **NOTE:** *Defining a point of origin and orientation for a new axis system is similar when the Y axis is privileged.*

CHANGE

Change current axis system.

Select: AXIS > CHANGE

Prompt: SEL AXS

Select any other available axis system.

SWAP

Swap axes of axis system.

Select: AXIS > SWAP

Prompt: SEL AXS

Select the axis system you wish to swap. The orientation of the axes will be reversed. Because only two axes are selectable in a 2D plane, the X axis will become the Y axis, and vice versa.

INVERT

Invert axes.

Select: AXIS > INVERT

Prompt: SEL AXS

Select axes. Axes pointing to the right will point to the left (and vice versa), and axes pointing downward will point upward (and vice versa).

FIXED

Fix axis previously unfixed.

Select: AXIS > FIXED

Prompt: SEL AXS

iSelect the axis. (Note that you can only select *unfixed* axes.)

✓ **NOTE:** *New axis systems are fixed by default. In brief, they will not move when a transformation is applied to the current view. When an axis is unfixed (see below), it will move with the transformation.*

UNFIXED

Unfix axis.

Select: AXIS > UNFIXED

Prompt: SEL AXS

Select axis in order to move the axis when a transformation is applied. (Note that only *fixed* axes are selectable.)

ANALYZE

Analyze axis system.

Select: AXIS > ANALYZE

Prompt: SEL AXS

Select axis system.The alphanumeric window displays the following information.

- Axis system type (direct or reverse).
- Axis system status (fixed or unfixed). If the axis is fixed, the workspace to which it belongs is displayed. If the axis system is unfixed, the set to which it belongs is displayed.
- Axis system status (current or not current).
- Coordinates of axis system origin and components of each axis.
- Euler angles with reference to the orientation of the analyzed axis system.

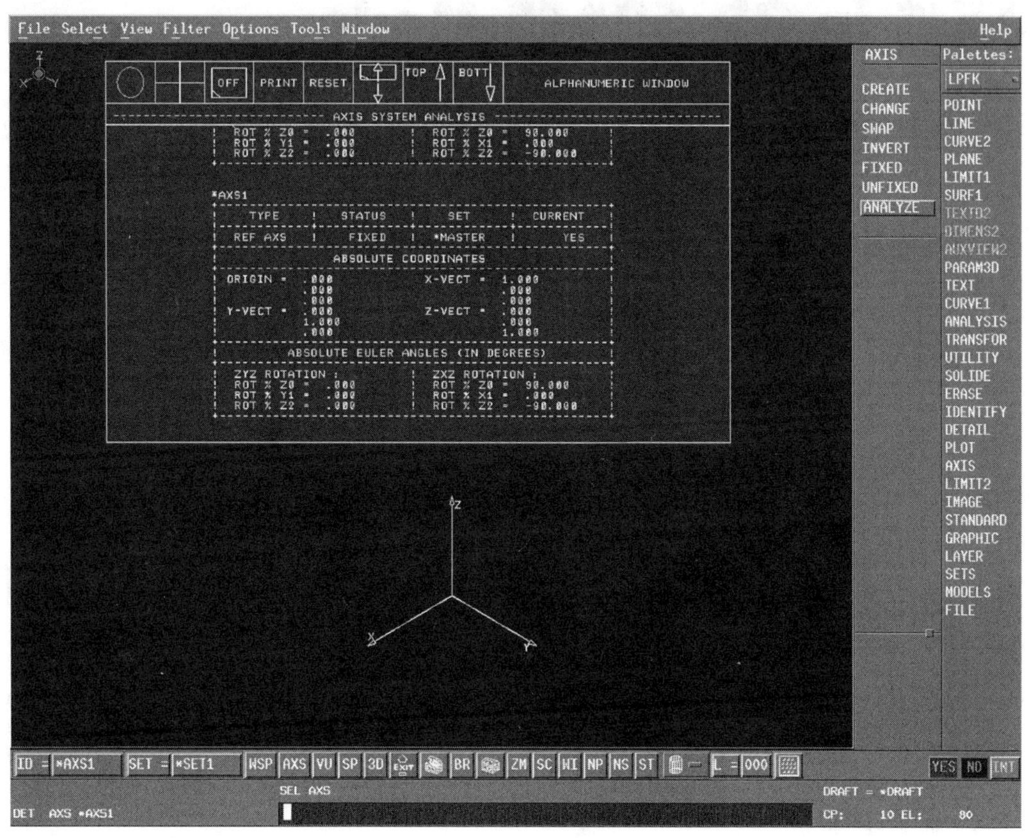

Axis system analysis display.

AXIS in 3D SPACE Mode

Chapter 13

The AXIS function is used to create and modify axis systems, and to switch the current axis system.

Option menu for AXIS in 3D SPACE mode.

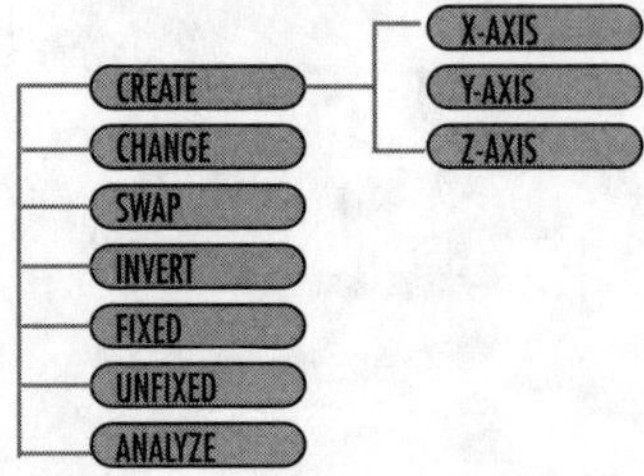

The four display modes of axis systems in SPACE mode are listed in the next table.

Type of axis system	Display mode
Current direct	Solid lines
Noncurrent direct	Dashed lines
Current reversed	Dotted lines
Noncurrent reversed	Dot-dashed lines

CREATE

Create new axis system.

Select: AXIS > CREATE | X-AXIS or Y-AXIS or Z-AXIS

Prompt: SEL PT/LN/PLN

When creating new axis systems you have a choice of creating an X, Y, or Z axis system (i.e., making it the privileged axis). The following descriptions assume that the Z axis is privileged. Axis systems can be created in two ways.

- Define the origin of the new system by selecting a PT (point) element. An axis system displays with the origin positioned on the point and axes parallel to those of the current axis system.

✓ **NOTE:** *If you do not require the new axis to be parallel to the current axis, it can be changed in one of the following three ways. (a) Key in the components of a vector (making* *the Z axis colinear with the vector). (b) Select an LN (line) element (making the Z axis colinear with the line). (c) Select a PLN (plane) element (making the Z axis perpendicular to the plane).*

- Select a LN (line) element, and an axis system displays with its origin as the projection of the current axis system origin point to the selected line. The Z axis will lie on the selected line and the X axis will be colinear with the intersection of the current axis system XY plane and the plane perpendicular to the selected line. You can also implicitly define an origin point for the new axis system by selecting a PLN (plane) element. A new axis will be created whose origin is the projection of the current axis system origin onto the selected plane. The Z axis will be perpendicular to the selected plane. The X axis will be colinear with the intersection of the current axis system XY plane and the selected plane.

✓ **NOTE:** *When a new axis system is created it automatically becomes the current axis system.*

If necessary, you can modify the other two axes in the axis system. Because the Z axis is privileged, the menu option only shows the X and Y axes, and the X axis is selected by default. You can modify it in one of three ways.

- Key in an angle value for the X axis; the axis system will be rotated about the Z axis.
- Select a LN (line) element; the X axis will become colinear with the projection of the line onto the XY plane.
- Select a PLN (plane) element; the X axis will be colinear with the intersection of this plane and the XY plane.

Once you have defined the system's origin point and orientation, click on the YES button to accept and make current the displayed axis system.

✓ **NOTE:** *Defining a point of origin and orientation for a new axis system is similar when the X or Y axis is privileged.*

CHANGE

Change current axis system.

Select: AXIS > CHANGE

Prompt: SEL AXS

Select any other available axis system.

SWAP

Swap two axes of axis system.

Select: AXIS > SWAP

First prompt: SEL 1ST AXIS VECT

Second prompt: SEL 2ND AXIS VECT

Select the axes in question. To complete the switch, select two axes (e.g., the X and Y axes) after the first and second prompts listed above.

✓ ***NOTE:*** *This function applies to all three axes in an axis system.*

INVERT

Invert axis of axis system.

Select: AXIS > INVERT

Prompt: SEL AXS

Select axis. An axis pointing to the right will point to the left (and vice versa), and an axis pointing downward will point upward (and vice versa).

FIXED

Fix axis.

Select: AXIS > FIXED

Prompt: SEL AXS

Select axis that has been axis. (Note that you can only select *unfixed* axes.)

✓ ***NOTE:*** *New axis systems are by default fixed. That is, they will not move when a transformation is applied to the current view. When an axis is unfixed (see below), it will move with the transformation.*

UNFIXED

Unfix axis.

Select: AXIS > UNFIXED

Prompt: SEL AXS

Select axis in order to move the axis when a transformation is applied. (Note that only *fixed* axes are selectable.)

ANALYZE

Analyze axis system.

Select: AXIS > ANALYZE

Prompt: SEL AXS

Select axis system. The alphanumeric window displays the following information.

- Axis system type (direct or reverse).
- Axis system status (fixed or unfixed). If the axis is fixed, the workspace to which it belongs is displayed. If the axis system is unfixed, the set to which it belongs is displayed.
- Axis system status (current or not current).
- Coordinates of axis system origin and components of each axis.
- Euler angles with reference to the orientation of the analyzed axis system.

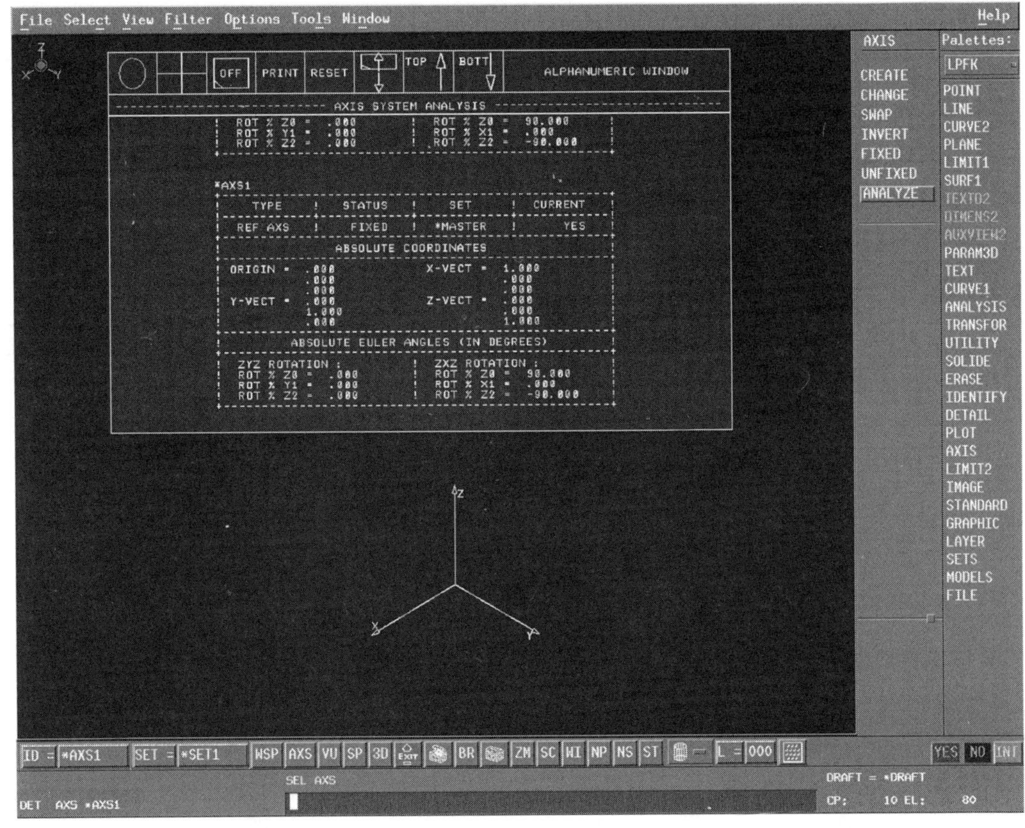

Axis system analysis display.

COMBIVU in DRAW Mode

Chapter 6

The COMBIVU function is used to generate geometry in a DRAW view using geometry in other DRAW views or from SPACE geometry.

Option menu for COMBIVU in DRAW mode.

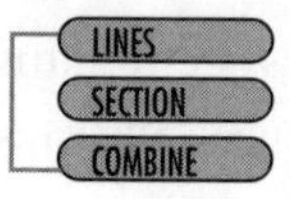

LINES

Create witness lines in the current DRAW view from elements selected in another DRAW view. When creating witness lines you sometimes have the choice of creating unlimited or limited lines.

✓ **NOTE:** *Before creating lines using COMBIVU verify that you are in the appropriate view using the AUXVIEW > CHANGE option or the VU button in the fixed menu area. Selecting either option will dim the current view.*

Create unlimited witness lines in current view

Select: COMBIVU > LINES

Prompt: SEL PT/LN/CRV // 1ST LIM: SEL LN

Choose one of the following procedures for creating unlimited witness lines using different element types.

- Select a PTD (point) element outside the current view. The line that results is the projection into the current view of the line passing through the selected point and perpendicular to the view to which the selected point belongs.
- Select an LND (line) element outside the current view, and then select an LND element from the same view as the first selected line. The line that results is the projection of the intersection for the two selected lines.
- Select an LND (line) element outside the current view. Click on the YES button to create lines passing through the end points of the selected line. (However, clicking on YES is unnecessary if the end points of the selected line yield the same line.)
- Select a CRV (curve) element outside the current view. The line that results is the projection of the apparent contour of the cylinder based on the selected curve and

perpendicular to the view containing the curve. If several solutions exist, the solution closest to the selection point is used. A temporary point displays to show the intersection point used.

Create limited witness lines in current view

Select: COMBIVU > LINES

Prompt: SEL PT/LN/CRV // 1ST LIM: SEL LN

Select two LND elements in the current view; the lines define the length of lines projected into the view. Once the limiting lines are selected, points, lines, or curves can be selected from other views as defined above for unlimited witness lines.

SECTION

Create cross sections of infinite cylinders by projecting them into a view; the projection is normal to the current view.

Select: COMBIVU > SECTION

Prompt: SEL VIEW

1. Select the view into which elements will be projected. Key in a depth value with respect to the view plane, or click on the YES button to accept the default value displayed in the message area.
2. Select a DRAW element in the current view to define its intersect with the cutting plane defined in the previous step. If necessary, key in a new depth value with respect to the view plane.
3. Click on the YES button to complete the process.

COMBINE

Generate elements in the current view by defining a construction plane or curve and selecting elements that lie on the plane or curve.

Generate elements in current view using planes and elements from other views

Select: COMBIVU > COMBINE

Prompt: SEL PLN // SEL CRV

1. Define the plane on which the geometry will lie by selecting an LND (line) element; this line cannot be in the current view.
2. Select a DRAW element from another view to be combined with the defined plane to create elements in the current view.

3. If necessary, click on the YES button to define another plane.

Generate elements in current view using curves and elements from other views

Select: COMBIVU > COMBINE

Prompt: SEL PLN // SEL CRV

1. Define the curve on which the desired geometry will lie by selecting a CRVD (curve) element outside the current view.
2. Select a DRAW element from another view to be combined with the selected curve to create elements in the current view.
3. If necessary, click on the YES button to select another defining curve.

CURVE2 in DRAW Mode

Chapters 2, 6, 9

The CURVE2 function in DRAW mode is used to create and modify 2D curve type elements (circles, conics, and splines).

Option menu for CURVE2 in DRAW mode.

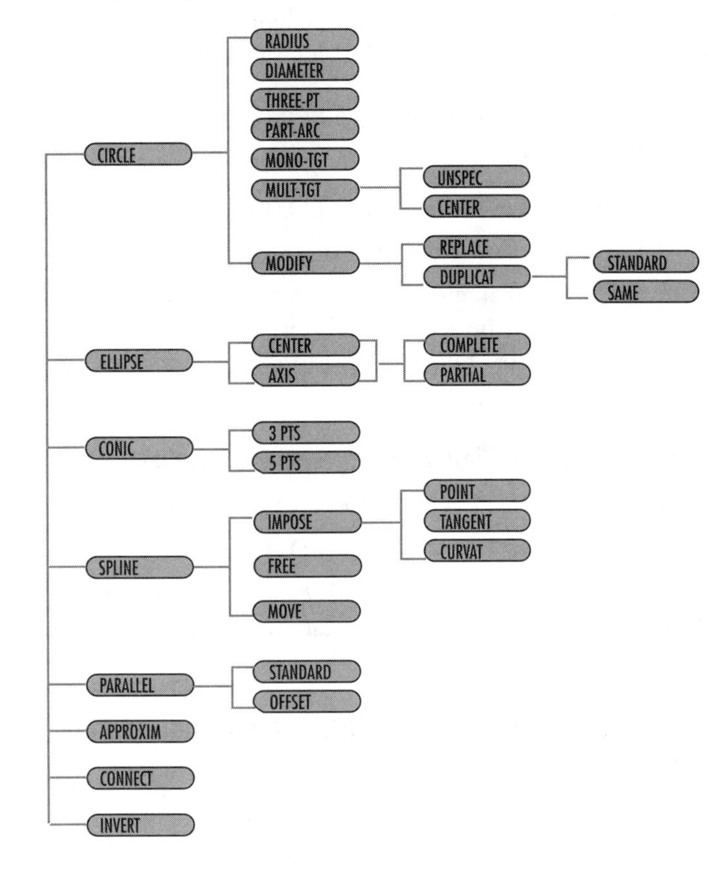

CIRCLE

Create circle.

Select: CURVE2 > CIRCLE | RADIUS or DIAMETER or THREE-PT or PART-ARC or MONO-TGT or MULT-TGT

Create circle using center and radius

Select: CURVE2 > CIRCLE | RADIUS

Prompt: CENTER: SEL PTD/CIRD

Take one of the following alternatives: (a) Select or indicate a point, or select a circle to define the center. Enter the circle radius, or select a point. A circle will be created on the defined center point passing through the selected point. (b) After center definition, select a line. A circle with radius equal to the length of the line is created. (c) Select a circle (a circle with same radius is created). (d) Click on YES to accept the standard radius value offered.

Create circle using center and diameter

Select: CURVE2 > CIRCLE | DIAMETER

Prompt: SEL LND / / SEL PTD/CIRD

To define the center, select or indicate a point, or select a circle.

Enter the circle diameter. Alternatively, after center definition, select a line (a circle with diameter equal to the length of the line is created), or select a circle (a circle with same diameter is created). Finally, you can click on YES to accept the standard diameter value provided.

Create circle via diametrically opposed points

Select: CURVE2 > CIRCLE | DIAMETER

Prompt: SEL LND / / SEL PTD/CIRD

Select a point or a circle (whose center will be selected). The created circle will pass through this point. Select or indicate a second point; a circle with diameter equal to the distance between the points is created.

Create circle using three points, or two points and radius

Select: CURVE2 > CIRCLE | THREE-PT

Prompt: SEL PTD1 <LND/CIRD>

Select or indicate three points; a circle passing through the three points is created. Alternatively, select or indicate two points, and then enter a radius value or click YES to accept the standard radius value offered. A circle of defined radius passing through the two points is created. Click on YES to obtain other possible solutions as necessary.

Create circular arc using three points, or two points and radius

Select: CURVE2 > CIRCLE | PART-ARC

Prompt: SEL PTD1 <LND/CRVD>

Select or indicate three points; an arc is created through the points. Click on YES to obtain the complementary arc as necessary. Alternatively, select or indicate two points, and then enter a radius. An arc is created. Click on YES to obtain other possible solutions as necessary.

Create circular arc with defined center and passing through point

Select: CURVE2 > CIRCLE | PART-ARC

Prompt: SEL PTD1 <LND/CRVD>

1. Select or indicate a point to create the first end of the arc.
2. Click on YES, and then select or indicate a center point.
3. Select or indicate a third point (an arc is created from the first point to a line from the center to the third point), or enter an angle value.
4. Click on YES to obtain the complementary arc as necessasry.

Create tangent circle defined by center

Select: CURVE2 > CIRCLE | MONO-TGT

Prompt: CENTER: SEL PTD/LND

1. Select a point or a line on which the center of the circle will lie.
2. Select a line or curve to which the circle will be tangent.
3. Indicate the region in which the center of the circle will be, and then enter a radius, or click on YES to accept the value offered. A tangent circle is created.
4. Click on YES to obtain other possible solutions as necessary.

Create circle tangent to other elements with or without defined center

Select: CURVE2 > CIRCLE | MULT-TGT | UNSPEC

Prompt: SEL 1ST PTD/LND/CIRD/CRVD

Choose one of the following alternatives.

- Select two point, line, circle, or curve elements and enter a radius value (which will become the standard value), or indicate the region in which the circle is to be created. Click on YES to obtain other possible solutions as necessary.
- Select two circle type elements, and enter a radius value. Click on YES to create the circle. Click on YES to obtain other possible solutions as necessary.
- Select three point, line, or circle type elements, or any combination of the same. A circle tangent to the three elements is created. Click on YES to obtain other possible solutions as necessary.

Create circle tangent to two other circles

Select: CURVE2 > CIRCLE | MULT-TGT | CENTER

Prompt: CENTER: SEL LND

1. Select a line; the center of the circle will be on this line.
2. Select two circles, and then indicate the region where the circle is to be created.
3. Click on YES to obtain other possible solutions as necessary.

Modify existing circle

Select: CURVE2 > CIRCLE | MODIFY | REPLACE or DUPLICAT | STANDARD or SAME

Prompt: SEL CIRD

Upon selection of DUPLICAT, you have the following options: with STANDARD selected, the duplicated circle will be on the current layer and contain the current standard graphical representation; and with SAME selected, the duplicated circle will be on the same layer as the selected circle, and will contain identical graphical representation.

Modify circle radius

Select a circle, and then enter a new radius value. Alternatively, select a point, and then click on YES to create the new circle.

Modify position of circle center

Enter x and y coordinates, or select two points to create a translation vector. The center of the circle will be moved or copied to the new position.

ELLIPSE

Create ellipses using different methods.

Select: CURVE2 > ELLIPSE | CENTER or AXIS

Create ellipse with known center

Select: CURVE2 > ELLIPSE | CENTER | COMPLETE

Prompt: CIR: SEL PTD <LND/CRVD>

Use one of the following alternatives.

- Select or indicate the center point of the ellipse. Enter lengths for the semi-axes, or enter an value for the angle between the x axis and the major axis of the ellipse, or select a line which defines the first semi-axis direction.

- Select or indicate the center point of the ellipse. Select or indicate two points as the ends of the semi-axes, and then select a point through which the ellipse will pass.

Create ellipse using center and vertex

Select: CURVE2 > ELLIPSE | CENTER | PARTIAL

Prompt: CIR: SEL PTD <LND/CRVD>

Select or indicate the center point of the ellipse. Select or indicate the end point of one of the semi-axes, and then select a point through which the partial ellipse will pass. Click on YES to obtain the complementary arc as necessary.

Create ellipse with known axis

Select: CURVE2 > ELLIPSE | AXIS | COMPLETE or PARTIAL

Prompt: SEL 1ST PTD 1ST AXIS / / SEL LND

Choose one of the following methods.

- Select or indicate two points to serve as the ends of one of the axes, and then select a point through which the ellipse will pass.
- Select or indicate two points as the ends of one of the axes, and then enter the length of the other axis.
- Select a line whose length will define one axis, and then select a point through which the ellipse will pass.
- Select a line whose length will define one axis, and then enter the length of the other axis.

With the PARTIAL option selected, click on YES to obtain the complementary arc as necessary.

CONIC

Create conic using three- or five-point method.

Select: CURVE2 > CONIC | 3 PTS or 5 PTS

Create conic using three points

Select: CURVE2 > CONIC | 3 PTS

Prompt: SEL 1ST PTD <LND/CRVD>

See the following illustration for the definition requirements of a conic arc.

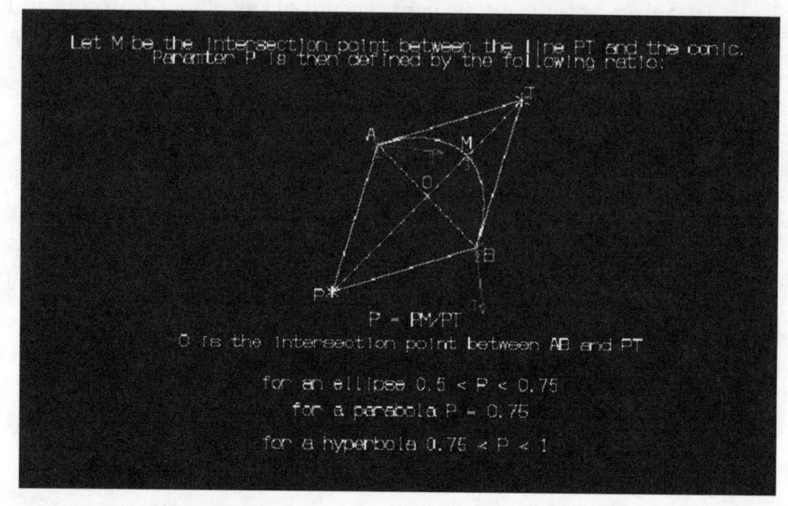

Definition of conic arc.

Create conic arc defined by two end points, intersection between tangents of points, and a specific condition

1. Select or indicate three points, and then enter a value for P. Click on YES to create a parabola.
2. Select a point through which the parabola will pass, or select a point or tangent symbol of one of the end points. Enter a value for the curvature radius.

Create conic arc defined by two end points, tangents of end points, and a specific condition

Nonparallel tangents

1. Select or indicate the first end point of the arc. Enter an angle value between the H axis and the first tangent, or select a line. The first tangent will be parallel to this line.
2. Select or indicate the second end point of the arc. Enter an angle value between the H axis and the second tangent, or select a line. The second tangent will be parallel to this line.
3. Enter a value for P and then click on YES to create a parabola. Alternatively, select a point through which the parabola will pass, or select a point or tangent symbol and enter a value for the curvature at this point.

Parallel tangents (ellipse arcs created)

1. Select or indicate the first end point of the arc. Enter an angle value between the H axis and the first tangent, or select a line. The tangent will be parallel to this line.

2. Select or indicate the second end point of the arc. Enter an angle value between the H axis and the second tangent, such that the defined tangent is parallel to the previously defined tangent. Select a point through which the ellipse will pass.

Create conic using five points

Select: CURVE2 > CONIC | 5 PTS

Prompt: SEL PTD (1ST EXTREMITY)

This method involves creating a conic arc using end points, plus one of the following: one point and two tangents; two points and one tangent; or three points.

1. Select a point.
2. Select a line to define the tangent, or enter an angle value between the H axis and the tangent as necessary.
3. Repeat to create the five constraints to define the arc.
4. Click on YES to obtain the complementary arc as necessary.

SPLINE

Create and modify curves with the possibility of imposing tangency and curvature at each point.

Select: CURVE2 > SPLINE | IMPOSE or FREE or MOVE

Impose points through which spline passes

Select: CURVE2 > SPLINE | IMPOSE | POINT

Prompt: SEL CRVD / /SEL PTD

1. Select or indicate points through which the spline must pass, and then click on YES to compute the spline.
2. If necessary, select the spline to be modified and define the current point.
3. Select a point, and click on YES to compute the spline.
4. Click on YES to replace the old spline with the new as necessary.

Impose tangent

Select: CURVE2 > SPLINE | IMPOSE | TANGENT

Prompt: SEL CRVD

1. Select the spline to be modified and define the current point as necessary.
2. Select a line (the tangent at the selected point will be parallel to this line), or enter an angle value for the tangent.

3. Select a point. The direction of the tangent is defined by a line passing through the selected point and the current point.
4. If necessary, select the tangent symbol to reverse direction. Click on YES to compute the spline.
5. Click on YES to replace the old spline with the new as necessary.

Impose curvature

Select: CURVE2 > SPLINE | IMPOSE | CURVAT

Prompt: SEL CRVD

Select the spline to be modified and define the current point as necessary. (The tangent at the current point must be imposed to define the curvature).

Enter a value for the curvature radius or select a circle (whose radius will be used as the curvature radius).

Click on YES to compute the spline.

Click on YES to replace the old spline with the new as necessary.

Free constraints at point on spline

Select: CURVE2 > SPLINE | FREE

Prompt: SEL CRVD

Select the spline to be modified, and then select a point symbol as necessary. The point and constraints are deleted. Alternatively, select a tangent, and the tangent and curvature constraints at the point are deleted. Another option would be to select a curvature symbol; the curvature constraint at the point is deleted. Click on YES to compute the spline. Click on YES to replace the old spline with the new as necessary.

Move current point

Select: CURVE2 > SPLINE | MOVE

Prompt: SEL CRVD

Select the spline to be modified and define the current point as necessary. Select or indicate a point; the current point will be moved to the defined point. Click on YES to compute the spline as necessary. Click on YES to replace the old spline with the new as necessary.

PARALLEL

Create curves parallel to other curves.

Select: CURVE2 > PARALLEL | STANDARD or OFFSET

Create standard parallel curves

Select: CURVE2 > PARALLEL | STANDARD

Prompt: SEL CRVD

Use one of the following alternatives.

- Select a curve, and then select a point. A curve is created parallel to the first curve which passes through the point. If desired, click on YES to create the parallel curve symmetrical to the selected curve.
- Select a curve, and indicate the region where the curve is to be created. Enter a distance or click on YES to accept the distance offered. Specify the number of desired parallel curves as necessary. If desired, click on YES to create the parallel curve(s) symmetrical to the selected curve.

Create curves parallel to other curves and offset by linear distance

Select: CURVE2 > PARALLEL | OFFSET

Prompt: SEL CRVD

Select a curve, and indicate the region in which the curve is to be created. Enter a distance or click on YES to accept the distance offered. Specify the number of parallel curves needed as necessary. Click on YES to create the curve(s) symmetrical about the selected curve as necessary.

APPROXIM

Approximate one curve by another curve with a given degree and tolerance value.

Select: CURVE2 > APPROXIM

Prompt: SEL CRVD

1. Select a curve. The following symbols display: a number 1 or 2 representing a locked point, and an arrow representing a locked tangent. The imposed limit conditions are highlighted.
2. If necessary, select the limit condition to be modified.
3. Click on YES to confirm the proposed values, or enter new values for degree and tolerance. The created curve displays.
4. Click on YES to replace the old curve with the new as necessary.

CONNECT

Create connection curve between two other curves.

Select: CURVE2 > CONNECT

Prompt: SEL 1ST CRVD

1. Select two curves. The end nearest the point of selection will be the connection end, and the following symbols display: a number 1 or 2 representing a locked point and an arrow representing a locked tangent. The imposed limit conditions are highlighted.
2. If necessary, select the limit condition to be modified.
3. Click on YES to confirm proposed values, or enter new values for degree and tolerance. The created curve displays.
4. Click on YES to replace the old curve with the new as necessary.

INVERT

Invert curves.

Select: CURVE2 > INVERT

Prompt: SEL CRVD

1. Select a spline. Symbols indicating the direction of the curve display at the ends.
2. Click on YES to invert the curve. Symbols arereversed.

CURVE2 in 2D SPACE Mode

INSIDE CATIA

Chapter 9

The CURVE2 function in 2D SPACE mode is used to create circles, ellipses, conic arcs, and curves. Tangency and curvature continuity can also be imposed at arc limits.

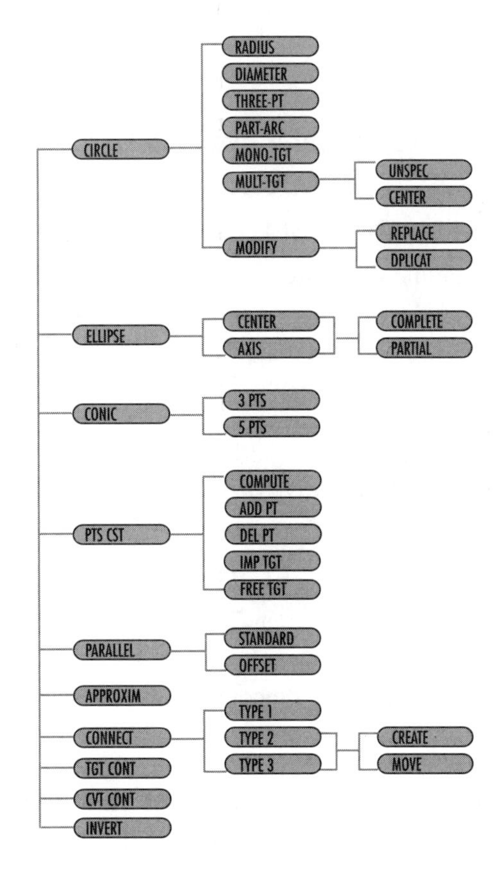

Option menu for CURVE2 in 2D SPACE mode.

CIRCLE

Create circles.

Select: CURVE2 > CIRCLE | RADIUS or DIAMETER or THREE-PT or PART-ARC or MONO TGT or MULT TGT or MODIFY

Create circle using center and radius

Select: CURVE2 > CIRCLE | RADIUS

Prompt: CENTER: SEL PT/CIR

1. To define the center, select or indicate a point, or select a circle.
2. To define the diameter, you have five alternatives: (a) Enter circle radius. (b) Select a point. (A circle is created on the defined center point passing through the selected point.) (c) Select a line. (A circle with radius equal to the length of the line is created.) (d) Select a circle. (A circle with same radius is created.) (e) Click on YES to accept the suggested standard radius value.

Create circle using center and diameter

Select: CURVE2 > CIRCLE | DIAMETER

Prompt: SEL LN / / SEL PT/CIR

1. To define the center, select or indicate a point, or select a circle.
2. To set the diameter, you have four alternatives: (a) Enter circle diameter. (b) Select a line. A circle with diameter equal to the length of the line is created. (c) Select a circle. A circle with same diameter is created. (d) Click on YES to accept the suggested standard diameter value.

Create circle by diametrically opposed points

Select: CURVE2 > CIRCLE | DIAMETER

Prompt: SEL LN / / SEL PT/CIR

1. Select a point or circle (i.e., circle center). The created circle will pass through this point.
2. Select or indicate a second point. A circle with diameter equal to the distance between the points is created.

Create circles using three points or two points and radius

Select: CURVE2 > CIRCLE | THREE-PT

Prompt: SEL PT1<LN/CRV>

1. Select or indicate three points; a circle passing through the three points is created. Alternatively, select or indicate two points and then enter a radius value or click YES to accept the standard suggested radius value. A circle of defined radius passing through the two points is created.
2. Click YES to obtain other possible solutions as necessary.

Create circular arc using three points or two points and radius

Select: CURVE2 > CIRCLE | PART-ARC

Prompt: SEL PT1<LN/CRV>

Select or indicate three points; an arc is created through the points. If required, click on YES to obtain the complementary arc as necessary.

Alternatively, select or indicate two points, and then enter a radius. An arc is created. Click on YES to obtain other possible solutions as necessary.

Create circular arc with defined center and passing through two points

Select: CURVE2 > CIRCLE | PART-ARC

Prompt: SEL PT1<LN/CRV>

1. Select or indicate two points to serve as arc end points.
2. Enter a radius value or click on YES to accept the suggested value.
3. Click on YES to obtain other possible solutions as necessary.

Create circular arc passing through point with defined center

Select: CURVE2 > CIRCLE | PART-ARC

Prompt: SEL PT1<LN/CRV>

1. Select or indicate a point to serve as the first arc end point.
2. Click on YES to confirm that the next entry will be the arc center.
3. Select or indicate the arc center point.
4. Select or indicate the second arc end point, or enter an angle.
5. Click on YES to obtain complementary arc as necessary.

Create tangent circle defined by center

Select: CURVE2 > CIRCLE | MONO-TGT

Prompt: CENTER: SEL LN

1. Select or indicate the center point of the circle to be created.
2. Select a line, circle, or curve. A tangent circle is created.
3. Click on YES to obtain other possible solutions as necessary.

Create circle tangent to two other elements with center on line

1. Select a line on which the center of the circle will lie.
2. Select a line, circle, or curve.
3. Enter a radius value or select another circle whose radius will be used.

4. Indicate the region where the circle is to be created. Click on YES to obtain other possible solutions as necessary.

Create circle tangent to other elements with or without defined center

Select: CURVE2 > CIRCLE | MULT-TGT | UNSPEC

Prompt: SEL 1ST PT/LN/CIR/CRV

1. Select two point, line, circle, or curve elements.
2. Enter a radius value (which will become the standard value) or indicate the region in which the circle is to be created. Click on YES to obtain other possible solutions as necessary.
3. Select two circle type elements, and enter a radius value.
4. Click on YES to create the circle. Click on YES to obtain other possible solutions as necessary.
5. Select three point, line, or circle type elements, or any combination of the same. A circle tangent to the three elements is created. Click on YES to obtain other possible solutions as necessary.

Create circle tangent to two other circles with defined center

Select: CURVE2 > CIRCLE | MULT-TGT | CENTER

Prompt: SEL CIR

1. Select a line; the center of the circle will be on this line.
2. Select two circles, and then indicate the region where the circle is to be created.
3. Click on YES to obtain other possible solutions as necessary.

Modify existing circle

Select: CURVE2 > CIRCLE | MODIFY | REPLACE or DUPLICAT

Prompt: SEL CIR

On selection of the DUPLICAT option, the following suboptions are accessible.

- STANDARD. Duplicated circle will be on the current layer with current standard graphical representation.
- SAME. Duplicated circle will be on the same layer as the selected circle with identical graphical representation.

Modify radius of circle

Select a circle, and then enter a new radius value. Alternatively, select a point, and then click on YES to create the new circle.

Modify position of circle center

Enter x and y coordinates, or select two points to create a translation vector. The center of the circle will be moved or copied to the new position.

ELLIPSE

Create ellipse.

Select: CURVE2 > ELLIPSE | CENTER or AXIS

Create ellipse with known center

Select: CURVE2 > ELLIPSE | CENTER | COMPLETE

Prompt: CTR: SEL PT<LN/CRV>

1. Select or indicate the center point of the ellipse.
2. Enter lengths for the semi-axes. Alternatively, enter a value for the angle between the x axis and the major axis of the ellipse, or select a line which defines the first semi-axis direction.
3. Select or indicate the center point of the ellipse.
4. Select or indicate two points as the ends of the semi-axes, and then select a point through which the ellipse will pass.

Create ellipse using center and vertex and passing through point

Select: CURVE2 > ELLIPSE | CENTER | PARTIAL

Prompt: CTR: SEL PT<LN/CRV>

1. Select or indicate the center point of the ellipse.
2. Select or indicate the end point of one of the semi-axes, and then select a point through which the partial ellipse will pass.
3. Click on YES to obtain the complementary arc as necessary.

Create ellipse with known axis

Select: CURVE2 > ELLIPSE | AXIS | COMPLETE or PARTIAL

Prompt: SEL 1ST PT 1ST AXIS / / SEL LN

1. Select or indicate two points that will serve as the ends of one of the axes, and then select a point through which the ellipse will pass.
2. Select or indicate two points as the ends of one of the axes, and then enter the length of the other axis.

3. Select a line whose length will define one axis, and then select a point through which the ellipse will pass.
4. Select a line whose length will define one axis, and then enter the length of the other axis.
5. When the PARTIAL option is selected, click on YES to obtain the complementary arc as necessary.

CONIC

Create conic.

Select: CURVE2 > CONIC | 3 PTS or 5 PTS

Create conic arc using three points

Select: CURVE2 > CONIC | 3 PTS

Prompt: SEL 1ST PT<LN/CRV>

See the next illustration for the definition requirements of a conic arc.

Definition of conic arc.

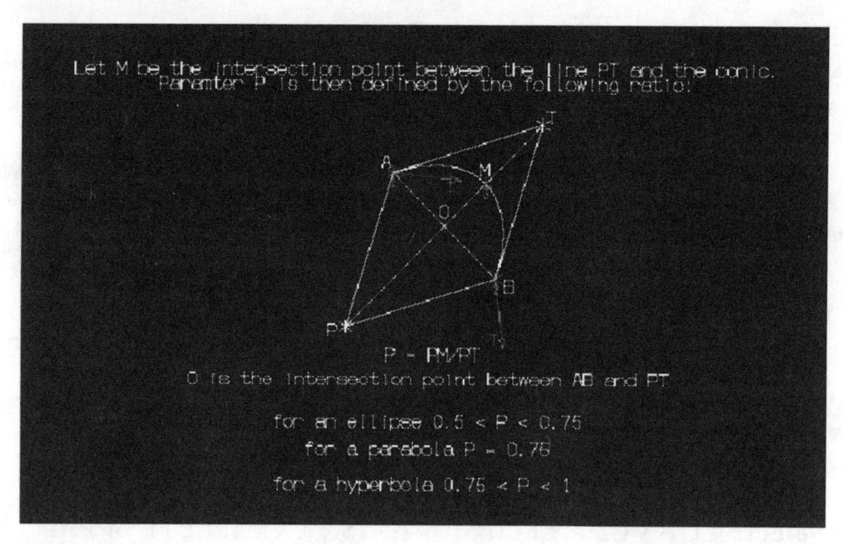

Create conic arc defined by two end points, intersection between point tangents, and specific condition

1. Select or indicate three points, and then enter a value for P.
2. Click on YES to create a parabola. Alternatively, select a point through which the parabola will pass, or select a point or tangent symbol of one of the end points, and then enter a value for the curvature radius.

Create conic arc defined by two end points, point tangents, and specific condition

Nonparallel tangents

1. Select or indicate the first end point of the arc. Enter an angle value between the H axis and the first tangent, or select a line. The first tangent will be parallel to this line.
2. Select or indicate the second end point of the arc. Enter an angle value between the H axis and the second tangent, or select a line. The second tangent will be parallel to this line.
3. Enter a value for P.
4. Click on YES to create a parabola. Alternatively, select a point through which the parabola will pass or select a point or tangent symbol, and then enter a value for the curvature at this point.

Parallel tangents

1. Select or indicate the first end point of the arc. Enter an angle value between the H axis and the first tangent, or select a line. The tangent will be parallel to this line.
2. Select or indicate the second end point of the arc. Enter an angle value between the H axis and the second tangent, such that the defined tangent is parallel to the previously defined tangent.
3. Select a point through which the ellipse will pass.

Create conic arc

Select: CURVE2 > CONIC | 5 PTS

Prompt: SEL PT(1ST EXTREMITY)

Create a conic arc using its end points, plus one of the following: (a) one point and two tangents; (b) two points and one tangent; or (c) three points.

1. Select a point.
2. If required, select a line to define the tangent, or enter an angle value between the H axis and the tangent.
3. Repeat the above steps to create the five constraints to define the arc.
4. Click on YES to obtain the complementary arc as necessary.

PTS CST

Create curves passing through specified constraint points.

Select: CURVE2 > PTS CST | COMPUTE or ADD PT or DEL PT or IMP TGT or FREE TGT

Suboptions can be selected to display information in the alphanumeric window. Constraint points are defined using the suboptions as follows.

- COMPUTE. Select a constraint (CST) element.
- ADD PT. Add points through which the curve must pass.
- DEL PT. Delete points through which the curve must pass.
- IMP TGT. Define tangency of point.
- FREE TGT. Remove tangency requirements of point.

The created curve will have arcs sufficient to respect tangency conditions, while passing through all points.

✓ **NOTE:** *When using the PTS CST option, the current point is the point on which a modification is carried out. The point is highlighted.*

PARALLEL

Create curve parallel to another curve.

Select: CURVE2 > PARALLEL | STANDARD or OFFSET

Create parallel curve symmetrical about another curve

Select: CURVE2 > PARALLEL | STANDARD

Prompt: SEL CRV

1. Select a curve, and then select a point. A curve is created parallel to the first curve and passing through the point.
2. Click on YES to create the parallel curve symmetrical about the selected curve.
3. Select a curve, and indicate the region where the curve is to be created.
4. Enter a distance or click on YES to accept the suggested distance.
5. If required, specify the number of parallel curves.
6. Click on YES to create the parallel curve(s) symmetrical about the selected curve as necessary.

Create parallel curve offset by linear distance

Select: CURVE2 > PARALLEL | OFFSET

Prompt: SEL CRV

1. Select a curve, and indicate the region in which the curve is to be created.

2. Enter a distance or click on YES to accept the suggested distance.
3. If required, specify the number of parallel curves.
4. Click on YES to create the curve(s) symmetrical about the selected curve as necessary.

APPROXIM

Approximate a curve by another curve at specific degree and tolerance value.

Select: CURVE2 > APPROXIM

Prompt: SEL CRV

1. Select a curve. The following symbols display: a number 1 or 2 representing a locked point, and an arrow representing a locked tangent. The imposed limit conditions are highlighted.
2. If desired, select the limit condition to be modified.
3. Click on YES to confirm suggested values, or enter new values for degree and tolerance. The created curve displays.
4. Click on YES to replace the old curve with the new.

CONNECT

Create connection curve between two other curves.

Select: CURVE2 > CONNECT | TYPE1 or TYPE2 or TYPE3

Prompt: SEL 1ST LN/CRV/CCV

Connect two curves in space

Select: CURVE2 > CONNECT | TYPE 1

Prompt: SEL 1ST LN/CRV/CCV

1. Select two curve elements; the ends nearest the point of selection will be connected. An arrow symbol indicating locked tangent, and an X symbol indicating locked curvature displays at the curve ends.
2. If required, select arrows to invert.
3. Click on YES to accept suggested degree value, or enter the desired value.

Connect two curves in space by planar fourth degree arc

Select: CURVE2 > CONNECT | TYPE 2 | CREATE

Prompt: SEL 1ST LN/CRV/CCV

1. Select a curve or line element. The end nearest the point of selection will be selected and a tangent vector displays. (Select the vector to invert as necessary.)
2. Select a second curve or line element. A connecting curve is computed.

Modify connecting curve

Reselect one of the previously selected curves, and take one of the following alternatives.

- Select a tangent vector to invert it.
- Select a point; the point is projected onto the connecting plane and then onto the nearest curve. This point becomes the new connecting point.
- Select a new curve or line element, and then enter the number of curve to be dropped.
- If the analysis window displays, click on YES to analyze, and then select the curve end to be retained.
- Click on YES to limit the two curves.

Make connecting curve pass through specified point

Select: CURVE2 > CONNECT | TYPE2 | MOVE

Prompt: SEL CORNER TO MOVE

1. Select a connecting curve, and then select or indicate a point. The connecting curve is modified to pass through the point.
2. Click on YES to iterate the modification.

Connect two curves by two fourth degree arcs

Select: CURVE2 > CONNECT | TYPE 3 | CREATE

Prompt: SEL 1ST LN/CRV/CCV

1. Select a curve or line element. The end nearest the point of selection will be selected and a tangent vector displays. (Select the vector to invert as necessary.)
2. Select a second curve or line element. The connecting points and tangents at the curve ends form the connecting plane, and a connecting curve is computed.

Modify connecting curve

1. Reselect one of the previously selected curves and then take one of the following alternatives.

   - Select a tangent vector to invert it.
   - Select a point; the point is projected onto the connecting plane and then onto the nearest curve. This point becomes the new connecting point.

- Select a new curve or line element, and then enter the number of curve to be dropped.
- Enter a new minimum radius value.
- Select or indicate a point inside the displayed AREA to define the point at which the curvature radius will be minimum.
- Select the vector tangent to the minimum radius point.

2. Select or indicate a point. The tangent at the minimum radius point is defined by a line joining the minimum radius point to the selected point.
3. Select a line element. The tangent is define by the direction of the selected line.
4. If an analysis window displays, click on YES to analyze the curve.
5. If the point of minimum radius is outside the AREA, click on YES to accept the suggested values.
6. Select curve to be retained, or click on YES to limit the curves.

Make connecting curve pass through specified point

Select: CURVE2 > CONNECT | TYPE3 | MOVE

Prompt: SEL CORNER TO MOVE

1. Select a connecting curve, and then select or indicate a point. The connecting curve is modified to pass through the point.
2. Click on YES to iterate the modification.

TGT CONT

Create curves with tangency continuity at arc limits.

Select: CURVE2 > TGT CONT

Prompt: SEL CRV

1. Select a curve element; symbols display at the arc limits. A symbol indicating the point of maximum deviation displays, and maximum deviation values are also displayed in the alphanumeric window.
2. Click on YES to select the created curve. Angle and tangent values display in the alphanumeric window.

CVT CONT

Create curves with curvature continuity at arc limits.

Select: CURVE2 > CVT CONT

Prompt: SEL CRV

✓ ***NOTE:*** *Only isolated curves can be accepted.*

1. Select a curve element; symbols display at the arc limits. A symbol indicating the point of maximum deviation and maximum deviation values display in the alphanumeric window.
2. If desired, click on YES to select the created curve. Angle and tangent values display in the alphanumeric window.

INVERT

Invert curves.

Select: CURVE2 > INVERT

Prompt: SEL LN/CCV/CST

1. Select a spline; symbols indicating the direction of the curve display at its ends.
2. Click on YES to invert the curve. Symbols are reversed.

CURVE2 in 3D SPACE Mode

Chapter 9

The CURVE2 function in SPACE mode is used to create curves in space. In addition, the following elements can be created using this function: circles, spines, helices, wireframe volumes, connecting curves, and parallel curves. Tangency and continuity can also be imposed at arc limits.

Option menu for CURVE2 in 3D SPACE mode.

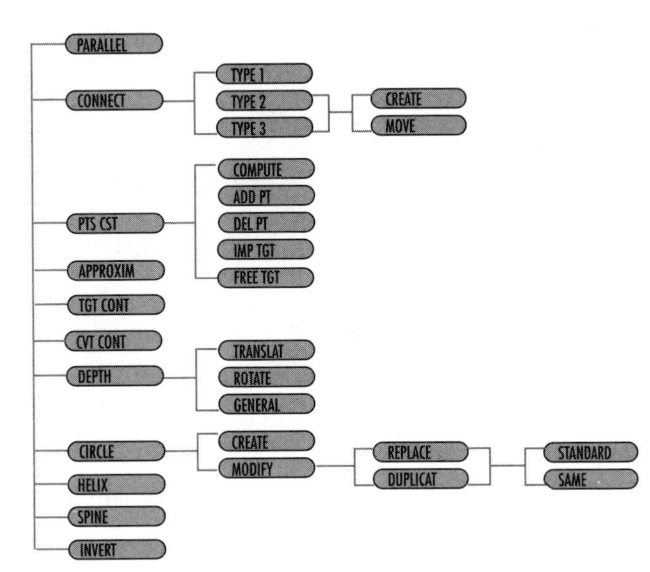

PARALLEL

Create planar curves parallel to another curve

Select: CURVE2 > PARALLEL

Prompt: SEL PLANAR CRV / / SEL SUR/FAC

1. Select a planar curve (CRV type element), and then indicate the region in which the parallel curve(s) (max. 20) is to be created.
2. Enter a distance and the number of curves (e.g., *5,20*) as necessary, or click on YES to accept the suggested values.
3. Click on YES to display the symmetrical option.

Create curves pseudo-parallel to curve on surface or face

1. Select a surface (SUR) or nonplanar face (FAC) element.
2. Select a curve element on the selected face or surface. An arrow indicating the relative position of the proposed curve displays. Select arrow to change direction in steps of 90 degrees.
3. Enter a distance and the number of curves as necessary, or click on YES to accept suggested values.
4. Click on YES to the symmetrical option.

✓ **NOTE:** *The creation of parallel curves will be refused if the tangency discontinuity is greater than 1, or if the distance proposed would result in degenerate geometry.*

CONNECT

Connect, modify, and move curves in space.

Select: CURVE2 > CONNECT | TYPE 1 or TYPE 2 or TYPE 3

Connect two curves in space

Select: CURVE2 > CONNECT | TYPE 1

Prompt: SEL 1ST LN/CRV/CCV

1. Select two curve elements; the ends nearest the point of selection will be connected. An arrow symbol indicating locked tangent and an X symbol indicating locked curvature display at curve ends.
2. Select arrows to invert as necessary.
3. Click on YES to accept the suggested degree value or enter desired value.

Create connecting curve

Select: CURVE2 > CONNECT | TYPE 2 | CREATE

Prompt: SEL 1ST LN/CRV/CCV

1. Select a curve or line element; the end nearest the point of selection will be selected. A tangent vector displays. (Select the vector to invert.)
2. Select a second curve or line element; a connecting curve is computed.

Modify connecting curve

Reselect a previously selected curve. At this point, proceed with one of the following alternatives:

- Select a tangent vector to invert it.

- Select a point. The point is projected onto the connecting plane and then onto the nearest curve. This point becomes the new connecting point.
- Select a new curve or line element, and then enter the number of curve to be dropped.
- Click on YES to analyze. If the analysis window displays, select the curve end to be retained.
- Click on YES to limit the two curves.

Make connecting curve pass through specified point

Select: CURVE2 > CONNECT | TYPE2 | MOVE

Prompt: SEL CORNER TO MOVE

Select a connecting curve, and then select or indicate a point. The connecting curve is modified to pass through the point. Click on YES to iterate the modification.

Create connecting curve

Select: CURVE2 > CONNECT | TYPE 3 | CREATE

Prompt: SEL 1ST LN/CRV/CCV

1. Select a curve or line element; the end nearest the point of selection will be selected. A tangent vector displays. (Select the vector to invert.)
2. Select a second curve or line element; the connecting points and tangents at the curve ends form the connecting plane. A connecting curve is computed.

Modify connecting curve

Reselect a previously selected curve. Proceed with one of the following alternatives.

- Select a tangent vector to invert it.
- Select a point; the point is projected onto the connecting plane and then onto the nearest curve. This point becomes the new connecting point.
- Select a new curve or line element, and then enter the number of the curve to be dropped.
- Enter a new minimum radius value.
- Select or indicate a point inside the displayed AREA to define the point at which the curvature radius will be minimum.
- Select the vector tangent to the minimum radius point.
- Select or indicate a point; the tangent at the minimum radius point is defined by a line joining the minimum radius point to the selected point.

- Select a line element; the tangent is defined by the direction of the selected line.
- Click on YES to analyze the curve if an analysis window displays.
- If the point of minimum radius is outside the AREA, click on YES to accept suggested values.
- Select curve to be retained.
- Click on YES to limit the curves.

Make connecting curve pass through specified point

Select: CURVE2 > CONNECT | TYPE3 | MOVE

Prompt: SEL CORNER TO MOVE

Select a connecting curve, and then select or indicate a point. The connecting curve is modified to pass through the point. Click on YES to iterate the modification.

PTS CST

Create curves passing through all specified constraint points.

Select: CURVE2 > PTS CST | COMPUTE or ADD PT or DEL PT or IMP TGT or FREE TGT

Prompt: SEL 1ST LN/CRV/CCV

1. Define the constraint points using the following menu options: add or delete points through which the curve must pass (ADD PT or DEL PT), and define tangency or not as desired at each point (IMP TGT or FREE TGT).
2. Compute the curve (COMPUTE) while taking into account the constraints in the previous steps.
3. If the alphanumeric window is activated, analytical data are included.
4. A submenu displays with the following options: POINT, TANGENT, NORMAL, and RADIUS. Any of these options can be selected to display information in the alphanumeric window.

✓ **NOTE:** *When using the PTS CST option, the current point is the point on which you carry out a modification displayed in highlight mode.*

APPROXIM

Approximate one curve by another of specified degree and tolerance.

Select: CURVE2 > APPROXIM

Prompt: SEL CRV

The following two options are offered in the panel activated upon selection of APPROXIM.

- STD. Standard approximation to comply with desired tolerance criteria.
- COARSE. Rougher approximation.

1. Select a curve element at each end of the selected element. A number 1 or 2 displays indicating locked point, as well as an arrow indicating locked tangent. The limit conditions display in highlighted mode.
2. If necessary, select the limit to be modified. Click on YES to accept suggested values for degree and tolerance, or enter new values.
3. Click on YES to replace the selected curve by the created curve.

TGT CONT

Create curves with tangency continuity at arc limits.

Select: CURVE2 > TGT CONT

Prompt: SEL CRV

1. Select a curve element; symbols display at the arc limits. Maximum deviation values and a symbol indicating the point of maximum deviation display in the alphanumeric window.
2. If necessary, click on YES the selected curve with the created curve.
3. If the alphanumeric window displays, angle and tangent values are included.

CVT CONT

Create curves with curvature continuity at arc limits.

Select: CURVE2 > CVT CONT

Prompt: SEL CRV

✓ ***NOTE:*** *Only isolated curves can be accepted.*

1. Select a curve element; symbols display at the arc limits. Maximum deviation values and a symbol indicating the point of maximum deviation display in the alphanumeric window.
2. If necessary, click on YES the selected curve with the created curve.
3. If the alphanumeric window displays, angle and tangent values are included.

DEPTH

Create wireframe by transforming chain of segments or curves.

Select: CURVE2 > DEPTH | TRANSLAT or ROTATE or GENERAL

Transform chain based on translation

Select: CURVE2 > DEPTH | TRANSLAT

Prompt: SEL LN/CRV/FAC / / YES:AUTO

1. Define the contour by selecting a face or segments to form the contour.

✓ **NOTE:** *Click on YES to activate the auto search facility before selecting an element.*

2. Click on YES to end contour selection.
3. Define the translation by taking one of the following alternatives: enter components of the translation vector, select two points to define the vector, or select a line to define direction.
4. Enter a distance value and a number as necessary, and then select the arrow to invert as necessary.

Transform chain based on rotation

Select: CURVE2 > DEPTH | ROTATE

Prompt: SEL LN/CRV/FAC / / YES: AUTO

1. Define the contour by selecting a face or segments to form the contour.
2. Define the contour by selecting a face or successively selecting segments. Alternatively, you can use the auto search as described above.
3. Click on YES to end contour selection.
4. Define the rotation by selecting an LN element as rotation axis, indicate the positive direction of rotation, then enter a rotation value and number as necessary.

Use stored transformation of chain

Select: CURVE2 > DEPTH | GENERAL

Prompt: SEL LN/CRV/FAC / / YES: AUTO

1. Define the contour by selecting a face or successively selecting segments, or use the auto search as described above.
2. Click on YES to end contour selection.
3. To select a transformation, enter its identifier, or click on YES and select from list.

CIRCLE

Create and modify circles.

Select: CURVE2 > CIRCLE | CREATE or MODIFY

Create circle

Select: CURVE2 > CIRCLE | CREATE

Prompt: SEL PT1/LN

Select a line to define the axis, and then select a point through which circle will pass. Alternatively, select three points in succession through which the circle must pass.

Modify existing circle

Select: CURVE2 > MODIFY | REPLACE or DUPLICAT | STANDARD or SAME

Prompt: SEL CIR

Suboptions are described below.

- DUPLICAT. Create duplicate circle.
- REPLACE. Replace original circle.
- STANDARD. Produce circle using current standard graphic and layer settings.
- SAME. Circle will have same graphic attributes and layer as selected circle.

Select a circle element to be modified. At this point, you have the following alternatives.

- Select another circle element; the selected circle will be modified to match the second selection.
- Enter a new radius value.
- Select a point element, and click on YES to confirm that the selected point is to be included in the modified circle. (If the point is not in the plane of the circle, it will be projected into the plane.)

To move the center of the circle, take the following steps.

1. Select the circle to be modified.
2. Define a translation vector by entering components, selecting points, or selecting a line and then entering a distance.

HELIX

Create revolution helix.

Select: CURVE2 > HELIX

Prompt: REVOLUTION AXIS: SEL PT1/LN

1. To define the helix axis, select a line element, or select two points.
2. Define the start point by selecting a point; an arrow displays which can be inverted by selection.
3. Click on YES to define a circle, or enter two angle values (pitch and variation of pitch per revolution).

SPINE

Create curve passing through a point on a plane normal to one or more other planes.

Select: CURVE2 > SPINE

Prompt: SEL PLN1

1. Select a plane and then select a point. (If the point is not on the selected plane, it will be projected onto the plane.)
2. Successively select plane elements to be intersected by the spine.
3. Click on YES to create the curve.

INVERT

Invert LN, CRV, and CCV type elements.

Select: CURVE2 > INVERT

Prompt: SEL LN/CCV/CST

Select a line or curve element. Symbols indicating direction of element display at element ends. Click on YES to invert selected element.

DETAIL in DRAW Mode

Chapters 5, 7, 18

The DETAIL function in DRAW mode is used to place elements in alternative workspaces which can be used on the master workspace as a ditto or copy of its elements. DETAIL is also used to perform transformations and analysis on dittos and to change between workspaces.

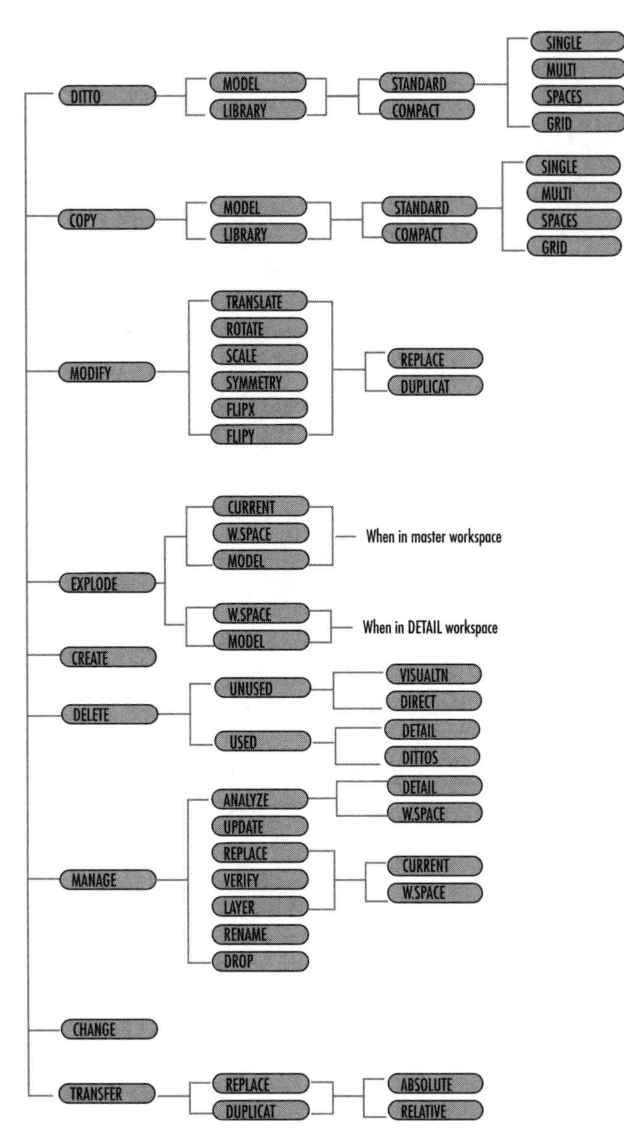

Option menu for DETAIL in DRAW mode.

DITTO

Create ditto.

Select: DETAIL > DITTO | MODEL or LIBRARY | STANDARD or COMPACT

Create dittos of detail existing in model

Select: DETAIL > DITTO | MODEL | STANDARD or COMPACT | SINGLE or MULTI or SPACES or GRID

Create single ditto of existing detail

Select: DETAIL > DITTO | MODEL | STANDARD or COMPACT | SINGLE

Prompt: KEY STRING / / ENTER LIST SEL DITTO / / YES: DISPLAY

Upon selecting STANDARD, dittos will be placed on the layer on which they were created. With COMPACT selected, dittos are placed on the current layer.

1. To select the detail, use one of the following three alternatives: (a) Click on YES to display available details, and then select the desired detail. (b) Enter a blank field to display the list of available details; select the desired detail. (c) Enter a string (up to 30 characters) to display the list of available details containing the string, and then select the desired detail.
2. Enter a new scale value if required, and then select or indicate a point. Alternatively, select two lines which intersect (the intersect point will be selected) to position the ditto.
3. Enter an orientation angle if required.
4. Click on YES to end orientation. You can now repeat the above steps to place additional dittos of the same detail or click on YES to select a new one.

Create multiple dittos of existing detail

Select: DETAIL > DITTO | MODEL | STANDARD or COMPACT | MULTI

Prompt: KEY STRING / / ENTER LIST SEL DITTO / / YES: DISPLAY

1. To select the detail, use one of the following three alternatives: (a) Click on YES to display available details, and then select the desired detail. (b) Enter a blank field to display the list of available details; select the desired detail. (c) Enter a string (up to 30 characters) to display the list of available details containing the string, and then select the desired detail.

2. Enter a new scale value if required.
3. Select or indicate a series of points or pairs of intersecting lines to position the ditto.
4. Enter an orientation angle if required.
5. Click on YES to end orientation. You can now repeat the above steps or click on YES to select a new detail.

Create spaced dittos of existing detail

Select: DETAIL > DITTO | STANDARD or COMPACT | SPACES

Prompt: KEY STRING / / ENTER LIST SEL DITTO / / YES: DISPLAY

1. To select the detail, use one of the following three alternatives: (a) Click on YES to display available details, and then select the desired detail. (b) Enter a blank field to display the list of available details; select the desired detail. (c) Enter a string (up to 30 characters) to display the list of available details containing the string, and then select the desired detail.
2. Enter a new scale value if required.
3. Use one of the following alternatives: (a) Select two points, and then enter the number of equally spaced dittos required. (b) Select a line or curve, and then select two points. Enter the number of equally spaced dittos required. (c) Key in the number of equally spaced dittos required along the selected element.
4. Click on YES, and then enter an approximate distance between dittos.
5. Enter an orientation angle if required.
6. You can now repeat the above steps or click on YES to select a new detail.

Create dittos in grid

Select: DETAIL > MODEL | STANDARD or COMPACT | GRID

Prompt: KEY STRING / / ENTER LIST SEL DITTO / / YES: DISPLAY

1. To select the detail, use one of the following three alternatives: (a) Click on YES to display available details, and then select the desired detail. (b) Enter a blank field to display the list of available details; select the desired detail. (c) Enter a string (up to 30 characters) to display the list of available details containing the string, and then select the desired detail.
2. Enter a new scale value if required.
3. Select a point type grid. Dittos will be placed on all points.
4. Enter an orientation angle if required.
5. You can now repeat the above or click on YES to select a new detail.

Create dittos using library details

Select: DETAIL > DITTO | LIBRARY | STANDARD or COMPACT | SINGLE or MULTI or SPACES or GRID

Prompt: KEY STRING / / ENTER:LIST SEL DITTO YES: DISPLAY

Upon selection of this option, a list of available details in the last used family display. If no family has been used, enter blank entry field to display list of available library families. Click on NO to select an alternative library if required, or select the desired family. From the displayed list, select the desired detail, and click on YES to confirm or NO to return to the list. Position the detail using the method described in MODEL options above.

✓ **NOTE:** *All MODEL options are also available in LIBRARY.*

COPY

Create copies of details in element form. Detail elements are copied into the workspace and the link with the detail is broken.

Select: DETAIL > COPY | MODEL or LIBRARY | STANDARD or COMPACT | SINGLE or MULTI or SPACES or GRID

Prompt: KEY STRING / / ENTER:LIST SEL DITTO YES: DISPLAY

✓ **NOTE:** *Methods of selection, placing, and orienting copies are identical to those described above for dittos.*

MODIFY

Modify dittos or copies.

Select: DETAIL > MODIFY | TRANSLATE or ROTATE or SCALE or SYMMETRY or FLIPX or FLIPY | REPLACE or DUPLICAT

Upon selection of the REPLACE option, the original ditto or copy will be replaced on execution of the modification. With the DUPLICAT option selected, the original ditto or copy will be retained.

✓ **NOTE 1:** *Copies can be modified immediately following placement only. Thereafter, they can be handled as elements.*

✓ **NOTE 2:** *If a ditto is incorrectly selected, click on YES to switch to a new one.*

Translate ditto or copy

Select: DETAIL > MODIFY | TRANSLATE | REPLACE or DUPLICAT

Prompt: SEL DITTO

1. Use one of the following alternatives to define the translation: select two points, select two parallel lines, enter x,y coordinates, or reselect a line and then a point or vice versa.
2. To define the direction, select two nonparallel lines, or select a line. Enter length.
3. Click on YES to select a new ditto.

Rotate ditto or copy

Select: DETAIL > MODIFY | ROTATE | REPLACE or DUPLICAT

Prompt: SEL DITTO

1. Use one of the following alternatives to define the rotation: (a) Select two points. (b) Select a point and then enter an angle. (c) Select two lines or select a line and then enter an angle.
2. Click on YES to select a new ditto.

Modify scale of ditto or copy

Select: DETAIL > MODIFY | SCALE | REPLACE or DUPLICAT

Prompt: SEL DITTO

1. Use one of the following alternatives to modify scale: (a) Enter a scale value. (b) Select two lines (LND1 and LND2) where the current scale is multiplied by the quotient of the two lengths. (c) Select two circles (CIRD1 and CIRD2) where the current scale is multiplied by the quotient of the two radii.
2. Click on YES to select a new ditto..

✓ **NOTE:** *Text items are not affected by scaling.*

Modify symmetry of ditto or copy

Select: DETAIL > MODIFY | SYMMETRY | REPLACE or DUPLICATE

Prompt: SEL DITTO

Define the symmetry by selecting a point or line. Click on YES to select a new ditto.

Modify symmetry around X or Y axis

Select: DETAIL > MODIFY | FLIPX or FLIPY | REPLACE or DUPLICAT

Prompt: SEL DITTO

The selected ditto or copy will be copied or replaced by symmetry around the X or Y axis.

EXPLODE

Transform ditto into copy.

Select: DETAIL > EXPLODE | CURRENT or W.SPACE or MODEL

Prompt: SEL DITTO

The CURRENT, W.SPACE, and MODEL options display when you are working in the master workspace. The W.SPACE and MODEL options display when you are working in a detail workspace. Options are described below.

- CURRENT. Explode ditto in current set.
- W.SPACE. Explode ditto in current workspace.
- MODEL. Explode ditto in model.

Select the ditto to be exploded. The ditto will be highlighted. Click on YES to confirm the explode task, and then click on YES again to explode all dittos of the same detail.

CREATE

Create new detail.

Select: DETAIL > CREATE

Prompt: 16 CHARACTERS FOR DETAIL ID KEY NEW DETAIL NAME

1. Enter a name for the new detail, which will become the current detail.

✓ ***NOTE:*** *The new name cannot exceed 16 characters, and must not be the same as an existing detail or symbol.*

2. Enter a comment of up to 40 characters as necessary, or click on YES if no comment is required. The newly created detail becomes the current workspace.

DELETE

Delete detail.

Select: DETAIL > DELETE

Delete an unused detail directly

Select the element for deletion, and then click on YES to confirm.

Select: DETAIL > DELETE | UNUSED | DIRECT

Prompt: KEY STRING / / ENTER: LIST YES: DISPLAY

Delete unused detail after visualization

Select an element for deletion, and then click on YES to confirm.

Select: DETAIL > DELETE | UNUSED | VISUALTN

Delete used detail and all associated dittos

Select an element for deletion, and then click on YES to confirm.

Select: DETAIL > DELETE | USED | DIRECT or VISUALTN | DETAIL

Delete all dittos of used detail

Select an element for deletion, and then click on YES to confirm.

Select: DETAIL > DELETE | USED | DIRECT or VISUALTN | DITTOS

MANAGE

Manage details.

Select: DETAIL > MANAGE | ANALYZE or UPDATE or REPLACE or VERIFY or LAYER or RENAME or DROP

Analyze details or workspaces

Select: DETAIL > MANAGE | ANALYZE | DETAIL or W.SPACE

Prompt: KEY STRING / / ENTER: LIST SEL DITTO / / YES: DISPLAY

To select the detail, use one of the following three alternatives: (a) Click on YES to display available details, and then select the desired detail. (b) Enter a blank field to display the list of available details; select the desired detail. (c) Enter a string (up to 30 characters) to display the list of available details containing the string, and then select the desired detail.

Analysis results display in the alphanumeric window in the form of the following lists: details using the selected detail and the number of times used, and details used by the selected detail and number of times used.

If the W.SPACE option is selected, click on YES and a panel showing the workspace tree structure of the current workspace displays.

Update external (library) detail vis-a-vis current library

Select: DETAIL > MANAGE | UPDATE

Prompt: KEY STRING / / ENTER: LIST SEL DITTO / / YES: DISPLAY

1. To select the detail, use one of the following three alternatives: (a) Click on YES to display available details, and then select the desired detail. (b) Enter a blank field to display the list of available details; select the desired detail. (c) Enter a string (up to 30 characters) to display the list of available details containing the string, and then select the desired detail.
2. Dates and times of the current and external detail display for comparison. Click on YES to update the selected detail to the external detail.

Replace ditto with another ditto

Select: DETAIL > MANAGE | REPLACE | W.SPACE or CURRENT

Prompt: SEL DITTO TO REPLACE

✓ **NOTE:** *This option is not available if the current workspace is an external detail.*

Upon selecting the W.SPACE option, the current workspace will be replaced. With the CURRENT option selected, the current set will be replaced.

1. To select the detail to be replaced (ditto 1), use one of the following three alternatives: (a) Click on YES to display available details, and then select the desired detail. (b) Enter a blank field to display the list of available details; select the desired detail. (c) Enter a string (up to 30 characters) to display the list of available details containing the string, and then select the desired detail.
2. Select the replacement ditto (ditto 2). Click on YES to confirm replacement.
3. Click on YES again if desired to replace all dittos 1 by ditto 2.

Verify nature of dittos

Select: DETAIL > MANAGE | VERIFY | CURRENT or W.SPACE

Prompt: SEL DITTO / / YES: ALL DITTOS

When the W.SPACE option is selected, the current workspace will be verified. With the CURRENT option selected, the current set will be verified.

1. Select a ditto. All dittos from the same detail will be highlighted. Alternatively, click on YES; all dittos are highlighted.
2. Reselect a ditto and all dittos at the same scale will be highlighted and the current scale shown in the message area. Click on NO and all dittos at a different scale will be highlighted.

Change standard dittos to compact dittos and vice versa

Select: DETAIL > MANAGE | LAYER | CURRENT or W.SPACE

Prompt: SEL DITTO TO TRANSFORM

When the W.SPACE option is selected, the current workspace will be modified. With the CURRENT option selected, the current set will be modified.

1. Select a ditto. Click on YES. If the selected ditto was standard it will be changed to compact and vice versa.
2. Click on YES to highlight all dittos of the same detail. Click on YES again to modify all if desired.

Rename detail or modify detail comment

Select: DETAIL > MANAGE | RENAME

Prompt: KEY STRING / /ENTER: LIST SEL DITTO / / YES: DISPLAY

Select a detail using one of the methods described under the DITTO heading. Enter a new name or click on YES to leave unchanged. Enter a new comment or click on YES to leave unchanged.

Break link between externally stored detail and library

Select: DETAIL > MANAGE | DROP

Prompt: KEY STRING / /ENTER: LIST SEL DITTO / / YES: DISPLAY

1. To select the detail, use one of the following three alternatives: (a) Click on YES to display available details, and then select the desired detail. (b) Enter a blank field to display the list of available details; select the desired detail. (c) Enter a string (up to 30 characters) to display the list of available details containing the string, and then select the desired detail.
2. Click on YES to change the selected detail from external to internal.

✓ ***NOTE:*** *On the right side of the list of available details, MODEL or LIBRARY is shown providing the status of each detail.*

CHANGE

Change current workspace.

Select: DETAIL > CHANGE

Prompt: KEY STRING / /ENTER: LIST SEL DITTO / / YES: DISPLAY

To select the detail, use one of the following three alternatives: (a) Click on YES to display available details, and then select the desired detail. (b) Enter a blank field to display the list of available details; select the desired detail. (c) Enter a string (up to 30 characters) to display the list of available details containing the string, and then select the desired detail.

The active workspace is changed to the selected detail workspace.

✓ **NOTE:** *Switching from master to detail workspace or vice versa can also be achieved by using the WSP button on the fixed menu bar.*

TRANSFER

Transfer elements from one workspace to another.

Select: DETAIL > TRANSFER | REPLACE or DUPLICAT | ABSOLUTE or RELATIVE

Prompt: KEY STRING / /ENTER: LIST SEL DITTO / / YES: DISPLAY

Upon selecting the REPLACE option, elements will be deleted from the sending workspace. With the DUPLICAT option selected, elements will be retained in the sending workspace.

To select the receiving workspace, take the following steps.

1. Choose one of the following alternatives: (a) Click on YES to display existing details, and then select the receiving detail. (b) Enter a blank field or string to display a list of details, and then select the receiving detail from the list. (c) Enter a new character string if it does not already exist; a new detail workspace with this name is proposed.
2. Click on YES to create the new receiving detail.
3. Enter a comment or click on YES to continue. You can now click on YES again to transfer all elements into the newly created detail, or select them by other means.
4. Select a ditto; the corresponding detail workspace becomes the receiving workspace.

To send elements to the receiving workspace, take the following steps.

1. Select the receiving workspace.
2. At this point, you proceed according to selection of ABSOLUTE or RELATIVE.
   - With the ABSOLUTE option selected, select elements and then click on YES to confirm the transfer. The elements will be transferred in the absolute position with respect to the axis of the receiving workspace.
   - With the RELATIVE option selected, define a transformation by entering coordinates or selecting geometry. Select the elements for transfer. Click on YES to transfer the selected elements to the receiving workspace with the transformation applied.

DETAIL in 3D SPACE Mode

Chapters 13, 18

The DETAIL function in 3D SPACE mode is used to place elements in alternative workspaces which can be used on the master workspace as a ditto or as a copy of its elements; to perform transformations and analysis on dittos; and to switch between workspaces.

Option menu for DETAIL in 3D SPACE mode.

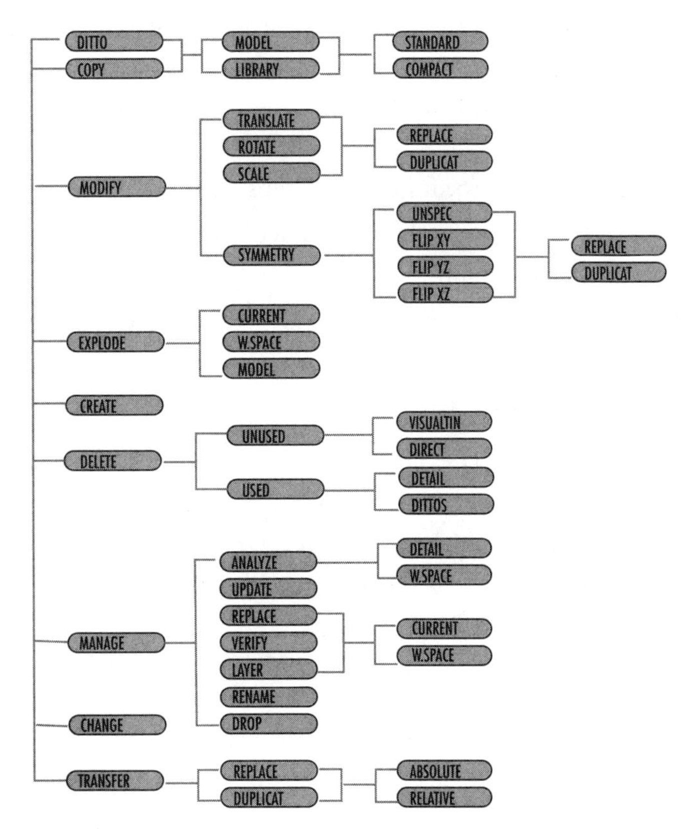

DITTO

Create dittos in models and libraries.

Select: DETAIL > DITTO | MODEL or LIBRARY

Create dittos of existing detail in model

Select: DETAIL > DITTO | MODEL | STANDARD or COMPACT

Prompt: KEY STRING / / ENTER LIST SEL DITTO / / YES: DISPLAY

✓ **NOTE:** *Upon selecting the STANDARD suboption, dittos will be placed on the layer on which they were created. With COMPACT selected, dittos will be placed on the current layer.*

Take one of the following alternatives.

1. Click on YES to display available details, and then select a detail.
2. Enter a blank field to display the list of available details, and then select a detail.
3. Enter a string (of up to 30 characters) to display the list of available details containing the string, and then select a detail.

To place the ditto, follow the steps below.

1. Enter a new scale value as necessary.
2. Enter a point; the ditto axis system will be placed on the selected point.
3. Enter a line, and then a second line or a plane; the ditto will be placed at the intersection of these elements. Click on YES to select a new detail.

Use library details

Select: DETAIL > DITTO | LIBRARY | STANDARD or COMPACT

Prompt: SEL DETAIL / / KEY STRING NO: FAMILY SELECTION

1. Upon selection of the LIBRARY option, a list of available details in the most recently used family displays. If no family has been used, enter a blank entry field to display the list of available library families.
2. If necessary, click on NO to select an alternative library, or select the desired family.
3. From the displayed list, select the desired detail.
4. Click on YES to confirm or NO to return to the list.
5. Position the detail using the method described in the previous section.

✓ **NOTE:** *All MODEL options are also available under LIBRARY.*

COPY

Create copies of details in elemental form. Upon using the COPY option, the detail elements are copied into the workspace and the link with the detail is broken.

Select: DETAIL > COPY | MODEL or LIBRARY | STANDARD or COMPACT

Prompt: KEY STRING / / ENTER: LIST SEL DITTO / / YES: DISPLAY

✓ ***NOTE:*** *Methods of copy selection, placement, and orientation are identical to those described above for dittos, except that copies can be modified immediately following placement only, after which point they are handled as elements alone.*

MODIFY

Modify dittos or copies.

Select: DETAIL > MODIFY | TRANSLAT or ROTATE or SCALE or SYMMETRY | STANDARD or COMPACT

Translate ditto or copy

Select: DETAIL > MODIFY | TRANSLAT | REPLACE or DUPLICAT

Prompt: SEL DITTO

When the REPLACE option is selected, the original ditto or copy will be replaced on execution of the modification. If the DUPLICAT option is selected, the original ditto or copy will be retained.

1. Take one of the following alternatives to define the translation: (a) Enter coordinates DX, DY, and DZ. The ditto's axis origin will be moved to the new coordinate. (b) Select a point. The ditto's axis origin will be moved to the new point. (c) Select a line, and then enter a distance.
2. Click on YES to select a new ditto or copy.

Rotate ditto or copy

Select: DETAIL > MODIFY | ROTATE | REPLACE or DUPLICAT

Prompt: SEL DITTO

1. Take one of the following alternatives to define the rotation: (a) Select a point; the rotation axis passes through the selected point. (b) Select a plane; the rotation axis will be perpendicular to the selected plane. Enter the rotation angle. (c) Select a line, and then enter a rotation angle.
2. Click on YES to modify.

Scale ditto or copy

Select: DETAIL > MODIFY | SCALE | REPLACE or DUPLICAT

Prompt: SEL DITTO

1. Enter a new scale value.

2. Select an existing ditto; the scale of the selected ditto will be applied.
3. Click on YES to modify.

Define symmetry for ditto or copy

Select: DETAIL > MODIFY | SYMMETRY | UNSPEC or FLIP XY or FLIP YZ or FLIP XZ | REPLACE or DUPLICATE

Prompt: SEL DITTO

Options in the SYMMETRY menu are described below.

- UNSPEC. Select a point, line, or plane. The symmetry is obtained about the selected element.
- FLIP XY. Symmetry obtained about the XY plane.
- FLIP YZ. Symmetry obtained about the YZ plane.
- FLIP XZ. Symmetry obtained about the XZ plane.

EXPLODE

Explode ditto.

Select: DETAIL > EXPLODE | CURRENT or W.SPACE or MODEL | W.SPACE or MODEL

Explode ditto in current set

Select: DETAIL > EXPLODE | CURRENT | W.SPACE or MODEL

Prompt: SEL DITTO

Select the ditto to be exploded; the ditto will be highlighted. Click on YES to confirm the explode, and then click on YES again to explode all dittos of the same detail.

Explode ditto in current workspace

Select: DETAIL > EXPLODE | W.SPACE | W.SPACE or MODEL

Prompt: SEL DITTO

Select the ditto to be exploded; the ditto will be highlighted. Click on YES to confirm the explode, and then click on YES again to explode all dittos of the same detail.

Explode ditto in model

Select: DETAIL > EXPLODE | MODEL | W.SPACE or MODEL

Prompt: SEL DITTO

Select the ditto to be exploded; the ditto will be highlighted. Click on YES to confirm the explode, and then click on YES again to explode all dittos of the same detail.

CREATE

Create new detail.

Select: DETAIL > CREATE

Prompt: 16 CHARACTERS FOR A DETAIL ID KEY NEW DETAIL NAME

1. Enter a name for the new detail, which will become the current detail.

✓ ***NOTE:*** *The new name must have between one and 16 characters and must not be the same as any existing detail or symbol.*

2. Enter a comment of up to 40 characters or click on YES if no comment is desired. The newly created detail becomes the current workspace.

DELETE

Delete detail.

Select: DETAIL > DELETE | UNUSED or USED

Prompt: KEY STRING / / ENTER: LIST SEL DITTO / / YES: DISPLAY

Delete unused detail directly

Select: DETAIL > DELETE | UNUSED | DIRECT

Select the element to be deleted, and then click on YES to confirm.

Delete unused detail after visualization

Select: DETAIL > DELETE | UNUSED | VISUALTN

Select the element to be deleted, and then click on YES to confirm.

Delete used detail and associated dittos

Select: DETAIL > DELETE | USED | DIRECT or VISUALTN | DETAIL

Select the element to be deleted, and then click on YES to confirm.

Delete dittos of used detail

Select: DETAIL > DELETE | USED | DIRECT or VISUALTN | DITTOS

Select the element to be deleted, and then click on YES to confirm.

MANAGE

Manage details and workspaces.

Select: DETAIL > MANAGE | ANALYZE or UPDATE or REPLACE

Analyzing details or workspaces

Select: DETAIL > MANAGE | ANALYZE | DETAIL or W.SPACE

Prompt: KEY STRING / / ENTER: LIST SEL DITTO / / YES: DISPLAY

1. Select a detail using one of the methods described under the DITTO option section. Analysis results display in the alphanumeric window in the form of lists of details using and used by the selected detail and number of times used, respectively.
2. If the W.SPACE option is selected, click on YES. A panel showing the workspace tree structure of the current workspace displays.

Update external (library) detail with respect to current library

Select: DETAIL > MANAGE | UPDATE

Prompt: SEL DITTO

1. Select a detail using one of the methods described under the DITTO option section. Dates and times of the current and external detail display for comparison.
2. Click on YES to update the selected detail to the external detail.

Replace ditto with another ditto

Select: DETAIL > MANAGE | REPLACE | W.SPACE or CURRENT

Prompt: SEL DITTO

✓ **NOTE:** *This option is not available if the current workspace is an external detail.*

When the W.SPACE option is selected, the current workspace will be replaced. If the CURRENT option is selected, the current set will be replaced.

1. Select a ditto using one of the methods described under the DITTO option section. Select the ditto to be replaced (ditto 1), and then select the replacement ditto (ditto 2).
2. Click on YES to confirm the replacement.
3. Click on YES again to replace all dittos 1 with ditto 2 as necessary.

Verify nature of dittos

Select: DETAIL > MANAGE | VERIFY | CURRENT or W.SPACE

Prompt: SEL DITTO / / YES: ALL DITTOS

When the W.SPACE option is selected, the current workspace will be verified. If the CURRENT option is selected, the current set will be verified.

1. Select a ditto. All dittos from the same detail will be highlighted. Alternatively, click on YES, and all dittos are highlighted.
2. Reselect a ditto; all dittos at the same scale will be highlighted and the current scale shown in the message area.
3. Click on NO and all dittos at a different scale will be highlighted.

Change standard dittos to compact dittos and vice versa

Select: DETAIL > MANAGE | LAYER | CURRENT or W.SPACE

Prompt: SEL DITTO

When the W.SPACE option is selected, the current workspace will be modified. If the CURRENT option is selected, the current set will be modified.

1. Select a ditto, and click on YES. If selected ditto was standard, it will be changed to compact, and vice versa.
2. Click on YES to highlight all dittos of the same detail.
3. Click on YES again to modify all dittos as necessary.

Rename detail or modify detail comment

Select: DETAIL > MANAGE | RENAME

Prompt: SEL DITTO

1. Select a detail using one of the methods described under the DITTO option section.
2. Enter a new name or click on YES to leave unchanged.
3. Enter a new comment or click on YES to leave unchanged.

Break link between externally stored detail and library

Select: DETAIL > MANAGE | DROP

Prompt: KEY STRING / / ENTER: LIST SEL DITTO / / YES: DISPLAY

1. Select a detail using one of the methods described in the DITTO option section.
2. Click on YES to change the selected detail from external to internal.

CHANGE

Change current workspace.

Select: DETAIL > CHANGE

Prompt: KEY STRING / / ENTER: LIST SEL DITTO / / YES: DISPLAY

Select a detail using one of the methods described in the DITTO option section. The active workspace is changed to the selected detail workspace.

TRANSFER

Transfer elements from one workspace into another.

Select: DETAIL > TRANSFER | REPLACE or DUPLICAT | ABSOLUTE or RELATIVE

Prompt: KEY STRING / / ENTER: LIST SEL DITTO / / YES: DISPLAY

When selecting the REPLACE option, elements will be deleted from the sending workspace. If the DUPLICAT option is selected, elements will be retained in the sending workspace.

✓ **NOTE:** *SPACE axis systems cannot be transferred between workspaces.*

Select receiving workspace

1. Click on YES to display existing details, and then select the receiving detail.
2. Enter a blank field or string to display a list of details, and then select a receiving detail from list.
3. Enter a new character string, if it does not already exist; a new detail workspace with this name is proposed. Click on YES to create the new receiving detail.
4. Enter a comment or click on YES to continue. You can now click on YES again to transfer all elements into the newly created detail, or select them by other means.
5. Select a ditto. The corresponding detail workspace becomes the receiving workspace.

Send elements to receiving workspace

With the receiving workspace selected, take one of the following alternatives.

1. Select the ABSOLUTE option. Select elements, and click on YES to confirm the transfer. Elements will be transferred in the absolute position with respect to the axis system of the receiving workspace.

2. Select the RELATIVE option. Define a transformation by entering coordinates or selecting geometry. Select the elements for transfer. Click on YES to transfer the selected elements to the receiving workspace with transformation applied.

DIMENS2 in DRAW Mode

Chapter 7

The DIMENS2 function is used to create and modify dimensions.

Option menu for DIMENS2 in DRAW mode.

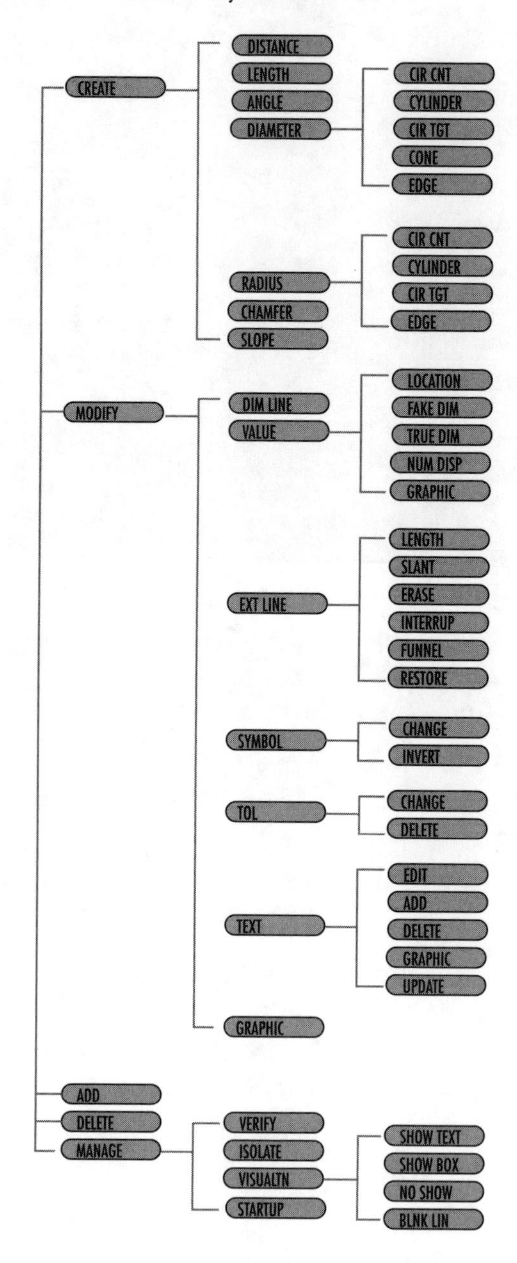

DIMENS2 Menu Structure

When the CREATE option is selected, the Management window is activated. This window first appears in a condensed form. To expand the window, select the MORE icon at the top right of the window, and to return the window to its original size select the LESS icon at the top right of the window. This window is used to define the presentation of dimensions created. Buttons in the following illustration are described below.

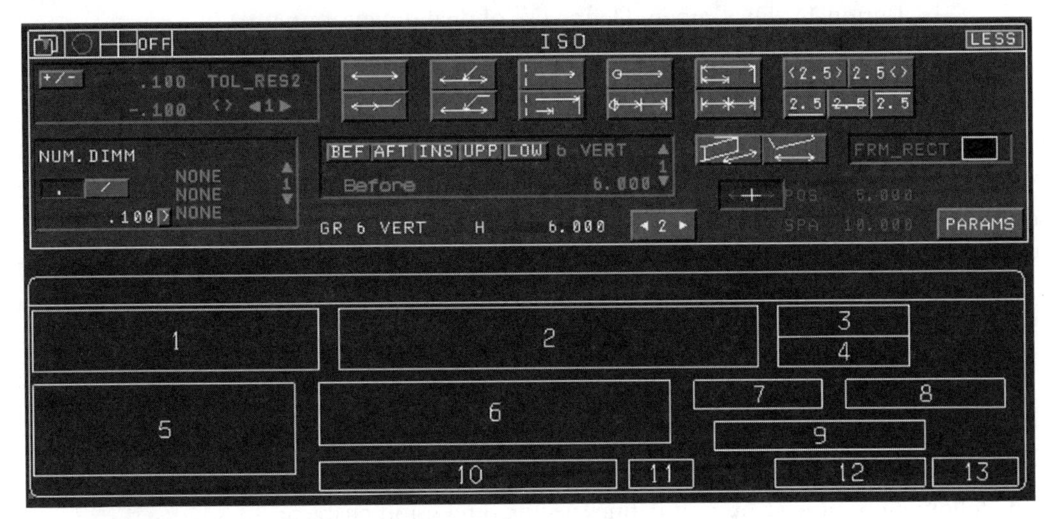

Typical DIMENS2 Expanded Management window.

✓ **NOTE:** *Your DIMENS2 Expanded Management window may vary from the window shown in the illustration. Because window configuration is set in the DRWSTD function, modification may be available only to your system administrator.*

- Button 1. Tolerancing mode. Switch the tolerancing mode on or off, and define tolerances.
- Button 2. Presentation mode. Define presentation of dimensions (i.e., single, cumulative, from a datum or stacked).
- Button 3. Dimension text location mode.
- Button 4. Score line. Define position of any score used. The score can be under, through, or over the text.
- Button 5. Numerical display. Define dimension text display standard to be set, and accuracy of displayed dimension.
- Button 6. Associated text. Define use and position of associated text.
- Button 7. Calculation mode.

- Button 8. Framing. Define required framing (i.e., box, circle, arrow, or any other frame defined in DRWSTD). To change the frame in use, select a frame symbol. For instance, for rectangular shapes, a list of available frames will be displayed.
- Button 9. Dimension position mode.
- Button 10. Graphism. Define font style and height of text.
- Button 11. Dual mode. Define use of dual dimensioning.
- Button 12. Spacing mode. Define distance for automatic spacing of dimension lines when using the stacking presentation option (see Button 2).
- Button 13. Access to parameters.

The following illustration shows the parameters that may be available to you. Your options may differ because availability and content of this window are typically controlled by system administrators via the DRWSTD function.

OFF ACCESS TO PARAMETERS RETURN

CURRENT STANDARD : ISO DISPLAY PARM UNIT : 1.000 (X MODEL UNIT)

DIM-LINE	DIM TEXT	SYMBOL	PREFIX	SCORE	CHAMFER	ARC LENGTH	HALF-DIM	DUAL
EXT-LINE	GRAPHIC	FRAME	TEXT	LEADER	CUMULATE	FORESHORTENED	FAKE DIM	SLOPE

Typical DIMENS2 Parameter Management window.

The following parameter options define specific dimension characteristics.

- DIM-LINE. Defines orientation, positioning, and length of dimension lines.
  - ONE PART & TWO PART. Position of dimension value relative to dimension line.
  - DIM LINE ORIENTATION. Orientation of dimension line relative to an existing line, current view, or the screen.
  - DIM LINE GAP. Gap between the dimension line and the dimension value.
  - DIM LINE OVERRUN. Amount dimension line overruns the extension line
  - DIM LINE DISPLAY. Show or no show dimension line.
  - RADIUS/DIAMETER DIM. How radius/diameter dimension lines are displayed.
- EXT-LINE. Orientation, positioning, and length of extension line.
  - BLANKING AND OVERRUN. Gap between extension line and element being dimensioned, and the overrun between the extension and dimension line.

- EXT ORIENTATION. Orientation of the extension line relative to the view or a set angle.
- LENGTH DIMENSION. Display of length dimension for linear, circular, and curvilinear dimensions.

- DIM TEXT. Orientation and positioning of dimension text in relation to dimension lines.
  - VERTICAL POSITIONING. Height of value above dimension line.
  - DIM TEXT ORIENTATION. Orientation of dimension value text relative to the current view, screen, or dimension line.
  - MASKING DIMENSION TEXT. Mask or no mask for dimension value text.
  - HORIZONTAL POSITIONING. Position of dimension value text along dimension line.
- GRAPHIC. Display, thickness, and color of all dimension elements.
  - DISPLAY FOR EACH DIMENSION. Display dimension text as text, box, or no display.
  - WRITING DIRECTION. Direction of text (left to right or right to left).
  - DIMENSION COMPONENTS. Thickness and color of dimension components.
  - SCALE. Whether text size is linked to the current view scale or free.
- SYMBOL. Define symbols used in dimensioning.
  - SYMBOL. Symbol used on dimension line for various dimension types.
  - CHAIN DIMENSIONING. Whether intermediate symbol is used, and the particular symbol used.
  - SYMBOL REVERSAL. When symbol is placed inside or outside extension line.
- FRAME. Size of framing to be used.
  - ENTITY. Select part of dimension text to be placed in the frame.
  - DUAL DIMENSIONING. Select part of dual dimension placed in frame.
- PREFIX. Prefix used for various dimension types (i.e., diameter, radius, slope, and so forth), position of prefix relative to dimension value, and graphism and font used for prefix.
- TEXT. Position text relative to dimension lines.
  - ASSOCIATED TEXTS POS. Position associated text.
- SCORE. Size and position of scores.

   - ENTITY. Select part of dimension text to be underscored, scored, or overscored.
   - POSITION AND LENGTH. Position and length of underscore, score, or overscore.
- LEADER. Leader end symbol, length, and orientation.
   - LEADER SYMBOL. Symbol used for leaders.
   - LEADER LENGTH. Length of leader.
   - LEADER ORIENTATION. Orientation of leader relative to dimension line to current view or the screen.
- CHAMFER. Display of chamfer dimensions.
   - IF 2 MEASURES. Separator used and height of separator.
   - ENTITY TO BE FRAMED. Part of chamfer dimension to be framed.
- CUMULATE. Display and orientation of cumulative dimensions.
   - VALUE SIGN DISPLAY. Display of cumulative dimension values.
   - ORIGIN. Display of origin symbol, scale, and extension, and whether a zero is displayed at the origin.
   - DIM TEXT ORIENTATION. Orientation of dimension value text relative to dimension or extension line.
   - DIM LINE. Length of dimension line, and position of dimension text relative to dimension line.
- ARC LENGTH. Display of arc dimensions.
   - SYMBOL DISPLAY. Whether symbol is used.
   - LENGTH. Display of dimension line.
- FORESHORTENED. Display of foreshortened radius dimensions.
   - SEGMENT ORIENTATION. Display of dimension line.
   - TEXT POSTION. Dimension text position.
   - BREAK POSITION. Dimension line break position.
   - ORIGIN. Symbol used for dimension origin.
- HALF-DIM. Display of half dimension mode.
   - INSIDE DIM. Display of dimension line and text for inside dimension.
   - OUTSIDE DIM. Display of dimension line and text for outside dimension.
- FAKE DIM. Display of fake dimensions.
   - FAKE DIM SYMBOLS. Display of symbol to highlight fake dimension.

- FAKE DIM SYMBOLS. Position of any symbol used.
- FAKE DIM SYMBOLS. Select symbol used and symbol height.
- DUAL. Display of dual dimensions.
  - DIM LINE. Place dimension text on top of each other or side by side.
  - FRACTION LINE. Whether to use a fraction line.
  - SIDE BY SIDE. Whether a separator is used and separator height.
  - TOP. Position of texts relative to dimension line.
  - JUSTIFICATION. Justification for dimension texts.
- SLOPE. Defines display of slope dimensions.
  - VALUE ORIENTATION. Orientation of dimension value (horizontal to the current view, horizontal to the screen, or parallel to slope).
  - VALUE PRESENTATION. Display of slope value.
  - JUSTIFICATION. Dimension text justification.
  - CALULATION MODE. Whether sin or tan values are used for calculating slope value.

✓ **NOTE:** *As mentioned previously, not all parameter options and suboptions discussed above will be available to you because they are controlled by the system administrator.*

CREATE

Create dimensions.

Select: DIMENS2 > CREATE | DISTANCE or LENGTH or ANGLE or DIAMETER or RADIUS or CHAMFER or SLOPE

Prior to creating dimensions, window options for the specific dimension style you require are necessary. Dimension style components include tolerance, tolerance style, accuracy, and dimension text. If such panels are not correctly set prior to creation of dimensions, results can be modified using the many DIMENS2 > MODIFY options.

Dimension appearance is also dependent on dimension standards set in the parameter options.

✓ **NOTE:** *All descriptions of CREATE options below apply to dimensioning elements created using the DRAW mode functions. When dimensioning elements created using AUXVIEW2, steps used may be different. For example, upon creating a dimension on a shaft using DIMENS2 > CREATE | DIAMETER | CYLINDER, you will be required only to select one edge because AUXVIEW2 retains the relationship between DRAW and SPACE elements.*

TIP: *When dimensions are created, they can normally be positioned by indicating a position. The first indicated position is often incorrect. If you then indicate a new position for the dimension while still using the CREATE option, the dimension line and value will be aligned with that position. However, if you select the dimension line before indicating a new position, only the dimension line will be aligned with the indicated position, and the value will remain in the same position relative to the dimension line as when the dimension was created. Similarly, if you select the dimension value before indicating a new position, dimension value alone will be aligned with the indicated position, and the dimension line will remain in the same position as when the dimension was created.*

Create linear dimensions between two elements

The style of the dimension is controlled by selections activated by Button 2 in the Management window.

Select: DIMENS2 > CREATE | DISTANCE

Prompt: SEL > ELEM

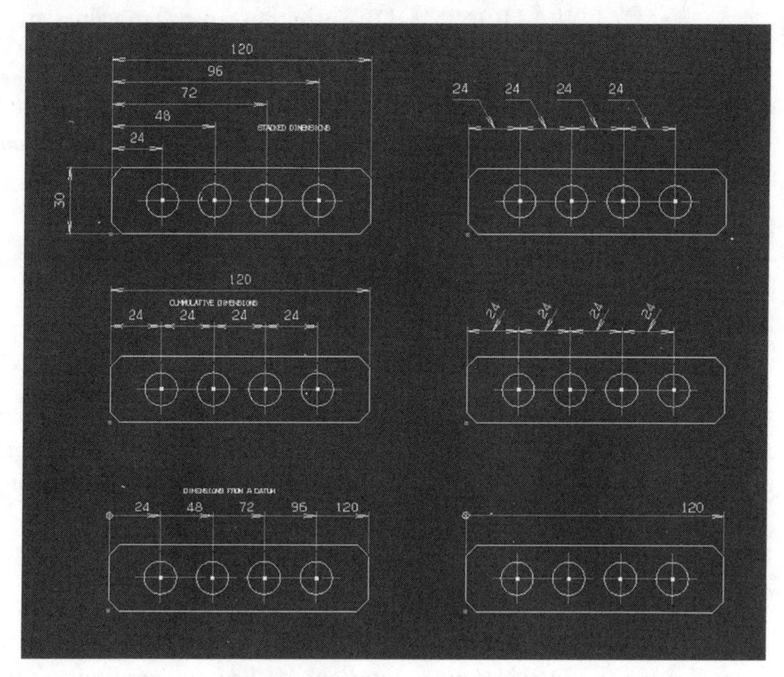

Typical selection of DISTANCE dimensioning styles.

Create distance dimension between two lines

1. Select two line elements.

2. Indicate or select a point to position the created dimension, or key in a number to the input information area and press <Enter>. The latter will be the distance from the first selected element to the dimension line.

Create distance dimension between line and point

1. Select a line and a point element.
2. Indicate or select a point to position the created dimension, or key in a number to the input information area and press <Enter>. The latter will be the distance from the first selected element to the dimension line.

If the line is selected first, the dimension will be normal to the selected line. If the point is selected first, the dimension will be directly between the point and the end of the line closest to the position of selection on the line.

Create distance dimension between two points

1. Select two point elements.
2. Indicate or select a point to position the created dimension, or key in a number to the input information area and press <Enter>. The latter will be the distance from the first selected element to the dimension line.

Create distance dimension between two circles

1. Select two circle elements. When the first circle is selected, a window appears providing you with the option of dimensioning to the circle center point or the circle edge.
2. Indicate or select a point to position the created dimension, or key in a number to the input information area and press <Enter>. The latter will be the distance from the first selected element to the dimension line.

Create distance dimension between line and circle

1. Select a line and a circle element.
2. Indicate or select a point to position the created dimension, or key in a number to the input information area and press <Enter>. The latter will be the distance from the first selected element to the dimension line.

If the line is selected first, the dimension will be normal to the selected line. If the circle is selected first, the dimension will be directly between the center point of the circle and the end of the line closest to the position of selection on the line.

Create distance dimension between point and circle

1. Select a point and a circle element.

2. Indicate or select a point to position the created dimension, or key in a number to the input information area and press <Enter>. The latter will be the distance from the first selected element to the dimension line.

Create dimensions of line or curve length

Create dimensions of line or curve length. Dimension style is controlled by selections activated by Button 2 in the Management window.

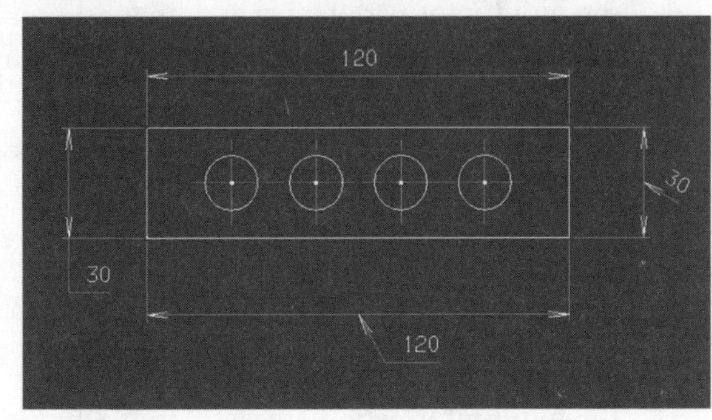

Typical selection of LENGTH dimensioning styles.

Select: DIMENS2 > CREATE | LENGTH

Prompt: SEL > ELEM

Create line length dimension

1. Select a line element.
2. Click on the YES button to dimension the complete line, or select or indicate points to define the line length to be dimensioned.
3. Indicate or select a point to position the created dimension.

Create length (circumference) dimension of curve

1. Select a curve or circle element.
2. Click on the YES button to dimension the complete curve, or select or indicate points to define the length of the curve to be dimensioned.
3. Indicate or select a point to position the created dimension.

Create angular dimensions between lines

The style of the dimension is controlled by selections activated by Button 2 in the Management window.

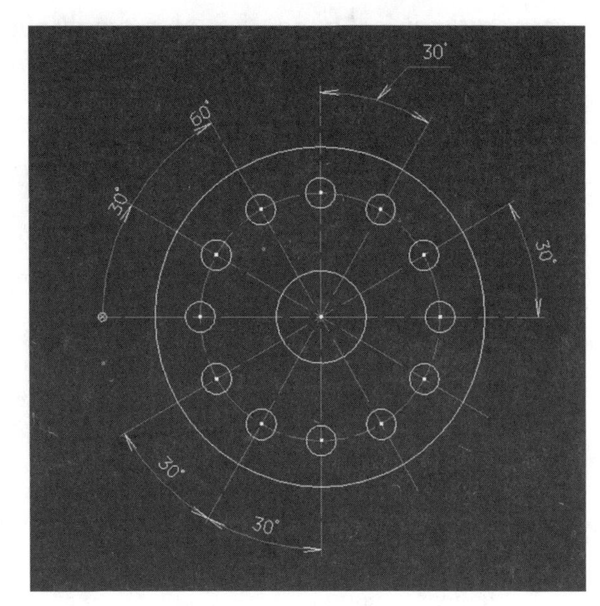

Typical selection of ANGLE dimensioning styles.

Select: DIMENS2 > CREATE | ANGLE

Prompt: 1ST_ELEM : SEL > ELEM

Create angular dimension between two lines

1. Select the first line element. If you wish the dimension to be on the opposite side of the line to which you selected, indicate to the other side of the line and the position of the temporary point element will change.
2. Select the second line element. If you wish the dimension to be on the opposite side of the line you selected, indicate to the other side of the line; the position of the temporary point element will change.
3. If the vector arrow is not pointing in the direction in which you wish to dimension, select the vector to change its direction.
4. Click on the YES button to continue.
5. Indicate or select a point to position the created dimension, or key in a number to the input information area and press <Enter>. The latter will be the distance from the first selected element to the dimension line.

Create diametrical curve dimensions

Dimension style is controlled by selections activated by Button 2 in the Management window.

Select: DIMENS2 > CREATE | DIAMETER | CIR CNT or CYLINDER or CIR TGTR or CONE or EDGE

Create diametrical dimension of curve through center point

Select: DIMENS2 > CREATE | DIAMETER | CIR CNT

Prompt: SEL > ELEM

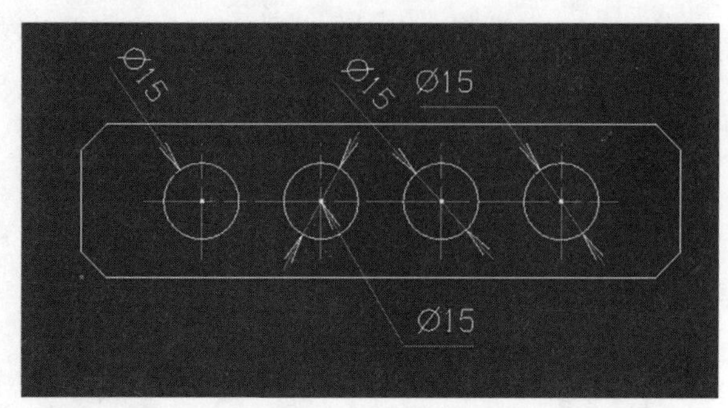

Typical example of DIAMETER | CIR CNT dimension.

1. Select a curve and the dimension will be created to the style set in the selections accessed through Button 2 in the Management window.
2. If a spherical diameter dimension is required, select the sphere icon from the Management window.

✓ **NOTE:** *The dimension line position and value can be modified as described in the previous TIP.*

Create diametrical cylinder dimension

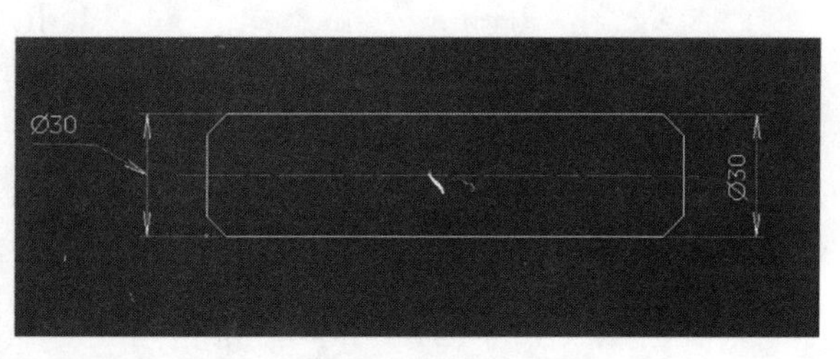

Typical example of DIAMETER | CYLINDER dimension.

Select: DIMENS2 > CREATE | DIAMETER | CYLINDER

Prompt: GENERATOR : SEL > ELEM

Select two lines to create the dimension. The two lines are assumed to be the cylinder edges.

✓ **NOTE:** *The dimension line position and value can be modified as described in the previous TIP.*

Create diametrical curve dimension through center point

Select: DIMENS2 > CREATE | DIAMETER | CIR TGT

Prompt: SEL > ELEM

Typical example of DIAMETER | CIR TGT dimension.

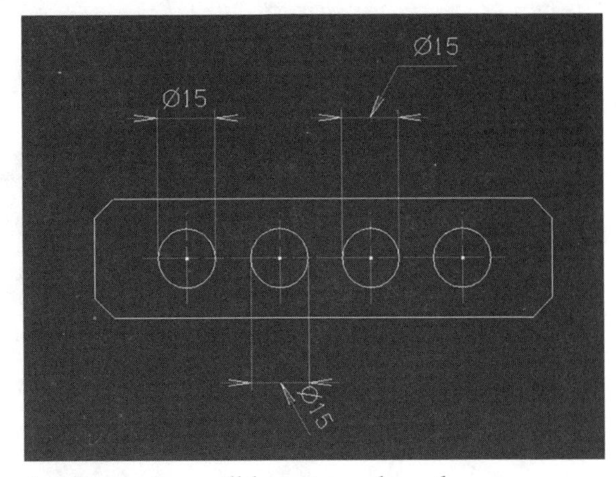

In the window that appears, define how the dimension will be oriented on the screen.

1. Select a curve.
2. Indicate or select a point to position the created dimension.
3. Click on the YES button to end the dimension definition.

Create conical dimension of two lines

Select: DIMENS2 > CREATE | DIAMETER | CONE

Prompt: GENERATOR : SEL > ELEM

Typical example of DIAMETER | CONE dimension.

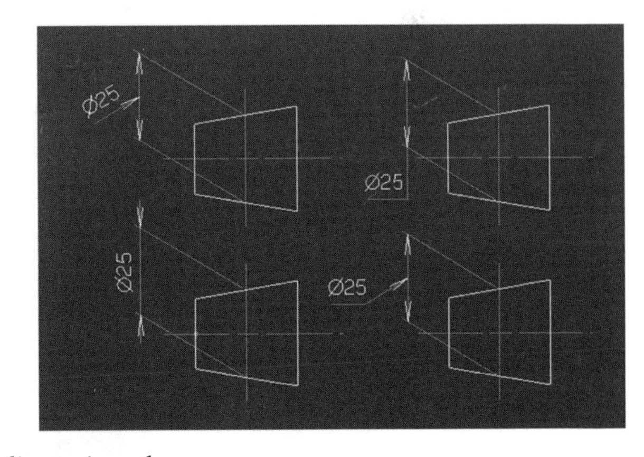

1. Select the two elements to be dimensioned.
2. Select the cutting line. The cutting line must be perpendicular to the cone bisector.
3. Indicate or select a point to position the created dimension.
4. Click on the YES button to end the dimension definition.

Create edge dimension of circle

Select: DIMENS2 > CREATE | DIAMETER | EDGE

Prompt: SEL > ELEM

Typical example of DIAMETER | EDGE dimension.

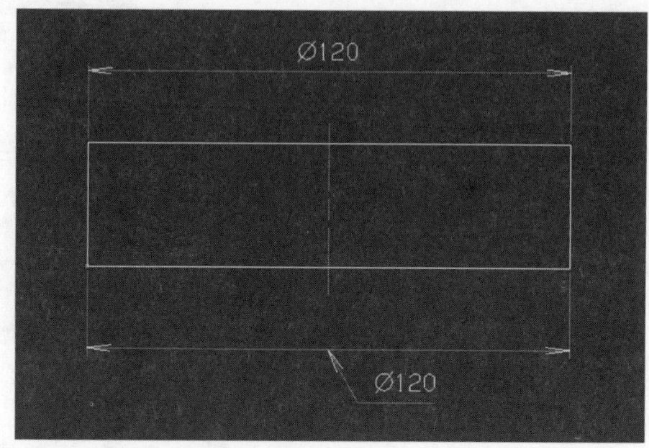

1. Select the element (circle edge) to be dimensioned.
2. Indicate or select a point to position the created dimension.
3. Click on the YES button to end the dimension definition.

Create radial curve dimensions

Select: DIMENS2 > CREATE | RADIUS | CIR CNT or CYLINDER or CIR TGT or EDGE

Dimension style is controlled by selections activated by Button 2 in the Management window.

Create radial curve dimension through center point

Typical example of RADIUS | CIR CNT dimension.

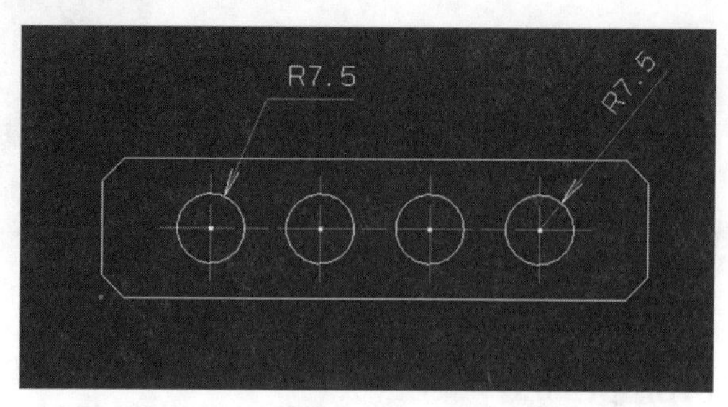

Select: DIMENS2 > CREATE | RADIUS | CIR CNT

Prompt: SEL > ELEM

1. Select a curve. The dimension is created to the style set in selections made upon activation of Button 2 in the Management window.
2. If a spherical radius dimension is required, select the sphere icon from the Management window.

✓ **NOTE:** *The dimension line position and value can be modified as described in the previous TIP.*

Create radial dimension of cylinder

Typical example of RADIUS | CYLINDER dimension.

Select: DIMENS2 > CREATE | RADIUS | CYLINDER

Prompt: GENERATOR : SEL > ELEM

Select two lines and the dimension will be created. The two lines are assumed to be the cylinder edges.

✓ **NOTE:** *The dimension line position and value can be modified as described in the previous TIP.*

Create radius dimension of curve through center point

Typical example of RADIUS | CIR CGT dimension.

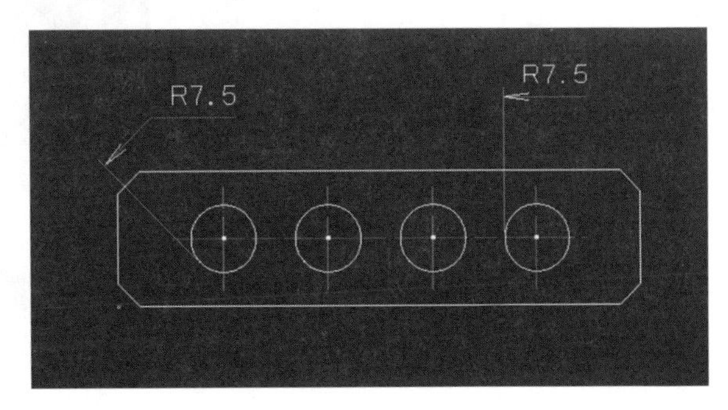

Select: DIMENS2 > CREATE | RADIUS | CIR TGT

Prompt: SEL > ELEM

Upon selecting this option a window appears in which you can define how the dimension will be oriented on the screen.

1. Select a curve.

2. Indicate or select a point to position the created dimension.
3. Click on the YES button to end the dimension definition.

Create edge dimension of circle

Typical example of RADIUS I EDGE dimension.

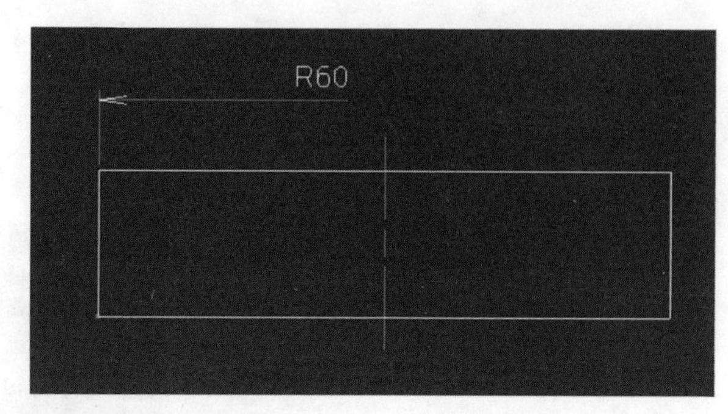

Select: DIMENS2 > CREATE | RADIUS | EDGE

Prompt: SEL > ELEM

1. Select the element (circle edge) to be dimensioned.
2. Indicate or select a point to position the created dimension.
3. Click on the YES button to end the dimension definition.

Create chamfer dimension

Dimension style is controlled by selections accessed through Button 2 in the Management window.

Typical selection of CHAMFER dimensioning styles.

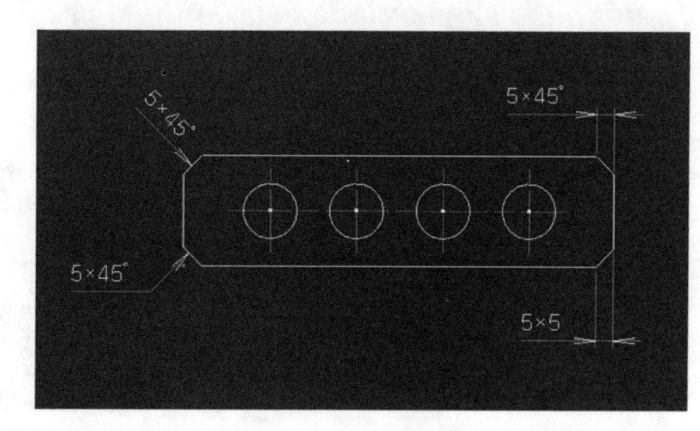

Select: DIMENS2 > CREATE | CHAMFER

Prompt: CHAMFER : SEL > ELEM

1. Select the chamfer to be dimensioned.
2. Select the line that will define the base of the chamfer.

3. Indicate or select a point to position the created dimension, or key in a number to the input information area and press <Enter>. This will be the distance from the first selected element to the dimension line.

Create slope dimension

Dimension style is controlled by selections accessed through Button 2 in the Management window.

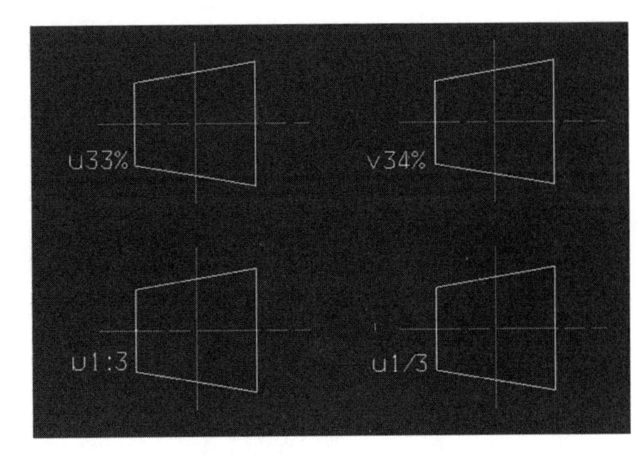

Typical selection of SLOPE dimensioning styles.

Select: DIMENS2 > CREATE | SLOPE

Prompt: REFERENCE_ELEMENT : SEL > ELEM

1. Select two lines between which the slope dimension will apply.
2. Indicate or select a point to position the created dimension, or key in a number to the input information area and press <Enter>. This will be the distance from the first selected element to the dimension line.
3. Click on the YES button to end the dimension.

MODIFY

Modify aspects of dimensions, including dimension line, dimension value, extension line, and symbol.

Select: DIMENS2 > MODIFY | DIM LINE or VALUE or EXT LINE or SYMBOL or TOL or TEXT or GRAPHIC

Modify dimension line position

Modify dimension line.

Select: DIMENS2 > MODIFY | DIM LINE

Prompt: DIMENSION : SEL ELEM

1. Select the dimension to be modified.
2. Indicate a new position for the dimension line, or select another dimension. The first selected dimension line will be aligned with the second selected dimension.
3. Select a point; the selected dimension line will be aligned with the selected point. Alternatively, key in a number to the input information area and press <Enter>. This will define the translation of the dimension line.
4. Click on the YES button to end modification of the selected dimension.

Modify dimension line spacing of stacked dimension system

Select the dimension system to be modified. The SPACING window appears in which you can modify the dimension line spacing.

Modify dimension value

With the VALUE suboption you can modify all aspects of a dimension value, such as changing its location or the value to a fake value, restoring a true value, and changing the graphics.

Select: DIMENS2 > MODIFY | VALUE | LOCATION or FAKE DIM or TRUE DIM or NUM DISP or GRAPHIC

Modify dimension value position

Select: DIMENS2 > MODIFY | VALUE | LOCATION

Prompt: MODIFY : SEL ELEM

Select the dimension to be modified, and then take one of the following alternatives.

- Indicate a new position for the dimension line.
- Select another dimension. The first selected dimension line will be aligned with the second selected dimension.
- Select a point. The selected dimension will be aligned with the selected point.
- Key in a number to the input information area and press <Enter>; the translation of the dimension line is defined.
- Click on the YES button to end modification of the selected dimension.

Change dimension value to fake value

Select: DIMENS2 > MODIFY | VALUE | FAKE DIM

Prompt: SEL DIMN / SYS_DIM / DIM

1. Select the dimension to be modified.

2. Enter a fake value (numerical or alphanumerical) in the activated window. Graphism and height of dimension text can also be changed in the window.
3. Click on the YES button to apply the changes (i.e., the fake dimension value).

✓ **NOTE:** *The appearance of the fake dimension will be dependent on the settings in the dimension parameter options (e.g., value to be underlined, value to be placed in parentheses, and so on).*

Restore true value of dimension

Use this option when the value has previously been changed using DIMENS2 > MODIFY | VALUE | FAKE DIM.

Select: DIMENS2 > MODIFY | VALUE | TRUE DIM

Prompt: SEL DIMN / SYS_DIM / DIM

1. Select the dimension(s) individually or use a multiselect option (e.g., *DIM*).
2. Click on the YES button to restore the true values.

Change graphical representation of dimension value

Select: DIMENS2 > MODIFY | VALUE |NUM DISP

Prompt: SEL MSELW DIMN // SEL DIM_VAL / DIM

Select the dimension(s) individually or by using a multiselect option (e.g., *DIM*). In the activated window you can modify graphism, height of value text, dimensioning style, and dimension value precision.

✓ **NOTE:** *Any changes to the window will be immediately applied to the selected dimension(s).*

Change graphical presentation of dimension value

Select: DIMENS2 > MODIFY | VALUE |GRAPHIC

Prompt: SEL DIMN / SYS_DIM / DIM

Select the dimension. In the activated window, you can apply under- or overscoring, and dimension value framing can be added or deleted.

✓ **NOTE:** *Any changes to the window will be applied immediately to the selected dimension(s).*

Modify dimension extension lines

Select: DIMENS2 > MODIFY | EXT LINE | LENGTH or SLANT or ERASE or INTERRUP or FUNNEL or RESTORE

Modify extension line length

Select: DIMENS2 > MODIFY | EXT LINE | LENGTH

Prompt: SEL DIMN / SYSD / DIM // SEL DIM_EXT

1. Select the dimension or extension line. When a dimension is selected, all extension lines associated with the dimension will be modified. If an extension line is selected, only the extension line selected will be modified.
2. At this point, you have the following alternatives: (a) Key in a new length to the input information area and press <Enter>. (b) Indicate a position for the length of the extension line to be modified. (c) Click on the YES button to restore a previously modified extension line to its original position from the dimensioned element.

Modify slant of extension line

Select: DIMENS2 > MODIFY | EXT LINE | SLANT

Prompt: CHG_ORIENTATION : SEL ELEM

1. Select the dimension to be modified.
2. Select a line to serve as the reference for the new orientation of extension lines, or key in a new angle to the input information area and press <Enter> to define the new orientation of the extension lines.

Delete extension line(s)

Select: DIMENS2 > MODIFY | EXT LINE | ERASE

Prompt: DIMN_SYSD_DIM_EXT : SEL ELEM

Select the dimension containing the extension lines to be erased.

Interrupt extension line

Select: DIMENS2 > MODIFY | EXT LINE | INTERRUP

Prompt: DIMN : SEL ELEM

1. Select the dimension extension line to be interrupted.
2. Indicate two positions along the extension line to define the break positions.

Create funnel-shaped extension line

Select: DIMENS2 > MODIFY | EXT LINE | FUNNEL

Prompt: DIM_EXT : SEL ELEM

1. In the activated window, you can set the parameters for the funnel-shaped extension line. The parameters to be defined are the angle and the length of the required funnel.
2. Select the extension line to be modified. Funnel window settings are applied.

Restore modified extension line

Select: DIMENS2 > MODIFY | EXT LINE | RESTORE

Prompt: DIM_EXT : SEL ELEM // SEL DIMN / SYSD

Select the extension line to be restored. The extension line will be restored to its appearance prior to modification.

Change or invert dimension line end symbol

Select: DIMENS2 > MODIFY | SYMBOL | CHANGE or INVERT

Change end symbol

Select: DIMENS2 > MODIFY | SYMBOL | CHANGE

Prompt: SYMB_TXTN_DIMN_SYSD : SEL ELEM

1. Select the dimension line end symbol to be changed.
2. In the activated window, you can select a replacement symbol. The symbols available for selection in the window will be shown in white. Available symbols are dependent on on standards set in the DRW STD function.

Invert end symbol

Select: DIMENS2 > MODIFY | SYMBOL | INVERT

Prompt: SEL DIMN/SYS_DIM / DIM

Select a dimension line. The end symbols are inverted, that is, arrows are inverted about the dimension extension line.

Change or delete tolerances

Select: DIMENS2 > MODIFY | TOL | CHANGE or DELETE

Change tolerances

Select: DIMENS2 > MODIFY | TOL | CHANGE

Prompt: SEL MSELW DIM // SEL SYS_DIM / DIM

1. Select the dimension(s) individually or use a multiselect option (e.g., **DIM*).
2. In the Tolerance window, you can modify the applied tolerance, apply a tolerance, and change tolerance presentation. Click on the YES button to apply changes.

✓ **NOTE:** *Tolerance presentation is set in the DRW STD function, which is normally controlled by a system administrator.*

Delete tolerances

Select: DIMENS2 > MODIFY | TOL | DELETE

Prompt: SEL DIMN / SYS_DIM / DIM

Select dimension.

✓ **NOTE:** *Applied tolerances can also be deleted using the DIMENS2 > MODIFY | TOL | CHANGE option.*

Edit, add, delete, or change graphical representation of applied text

Select: DIMENS2 > MODIFY | TEXT | EDIT or ADD or DELETE or GRAPHIC or UPDATE

Edit tolerance applied to dimension

Select: DIMENS2 > MODIFY | TEXT | EDIT

Prompt: TXT_REF : SEL TXT_TXT

Select the dimension text to be edited. Two windows are activated in which you can change text graphical attributes and edit selected text, respectively. When either of the windows are modified and <Enter> is pressed, the selected text will be modified.

✓ **NOTE:** *Dimension text can also be edited using TEXTD2 > MODIFY | TEXT | EDIT.*

Add text to dimension

Select: DIMENS2 > MODIFY | TEXT | ADD

Prompt: TXT_REF : SEL TXT_TXT // SEL DIM_VAL

Select the dimension text to which you wish to add text. A window is activated allowing you to add text in various positions (before, after, insert, above, lower) relative to the dimension value.

The text is created for the relevant positions by selecting the line beneath the position icon and keying in the required text to the text box in the window. The text is then applied by selecting the relevant icon. Graphism and text height can also be modified using this window.

Delete applied text

Select: DIMENS2 > MODIFY | TEXT | DELETE

Prompt: SEL TXT_TXT / TXTN / DIM

Select the text to be deleted.

✓ **NOTE:** *Applied text can also be switched off using the DIMENS2 > MODIFY | TEXT | ADD option.*

Modify graphical attributes of text

Select: DIMENS2 > MODIFY | TEXT | GRAPHIC

Prompt: MSELW TXTN / DIMN SYSD // SEL TXT_TXT

Select the dimension(s) whose attributes are to be modified individually or by using a multiselect option (e.g., **DIM*, any dimensions affected by a change to a program, attributes, or identifer). In the window which activates, you can change the attributes listed below.

- Graphism
- Height and width of text
- Thickness and color of text
- Underscore, score or overscore
- Frames
- Spacing between lines of text

Changes made in the window will be automatically applied.

Alternatively, you can select the text to be modified and then click on the YES button. Select another piece of text to serve as the reference for the first selected text (i.e., the graphical attributes applied to the second selected piece of text will be applied to the first selected text).

Update dimensions

Select: DIMENS2 > MODIFY | TEXT | UPDATE

Prompt: MSELW TXTN / TXTD / DIMN / SYSD

1. Select dimension(s) individually or by using a multiselect option (e.g., **DIM*).
2. Click on the YES button to confirm the update.

Change graphical attributes

Select: DIMENS2 > MODIFY | GRAPHIC

Prompt: SEL MSELW DIMN

Change graphical attributes of dimension

Select dimension(s) whose graphical attributes are to be changed. You can use individual or multiselection (e.g., **DIM*). In the window which activates, you can change the attributes listed below.

- Dimension line thickness and color
- Extension line thickness and color
- Symbol thickness and color
- Value thickness and color
- Frame thickness and color
- Score thickness and color

ADD

Add new dimensions to an existing dimension system.

Select: DIMENS2 > ADD

Prompt: SEL DIMN / SYSD / DIM

1. Select a dimension to which a new cumulative dimension is to be added.
2. Select element(s) to be dimensioned.
3. Click on the YES button to end selection.

DELETE

Delete dimensions or dimension systems.

Select: DIMENS2 > DELETE

Prompt: EXT LINE : SEL DIM_EXT SEL DIMN // SEL SYSD // SEL DIM

Delete single dimension

Select the required dimension or dimension extension line and the dimension will be deleted.

Delete extension line from set of cumulative dimensions

1. Select the required dimension.
2. Indicate near the extension to be deleted. The extension line will be deleted and the associated dimension will be adjusted to suit.

MANAGE

Manage dimensions.

Select: DIMENS2 > MANAGE | VERIFY or ISOLATE or VISUALTN or STARTUP

Verify dimensions

Select: DIMENS2 > MANAGE | VERIFY

Prompt: MSELW DIMN / SYSD / DIM

1. In the activated window, you have three verification choices: FAKE DIM, TRUE LENGTH, and ASSOCIATIVITY. Select the desired option.
2. Key in **DIM* to the input information area and press <Enter>.
3. All dimensions pertaining to the required option will be highlighted.

Isolate dimension(s)

Select: DIMENS2 > MANAGE | ISOLATE

Prompt: MSELW TXTN / DIMN / SYSD / TXTD /DIM

1. Select the dimension(s) individually or use a multiselect option (e.g., **DIM*).
2. Click on the YES button to confirm isolation of selected dimension(s).

✗ **WARNING:** *Once a dimension has been isolated, it cannot be reattached to its creation elements.*

Modify visualization of dimension value and text

Select: DIMENS2 > MANAGE | VISUALTN | SHOW TXT or SHOW BOX or NO SHOW or BLNK LIN

Show dimension values and text

Select: DIMENS2 > MANAGE | VISUALTN | SHOW TXT

Prompt: MSELW TXTN / DIMN / SYSD / DIM / TXD

Select dimension(s) previously modified using DIMENS2 > MODIFY | VISUALTN | SHOW BOX or NO SHOW individually or use a multiselect option (e.g., **DIM*). The dimension value or text will then be restored to its normal display mode.

Show dimension values and text as boxes

Select: DIMENS2 > MANAGE | VISUALTN | SHOW BOX

Prompt: MSELW TXTN / DIMN / SYSD / DIM / TXD

Select dimension(s) to be shown as boxes individually or use a multiselect option (e.g., **DIM*).

Hide dimension values and text

Select: DIMENS2 > MANAGE | VISUALTN | NO SHOW

Prompt: MSELW TXTN / DIMN / SYSD / DIM / TXD

Select dimension(s) to be hidden individually or use a multiselect option (e.g., **DIM*).

Identify all blank lines or blank subtexts

Select: DIMENS2 > MANAGE | VISUALTN | BLNK LIN

Prompt: MSELW TXTN / DIMN / SYSD / DIM / TXD

Select dimension(s) to be identified individually or use a multiselect option (e.g., **DIM*).

Reset original status of modal parameters

Select: DIMENS2 > MANAGE | STARTUP

Reset modal parameters under DIMENS2 and TEXTD2 modified during the creation of a model with this option.

In the STARTUP LIST window, select from available startup customization models. The selection becomes the current customization.

DRAFT in DRAW Mode

Chapter 6

The DRAFT function in DRAW mode is used to create new drafts and modify existing drafts.

Option menu for DRAFT in DRAW mode.

CHANGE

Change current draft when more than two drafts are available in the model.

Select: DRAFT > CHANGE

Prompt: KEY DRAFT ID // YES: LIST

If known, key in the draft identification. Alternatively, click on the YES button to view the list of available drafts and select a different one.

Change current draft when no alternative drafts exist

If no alternative draft exists, the prompt is SEL MENU.

CREATE

Create draft from existing draft.

Create new draft when only one draft exists

Select: DRAFT > CREATE

Prompt: KEY NEW DRAFT ID

When only one draft is available in the model, you must duplicate the draft in order to create a new one. You will be prompted to key in the new draft identity.

Create new draft when more than one draft exists

Select: DRAFT > CREATE

Prompt: KEY DRAFT TO DUPLICATE // YES: LIST

When more than one draft is available, you will be prompted to either key in the draft identification (if known) or click on the YES button to view alternative drafts and select a draft to duplicate. After selecting a draft, you must key in the identification for the new draft.

DELETE

Delete drafts from model.

Select: DRAFT > DELETE

Prompt: KEY DRAFT ID // YES: LIST

1. Key in the identification of the draft you wish to delete, or click on the YES button and select a draft to delete.
2. Click on the YES button to confirm deletion.

RENAME

Change the name of draft.

Select: DRAFT > RENAME

Prompt: KEY DRAFT ID // YES: LIST

1. Key in the identification of a draft, or click on the YES button and select a draft to rename.
2. Key in the new name for the draft.

TRANSFER

Transfer draw views from one draft to another.

Select: DRAFT > TRANSFER

Prompt: KEY VIEW ID // YES: LIST

1. Key in the name of the view to transfer into the current draft, or click on the YES button to select one or more views from the list of available views.
2. Click on the YES button to end the selection. The views are transferred into the current draft.

COPY

Copy views from one draft into another.

Select: DRAFT > COPY

Prompt: KEY VIEW ID // YES: LIST

1. Key in the name of the view to copy into the current draft, or click on the YES button to select one or more views from the list of available views.
2. Click on the YES button to end the selection. The views are copied into the current draft.

DRW>SPC in SPACE Mode

Chapter 14

The DRW>SPC function is a SPACE mode function used to create geometry in SPACE mode using elements from DRAW mode. The only option is CREATE.

CREATE

Create geometry in SPACE mode using elements from DRAW mode.

Select: DRW>SPC > CREATE

Prompt: SEL PLN // SEL CRV

Two ways in which elements can be created in the SPACE work area using elements from the DRAW work area are described below.

- Define a plane in SPACE and duplicate the DRAW elements onto the selected plane.

1. Select a plane to define the plane on which the DRAW geometry is to be duplicated.
2. Select the DRAW geometry to be duplicated, either by selecting individual elements or by using one of the many available multiselection options. (See Appendices A and B.)
3. Click on the YES button to end the selection process when all required SPACE elements have been created.

- Select a curve in the DRAW work area to be duplicated in the SPACE work area, and then define its receiving plane by using a line in an associated view.

1. Select the curve to be duplicated.
2. Select a line(s) in an associated view to define the receiving plane in the SPACE work area.
3. After the curve has been duplicated onto all required planes, click on the YES button to end the selection process.

DRW STD

The DRW STD function is used to manage and create descriptions and standards used in the TEXTD2 and DIMENS2 functions. It is also used to create patterns used in the PATTERN function.

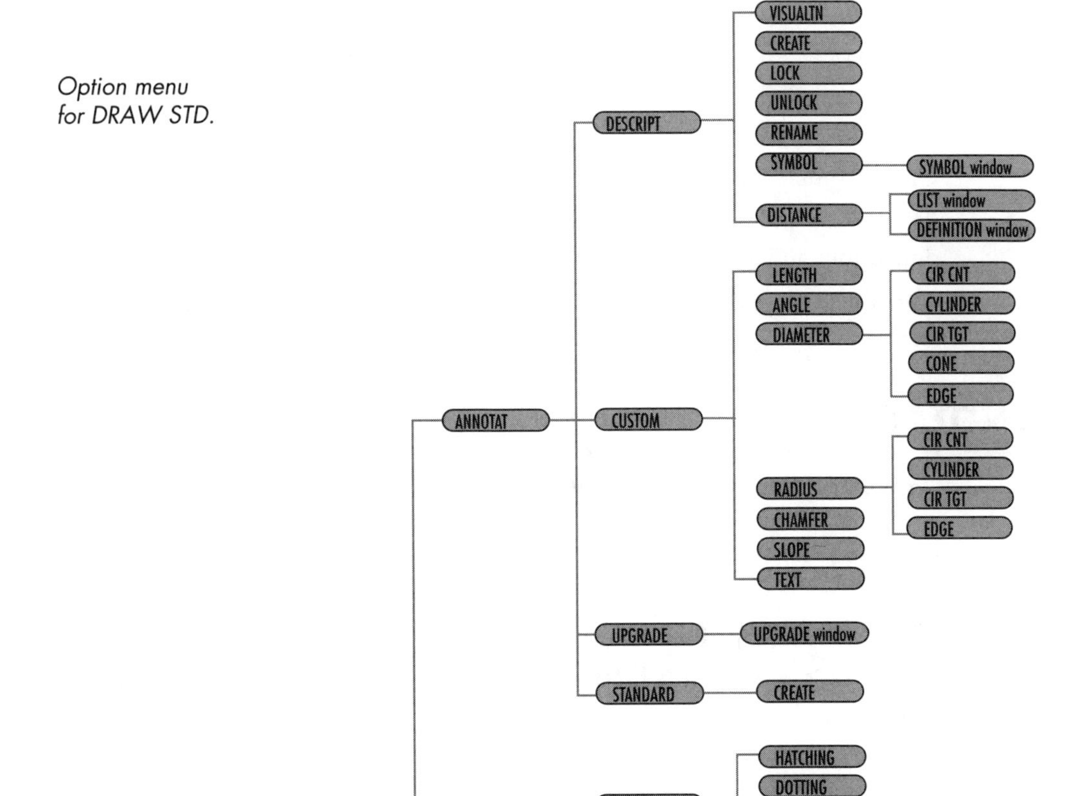

Option menu for DRAW STD.

The DRW STD function is often controlled by a system administrator in order to impose and enforce company standards. Because the purpose of this function is to enable an administrator to control the items listed below, available options will not be discussed in depth.

- Visualize existing descriptions used in the TEXTD2 and DIMENS2 functions and existing patterns used in the PATTERN function.
- Create new descriptions and patterns.
- Lock and unlock descriptions and patterns.
- Rename existing descriptions and patterns.
- Assign values to the various end symbols used in the TEXTD2 and DIMENS2 functions.
- Define contents of the MANAGEMENT windows used in the DIMENS2 function.
- Create company standards used in the TEXD2 and DIMENS2 functions.
- Enable users to perform the conversion of, and if needed modifications on, old dimensions and annotations.
- Define hatching, dotting, coloring, and cell patterns, and apply them to SHAP elements.

ANNOTAT

Manage and create descriptions and standards used in the TEXTD2 and DIMNES2 functions.

Select: DRW STD > ANNOTAT | DESCRIPT

Manage and create descriptions

Select: DRW STD > ANNOTAT | DESCRIPT | VISUALTN or CREATE or LOCK or UNLOCK or RENAME or SYMBOL

Visualize available descriptions

Select: DRW STD > ANNOTAT | DESCRIPT | VISUALTN

Prompt: SEL DESCRIPTION TYPE

The DESCRIPTION window in the next illustration displays.

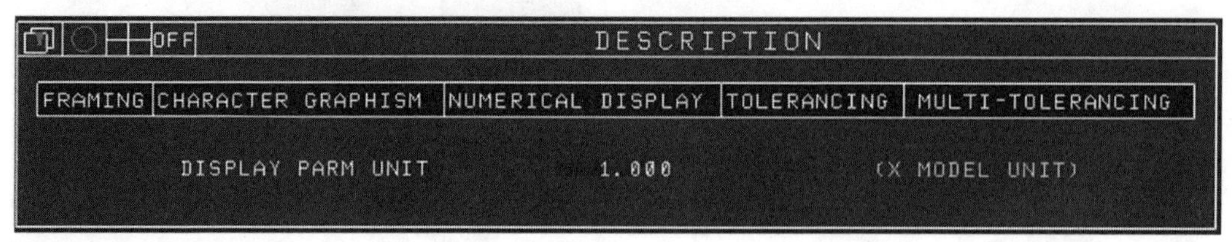

DESCRIPTION window.

You can make the following selections in the window.

- FRAMING
- CHARACTER GRAPHISM
- NUMERICAL DISPLAY
- TOLERANCING
- MULTI-TOLERANCING

Upon selecting one of the above topics, a list of available options displays. A typical example is shown in the following illustration.

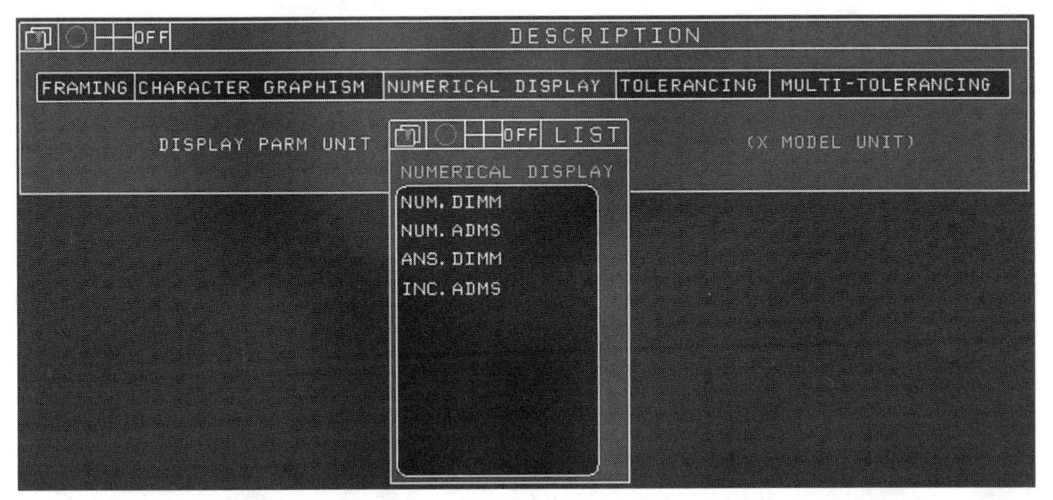

List of available NUMERICAL DISPLAY descriptions.

If an additional selection is made from the above list (e.g., NUM.ADMS), the characteristics of the selected description will be displayed as shown in the next illustrations.

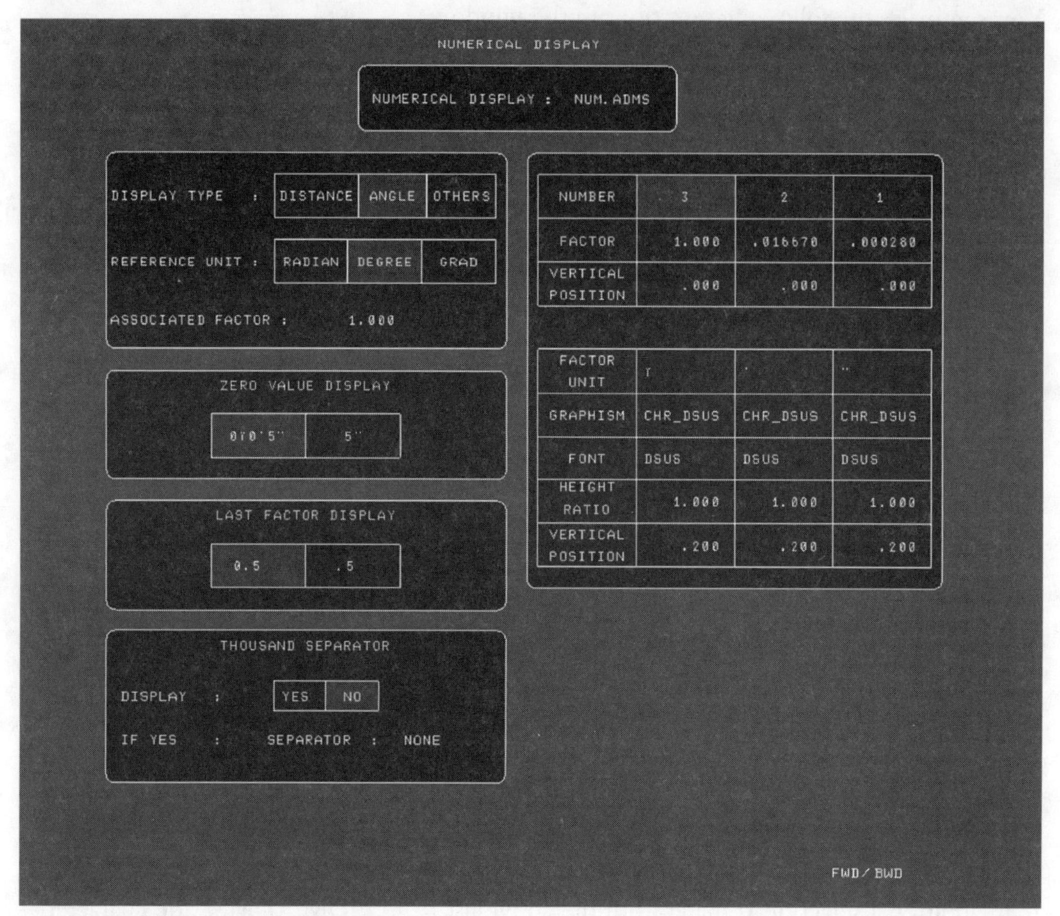

Page 1 of NUM.ADMS characteristics.

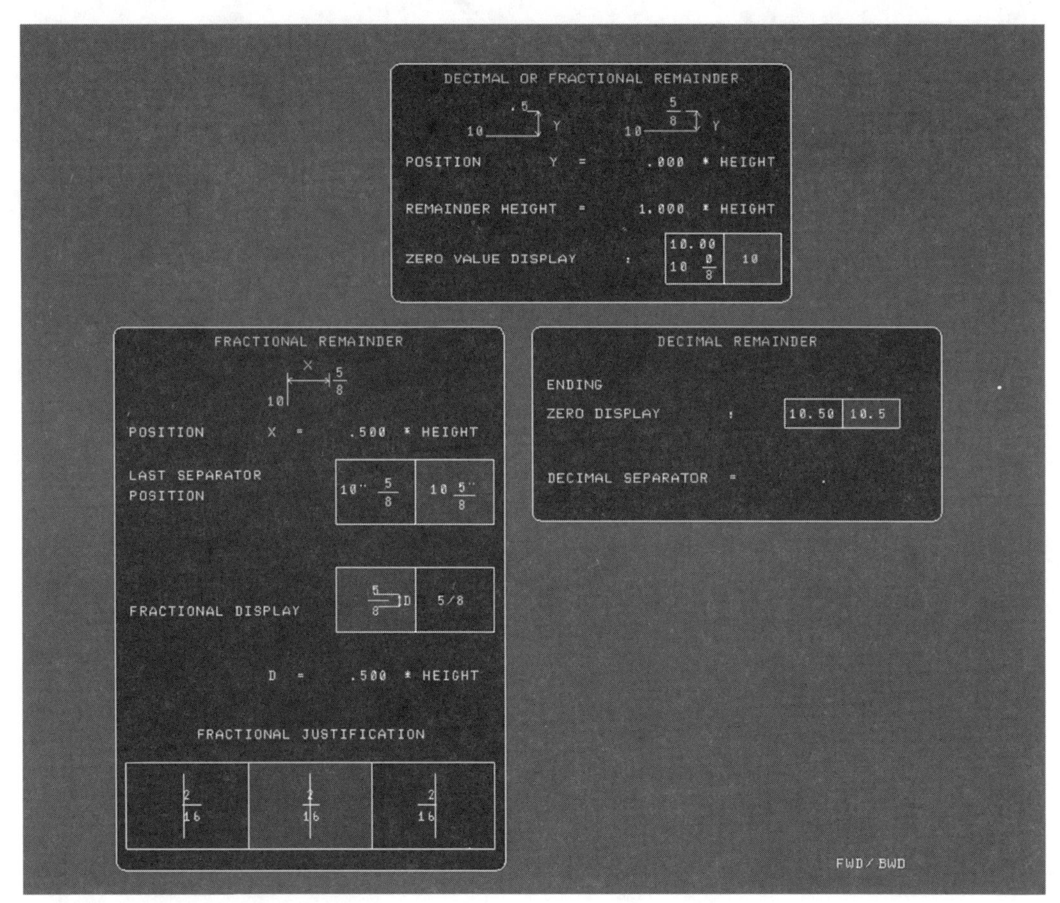

Page 2 of NUM.ADMS characteristics.

✓ **NOTE:** *This procedure can be used to visualize descriptions available in the DESCRIPTION window.*

Create definition of descriptions

Select: DRW STD > ANNOTAT | DESCRIPT | CREATE

Because this option is typically available to system administrators only, details are not provided in this book.

Lock availability of descriptions

Select: DRW STD > ANNOTAT | DESCRIPT | LOCK

Because this option is typically available to system administrators only, details are not provided in this book.

Unlock availability of descriptions

Select: DRW STD > ANNOTAT | DESCRIPT | UNLOCK

Because this option is typically available to system administrators only, details are not provided in this book.

Rename available descriptions

Select: DRW STD > ANNOTAT | DESCRIPT | RENAME

Because this option is typically available to system administrators only, details are not provided in this book.

Control symbol size parameters and current standards

Select: DRW STD > ANNOTAT | DESCRIPT | SYMBOL

Prompt: SEL OPTION

The window appearing in the next illustration displays.

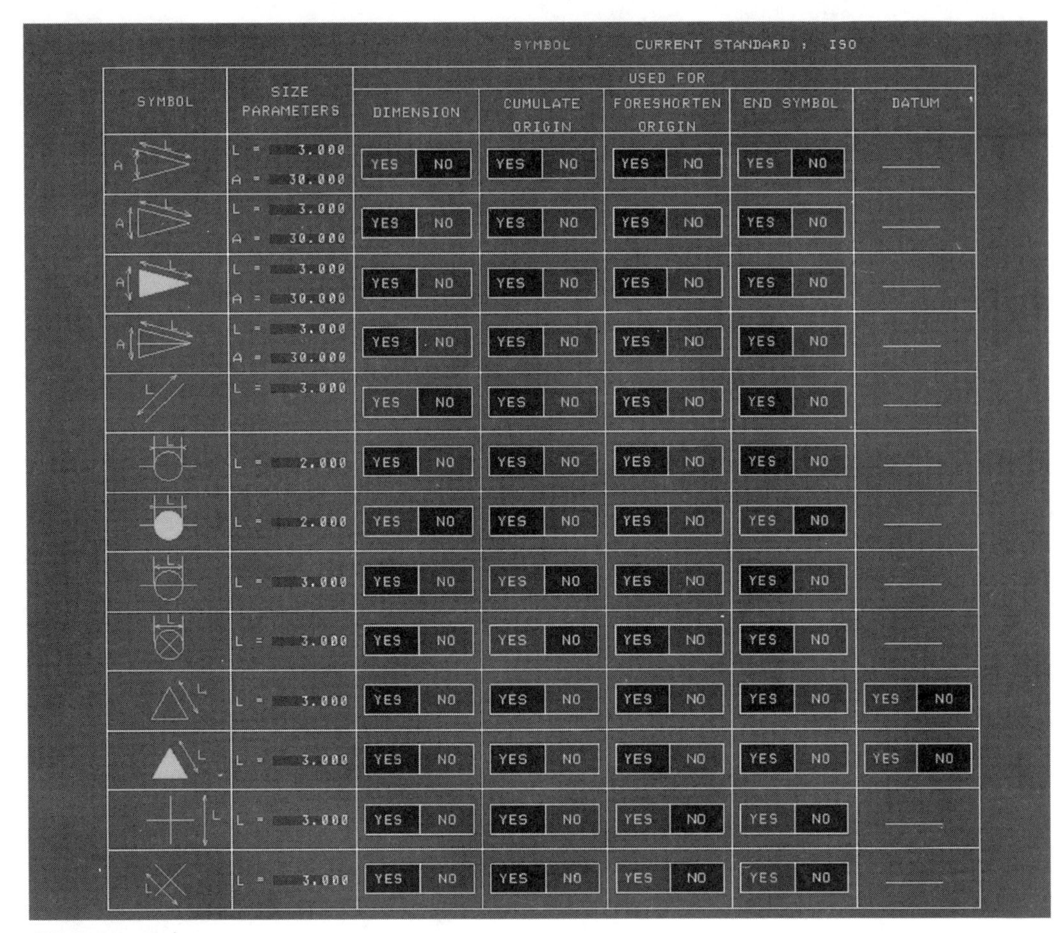

SYMBOL window.

The following parameters are defined in the above window.

- Current symbol standard (selectable).
- Size parameters of symbols (selectable).
- Where and for what reason the symbols are used (nonselectable; dependent on the current standard).

Change current standard

1. Select the white text (e.g., ISO in the above illustration).
2. In the ensuing window, select the required standard.

Change size parameter

Select the required dimension and key in the new value.

✓ **NOTE:** *Changing size parameters will not alter existing symbols. Only subsequently created symbols will bear the new size.*

CUSTOM

Customize MANAGEMENT windows accessed in the DIMENS2 function.

Select: DRW STD > CUSTOM | DISTANCE or LENGTH or ANGLE or DIAMETER or RADIUS or CHAMFER or SLOPE or TEXT

Buttons (icons) pertaining to specific windows can be modified and moved by selecting respective CUSTOM suboptions. All MANAGEMENT windows are customized in the same way as the following example for DISTANCE.

Customize Distance window

Select: DRW STD > ANNOTAT | CUSTOM | DISTANCE

Prompt: SEL PARAMETER // SEL CORNER MOVE : SEL POSITION // YES : DELETE

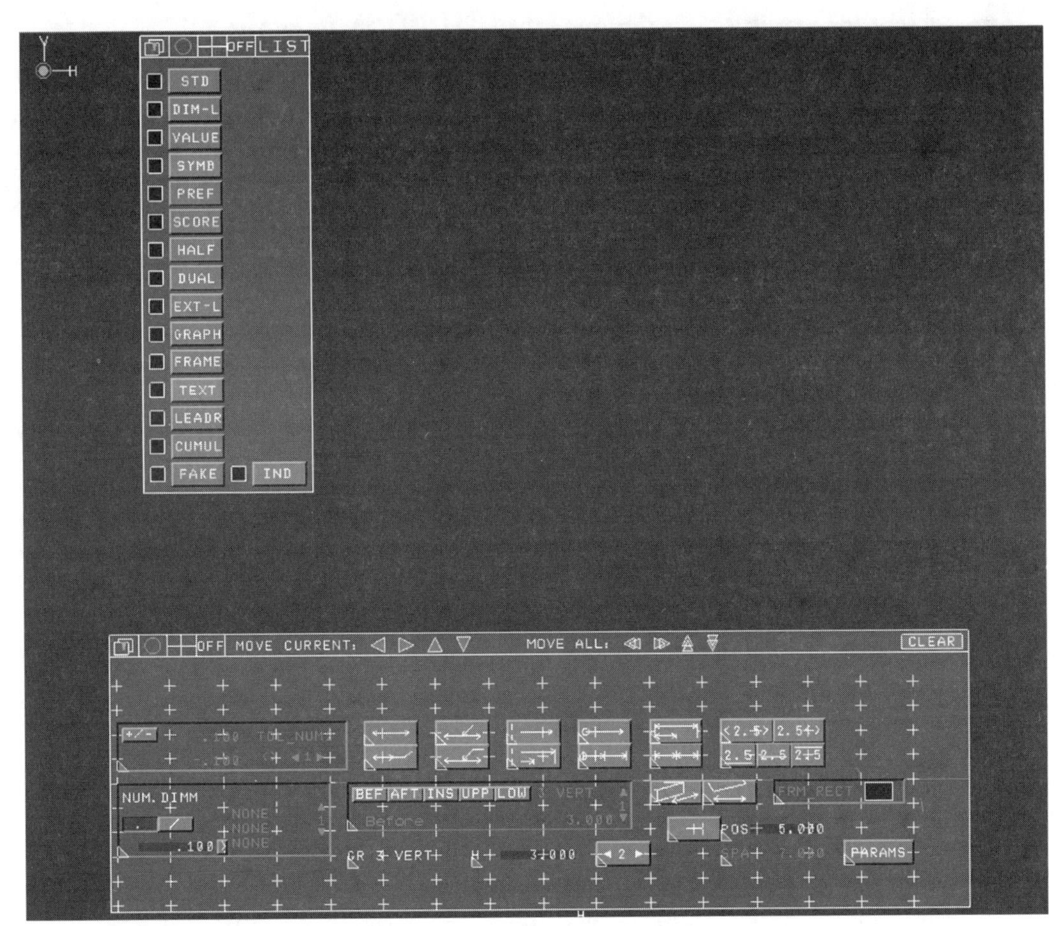

LIST and DEFINITION dialogs for DISTANCE option under DIMENS2 function.

Move default parameters from DEFINITION window into LIST window

1. Select the check box associated with the parameter to be removed from the DEFINITION window.
2. Click on the YES button to delete selected parameter from the window. The selected parameter now appears in the LIST window.

Move all default parameters from DEFINITION window into LIST window

Select the CLEAR command from the DEFINITION window. The window is cleared of all parameters. The parameters are now displayed in the LIST window.

Move parameters from LIST window into DEFINITION window

1. Select the check box associated with the parameter to be moved into the DEFINITION window.
2. Select a cross in the DEFINITION window corresponding to the anchor point to which the selected parameter will be moved.
3. Click on the YES button to leave the parameter selected in the LIST window, or if desired, use the MOVE CURRENT or MOVE ALL commands to reposition the required parameters. The MOVE CURRENT command moves the selected parameter only, and the MOVE ALL command simultaneously moves all parameters.

Move parameters within Definition window

1. Select the corner of the parameter to be moved.
2. Select a cross in the DEFINITION window corresponding to the anchor point to which the selected parameter will be moved.
3. If required, use the MOVE CURRENT or MOVE ALL commands to reposition the required parameters. The MOVE CURRENT command moves the selected parameter only, and the MOVE ALL command simultaneously moves all parameters.

✓ **NOTE:** *Modifications executed with the CUSTOM option affect only the current model. If the modifications are to be applied to subsequent models, the current model must be saved as a startup model.*

Control upgrade of old text and dimensions

Select: DRW STD > ANNOTAT | UPGRADE

Prompt: SEL OPTION

The following UPGRADE window displays.

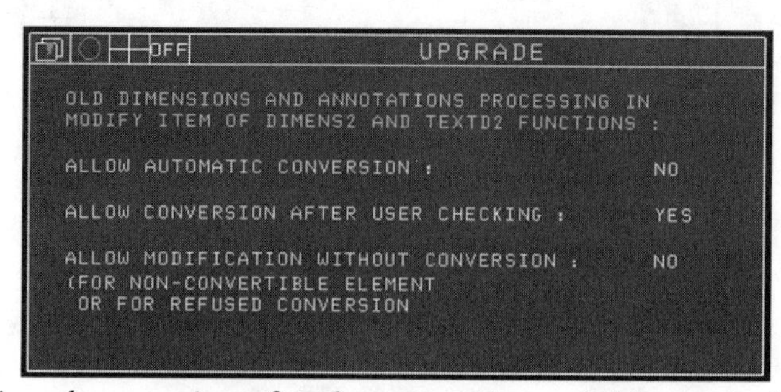

DRW STD UPGRADE window.

The window contains the following three options for the conversion of old text and dimensions.

- Automatic conversion
- Conversion after user checking
- Modification without conversion

Selections within these options define how old text or dimensions are dealt with. For instance, if selections match those in the next illustration, you cannot convert old text and dimensions, but you can modify them.

Old text and dimensions transparent modification.

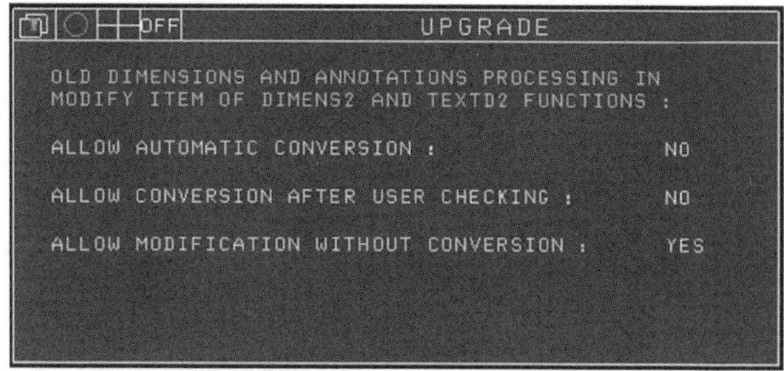

If selections match the next illustration, the conversion of old text and dimensions occurs only after user checking. When the user refuses conversion, modification of old text and dimensions takes place.

Controlled conversion, and if necessary, transparent modification.

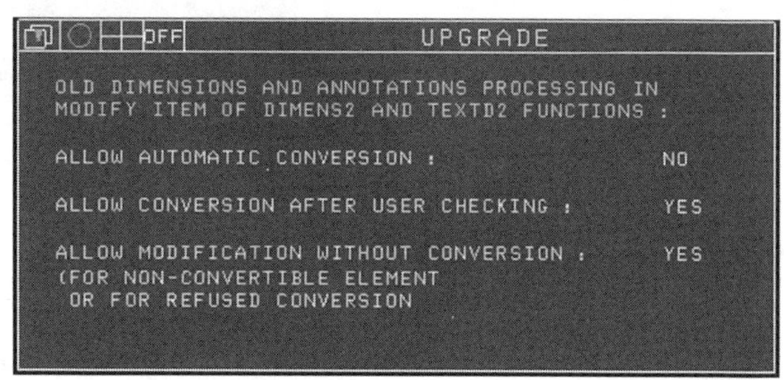

When selections match the following illustration, the conversion of text and dimensions takes place only after user checking. If the user does not convert old text and dimensions upon checking, modification of same does not take place.

Controlled conversion and no transparent modification.

OFF
UPGRADE
OLD DIMENSIONS AND ANNOTATIONS PROCESSING IN
MODIFY ITEM OF DIMENS2 AND TEXTD2 FUNCTIONS :
ALLOW AUTOMATIC CONVERSION : NO
ALLOW CONVERSION AFTER USER CHECKING : YES
ALLOW MODIFICATION WITHOUT CONVERSION : NO
(FOR NON-CONVERTIBLE ELEMENT
OR FOR REFUSED CONVERSION

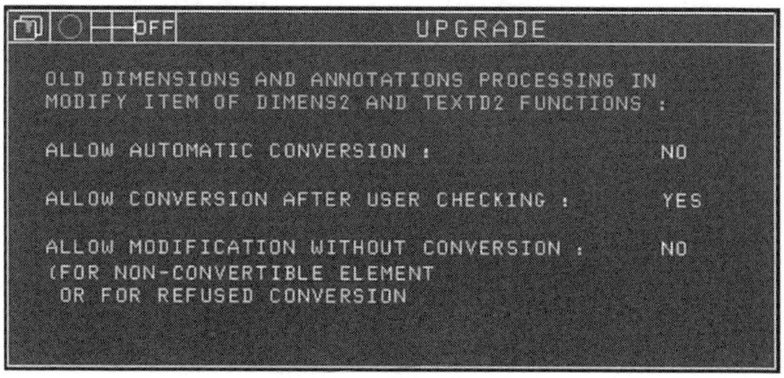

When selections match the next illustration, the conversion of old text and dimensions is automatic and cannot be refused by the user. Modification of old text and dimensions occurs if the same cannot be converted.

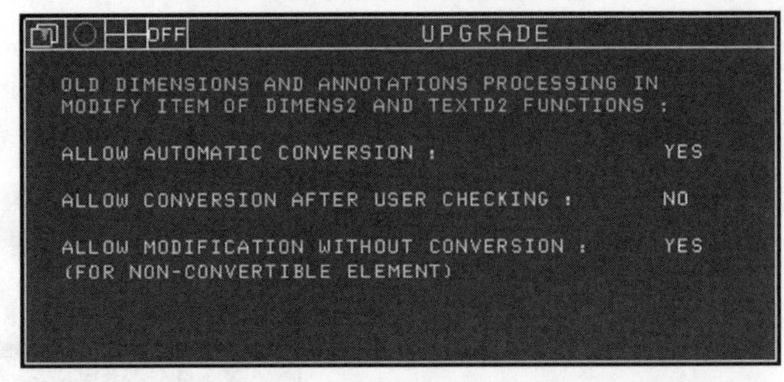

Automatic conversion and transparent modification.

When selections match the following illustration, the conversion of old text and dimensions is automatic and cannot be refused by the user, and modification of the same does not occur.

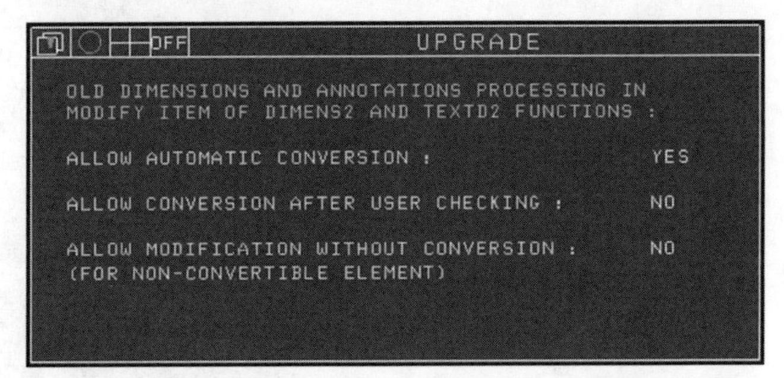

Automatic conversion exclusively.

✓ **NOTE 1:** *The above upgrade options are used when text or dimensions are modified in the TEXD2 and DIMENS2 functions.*

✓ **NOTE 2:** *Modifications with the UPGRADE option will affect only the current model. If the modifications are to be applied to subsequent models, the current model must be saved as a startup model.*

Create new standards for use with the DIMENS2 and TEXTD2 functions

Select: DRW STD > ANNOTATE | STANDARD

Because this option is normally available only to system administrators, it is not discussed in this book.

PATTERN

Manage and create patterns used in the PATTERN function. Options in this menu are typically available only to the system administrator. Consequently, they are not discussed in detail in this book.

Select: DRW STD > PATTERN | CREATE or LOCK or UNLOCK or RENAME

- CREATE. Create definition of patterns.
- LOCK. Lock pattern availability.
- UNLOCK. Unlock pattern availability.
- RENAME. Rename available patterns.

ERASE in DRAW and SPACE Modes

Chapters 3 to 11

The ERASE function in DRAW and SPACE modes is used to delete, hide, or deselect elements, and to pack the model.

Option menu for ERASE in DRAW and SPACE modes.

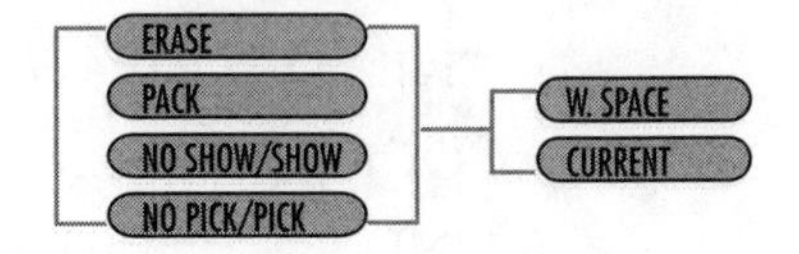

ERASE

Delete elements from model.

Select: ERASE > ERASE | W.SPACE or CURRENT

Prompt: WSP: MULTI-SELSEL: ELEM

When you choose W.SPACE, all elements in the model are selectable. Upon choosing CURRENT, only SPACE elements in the current set and DRAW elements in the current view and set are selectable. The element or elements you select are deleted.

✗ ***WARNING:*** *You only have one chance to say "no." If you take any other action at this point, deleted elements cannot be recovered in the current model.*

✓ ***NOTE:*** *You cannot delete the current axis systems, elements geometrically linked to other elements, or elements not included in the current model (e.g., an element in a passive overlaid model).*

Upon selecting a dimensioned element or element with an associated note, the following message displays: MESSAGE. NECESSARY ELEMENT FOR DIM/NOTE. Click on the YES button to confirm the deletion.

PACK

Remove unoccupied space from model.

Select: ERASE > PACK | W.SPACE or CURRENT

Prompt: YES: PACK

The Model Status window contains the following information.

- Index filling ratio. Size of occupied part of INDEX table in kilobytes.
- Data filling ratio. Size of occupied part of DATA table in kilobytes.

- Ext. data filling ratio. Size of occupied part of EXTENDED DATA table in kilobytes.

Click on the YES button to PACK the model.

✓ **NOTE:** *Model visualization is unaffected.*

NO SHOW/SHOW

Permanently hides elements from the standard display mode (NO SHOW) or redisplays previously hidden elements (SHOW).

Select: ERASE > NO SHOW | W.SPACE or CURRENT

Prompt: WSP: MULTI-SELSEL: ELEM // YES: SWAP

1. Select one or more elements to hide.
2. Click on the YES button to switch to the NO SHOW area, and then select one or more elements to redisplay.
3. Click on the YES button to return to the normal work area, or select another function.

NO PICK/PICK

Make elements unselectable (NO PICK) or make previously unselectable elements available (PICK).

Select: ERASE > NO PICK | W.SPACE or CURRENT

Prompt: WSP: MULTI SEL SEL: ELEM // YES: SWAP

1. Select one or more elements to make them unselectable.
2. Click on the YES button to switch to the NO PICK area, and then select one or more elements to make them selectable again.
3. Click on the YES button to return to the normal work area or select any other function.

FILE in DRAW and SPACE Modes

FILE function used throughout

Option menu for FILE in DRAW and SPACE modes.

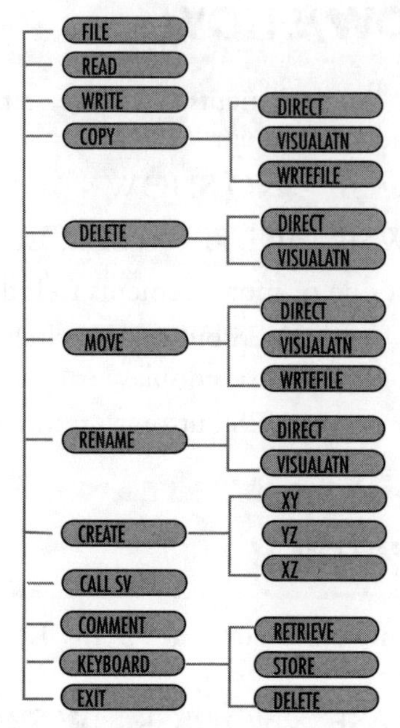

The FILE function in DRAW and SPACE modes is used to define the current file; read, write, delete, copy, or rename models; and create a start model or restore previously saved models. This function is also used to define palettes, write comments (of up to 500 lines) in a model, and end a CATIA session.

FILE

Define current model file. Choose one of the following alternatives to select a file.

Select: FILE > FILE

Prompt: KEY STRING

- Key in a string of up to 24 characters and press <Enter> to view all files containing the string. Select a file.
- Press <Enter> with a blank entry field to view a list of all available files. Select a file.

READ

Display model from current model file.

Select: FILE > READ

Prompt: KEY STRING

To read a model, key in a string contained within the model name you seek and select the model from the displayed list. Alternatively, press <Enter> with a blank entry field to choose from a list of available files.

Once the model is read, if it is incorrect or an alternative is required, click on the NO button to return to the list. The last model you read remains highlighted.

WRITE

Write or save model into current model file.

Select: FILE > WRITE

Prompt: KEY MODEL ID

1. To write a model into the current file, key in the name for the model and press <Enter>, or click on the YES button to accept the current name.
2. If the model already exists, the prompt YES REPLACE EXISTING MODEL displays. Click on YES to overwrite the model.

COPY

Copy model from one model file to another.

Select: FILE > COPY | DIRECT or VISUALATN or WRTEFILE

Copy model to defined receiving file

Select: FILE > COPY | DIRECT

Prompt: KEY OUTPUT FILE STRING

Select the model to be copied. Click on the YES button to end the selection and again to confirm the copy.

If a model with an identical name already exists in the receiving file, you can enter a new name. Click on the YES button to overwrite the name, or click on the NO button to refuse the copy.

View model before copying

Select: FILE > COPY | VISUALATN

Prompt: KEY OUTPUT FILE STRING

1. Select the sending file, and then select one or more models to be copied.
2. Click on the YES button to end the selection; the model will display.
3. Click on the YES button to confirm the copy.

Copy model without defining receiving file

Select: FILE > COPY | WRTEFILE

1. Key in the file name or string (up to 24 characters), or press <Enter> with a blank entry field to view a complete list of files.
2. Select the receiving file.

DELETE

Delete one or more models from file.

Select: FILE > DELETE | DIRECT or VISUALATN

Delete model from particular model file

Select: FILE > DELETE | DIRECT

Prompt: KEY STRING

1. Select a model file.
2. Select one or more models from the list, and then click on the YES button to confirm deletion.

View model before deletion

Select: FILE > DELETE | VISUALATN

Prompt: KEY STRING

1. Select a model file and then select a model to display.
2. Click on the YES button to delete the model or NO to cancel.

MOVE

Move model from one file to another. Suboptions work in the same manner as COPY suboptions (see previous description).

Select: FILE > MOVE | DIRECT or VISUALATN or WRTE FILE

Prompt: KEY OUTPUT FILE STRING

RENAME

Change name or view model before name change.

Select: FILE > RENAME | DIRECT or VISUALATN

Change model name

Select: FILE > RENAME | DIRECT

Prompt: KEY STRING

1. Select the model to be renamed, key in the new name, and press <Enter>.
2. If a model with the same name already exists, you can change the name, click on the YES button to overwrite the model, or click on NO to cancel the operation.

View model before changing name

Select: FILE > RENAME | VISUALATN

Prompt: KEY STRING

1. Select a model to view.
2. Key in the new name for the model and press <Enter>.
3. Click on the YES button to replace the model if a model with the same name is found. Key in a new name, or click on the NO button to cancel the operation.

CREATE

Select: FILE > CREATE | XY or YZ or XZ

Prompt: YES: CONFIRM

Create a basic empty model or access a start model (if your system administrator has made this option available). Creating models differs slightly in DRAW and SPACE modes.

- In DRAW mode, click on the YES button to confirm the model creation. If no start model has been established, an empty model with a draw axis is created.
- In SPACE mode, select one of the three planes offered to determine the initial draw view. If no start model has been established, an empty model with a space axis is created.

CALL SV

Restore model previously saved using the SV button on fixed menu.

✓ ***NOTE:*** *You can view the definition and purpose of fixed menu buttons in the message area by positioning the cursor over a button.*

Select: FILE > CALL SV

Prompt: YES: CONFIRM

Click on the YES button to restore the previously saved model.

✗ **WARNING:** *Take care not to reselect the SV button. If you do, the CALL SV option will be lost.*

✓ **NOTE:** *The SV button on the fixed menu has the same function as the CALL SV option in the FILE menu.*

COMMENT

Select: FILE > COMMENT

Prompt: SEL TEXT

Add textual information to a model. This option is available for the current model or by selecting a model from a file list. When COMMENT is selected, a full screen panel displays four windows, as shown in the next illustration.

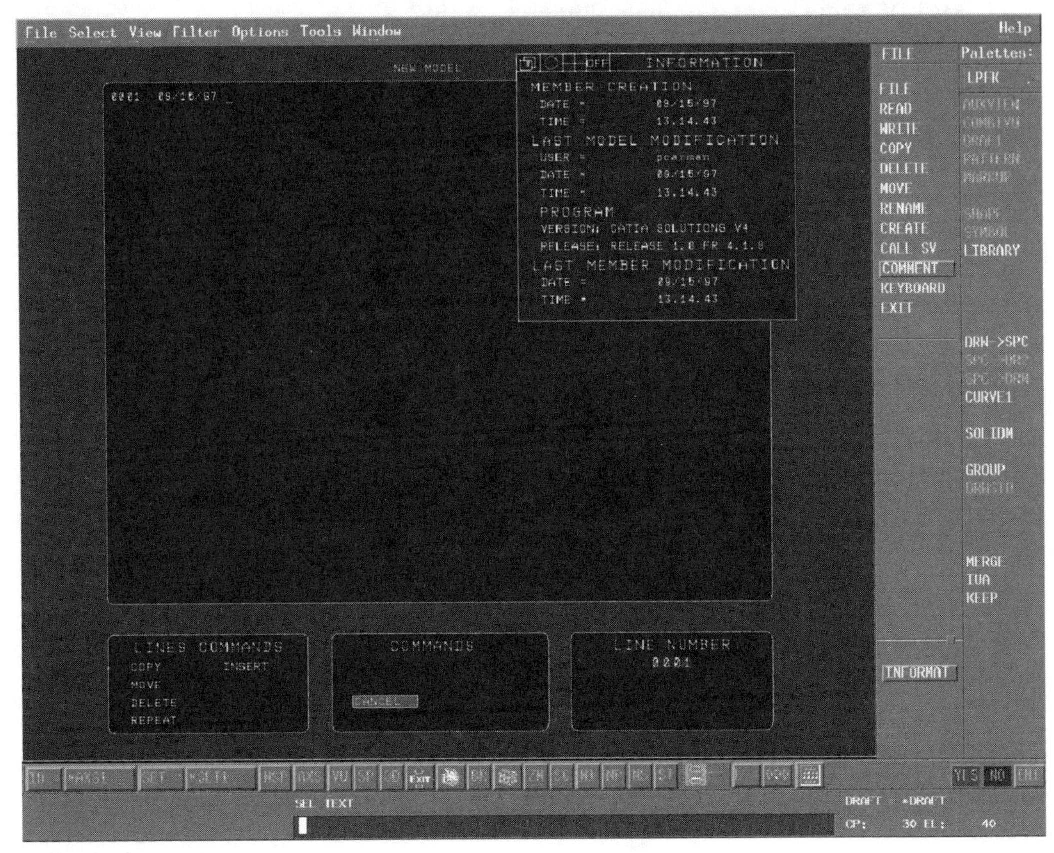

COMMENT screen.

INFORMATION window

The INFORMATION window displays the following data for a saved model.

- Date and time of model creation
- User name, date, and time of last modification
- Software version and release
- Date and time of last member modification

LINES COMMANDS window

Prompt: SEL TEXT // YES: RESET LINE COMMAND

The LINES COMMANDS window contains five major options described below. The prompt for all options is the same.

•COPY. Copy text from one position to another.

- •MOVE. Move text from one position to another.
- •DELETE. Delete line.
- •REPEAT. Repeat line of text.
- •INSERT. Insert next line of text.

COMMANDS window

Prompt: SEL TEXT

The COMMANDS window offers a single option, CANCEL, which allows you to cancel the entire comment.

LINE NUMBER window

The LINE NUMBER window simply indicates the current line number in the comment.

KEYBOARD

Define and manage palettes and LPFK keyboards.

Select: FILE > KEYBOARD | RETRIEVE or STORE or DELETE

Prompt: SEL KEYBOARD SEL FUNCTION // SEL FUNCTION KEY

When the previous commands and above prompt are issued, the Palette Keyboard Creation screen in the next illustration appears.

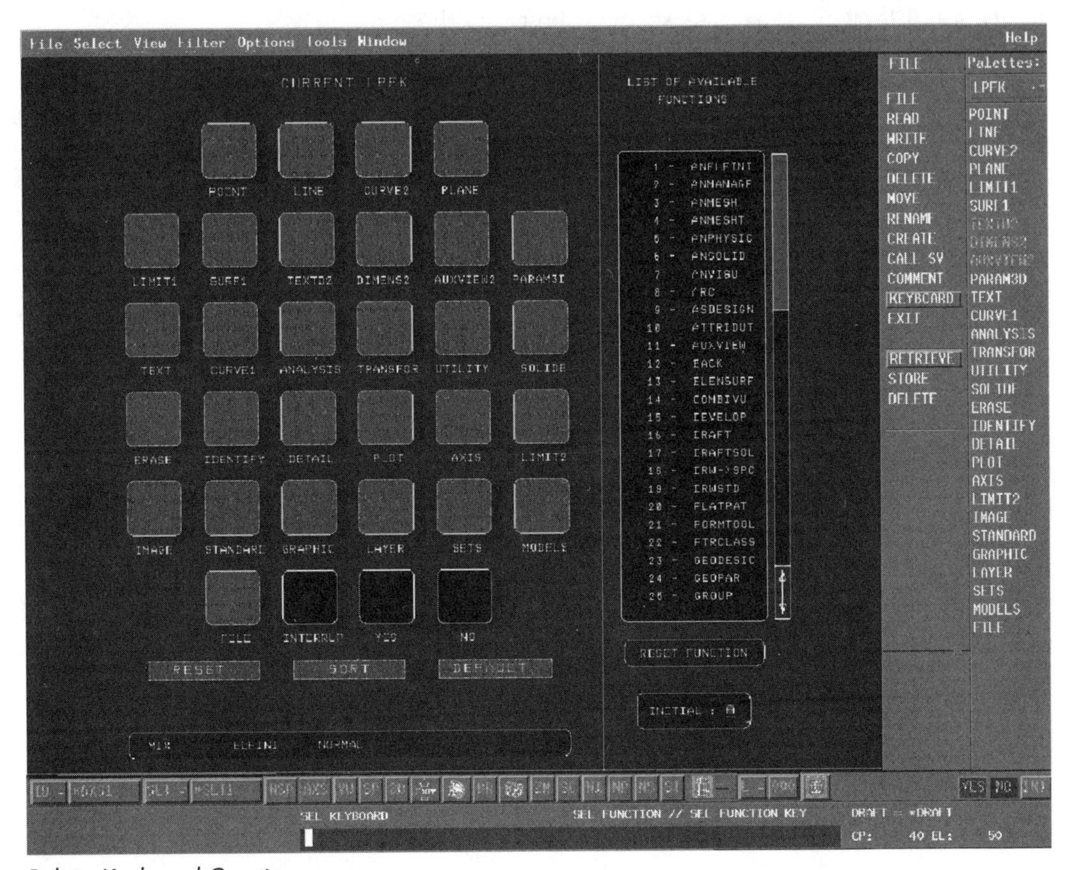

Palette Keyboard Creation screen.

Retrieve keyboard

Select: FILE > KEYBOARD | RETRIEVE

The following list describes the options available when the above command sequence is issued.

- •RESET. Remove all function from LPFK keys and prepares the system for the definition of a new keyboard.
- •DEFAULT. Restore default LPFK keyboard.
- •SORT. Sort allocated functions into alphabetical order.
- •RESET FUNCTION. Reset individual keys.

Retrieve a previously stored keyboard by selecting it from the list.

Store keyboard

Select: FILE > KEYBOARD | STORE

Delete stored keyboard

Select: FILE > KEYBOARD | DELETE

EXIT

End CATIA session.

Select: FILE > EXIT

Prompt: YES: END OF WORK SESSION

Click on the YES button to end the work session and again to end the execution.

GRAPHIC in DRAW Mode

Chapters 3, 5, 6, 7

The GRAPHIC function in DRAW mode is used to modify, verify, and analyze the display attributes of DRAW elements. SPACE elements displayed in DRAW views can also be modified.

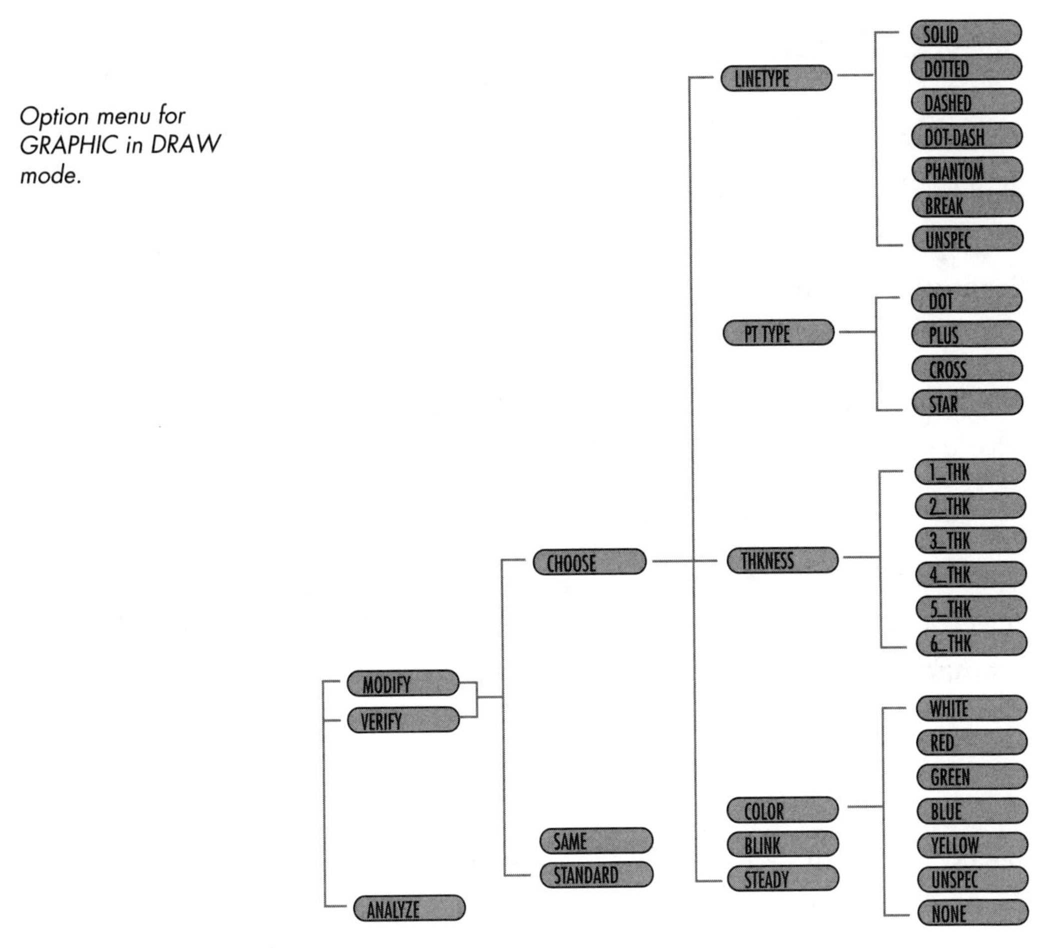

Option menu for GRAPHIC in DRAW mode.

✓ **NOTE:** *The thickness values assigned to THKNESS options are defined with the STANDARD function, where you have access to up to 32 complex linetypes. Linetype visualization may vary depending on your setup. DRESS UP options are available only when SPACE elements are displayed. The No Show/Show and No Pick/Pick options* *are toggle switches (e.g., if the No Show option is selected, the Show option displays but the working mode is No Show).*

MODIFY

Modify general display attributes.

Select: GRAPHIC > MODIFY | CHOOSE or SAME or STANDARD

Choose attribute style

Select: GRAPHIC > MODIFY | CHOOSE | LINETYPE or PT TYPE or THKNESS or COLOR or BLINK or STEADY

Modify display attributes of vector type elements

Select: GRAPHIC > MODIFY | CHOOSE | LINETYPE | Various

Prompt: WSP MULTI-SEL SEL ELEM

Use one of the following alternatives.

Select any item except UNSPEC among LINETYPE options. Select any suitable element; the element's display attributes will change to the selected linetype.

Select the UNSPEC item in the LINETYPE menu. Click on YES to display available linetypes or enter the required linetype number. Select any suitable element; the element's display attributes will change to the specified linetype.

Modify display attributes of point type elements

Select: GRAPHIC > MODIFY | CHOOSE | PT TYPE | Various

Prompt: WSP MULTI-SEL SEL SPACE ELT / / YES: SWAP

Select any item. Select a point; the element's display attributes will change to the selected point type.

Modify thickness of vector type elements

Select: GRAPHIC > MODIFY | CHOOSE | THKNESS | Various

Prompt: WSP MULTI-SEL SEL SPACE ELT / / YES: SWAP

✓ **NOTE:** *The actual thickness values assigned here are defined in the STANDARD function. Elements will appear dimmed if respective thickness is equal to or less than the visibility index set in the STANDARD function. The real thickness of elements can be viewed on screen when the THICKNESS option in the Display and Manipulation window is switched on.*

Select any item, and then select any suitable element. The thickness attribute changes.

Modify element color

Select: GRAPHIC > MODIFY | CHOOSE | COLOR

Prompt: MSELW ELEM

In the Color Chooser panel, select a color or enter a color number in the panel entry field. Select an element; the element's color will change to the one selected.

✓ **NOTE:** *The Color Chooser Panel can also be displayed by using the Change Color option in the Tools pull-down menu.*

Highlight or make elements blink

Select: GRAPHIC > MODIFY | CHOOSE | BLINK

Prompt: WSP MULTI-SEL SEL ELEM

✓ **NOTE:** *In Motif CATIA, items can no longer be made to blink. Elements selected using this option will be highlighted.*

Select any element; the element will be highlighted.

Undo blink (highlight) of elements

Select: GRAPHIC > MODIFY | CHOOSE | STEADY

Prompt: WSP MULTI-SEL SEL ELEM

Select any element previously selected using the BLINK option. The element will return to steady or unhighlighted mode.

Modify display attributes of elements by selecting other elements with desired attributes

Select: GRAPHIC > MODIFY | SAME

Prompt: REFERENCE: SEL ELEM

1. Select any element; this element's display attributes will be applied to the elements subsequently selected.
2. Select any element. Display attributes of element will be changed to be the same as the first selected element.
3. Click on YES to choose a new defining element as necessary.

Apply standard display attributes to elements

Select: GRAPHIC > MODIFY | STANDARD

Prompt: WSP MULTI-SEL SEL ELEM

Select any element. The display attributes of the selected element will be restored to attributes defined in the STANDARD function.

VERIFY

Verify display attributes.

Select: GRAPHIC > VERIFY | CHOOSE or SAME or STANDARD

Verify element display attributes

Select: GRAPHIC > VERIFY | CHOOSE | LINETYPE or PT TYPE or THKNESS or COLOR

Verify elements with same linetype display attributes

Select: GRAPHIC > VERIFY | CHOOSE | LINETYPE | Various

Prompt: YES: CONFIRM

Take one of the following alternatives.

Select any item, except UNSPEC, to define the linetype to be verified. Click on YES to display only elements with the selected linetype.

Select the UNSPEC item. Click on YES to display available linetypes, or enter linetype number. Click on YES to display only elements with selected linetype.

Verify elements with same point type display attributes

Select: GRAPHIC > VERIFY | CHOOSE | PT TYPE | Various

Prompt: YES: CONFIRM

Select any item to define the point type to be verified. Click on YES to display only points with the selected display attributes.

Verify elements with same thickness

Select: GRAPHIC > VERIFY | CHOOSE | THKNESS | Various

Prompt: YES: CONFIRM

Select an item to define the thickness to be verified. Click on YES to display only the elements with the selected thickness.

Verify elements with same color

Select: GRAPHIC > VERIFY | CHOOSE | COLOR | Various

Prompt: YES: CONFIRM

Select any item to define the color to be verified. (If UNSPEC is selected, enter color number.) Click on YES to display only the elements of the selected color.

Verify elements displayed in blinking or highlighted mode

Select: GRAPHIC > VERIFY | CHOOSE | BLINK

Prompt: YES: CONFIRM

Click on YES to display only the elements in blinking or highlighted mode.

Verify elements displayed in steady or unhighlighted mode

Select: GRAPHIC > VERIFY | CHOOSE | STEADY

Prompt: YES: CONFIRM

Click on YES to display only the elements in steady or unhighlighted mode.

Verify elements with same display attributes

Select: GRAPHIC > VERIFY | SAME

Prompt: SEL ELEM

Select any element; this element's display attributes will define the display attributes to be verified. Click on YES to display only the elements with the same display attributes as the first selected element.

Verify elements with standard display attributes

Select: GRAPHIC > VERIFY | STANDARD

Prompt: SEL ELEM

Click on YES to display only the elements with standard display attributes as defined in the STANDARD function.

ANALYZE

Present display attributes of elements in alphanumeric window.

Select: GRAPHIC > ANALYZE

Prompt: SEL ELEM

Display the alphanumeric window by pressing <Alt>+<+>, or by using the Alpha window on option in the Display and Manipulation window. Select any element; the display attributes of the selected element will appear in the alphanumeric window.

GRAPHIC in SPACE Mode

Chapter 9

The GRAPHIC function in SPACE mode is used to modify, verify, and analyze the display attributes of SPACE elements.

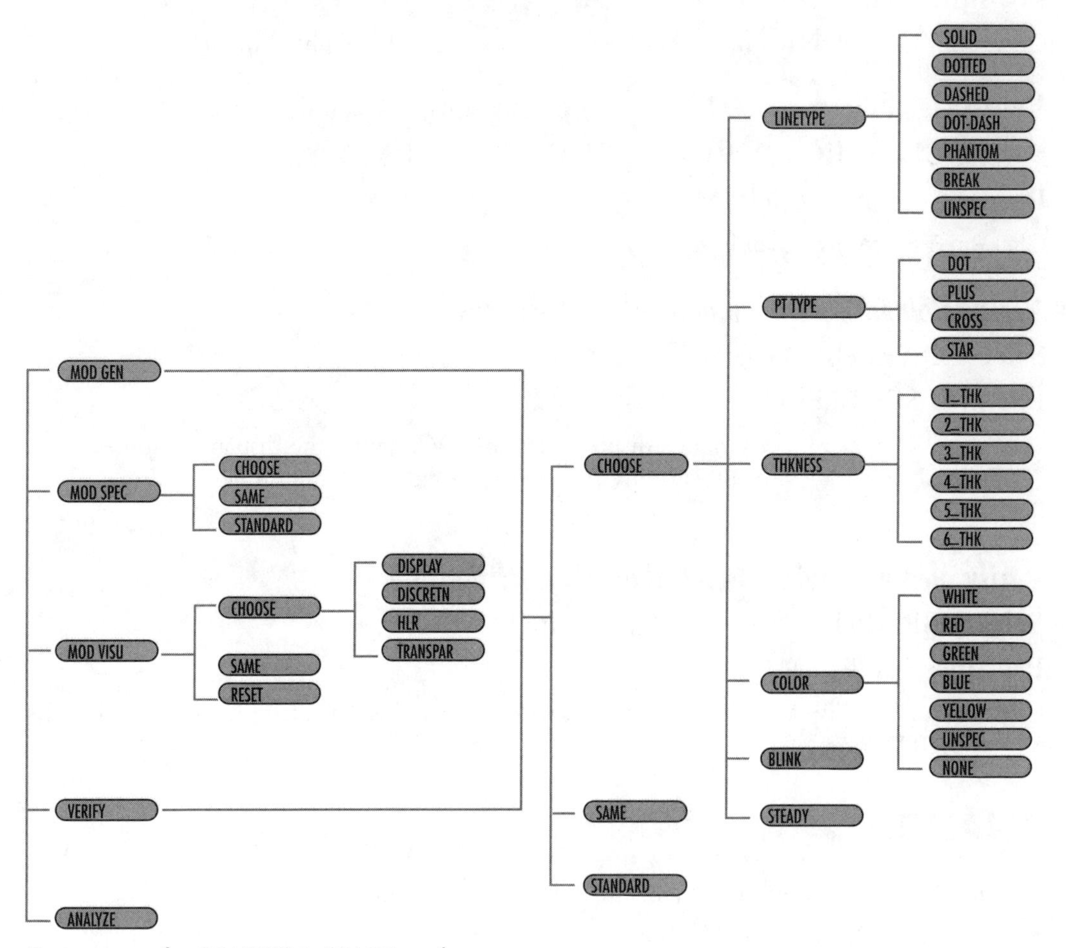

Option menu for GRAPHIC in SPACE mode.

✓ ***NOTE:*** *The thickness values assigned to MODIFY and VERIFY suboptions are defined using the STANDARD function, where you can also access up to 32 complex linetypes. Linetype visualization may vary depending on your setup. DRESS UP options are available when SPACE elements are displayed. The No Show/Show and No Pick/Pick* *options are toggle switches. For instance, if the No Show option is selected, the Show option displays but the working mode is No Show.*

MOD GEN

Modify general display attributes.

Select: GRAPHIC > MOD GEN | CHOOSE or SAME or STANDARD

Choose attribute style

Select: GRAPHIC > MOD GEN | CHOOSE | LINETYPE or PT TYPE or THKNESS or COLOR or BLINK or STEADY

Modify display attributes of vector type elements

Select: GRAPHIC > MOD GEN | CHOOSE | LINETYPE | Various

Prompt: WSP MULTI-SEL SEL ELEM

Use one of the following alternatives.

Select any item except UNSPEC among LINETYPE options, and then select any suitable element. The element's display attributes will change to the selected linetype.

Select the UNSPEC item in the LINETYPE menu. Click on YES to display available linetypes or enter the required linetype number. Select any suitable element; the element's display attributes will change to the specified linetype.

Modify display attributes of point type elements

Select: GRAPHIC > MOD GEN | CHOOSE | PT TYPE | Various

Prompt: WSP MULTI-SEL SEL SPACE ELT / / YES: SWAP

Select any item in the PT TYPE menu, and then select a point. The element's display attributes will change to the selected point type.

Modify thickness of vector type elements

Select: GRAPHIC > MOD GEN | CHOOSE | THKNESS | Various

Prompt: WSP MULTI-SEL SEL SPACE ELT / / YES: SWAP

✓ ***NOTE:*** *Thickness values assigned here are defined in the STANDARD function. Elements will appear dimmed if respective thicknesses are equal to or less than the visibility index set in the STANDARD function. The real thickness of elements can be viewed on screen when the THICKNESS option in the Display and Manipulation window is switched on.*

Select any item in THKNESS menu, and then select any suitable element. The thickness attribute changes.

Modify element color

Select: GRAPHIC > MOD GEN | CHOOSE | COLOR

Prompt: MSELW ELEM

In the Color Chooser panel, select a color or enter a color number in the panel entry field. Select an element; the element's color will change to the one selected.

✓ **NOTE:** *The Color Chooser Panel can also be activated by using the Change Color option in the Tools pull-down menu.*

Highlight or make elements blink

Select: GRAPHIC > MOD GEN | CHOOSE | BLINK

Prompt: WSP MULTI-SEL SEL ELEM

✓ **NOTE:** *In Motif CATIA, items can no longer be made to blink. Elements selected using this option will be highlighted.*

Select any element; the element will be highlighted.

Undo blink (highlight) of elements

Select: GRAPHIC > MOD GEN | CHOOSE | STEADY

Prompt: WSP MULTI-SEL SEL ELEM

Select any element previously selected using the BLINK option. The element will return to steady or unhighlighted mode.

Modify element display attributes by selecting other elements with desired attributes

Select: GRAPHIC > MOD GEN | SAME

Prompt: REFERENCE: SEL ELEM

1. Select any element; this element's display attributes will be applied to the elements subsequently selected.
2. Select any element. Element display attributes become the same as the first selected element.
3. Click on YES to choose a new defining element as necessary.

Apply standard display attributes to elements

Select: GRAPHIC > MOD_GEN | STANDARD

Prompt: WSP MULTI-SEL SEL ELEM

Select any element. The display attributes of the selected element will be restored to attributes defined in the STANDARD function.

MOD SPEC

Modify specific element attributes.

Select: GRAPHIC > MOD SPEC | CHOOSE or SAME or STANDARD

Display graphic attributes of elements

Select: GRAPHIC > MOD SPEC | CHOOSE

Prompt: MSELW ELEM

Select a single element or multiselect elements to activate a list of available element types. Select an element from the list. Another window containing element attributes as set in the STANDARD function displays.

Apply display attributes to elements using other elements

Select: GRAPHIC > MOD SPEC | SAME

Prompt: REFERENCE: SEL ELEM

Select a reference element. Click on YES to display attributes as necessary. Select one or more elements whose attributes will be modified to match the reference element.

Apply standard display attributes to elements

Select: GRAPHIC > MOD SPEC | STANDARD

Prompt: MSELW ELEM

Select one or more elements. Standard display attributes will be applied.

MOD VISU

Modify visualization mode.

Select: GRAPHIC > MOD VISU | CHOOSE or SAME or RESET

Choose specific entities for display mode modification

Select: GRAPHIC > MOD VISU | CHOOSE | DISPLAY or DISCRETN or HLR or TRANSPAR

Modify display mode type

Select: GRAPHIC > MOD VISU | CHOOSE | DISPLAY

Prompt: MSELW ELEM

The Display Mode window appears. With NHR selected, selected elements will be displayed in no hidden line removal mode. With CURRENT DISPLAY MODE SENSITIVE selected, selected elements display at current display mode settings.

Modify discretization

Select: GRAPHIC > MOD VISU | CHOOSE | DISCRETN

The Discretization window displays. Enter a value for SAG (deviation of faceted curves from true geometry), and then enter a value for the STEP (maximum length of facet lines).

Modify hidden line removal

Select: GRAPHIC > MOD VISU | CHOOSE | HLR

The Process Option window displays. When CHOOSE PARAMATERS is selected, the Specific Parameters window appears. In the latter window, special HLR display features can be defined and applied (but only in nondynamic display modes). Click on the BR button to apply new settings.

When RESET_TO_STANDARD is selected, selected elements will be restored to standard HLR settings after clicking on the BR button.

With SAME_ELEMENT selected, proceed to select an element with special HLR display features as the reference. Select other elements to which these features will be applied.

When RESTORE_SPECIFIC is selected, select an element made transparent to the current standard. Special display attributes will be restored.

Modify transparency

Select: GRAPHIC > MOD VISU | CHOOSE | TRANSPAR

✓ ***NOTE:*** *This option is useful only in shaded mode.*

With the YES option selected, selected elements will appear transparent.

When the NO option is selected, elements previously made transparent are made nontransparent.

Apply display parameters of element to other elements

Select: GRAPHIC > MOD VISU | SAME

Prompt: REFERENCE: SEL ELEM

Select a reference element. Select other elements whose display parameters will be modified to match the reference element.

Reset all display parameters to default settings

Select: GRAPHIC > MOD VISU | RESET

Prompt: MSELW ELEM

Select an element. All display parameters are reset.

VERIFY

Verify display attributes.

Select: GRAPHIC > VERIFY | CHOOSE or SAME or STANDARD

Verify element display attributes

Select: GRAPHIC > VERIFY | CHOOSE | LINETYPE or PT TYPE or THKNESS or COLOR

Verify elements with same linetype display attributes

Select: GRAPHIC > VERIFY | CHOOSE | LINETYPE | Various

Prompt: YES: CONFIRM

Use one of the following alternatives.

Select any item among LINETYPE options except UNSPEC to define the linetype to be verified. Click on YES to display only elements with the selected linetype.

Select the UNSPEC item in LINETYPE menu. Click on YES to display available linetypes, or enter linetype number. Click on YES to display only elements with the selected linetype.

Verify elements with same point type display attributes

Select: GRAPHIC > VERIFY | CHOOSE | PT TYPE | Various

Prompt: YES: CONFIRM

Select an item in the PT TYPE menu to define the point type to be verified. Click on YES to display only points with the selected display attributes.

Verify elements with same color

Select: GRAPHIC > VERIFY | CHOOSE | COLOR | Various

Prompt: YES: CONFIRM

Select any item in the COLOR menu to define the color to be verified. (If UNSPEC is selected, enter color number.) Click on YES to display only the elements of the selected color.

Verify elements with same thickness

Select: GRAPHIC > VERIFY | CHOOSE | THKNESS | Various

Prompt: YES: CONFIRM

Select any item in the THKNESS menu to define the thickness to be verified. Click on YES to display only the elements with the selected thickness.

Verify elements displayed in blinking or highlighted mode

Select: GRAPHIC > VERIFY | CHOOSE | BLINK

Prompt: YES: CONFIRM

Click on YES to display only the elements in blinking or highlight mode.

Verify elements displayed in steady or unhighlighted mode

Select: GRAPHIC > VERIFY | CHOOSE | STEADY

Prompt: YES: CONFIRM

Click on YES to display only the elements in steady or unhighlighted mode.

Verify elements with same display attributes

Select: GRAPHIC > VERIFY | SAME

Prompt: SEL ELEM

Select any element; this element's display attributes will define the display attributes to be verified. Click on YES to display only the elements with the same display attributes as the first selected element.

Verify elements with standard display attributes

Select: GRAPHIC > VERIFY | STANDARD

Prompt: SEL ELEM

Click on YES to display only the elements with standard display attributes as defined in the STANDARD function.

ANALYZE

Present display attributes of elements in alphanumeric window.

Select: GRAPHIC > ANALYZE

Prompt: SEL ELEM

Display the alphanumeric window by pressing <Alt>+<+>, or by using the Alpha window on option in the Display and Manipulation window. Select any element; the display attributes of the selected element will be displayed in the alphanumeric window.

GROUP in DRAW Mode

Chapter 5

The GROUP function in DRAW mode is used to group elements for common processing.

Option menu for GROUP in DRAW mode.

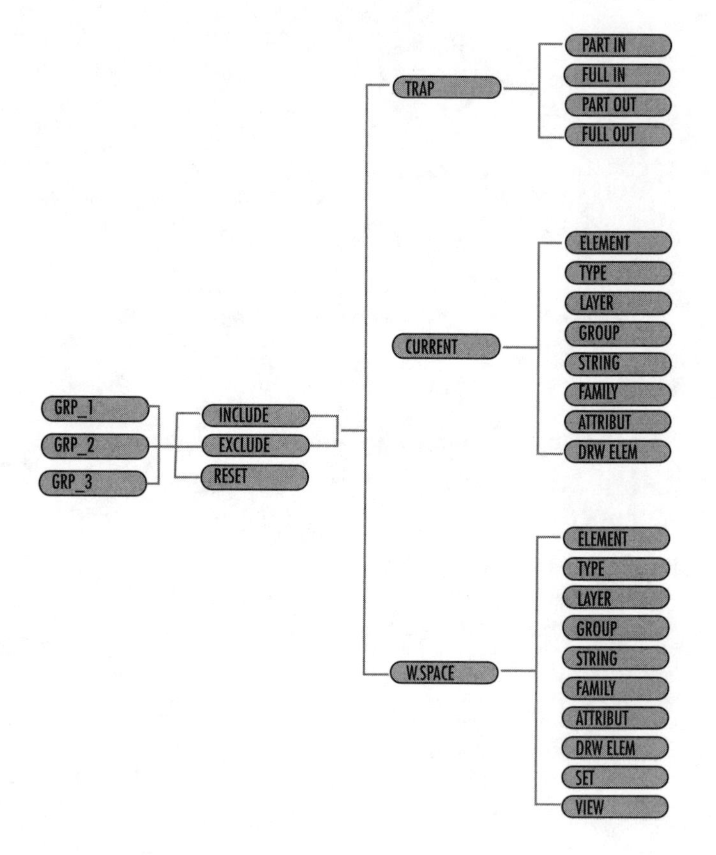

GRP_1, GRP_2, GRP_3

Specify group.

Select: GROUP > GRP_1 or GRP_2 or GRP_3 | INCLUDE or EXCLUDE or RESET

Include elements in selected group

Select: GROUP > GRP_1 or GRP_2 or GRP_3 | INCLUDE | TRAP or CURRENT or W.SPACE

Exclude elements in selected group

Select: GROUP > GRP_1, 2 or 3 | EXCLUDE | TRAP or CURRENT or W.SPACE

✓ **NOTE:** *All methods of selecting elements that follow are available whether the INCLUDE or EXCLUDE option is selected.*

Enclose or trap group of elements by indicating points around group

Select: GROUP > GRP_1 or GRP_2 or GRP_3 | INCLUDE or EXCLUDE | TRAP | PART IN or FULL IN or PART OUT or FULL OUT

1. Key in two or more points to form the trap. Two points will create a rectangular trap (the points represent the rectangle's diagonal corners). Keying in more than two points generates a polygonal trap. If a point is indicated incorrectly, click on the NO button to cancel it.
2. Click on the YES button once to close the trap and once to transfer the elements into the selected group. The elements will temporarily disappear.

The following sections describe specific types of traps.

Select elements wholly inside or overlapping trap boundary

Select: GROUP > GRP_1 or GRP_2 or GRP_3 | INCLUDE or EXCLUDE | TRAP | PART IN

Prompt: IND 1ST PT

Select elements wholly inside trap

Select: GROUP > GRP_1 or GRP_2 or GRP_3 | INCLUDE or EXCLUDE | TRAP | FULL IN

Prompt: IND 1ST PT

Select elements wholly outside or overlapping trap boundary

Select: GROUP > GRP_1 or GRP_2 or GRP_3 | INCLUDE or EXCLUDE | TRAP | PART OUT

Prompt: IND 1ST PT

Select elements wholly outside trap boundary

Select: GROUP > GRP_1 or GRP_2 or GRP_3 | INCLUDE or EXCLUDE | TRAP | FULL OUT

Prompt: IND 1ST PT

Select current set and view or entire workspace

Select: GROUP > GRP_1 or GRP_2 or GRP_3 | INCLUDE or EXCLUDE | CURRENT or W.SPACE

Transfer individual elements

Select: GROUP > GRP_1 or GRP_2 or GRP_3 | INCLUDE or EXCLUDE | CURRENT or W.SPACE | ELEMENT

Prompt: SEL ELEM

Transfer elements of specified types

Select: GROUP > GRP_1 or GRP_2 or GRP_3 | INCLUDE or EXCLUDE | CURRENT or W.SPACE | TYPE

Prompt: SEL ELEM // KEY TYPE

Transfer elements in selected layer

Select: GROUP > GRP_1 or GRP_2 or GRP_3 | INCLUDE or EXCLUDE | CURRENT or W.SPACE | LAYER

Prompt: YES: CUR LAYERSEL ELEM // KEY LAYER

Transfer elements in group containing selected element

Select: GROUP > GRP_1 or GRP_2 or GRP_3 | INCLUDE or EXCLUDE | CURRENT or W.SPACE | GROUP

Prompt: SEL ELEM // KEY NUM

Transfer elements with identifier containing specified string

Select: GROUP > GRP_1 or GRP_2 or GRP_3 | INCLUDE or EXCLUDE | CURRENT or W.SPACE | STRING

Prompt: KEY STRING

Transfer elements of selected element family

Select: GROUP > GRP_1 or GRP_2 or GRP_3 | INCLUDE or EXCLUDE | CURRENT or W.SPACE | FAMILY

Prompt: SEL ELEM

Transfer elements sharing attributes of selected element

Select: GROUP > GRP_1 or GRP_2 or GRP_3 | INCLUDE or EXCLUDE | CURRENT or W.SPACE | ATTRIBUT

Prompt: SEL ELEM // YES: CRITERIA

Other selections are necessary but are beyond the scope of this book.

Transfer DRAW elements

Select: GROUP > GRP_1 or GRP_2 or GRP_3 | INCLUDE or EXCLUDE | CURRENT or W.SPACE | DRW ELEM

Prompt: YES: CONFIRM

Transfer elements of selected set

Select: GROUP > GRP_1 or GRP_2 or GRP_3 | INCLUDE or EXCLUDE | W.SPACE | SET

Prompt: SEL SET

Transfer elements of view containing selected set

Select: GROUP > GRP_1 or GRP_2 or GRP_3 | INCLUDE or EXCLUDE | W.SPACE | VIEW

Prompt: SEL VIEW

Reset or empty selected group

Select: GROUP > GRP_1, 2 or 3 | RESET

GROUP in SPACE Mode

Chapter 13

The GROUP function in SPACE mode is used to group elements for common processing.

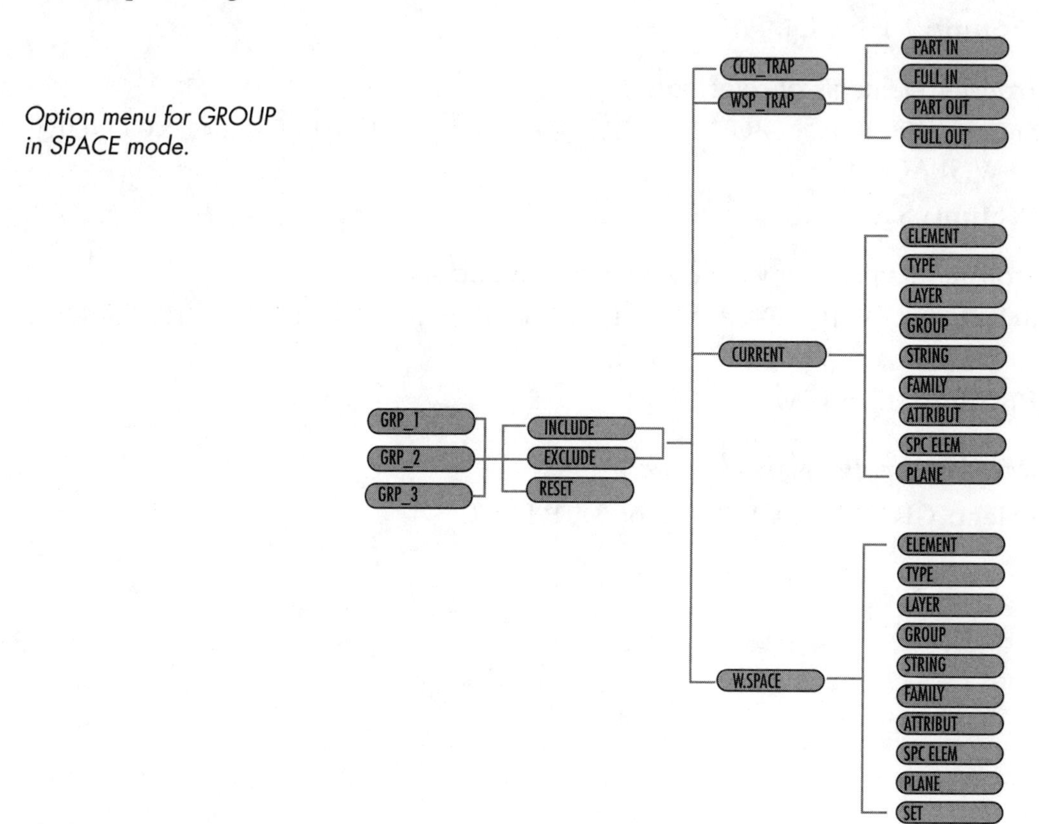

Option menu for GROUP in SPACE mode.

GRP_1, GRP_2, GRP_3

Specify group.

Select: GROUP > GRP_1 or GRP_2 or GRP_3 | INCLUDE or EXCLUDE or RESET

Include elements in selected group

Select: GROUP > GRP_1 GRP_2 or GRP_3 | INCLUDE | CUR_TRAP or WSP_TRAP or CURRENT or W.SPACE

Exclude elements in selected group

Select: GROUP > GRP_1, 2 or 3 | EXCLUDE | CUR_TRAP or WSP_TRAP or CURRENT or W.SPACE

✓ **NOTE:** *All methods of selecting elements that follow are available whether the INCLUDE or EXCLUDE option is selected.*

Enclose or trap group of elements by indicating points around group

Select: GROUP > GRP_1 or GRP_2 or GRP_3 | INCLUDE or EXCLUDE | CUR_TRAP or WSP_TRAP | PART IN or FULL IN or PART OUT or FULL OUT

Use CUR_TRAP to select elements only in the current set. In contrast, use WSP_TRAP if you wish to select all SPACE elements. (Note that VOL, DIT, and SOL elements are not always selectable using this method; other selection methods are available.)

1. Key in two or more points to form the trap. Selecting two points creates a rectangular trap (the points represent the rectangle's diagonal corners). Keying in more than two points generates a polygonal trap. If a point is indicated incorrectly, click on the NO button to cancel it.
2. Click on the YES button once to close the trap and once to transfer the elements into the selected group. The elements will temporarily disappear.

The following sections describe specific types of traps.

Select elements wholly inside or overlapping trap boundary

Select: GROUP > GRP_1 or GRP_2 or GRP_3 | INCLUDE or EXCLUDE | CUR_TRAP or WSP_TRAP | PART IN

Prompt: IND 1ST PT

Select elements wholly inside trap

Select: GROUP > GRP_1 or GRP_2 or GRP_3 | INCLUDE or EXCLUDE | CUR_TRAP or WSP_TRAP | FULL IN

Prompt: IND 1ST PT

Select elements wholly outside or overlapping trap boundary

Select: GROUP > GRP_1 or GRP_2 or GRP_3 | INCLUDE or EXCLUDE | CUR_TRAP or WSP_TRAP | PART OUT

Prompt: IND 1ST PT

Select elements wholly outside trap boundary

Select: GROUP > GRP_1 or GRP_2 or GRP_3 | INCLUDE or EXCLUDE | CUR_TRAP or WSP_TRAP | FULL OUT

Prompt: IND 1ST PT

Select current set and view or entire workspace

Select: GROUP > GRP_1 or GRP_2 or GRP_3 | INCLUDE or EXCLUDE | CURRENT or W.SPACE

Transfer individual elements

Select: GROUP > GRP_1 or GRP_2 or GRP_3 | INCLUDE or EXCLUDE | CURRENT or W.SPACE | ELEMENT

Prompt: SEL ELEM

Transfer elements of specified types

Select: GROUP > GRP_1 or GRP_2 or GRP_3 | INCLUDE or EXCLUDE | CURRENT or W.SPACE | TYPE

Prompt: SEL ELEM // KEY TYPE

Transfer elements in selected layer

Select: GROUP > GRP_1 or GRP_2 or GRP_3 | INCLUDE or EXCLUDE | CURRENT or W.SPACE | LAYER

Prompt: YES: CUR LAYERSEL ELEM // KEY LAYER

Transfer elements in group containing selected element

Select: GROUP > GRP_1 or GRP_2 or GRP_3 | INCLUDE or EXCLUDE | CURRENT or W.SPACE | GROUP

Prompt: SEL ELEM // KEY NUM

Transfer elements with identifier containing specified string

Select: GROUP > GRP_1 or GRP_2 or GRP_3 | INCLUDE or EXCLUDE | CURRENT or W.SPACE | STRING

Prompt: KEY STRING

Transfer elements of selected element family

Select: GROUP > GRP_1 or GRP_2 or GRP_3 | INCLUDE or EXCLUDE | CURRENT or W.SPACE | FAMILY

Prompt: SEL ELEM

Transfer elements sharing attributes of selected element

Select: GROUP > GRP_1 or GRP_2 or GRP_3 | INCLUDE or EXCLUDE | CURRENT or W.SPACE | ATTRIBUT

Prompt: SEL ELEM // YES: CRITERIA

Other selections are necessary but are beyond the scope of this book.

Transfer SPACE elements

Select: GROUP > GRP_1 or GRP_2 or GRP_3 | INCLUDE or EXCLUDE | CURRENT or W.SPACE | SPC ELEM

Prompt: YES: CONFIRM

Transfer elements lying on selected 3D plane

Select: GROUP > GRP_1 or GRP_2 or GRP_3 | INCLUDE or EXCLUDE | W.SPACE | PLANE

Prompt: SEL PLN / PLANAR FACE

Transfer elements of selected set

Select: GROUP > GRP_1 or GRP_2 or GRP_3 | INCLUDE or EXCLUDE | W.SPACE | SET

Prompt: SEL SET

Reset or empty selected group

Select: GROUP > GRP_1, 2 or 3 | RESET

IDENTIFY in DRAW and SPACE Modes

Chapter 18

The IDENTIFY function in DRAW and SPACE modes is used to manage element identifiers.

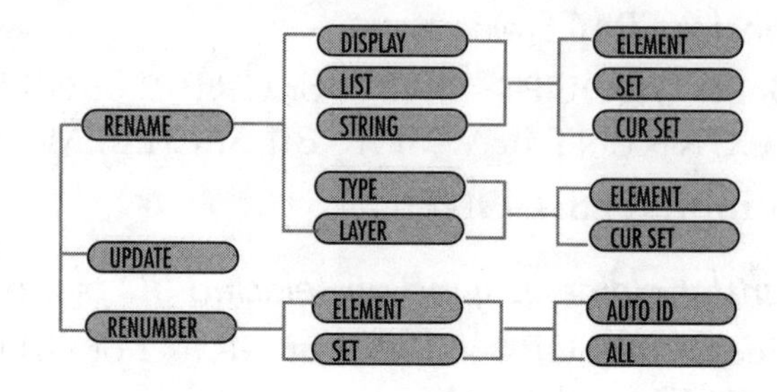

Option menu for IDENTIFY in DRAW and SPACE modes.

RENAME

Rename element identifiers.

Select: IDENTIFY > RENAME | DISPLAY or LIST or STRING

Rename selected identifiers

Select: IDENTIFY > RENAME | DISPLAY | ELEMENT or SET or CUR SET

Prompt: SEL ELEM

1. Select an element or set; the identifier of the element or set is displayed in the information entry area.
2. Key in a new name and press <Enter>.

Rename identifiers selected from list

Select: IDENTIFY > RENAME | LIST | ELEMENT or SET or CUR SET

Prompt: SEL ELEM

1. Select an element or set from the list; the identifier of the element or set is displayed in the information entry area.
2. Key in a new name and press <Enter>.

Rename identifiers according to string

Select: IDENTIFY > RENAME | STRING | ELEMENT or SET or CUR SET

Prompt: KEY STRING

1. Key in a string pertaining to an element identifier; the list of identifiers including the string you specified displays. Select from the list. The selected name displays in the information entry area.
2. Key in a new name and press <Enter>.

Rename identifiers by type

Select: IDENTIFY > RENAME | TYPE | ELEMENT or CUR SET

Prompt: SEL ELEM // KEY TYPE

1. Select an element or key in the name of an element type (e.g., LN or CRV). The list of identifiers of that element type or set displays. Select from the list. The selected name displays in the information entry area.
2. Key in a new name and press <Enter>.

Switch layers

Select: IDENTIFY > RENAME | LAYER | ELEMENT or CUR SET

Prompt: SEL ELEM KEY LAYER // YES: CUR LAYER

1. Select an element within a layer of your choice or enter the layer number.
2. Click on YES to work in the selected layer.

UPDATE

Update element identifiers after merging.

Select: IDENTIFY > UPDATE

Prompt: YES: ACCEPT

Click on YES to update element identifiers merged from extended to standard form.

RENUMBER

Renumber entire list of identifiers in model.

Select: IDENTIFY > RENUMBER | ELEMENT or SET | AUTO ID or ALL

Renumber identifiers

Select: IDENTIFY > RENUMBER | ELEMENT or SET | AUTO ID

Prompt: YES: ACCEPT

Click on YES to renumber all elements.

Retrieve identifiers from deleted elements

Select: IDENTIFY > RENUMBER | ELEMENT or SET | ALL

Prompt: YES: ACCEPT

Click on YES to retrieve all identifiers that became available when elements were deleted.

IMAGE in DRAW and SPACE Modes

Chapters 5, 13, 14

The IMAGE function in DRAW and SPACE modes is used to manage the display of elements in a model through the use of windows and groups of windows called "screens."

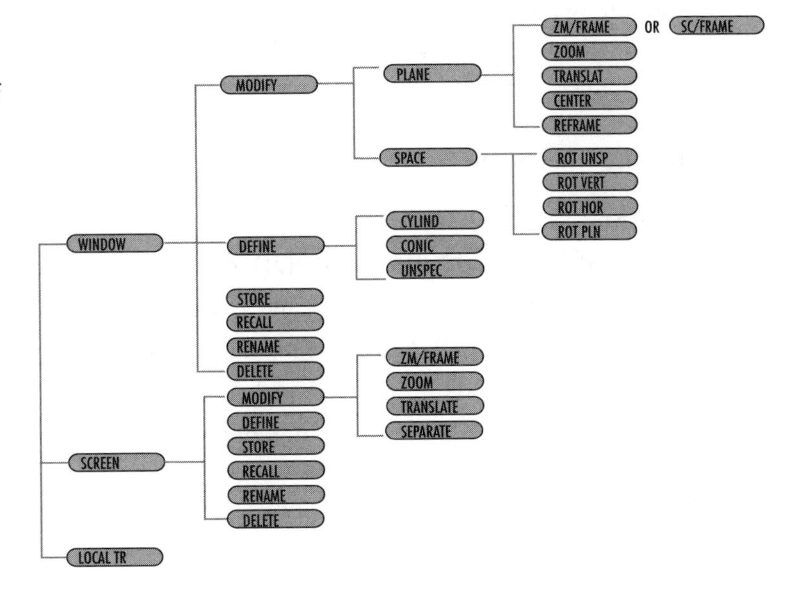

Option menu for IMAGE function in DRAW and SPACE modes.

WINDOW

Manage windows.

Select: IMAGE > WINDOW | MODIFY or DEFINE or STORE

Modify windows

Select: IMAGE > WINDOW | MODIFY | PLANE or SPACE

Modify window display in screen plane

Select: IMAGE > WINDOW | MODIFY | PLANE | SC/FRAME or ZM/FRAME or ZOOM or TRANSLAT or CENTER or REFRAME

Zoom window frame

Select: IMAGE > WINDOW | MODIFY | PLANE | SC/FRAME or ZM/FRAME

Prompt: IND 1st PT // SEL ELEM // KEY ZOOM

If there is only one window in the current screen, ZM/FRAME will be the only available option. When multiple windows display, the SC/FRAME option is available. In this instance, select the desired window.

1. Indicate two points; the image will zoom so that these points are at the edge of the visible area and arranged symmetrically about the center.
2. Select an element; the image will move so that the median point of the selected element is at the center of the screen.
3. Key in a new zoom factor and press <Enter>.

Zoom window

Select: IMAGE > WINDOW | MODIFY | PLANE | ZOOM

Prompt: KEY ZOOM // IND ZOOM // SEL ELEM

When multiple windows display, select the desired window.

1. Key in a new scale or zoom factor and press <Enter>.
2. Indicate a point above the center line of the screen to zoom out. Indicate a point below the center line of the screen to zoom in.
3. Select an element; the image will move so that the median point of the selected element is at the center of the screen.

Move display in window

Select: IMAGE > WINDOW | MODIFY | PLANE | TRANSLATE

Prompt: IND POINT // SEL ELEM

1. Indicate two points. The image will be moved along the translation defined by the two points.
2. Select two elements. The median point of the first selected element will be orthographically projected onto the second selected element defining a translation. The image will be moved along the translation.

Center window axis

Select: IMAGE > WINDOW | MODIFY | PLANE | CENTER

Prompt: SELELEM // IND PT // YES: CENTER

- If there is only one window, click on the YES button to center the image.

- If there are several windows, click on the YES button to center the axis of each window into the center of the window. Alternatively, define the window and then click on the YES button to center only the selected window.

Select an element in any window; the image will be centered on the median point of the selected element.

Reframe window

Select: IMAGE > WINDOW | MODIFY | PLANE | REFRAME

Prompt: YES: REFRAME

Click on the YES button to reframe the window to extents.

✓ ***NOTE:*** *This option is not available for a conic window.*

Manipulate window in space

Select: IMAGE > WINDOW | MODIFY | SPACE | ROT UNSP or ROT VERT or ROT HOR or ROT PLN

Unspecified rotation

Select: IMAGE > WINDOW | MODIFY | SPACE | ROT UNSP

Select an element, and then key in an angle and press <Enter>. The image will be rotated by the specified angle around the selected element. Click on the YES button to repeat.

Vertical rotation

Select: IMAGE > WINDOW | MODIFY | SPACE | ROT VERT

Select a point, and then key in an angle and press <Enter>. The image is rotated about a line passing through the selected point parallel to the Y axis of a virtual axis system and parallel to the edges of the screen. Click on the YES button to repeat.

Horizontal rotation

Select: IMAGE > WINDOW | MODIFY | SPACE | ROT HOR

Select a point, and then key in an angle and press <Enter>. The image is rotated about a line passing through the selected point parallel to the X axis of a virtual axis system and parallel to the edges of the screen. Click on the YES button to repeat.

Rotate plane

Select: IMAGE > WINDOW | MODIFY | SPACE | ROT PLN

Key in the rotation angle and press <Enter>. Alternatively, indicate or select a point. The image rotates until the selected point coincides with the right side of the horizontal passing through the center of the window.

Define windows

Select: IMAGE > WINDOW | DEFINE | CYLIND or CONIC or UNSPEC

Define cylindrical projection window

Select: IMAGE > WINDOW | DEFINE | CYLIND

1. If there are several windows in the current screen, define the window to be modified.
2. Key in one of the following window identifiers: YZ, XZ, or XY from the 3D axis system; P for a window parallel to the screen; or D for a single draw window.
3. Select a PLN type element. The window will be the view along a line normal to the plane and its center will be the center of the selected plane.
4. Select an LN CRV type element. The line of sight will coincide with the selected vector.
5. Key in components of the line of sight. In each instance, click on the YES button to invert the line of sight.

Define conic projection window

Select: IMAGE > WINDOW | DEFINE | CONIC

1. If there are several windows in the current screen, define the window to be modified.
2. Select a point to be the observer's eye, and then select a plane. The plane moves so as to be parallel to the screen and at center.
3. Select a line or curve. The line of sight will coincide with the selected vector.
4. Key in components of the line of sight. In each instance, click on the YES button to invert the line of sight.

✓ **NOTE:** *The projection plane is 500 mm from the observer's eye.*

Define unspecified projection window

Select: IMAGE > WINDOW | DEFINE | UNSPEC

1. Define the window to be modified as necessary.
2. Key in length for the NX, NY, and NZ vectors, and angles for the axis system AX, AY, and AZ.
3. Click on the YES button to iterate.

Store windows

Select: IMAGE > WINDOW | STORE

If the screen consists of several windows, define the window to be stored. Key in an alpha-numeric name for the window and press <Enter>.

Recall windows

Select: IMAGE > WINDOW | RECALL

Key in a string or enter a blank field to display list of existing windows. Select a window from the list.

Rename windows

Select: IMAGE > WINDOW | RENAME

A list of existing windows is displayed.

Select a window from the list. Key in a new name for the window and press <Enter>.

Delete windows

Select: IMAGE > WINDOW | DELETE

A list of available windows is displayed.

- Select a window from the list; the window will be deleted.
- Click on the YES button to delete all windows.

SCREEN

Manage groups of windows called screens.

Select: IMAGE > SCREEN | MODIFY or DEFINE or STORE or RECALL or RENAME or DELETE

Modify screens

Select: IMAGE > SCREEN | MODIFY | ZM/FRAME or ZOOM or TRANSLAT or SEPARATE

Zoom screen frame

If there is only one window in the current screen, the available options are the same as for WNDOW | MODIFY | PLANE.

Select: IMAGE > SCREEN | MODIFY | ZM/FRAME (zoom frame)

Prompt: IND 1st PT // SEL ELEM // KEY ZOOM

1. Indicate two points. The model will zoom so that the points are at the edge of the visible area symmetrically about the center.
2. Select an element. The model moves so that the median point of the selected element is at the center of the screen.
3. Key in a new zoom factor and press <Enter>.

Zoom screen

Select: IMAGE > SCREEN | MODIFY | ZOOM

Prompt: KEY ZOOM // IND ZOOM | // SEL ELEM

1. Key in a new zoom factor and press <Enter>.
2. Indicate a point above the center line of the screen to zoom out, and then indicate a point below the center line of the screen to zoom in.
3. Select an element. The model moves so that the median point of the selected element is at the center of the screen.

Translate screen

Select: IMAGE > SCREEN | MODIFY | TRANSLAT

Prompt: IND PT // SEL ELEM

1. Indicate two points. The model will be moved along the translation defined by the two points.
2. Select two elements. The median point of the first selected element will be orthographically projected onto the second selected element defining a translation. The model will be moved along the translation.

Separate screen

Select: IMAGE > SCREEN | MODIFY | SEPARATE

Prompt: IND PT // SEL ELEM // YES: MID SCR

1. Indicate a point. The separation lines move to the selected point.
2. Select an element. The separation lines move to the median point of the selected element.
3. Click on the YES button to center the seperation lines.

Define screen

Select: IMAGE > SCREEN | DEFINE

Prompt: SEL CONFIG // SEL STD SCREEN YES: DISPLAY CURRENT SCREEN

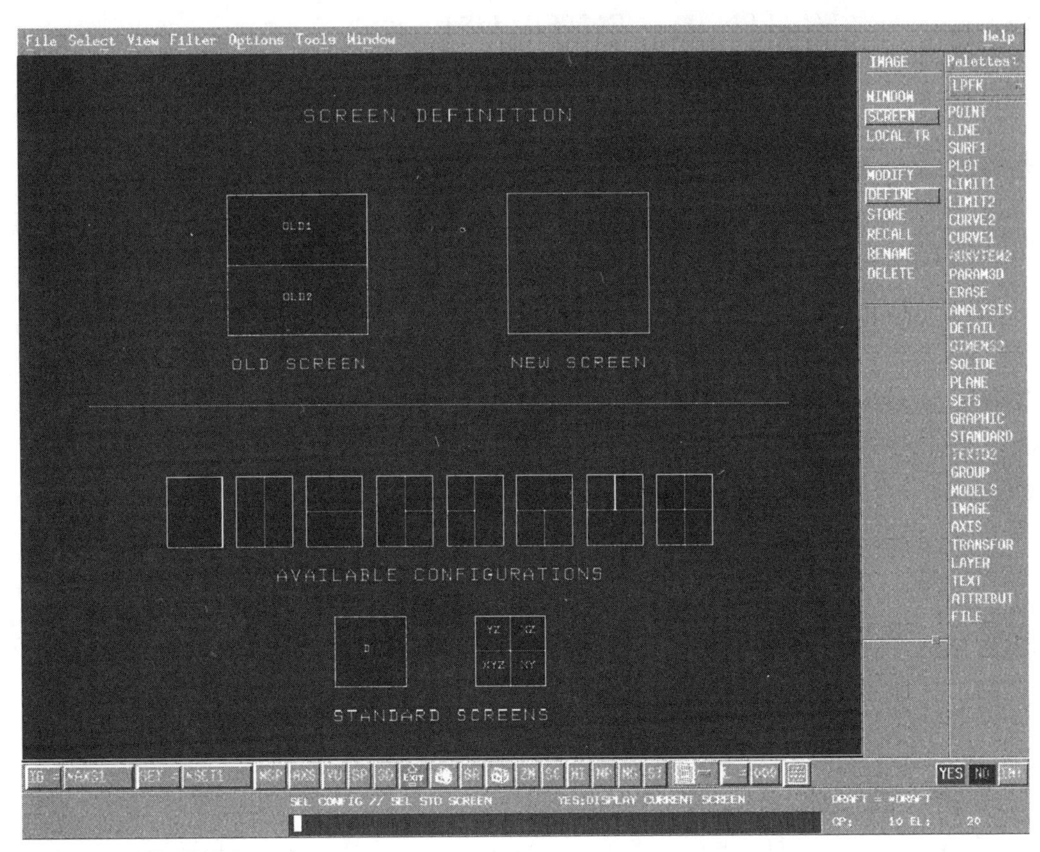

SCREEN DEFINITION window.

1. In the SCREEN DEFINITION window, select from the displayed standard screen configurations.
2. Select stored windows or standard windows for each window in your chosen screen layout.
3. Click on the YES button to confirm the screen creation.

Store screen

Select: IMAGE > SCREEN | STORE

Prompt: KEY SCREEN ID

Key in a name for the screen storage file and press <Enter>.

Recall screen

Select: IMAGE > SCREEN | RECALL

Prompt: KEY SCREEN ID // ENTER: LIST

Key in the name of the screen you want to recall. Enter a blank field to view the list of stored screens, and then select from the list.

Rename screen

Select: IMAGE > SCREEN | RENAME

Prompt: SEL SCREEN

From the displayed list select the screen to be renamed. Key in the new name and press <Enter>.

Delete screen

Select: IMAGE > SCREEN | DELETE

Prompt: SEL SCREEN // YES: ALL

From the displayed list select the screen to be deleted. Alternatively, click on the YES button to delete all screens.

LOCAL TR

Perform local transformations.

Select: IMAGE > LOCAL TR

Prompt: LOCAL TRANSFORMATIONS

The facility to manipulate the model using the local transformation tools (dials, spaceball, arrow keys, and so forth) is locked upon selection of the IMAGE function. The facility can be made available without leaving the function by selecting the LOCAL TR option from the menu.

IUA in DRAW and SPACE Modes

Interactive user access (IUA) commands are procedures that access model interface routines. IUA commands can be transparently selected (i.e., while working within a CATIA function) by keying in */M* followed by an IUA command name (e.g., */M AND2PT* to perform a relative analysis between two points), or they can be executed from the IUA function.

Before IUA commands can be selected transparently, certain operations (see FILE option below) must be performed from the IUA function. The IUA function is used to initiate the execution of an IUA command, and to display, modify, and define an IUA procedure.

Option menu for IUA function.

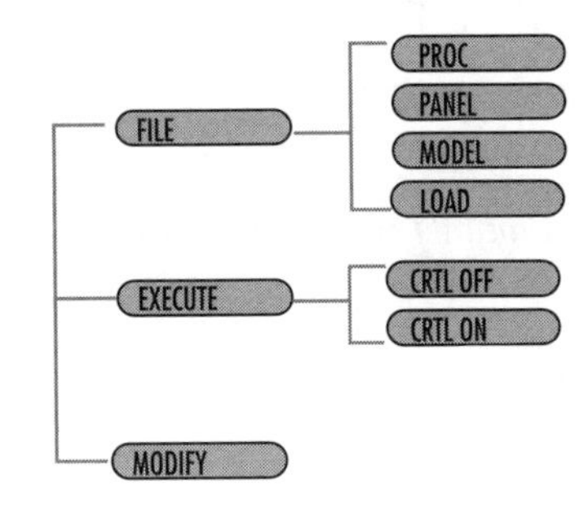

FILE

Define user files. User files must be loaded before an IUA command can be executed.

Select: IUA > FILE | PROC or PANEL or MODEL or LOAD

Define user procedure file

Select: IUA > FILE | PROC

Prompt: SEL FILE ID // NO : RESET

From the list of user procedure files, select the desired file to allocate it, or click on the NO button to deallocate the user file.

✓ ***NOTE:*** *Allocated files will be displayed in yellow, and cannot be selected.*

Define user panel file

Select: IUA > FILE | PANEL

Prompt: SEL FILE ID // NO : RESET

From the list of user panel files, select the desired file to allocate it, or click on the NO button to deallocate the user file.

✓ **NOTE:** *Allocated files will be displayed in yellow, and cannot be selected.*

Define user CATIA model file

Select: IUA > FILE | MODEL

Prompt: SEL FILE ID // NO : RESET

From the list of user model files select the required file.

Define user load modules file

Select: IUA > FILE | LOAD

Prompt: SEL FILE ID // NO : RESET

From the list of user load files, select the required file to allocate it, or click on the NO button to deallocate the user file.

✓ **NOTE:** *Allocated files will be displayed in yellow and cannot be selected.*

EXECUTE

Execute required IUA procedure. Options are described below.

Select: IUA > EXECUTE | CRTL OFF or CRTL ON

Prompt: KEY STRING

- CTRL OFF. The source program will not be displayed on the screen when an error is detected in the procedure.
- CTRL ON. The source program will be displayed on the screen when an error is detected in the procedure.

1. Key in a character string to display the commands containing the required string, or press <Enter> with the input information area empty to display the complete list of executable procedures.
2. Select the procedure to be executed, and follow the prompts in the message area. Alternatively, if there is a current procedure, click on the YES button to initiate execution of the procedure.
3. Click on the NO button as necessary to refuse creation of geometric elements if they were created by the procedure.

✓ **NOTE:** *When the list of procedures is displayed, a short description of usage use follows the procedure identifier.*

MODIFY

Modify IUA procedure.

Select: IUA > MODIFY

Because this option is typically available only to the system administrator, details are not provided in this book.

Using IUA Commands Transparently

IUA command procedures can be used transparently by keying in */M* followed by the command name. The list of standard available commands can be obtained from the "Interactive User Access Commands Quick Reference" book. A list of available commands can also be obtained interactively by taking the following steps.

1. Key in */M HELP* to the input information area. The list shown in the next illustration appears.

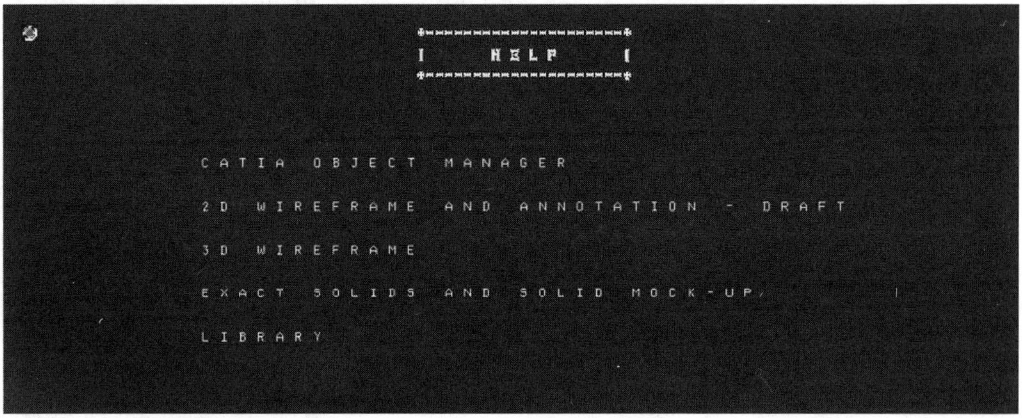

IUA Help panel opening page.

2. From the displayed list you can select a heading to narrow the search field, such as *2D WIREFRAME AND ANNOTATION - DRAFT*. The following list displays.

IUA commands available under 2D WIREFRAME AND ANNOTATION - DRAFT heading.

```
+--------------------+
| HELP   D R A F T   |
+--------------------+

A N A L Y S I S              A X I S
C U R V E 2                  D I M E N S 2
L I N E                      P A T T E R N
P O I N T                    S H A P E
T E X T D 2                  T R A N S F O R
```

3. From the displayed list you can select another heading to narrow the search field, such as *ANALYSIS*. The following list displays. When the list of procedures displays, a short description of usage follows each procedure identifier.

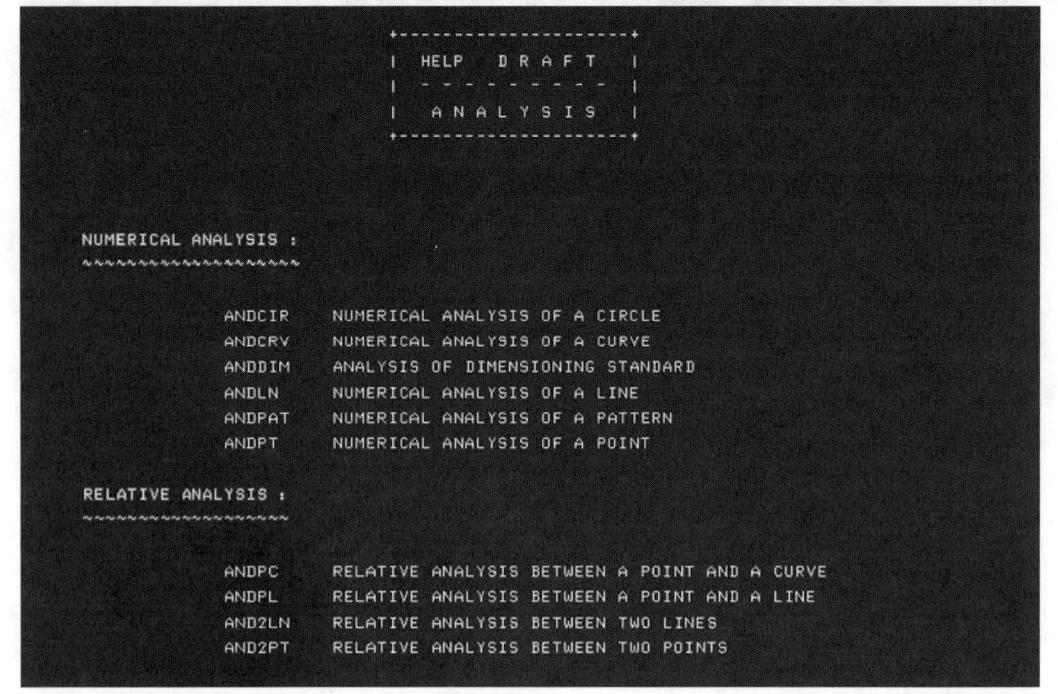

HELP D R A F T
A N A L Y S I S
NUMERICAL ANALYSIS :
ANDCIR NUMERICAL ANALYSIS OF A CIRCLE
ANDCRV NUMERICAL ANALYSIS OF A CURVE
ANDDIM ANALYSIS OF DIMENSIONING STANDARD
ANDLN NUMERICAL ANALYSIS OF A LINE
ANDPAT NUMERICAL ANALYSIS OF A PATTERN
ANDPT NUMERICAL ANALYSIS OF A POINT
RELATIVE ANALYSIS :
ANDPC RELATIVE ANALYSIS BETWEEN A POINT AND A CURVE
ANDPL RELATIVE ANALYSIS BETWEEN A POINT AND A LINE
AND2LN RELATIVE ANALYSIS BETWEEN TWO LINES
AND2PT RELATIVE ANALYSIS BETWEEN TWO POINTS

IUA commands available under ANALYSIS heading.

4. Select the required command procedure to be executed, and follow the prompts in the message area. Once the procedure is complete, you will return to the function selected prior to execution of the IUA command procedure.

KEEP in DRAW and SPACE Modes

Chapter 18

The KEEP function is used to merge passive models into the active model, and to delete elements, sets, details, or applications no longer required in the current model.

Option menu for KEEP function in DRAW and SPACE modes.

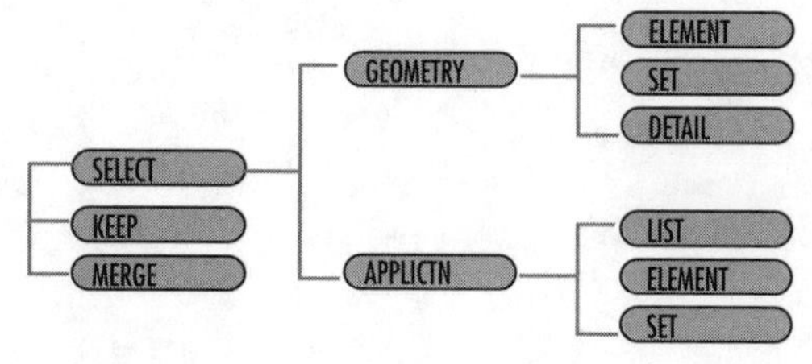

The KEEP option of the KEEP function is used to retain selected elements, sets, and details, and delete all others. Before using the KEEP option, however, selecting entities to be retained via the SELECT option is necessary. (The KEEP option is discussed after SELECT.)

Prior to selecting entities, certain general rules that apply to the entities being retained should be reviewed. The rules are listed below.

- All elements logically linked to the selected elements will be retained.
- The set to which the selected element entity belongs will be retained.
- The view to which the selected DRAW element belongs will be retained, as well as the draft to which the selected view belongs.
- When an occurrence of a detail or symbol is selected, the detail workspace and contents will be retained.
- When a set is selected the workspace to which it belongs will be retained.

SELECT

Select entities to retain.

Select: KEEP > SELECT | GEOMETRY or APPLICTN

Upon using the SELECT option, a MANAGE window appears. The window is used to control the display of the list of entities with identifiers. The following options can be selected from the window.

- Sort entities by creation date or alphabetical order.
- Display all elements, only selected elements, or only those that can be selected.
- Refine the list by using a specific search string.

- Another window displays when a list of entities is requested by clicking on the YES button as requested in the following options. Once this window displays, a SELECT ALL icon becomes available in the MANAGE window. The icon makes possible the selection of all entities displayed in the additional windows.

Select geometric elements, sets, or details

Select: KEEP > SELECT | GEOMETRY | ELEMENT or SET or DETAIL

Select elements to retain

Select: KEEP > SELECT | GEOMETRY | ELEMENT

Prompt: MSELW ELEM // YES : LIST

1. Select element(s) to retain individually or through a multiselect option (e.g., **LND* to select all DRAW mode lines). Alternatively, click on the YES button to display the list of elements and select the required elements from the window. All selected elements will be highlighted.
2. Once the elements to be retained have been selected, you can proceed to make further selections or select the KEEP > KEEP option.

Select sets to retain

Select: KEEP > | SELECT | GEOMETRY | SET

Prompt: SEL SET // YES : LIST

1. Select the set(s) to be kept. Alternatively, click on the YES button to display the list of sets, and select the required details from the window. All selected sets will be highlighted.
2. Once the set(s) to be kept have been selected, you can proceed to make further selections or to the KEEP > KEEP option.

Select details to retain

Select: KEEP > SELECT | GEOMETRY | DETAIL

Prompt: SEL WSP_SP / WSP_DR // YES : LIST

1. Select the detail(s) to be kept. Alternatively, click on the YES button to display the list of details, and select the required details from the window. All selected details will be highlighted.
2. Once the detail(s) to be kept have been selected, you can proceed to make further selections or to the KEEP > KEEP option.

✓ **NOTE:** *All above options can be used to select a combination of elements, sets, and details to be kept using KEEP > KEEP.*

Select applications and application elements and sets

Select: KEEP > SELECT | APPLICTN | LIST or ELEMENT or SET

Select applications

Select: KEEP > SELECT | APPLICTN | LIST

Prompt: SEL ITEM

In the Applications window, select application(s) from the current workspace.

✓ **NOTE:** *Applications include kinematics, robotics, numerical control, and user applications.*

Select application elements to keep

Select: KEEP > SELECT | APPLICTN | ELEMENT

Prompt: SEL ITEM

In the Applications window, select an element or elements from one or more applications selected via the SELECT | APPLICTN | LIST option.

Select application set to retain

Select: KEEP > SELECT | APPLICTN | SET

Prompt: SEL ITEM

In the Applications window, select a set or sets from one or more applications selected via the SELECT | APPLICTN | LIST option. Upon selecting an application set, all respective elements are selected.

✓ **NOTE:** *All the above options can be used to select a combination of elements, sets, and details to be retained using KEEP > KEEP.*

KEEP

Retain items selected using one or more of the above options.

Select: KEEP > KEEP

Prompt: YES : CONFIRM (KEEP)

Click on the YES button. All elements, details, sets, and applications selected via usage of the SELECT options will be retained, and all other elements, details, sets, and applications will be deleted from the model.

✓ **NOTE:** *Stored screens, windows, translations, and so forth will be lost when the KEEP option is used.*

✘ **WARNING:** *Once the YES button has been selected to retain selected entities and delete all other entities, there is no opportunity to select the NO button and undo the last command.*

MERGE

Merge (copy) passive model into current active model. You must therefore overlay a passive model in order to use this option.

Select: KEEP > MERGE

Prompt: MODEL : SEL ELEM // KEY (STRING)

Before selecting a passive model to be merged, the following general rules are worthy of review.

- The created model will take the name of the active model.
- The created model takes the standards (e.g., dimensions, colors, units) of the active model.
- If a different unit is used for each model, the CATUNIT utility restrictions apply to the merging process.
- Only the **MASTER* workspaces of each model will be merged. Details and sets will remain separate.
- If an element or set from the passive model has an automatic identifier [e.g., beginning with the asterisk (*) character], the identifier of the element or set will be modified.
- If two elements, sets, details, and so forth have the same manual identifier, the identifier from the passive model will be modified. A dollar sign ($) character followed by an integer ranging from 1 to 9 will be placed before the original identifier.
- If two views have the same identifier and are linked to the same SPACE background plane, the view belonging to the passive model will be deleted and the related elements will pertain to the corresponding view of the active model.

1. Select an element of the passive model to be merged. The selected model will be highlighted. Alternatively, press <Enter> to display the list of available models, and select the required model. The selected model will be highlighted.
2. Click on the YES button to confirm that you wish to merge the contents of the selected passive model into the current active model. The contents of the passive model will be merged (copied) into the current active model; the original passive model will not be changed in any way.

LAYER in DRAW and SPACE Modes

Chapters 5, 13, 14, 15, 17

The LAYER function in DRAW and SPACE modes is used to create and apply layer filters, transfer elements between layers, and analyze and manage layers and filters.

Option menu for LAYER in DRAW and SPACE modes.

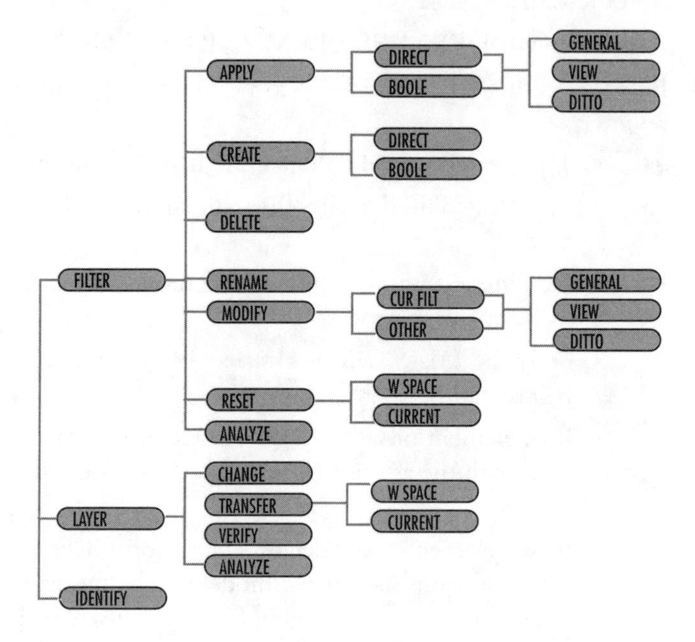

FILTER

Create, apply, and manage layer filters.

Select: LAYER > FILTER | APPLY or CREATE or DELETE or RENAME or MODIFY or RESET or ANALYZE

Apply layer filters

Select: LAYER > FILTER | APPLY | DIRECT or BOOLE

Apply layer filters in direct fashion

Select: LAYER > FILTER | APPLY | DIRECT | GENERAL or VIEW or DITTO

Apply layer filter to screen display

Select: LAYER > FILTER | APPLY | DIRECT | GENERAL

Prompt: SEL FILTER

The list of available filters displays. If none have been created, the two default filters listed below are available.

- ALL displays all 254 layers.
- LAYCUR displays elements in current layer.

Apply layer filter to view

Select: LAYER > FILTER | APPLY | DIRECT | VIEW

Prompt: SEL FILTER

The list of available filters displays. If none have been created, the three default filters listed below are available.

- ALL displays all 254 layers in selected view.
- LAYCUR displays elements in current view on current layer.
- NONE restores view filter to same as GENERAL layer filter.

Apply layer filter to ditto

Select: LAYER > FILTER | APPLY | DIRECT | DITTO

Prompt: SEL FILTER

The list of available filters displays. If none have been created, the three default filters listed below are available.

- ALL displays all 254 layers in selected ditto.
- LAYCUR displays elements in ditto on current layer.
- NONE restores view filter to same as GENERAL layer filter.

✓ ***NOTE:*** *For each filter above, you can choose DESCRIPTION mode ON or OFF. If the DESCRIPTION mode is ON, the status of the filter displays on the screen.*

Apply layer filter via Boolean operations

Select: LAYER > FILTER | APPLY | BOOLE | GENERAL or VIEW or DITTO

Apply layer filter created by combining other filters via Boolean operations

Select: LAYER > FILTER | APPLY | BOOLE | GENERAL

Prompt: SEL FILTER or SEL COMMANDS

1. Create a filter using Boolean operators and filters selected from the lists.
2. Click on the YES button to confirm creation. The status of the layers of the filter you created displays.
3. Click on YES to apply the filter.

Apply layer filter created via combining view filters in DRAW mode

Select: LAYER > FILTER | APPLY | BOOLE | VIEW

Prompt: SEL FILTER or SEL COMMANDS

1. Create a filter using the Boolean operators and filters selected from the lists.
2. Click on the YES button to confirm creation. The status of the layers of the filter you created displays.
3. Click on YES to apply the filter.

Apply layer filter created via combining ditto filters

Select: LAYER > FILTER | APPLY | BOOLE | DITTO

Prompt: SEL FILTER or SEL COMMANDS

1. Create a filter using the Boolean operators and filters selected from the lists.
2. Click on the YES button to confirm creation. The status of the layers of the filter you created displays.
3. Click on YES to apply the filter.

Create layer filter

Select: LAYER > FILTER | CREATE | DIRECT or BOOLE

Create layer filter from other layer filters stored in model

Select: LAYER > FILTER | CREATE | DIRECT

Prompt: SEL FILTER // YES: GENERAL

1. Select GENERAL, VIEW, or DITTO depending on the type of filter required.
2. Select a filter from the list of existing filters in the model. Upon selection of a filter from the list, the LIST OF LAYERS screen displays. In the screen, layers visible in the selected filter are shown in white, and invisible layers in green.
3. In the list, select layers to make them either visible or invisible.
4. Click on YES to create the modified filter, and then enter a new name for the filter.

Create layer filter by combining other filters using Boolean operators

The steps required in creating a filter are the same used in applying a Boolean filter (see previous description).

Delete filters

Select: LAYER > FILTER | DELETE

Prompt: SEL FILTER

Select a filter from the list, and then click on the YES button.

Rename filters

Select: LAYER > FILTER | RENAME

Prompt: SEL FILTER

1. Select a filter from the list.
2. Key in a new name for the filter, and then press <Enter>.

Modify filter status

Select: LAYER > FILTER | MODIFY | CUR.FILT or OTHER

Modify status of current filter

Select: LAYER > FILTER | MODIFY | CUR.FILT | GENERAL or VIEW or DITTO

Modify status of general applied filter

Select: LAYER > FILTER | MODIFY | CUR.FILT | GENERAL

Prompt: SEL ELEM

1. Select an element; the status of the general applied filter displays.
2. Select a layer or key in a combination of layers.
3. Click on the YES button to accept the modifications.

Modify status of filter applied to current view

Select: LAYER > FILTER | MODIFY | CUR.FILT | VIEW

Prompt: SEL VIEW // YES: CURRENT VIEW

1. Click on the YES button to modify the filter applied to the current view.
2. Select a view, and the filter applied to the view displays.
3. Select a layer or key in a combination of layers; the status of the layers resulting from this combination is inverted.
4. Click on the YES button to confirm the modification.

Modify status of filter applied to ditto or macroprimitive

Select: LAYER > FILTER | MODIFY | CUR.FILT | DITTO

Prompt: SEL DITTO/DITD/SOL

1. Select a ditto or a macroprimitive; the filter applied to the element displays.
2. Select a layer or key in a combination of layers; the status of the layers resulting from the combination is inverted.
3. Click on the YES button to confirm the modification.

Modify status of other filters

Select: LAYER > FILTER | MODIFY | OTHER | GENERAL or VIEW or DITTO

Prompt: SEL FILTER

1. A list of filters displays. (The ALL, LAYCUR, and NONE filters cannot be modified.)
2. Select a filter; the status of the filter displays.
3. Select a layer or key in a combination of layers; the status of the layers resulting from the combination is inverted.
4. Click on the YES button to confirm the modification.

Reset filters

Select: LAYER > FILTER | RESET | W.SPACE or CURRENT

Reset filters applied in any workspace

Select: LAYER > FILTER | RESET | W.SPACE

Prompt: YES: RESET WSP

Click on the YES button to reset the filters to their default status.

Reset filters applied in current workspace

Select: LAYER > FILTER | RESET | CURRENT

Prompt: YES: RESET CURRENT

Click on the YES button to reset the filters to their default status in the current workspace.

Analyze status of filter applied to screen, view, or ditto

Select: LAYER > FILTER | ANALYZE

Prompt: SEL FILTER // YES: GENERAL

1. Select GENERAL, VIEW, or DITTO from the menu. A list of filters displays. Select a filter to view its status, or select a view or ditto to view the status of the filter applied to the element.

2. Click on the YES button to display the status of the current GENERAL filter, and then click on YES to continue the analysis.

LAYER

Manipulate layers independently of filters.

Select: LAYER > LAYER | CHANGE or TRANSFER or VERIFY or ANALYZE

Change current layer

Select: LAYER > LAYER | CHANGE

Prompt: SEL ELEM // SEL LAYER // KEY LAYER

Use one of the following alternatives: (a) Select a layer from the list. (b) Key in the number of the desired layer. (c) Select an element. The current layer displays in red in the list.

Transfer elements between layers

Select: LAYER > LAYER | TRANSFER | W.SPACE or CURRENT

Prompt: SEL ELEM // SEL LAYERKEY LAYER // YES: CURRENT

1. A list of layers displays. Select a layer from the list, key in a layer number, or select an element on the required layer. This layer becomes the receiving layer.
2. Click on the YES button to confirm layer selection.
3. Select one or more elements, and then click on the YES button to complete the transfer.

✓ **NOTE:** *In order to transfer standard ditto elements to a different layer, either convert them to compact elements or transfer them from the detail workspace.*

Display elements on particular layer

Select: LAYER > LAYER | VERIFY | PICK or NO PICK

Prompt: SEL ELEM // SEL LAYERKEY LAYER // YES: CURRENT

1. A list of layers displays. In the Verify Mode window, select the desired display type.
2. Select a layer from the list, key in the required layer number, or select an element on the required layer. The layer displays.
3. Click on the YES button to display the elements on the selected layer. A Layer Verify window displays.

✓ **NOTE:** *Standard dittos are not displayed when using this method.*

Analyze layers in use in the current workspace

Select: LAYER > LAYER | ANALYZE

Prompt: SEL ITEM

The List of Layers screen highlights the layers currently in use. To change the current layer, select a layer from the list or key in the layer number. Select an element on the required layer.

IDENTIFY

Create or modify layer table in project file.

✓ **NOTE:** *This option may be accessible only by your system administrator.*

Select: LAYER > IDENTIFY

Prompt: YES: CONTINUE

1. Click on the YES button. The List of Layers screen displays.
2. Select a layer or click on the YES button to accept the current layer.
3. Enter a new identifier for the selected layer.

LIBRARY in DRAW and SPACE Modes

Chapter 5

The LIBRARY function in DRAW and SPACE modes is used to define the current library and family, write objects which become details into a library, and read library details into a model. The use and availability of the LIBRARY function largely depends on your site setup. Consult your system administrator for further information.

Option menu for LIBRARY in DRAW and SPACE modes.

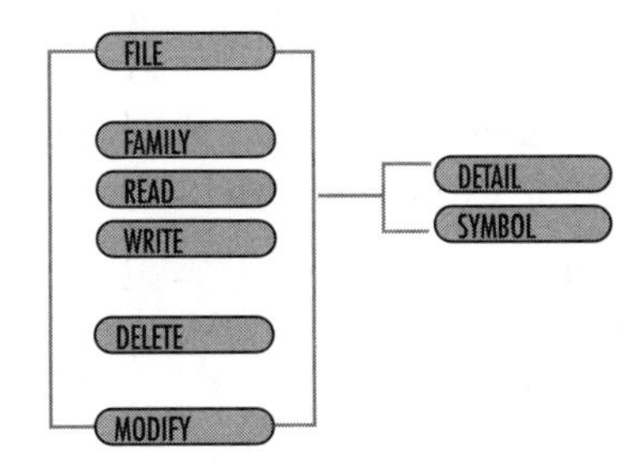

✓ ***NOTE:*** *The NCM TOOL, SECTION, and NCL TOOL options are available only if the Fixed Axis Milling, Multiple Axis Milling, and Lathe products are installed.*

FILE

Select library containing objects to be handled.

Select: LIBRARY > FILE

Prompt: KEY STRING (if LIBRARY has already been selected)

Prompt: SEL KEYWORD / / YES: END KEY OBJECT NAME (if LIBRARY has not previously been selected)

If a library has already been accessed, upon selection of LIBRARY function the last used library and family will be selected in order for an object to be read. If a library has not been accessed, enter a blank field to display a list of available libraries. If the automatically selected library is not the desired library, reselect the FILE option and press <Enter> to display the list.

Select the desired library from the list, or enter a character string contained in the name of the desired library to shorten the list.

FAMILY

Select family containing objects to be handled.

Select: LIBRARY > FAMILY | DETAIL or SYMBOL

Prompt: KEY STRING (if LIBRARY has already been selected)

Prompt: SEL KEYWORD / / YES: END KEY OBJECT NAME (if LIBRARY has not previously been selected)

If a family has already been accessed, upon selection of LIBRARY function the last used family will be selected. If a family has not been accessed, enter a blank field to display a list of available families. If the automatically selected family is not the desired family, reselect the FAMILY option and press <Enter> to display the list.

Select the desired family from the list, or enter a character string contained in the name of the family to shorten the list.

READ

Read an object from a library family into the current model.

Select: LIBRARY > READ | DETAIL or SYMBOL

Prompt: SEL KEYWORD / / YES: END KEY OBJECT NAME

✓ ***NOTE:*** *Before attempting to read a library object into a model, verify that the correct mode (DRAW or SPACE) has been selected and that the detail name does not already exist in the model.*

1. After the correct library and family have been selected, a window containing four types of keywords is displayed which can be used (depending on your setup) to help you search for the desired object. If necessary, select keywords from the displayed window, or click on YES to display a full list of objects in the selected family.
2. Click on YES to accept the selection or NO to refuse. Select an object from the list, and then click on YES to confirm or NO to refuse.
3. After making the selection, click on YES to accept it or NO to refuse.

WRITE

Write objects from a model into a library.

Select: LIBRARY > WRITE | DETAIL or SYMBOL

Prompt: KEY STRING / / ENTER: LIST
SEL DITTO / / YES: DISPLAY

✓ **NOTE:** *Any element can be written into a library as long as the object does not already exist in a library.*

The use of this option will depend on your level of access. A model detail or symbol must exist to be written into a library.

1. Using the method described in the previous entry, select the desired library and family. Enter a character string or a blank field to view a list of available details or symbols, or click on YES to display available details or symbols.
2. Select the desired detail or symbol from the list or display. If necessary, select keywords to assign to the selected object(s), and then click on YES to end. Click on YES again to confirm or NO to refuse.

DELETE

Delete object from library.

Select: LIBRARY > DELETE

Prompt: SEL KEYWORD / / YES: END KEY OBJECT NAME

The use of this option will depend on your level of access.

1. Use the method described in previous entries to select the desired library and family.
2. Enter a character string or a blank field to display a list of available details or symbols, or click on YES to display available details or symbols.
3. Select detail or symbol to be deleted (which will be displayed), and then click on YES to confirm deletion or NO to refuse.

MODIFY

Modify keywords of library object.

Select: LIBRARY > MODIFY | DETAIL or SYMBOL

Prompt: SEL KEYWORD / / YES: END KEY OBJECT NAME

1. Use one of the following alternatives: (a) Enter a keyword. (b) Enter an object name. (c) Click on YES to view a complete list.
2. Select an object from the displayed list. If desired, select keywords to be assigned.
3. Click on YES to end, and then click on YES again to confirm or NO to refuse.

UPDATE

Update library detail to reflect changes to object in model.

Select: LIBRARY > UPDATE | DETAIL or SYMBOL

Prompt: KEY STRING / / ENTER: LIST

SEL DITTO / / YES: DISPLAY

Use of this option will depend on your level of access.

1. Select the object to be updated by entering a string or a blank field.
2. Use one of the following alternatives: (a) Select from the displayed list. (b) Select a ditto. (c) Click on YES to display available dittos.
3. Click on YES to replace or NO to refuse.

LIMIT1 in DRAW Mode

Chapters 3, 6, 7

The LIMIT1 function in DRAW mode is used to modify the limits of DRAW elements.

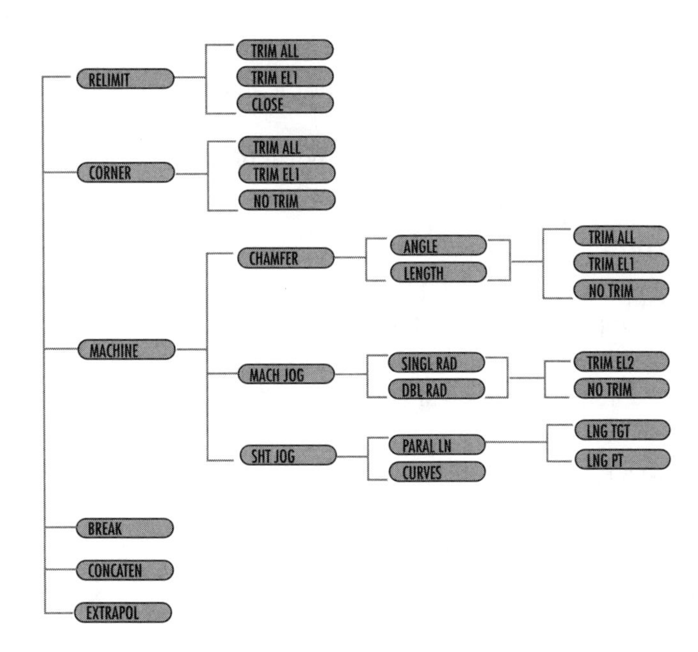

Option menu for LIMIT1 in DRAW mode.

RELIMIT

Modify limits of DRAW elements.

Select: LIMIT1 > RELIMIT | TRIM ALL or TRIM EL1 or CLOSE

Trim or extend two elements simultaneously to intersection point

Select: LIMIT1 > RELIMIT | TRIM ALL

Prompt: SEL PT/LN1/CRV1

1. Select a point, and then select a line or curve in the region to be retained. The line or curve is relimited to the point or the projection point.
2. Select a line or curve, and then select a point, line, or curve. The line or curve is relimited to the point of actual or projected intersection.
3. Indicate in another quadrant for an alternative solution as necessary.

Relimit one end of DRAW element

Select: LIMIT1 > RELIMIT | TRIM EL1

Prompt: SEL PT/LN1/CRV1

1. Select elements individually or by multiselect in the region to be retained.
2. Select a line, point, or curve. If the first selection was via keyboard entry or multiselect, indicate the region to be retained. The first selected elements are relimited to the point of the actual or projected intersection with the second selected element.
3. Indicate in another quadrant for an alternative solution as necessary.

Restore default limits of DRAW elements

Select: LIMIT1 > RELIMIT | CLOSE

Prompt: SEL LN | CRV

1. Select a line (LND), curve (CRVD), circle (CIRD), or ellipse (ELLD). Numbers are displayed at the ends of selected elements.
2. Select a number if one end is to be modified, or click on YES for both ends. A partial circle or ellipse will close, and a line will change from segment to unlimited.

CORNER

Create circular arcs between DRAW elements.

Select: LIMIT1 > CORNER | TRIM ALL or TRIM EL1 or NO TRIM

Limit two DRAW elements with circular arc

Select: LIMIT1 > CORNER | TRIM ALL

Prompt: SEL 1st LN/CRV / / KEY RAD

1. Enter a new radius value as necessary.
2. Select a line or curve on the side to be retained. (A new radius value can also be entered at this stage.)
3. Select a line or curve. A concave connecting arc is created.
4. Click on YES to view the convex solution. Click on YES again to create a line joining the tangent points.
5. Indicate in another region as necessary, and the corner will be recomputed.

Create circular arc joining two elements while relimiting single element

Select: LIMIT1 > CORNER | TRIM EL1

Prompt: SEL 1st LN/CRV / / KEY RAD

1. If desired, enter a new radius value.
2. Select a line or curve on the side to be retained. This element will be relimited. (A new radius value can also be entered at this stage.)
3. Select a line or curve. This element will not be relimited. A concave connecting arc is created.
4. Click on YES to view the convex solution. Click on YES again to create a line joining the tangent points.
5. Indicate in another region as necessary, and the corner will be recomputed.

Create circular arc joining two elements without relimiting

Select: LIMIT1 > CORNER | NO TRIM

Prompt: SEL 1st LN/CRV / / KEY RAD

1. Enter a new radius value as necessary.
2. Select a line or curve on the side to be retained. (A new radius value can also be entered at this stage.)
3. If necessary, enter values for leg length LNG1 and LNG2. Alternatively, click on YES to view the convex solution. Click on YES again to create a line joining the tangent points.
4. If required, indicate in another region and the corner will be recomputed.

MACHINE

Create chamfers and connecting curves.

Select: LIMIT1 > MACHINE | CHAMFER or MACH JOG or SHT JOG

Create chamfers

Select: LIMIT1 > MACHINE | CHAMFER | ANGLE or LENGTH

Create chamfer by specifying angle and one leg length

Select: LIMIT1 > MACHINE | CHAMFER | ANGLE | TRIM ALL or TRIM EL1 or NO TRIM

Prompt: SEL 1st LN/CRV

1. Select a line or curve in the region to be retained.
2. Select a second line in the region to be retained.

3. If required, enter length and angle values. A 45-degree angle is assumed if only the length value is entered. Alternatively, click on YES to accept suggested values. The chamfer is created.
4. Indicate in another quadrant for an alternative solution as necessary.

TRIM ALL, TRIM EL1, and NO TRIM work as described under the CORNER option section.

Create chamfer by specifying two leg lengths

Select: LIMIT1 > MACHINE | CHAMFER | LENGTH | TRIM ALL or TRIM EL1 or NO TRIM

Prompt: SEL 1st LN/CRV

1. Select a line or curve in the region to be retained.
2. Select a second line or curve in the region to be retained.
3. If required, enter values for leg lengths LNG1 and LNG2. Click on YES to accept suggested values. Chamfers are created.
4. Indicate in another quadrant for an alternative solution as necessary.

TRIM ALL, TRIM EL1, and NO TRIM work as described under the CORNER option section.

Create connecting arcs between two parallel lines

Select: LIMIT1 > | MACHINE | MACH JOG | SINGL RAD or DBL RAD | TRIM EL2 or NO TRIM

Prompt: REFERENCE LINE: SEL LN1

Create single connecting curve

Select: LIMIT1 > | MACHINE | MACH JOG | SINGL RAD | TRIM EL2 or NO TRIM

Prompt: REFERENCE LINE: SEL LN1

1. Select the first line in the region to be retained.
2. Select the second line parallel to the first in the region to be retained.
3. Indicate the region in which the connection is to be created, or select a point.
4. Enter a radius or click on YES to accept the suggested value. The connecting curve is created.

TRIM EL2 and NO TRIM work as described in the CORNER option section.

Create two connecting curves

Select: LIMIT1 > | MACHINE | MACH JOG | DBL RAD | TRIM EL2 or NO TRIM

Prompt: REFERENCE LINE: SEL LN1

1. Select the first line in the region to be retained.
2. Select a second line parallel to the first in the region to be retained.
3. Indicate the region in which the connecting arc is to be created, or select a point.
4. Enter values for RAD1 and RAD2 (the sum of which must be greater than or equal to the distance between the parallel lines), or click on YES to accept the suggested values.

TRIM EL2 and NO TRIM work as described under the CORNER option section.

Create connecting arcs between pairs of parallel lines or curves

Select: LIMIT1 > MACHINE | SHT JOG | PARAL LN or CURVES

Connection length provides tangency points of connecting curves

Select: LIMIT1 > MACHINE | SHT JOG | PARAL LN | LNG TGT or LNG PT

Prompt: REFERENCE LINE: SEL 1st LN

1. Select the first line in the region to be retained.
2. Select the second line parallel to the first in the region to be retained.
3. Select two more lines parallel to each other at the same distance apart as the first two.
4. Indicate the region in which the connecting curves are to be created.
5. Enter values for inner radius and connection length, or click on YES to accept suggested values. Connecting curves are created.

When the LNG PT option is selected, the connecting length provides the points of intersection between connecting and selected lines.

Create connecting curves between pairs of parallel curves

Select: LIMIT1 > MACHINE | SHT JOG | CURVES

Prompt: REFERENCE: SEL 1ST LN/CRV

1. Select the first line or curve in the region to be retained.
2. Select the line or curve parallel to the first in the region to be retained.
3. Select two more lines or curves parallel to each other at the same distance apart as the first two.

4. Indicate the region in which the connecting curves are to be created, or select a point.
5. Enter values for inner radius and connection length, or click on YES to accept suggested values. Connect curves are created.

BREAK

Create breaks in DRAW elements.

Select: LIMIT1 > BREAK

Prompt: CUR MULTI SEL ELEM TO BE BROKEN: SEL ELEM

On selection of the BREAK option, a panel containing the following options displays.

- NO GAP. Create gap at break point.
- GAP. Create gap at break point depending on NORMAL or OBLIQUE.
- NORMAL. Create gap if element to be broken is less than displayed distance from second element.
- OBLIQUE. Create gap at point of projection of second element onto first.

1. Select a point. The break is made at the point if it is on the line or at the projection point if not. Alternatively, select a line or curve. Proposed break points at intersections are displayed with symbols and numbers.
2. Enter the numbers of break points, or click on YES to break at all points.

To break a curve into elemental arcs, take the following steps.

1. Select a curve.
2. Select a temporary symbol, or click on YES to accept all break points.

CONCATEN

Concatenate element ends into single element.

Select: LIMIT1 > CONCATEN

Prompt: SEL LN/CRV

1. Select a line or curve.
2. Select a second line or curve connected to the first. The concatenation is executed upon selection of the second element, and tangent discontinuity is indicated by a double arrow symbol.
3. If automatic concatenation of additional elements is possible, click on YES to execute or NO to refuse.

✓ **NOTE:** *Elements in line but not joined can also be concatenated.*

EXTRAPOL

Extrapolate line or curve.

Select: LIMIT1 > EXTRAPOL

Prompt: SEL LN/CRV

1. Select a line or a curve. Numbers are displayed at the ends.
2. Select the end to extrapolate by selecting or entering a number.
3. Depending on element type, enter a value for length or percentage, or click on YES to accept the suggested value.
4. If required, click on YES again to iterate the extrapolation.

✓ ***NOTE:*** *If the message SUPPRESS LIMIT CURVE BEFORE EXTRAPOLATION is encountered, use LIMIT1 > RELIMIT | CLOSE to relimit one or both ends of element.*

LIMIT1 in 2D SPACE Mode

The LIMIT1 function in 2D SPACE mode is used to modify the limits of 2D SPACE elements.

Option menu for LIMIT1 in 2D SPACE mode.

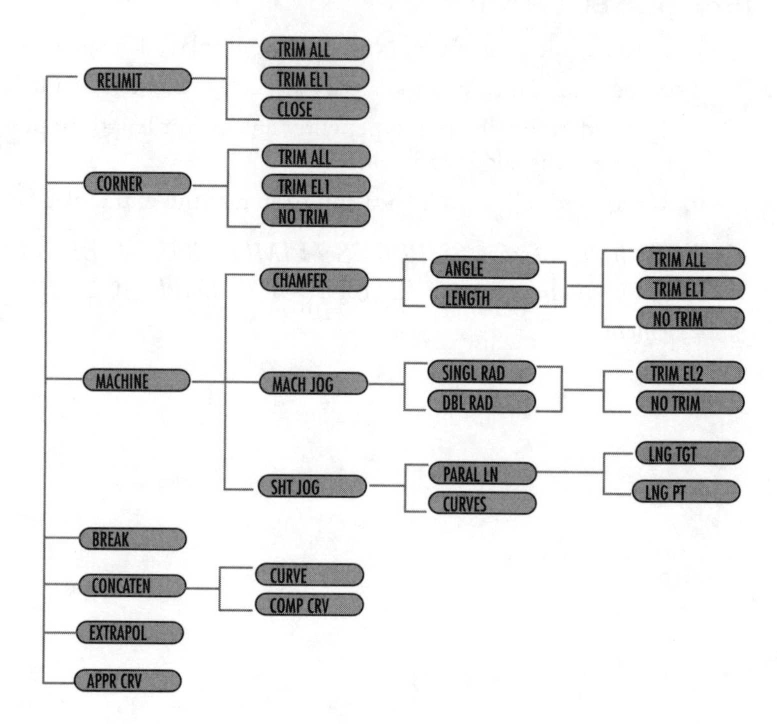

RELIMIT

Modify element limits.

Select: LIMIT1 > RELIMIT | TRIM ALL or TRIM EL1 or CLOSE

The Tolerance panel displays the following options.

- STD. Relimit using default intersection projection tolerance.
- COARSE. Relimit using a different value of intersection projection tolerance.

Trim or extend two elements simultaneously to intersection point

Select: LIMIT1 > RELIMIT | TRIM ALL

Prompt: SEL PT/LN1/CRV1/NRBC1

Select a point, and then select a line or curve in the region to be retained. The line or curve is relimited to the point or the projection point.

Alternatively, select a line or curve, and then select a point, line, or curve. The line or curve is relimited to the point of actual or projected intersection. If desired, indicate in another quadrant for an alternative solution.

Relimit one end of DRAW element

Select: LIMIT1 > RELIMIT | TRIM EL1

Prompt: WSP MULTI-SEL SEL 1ST LN/CRV/NRBC

1. Select elements individually or by multiselect in the region to be retained. If multiselect was used, click on YES to end selection.
2. Select a line point or curve.
3. If the first selection was made through keyboard entry or multiselect, indicate the region to be retained. The first selected elements are relimited to the point of actual or projected intersection with the second selected element.
4. If desired, indicate in another quadrant for an alternative solution.

Restore default limits of DRAW elements

Select: LIMIT1 > RELIMIT | CLOSE

Prompt: SEL LN / CRV/ NRBC

1. Select a line (LN) or curve (CRV). If a part circle is selected, it will close automatically.
2. If a line is selected, numbers are displayed at the ends of selected element.
3. Select a number if one end is to be modified, or click on YES for both ends. A part circle or ellipse will close, whereas a line will change from segment to unlimited.

CORNER

Select: LIMIT1 > CORNER | TRIM ALL or TRIM EL1 or NO TRIM

Create circular arcs between DRAW elements.

Limit two DRAW elements with circular arc

Select: LIMIT1 > CORNER | TRIM ALL

Prompt: SEL 1st LN/CRV / / KEY RAD

1. If desired, enter a new radius value.

2. Select a line or curve on the side to be retained. (A new radius value can be entered at this stage as well.)
3. Select a line or curve. A concave connecting arc is created.
4. Click on YES to view the convex solution. Click on YES again to create a line joining the tangent points.
5. If required, indicate in another region and the corner will be recomputed.

Create circular arc joining two elements while relimiting one element

Select: LIMIT1 > CORNER | TRIM EL1

Prompt: SEL 1st LN/CRV / / KEY RAD

1. If desired, enter a new radius value.
2. Select a line or curve on the side to be retained. This element will be relimited. (A new radius value can be entered at this stage as well.)
3. Select a line or curve. This element will not be relimited. A concave connecting arc is created.
4. Click on YES to view the convex solution. Click on YES again to create a line joining the tangent points.
5. If required, indicate in another region and the corner will be recomputed.

Create circular arc joining two elements without relimiting

Select: LIMIT1 > CORNER | NO TRIM

Prompt: SEL 1st LN/CRV / / KEY RAD

1. If desired, enter a new radius value.
2. Select a line or curve on the side to be retained. (A new radius value can also be entered at this stage.)
3. Select a line or curve. A concave connecting arc is created.
4. Click on YES to view the convex solution. Click on YES again to create a line joining the tangent points.
5. If required, indicate in another region and the corner will be recomputed.

MACHINE

Create chamfers and connecting curves.

Select: LIMIT1 > MACHINE | CHAMFER or MACH JOG or SHT JOG

Create chamfer

Select: LIMIT1 > MACHINE | CHAMFER | ANGLE or LENGTH

Create chamfer by specifying angle and one leg length

Select: LIMIT1 > MACHINE | CHAMFER | ANGLE | TRIM ALL or TRIM EL1 or NO TRIM

Prompt: SEL 1ST LN/CRV

TRIM ALL, TRIM EL1, and NO TRIM options work as described under the CORNER option above.

1. Select a line or curve in the region to be retained.
2. Select a second line in the region to be retained.
3. If desired, enter length and angle values. If a length value alone is entered, a 45-degree angle is assumed. Alternatively, click on YES to accept suggested values. The chamfer is created.

Create chamfer by specifying two leg lengths

Select: LIMIT1 > MACHINE | CHAMFER | LENGTH | TRIM ALL or TRIM EL1 or NO TRIM

Prompt: SEL 1st LN/CRV

The TRIM ALL, TRIM EL1, and NO TRIM options work as described under the CORNER option above.

1. Select a line or curve in the region to be retained.
2. Select a second line or curve in the region to be retained.
3. If desired, enter values for leg lengths LNG1 and LNG2, or click on YES to accept suggested values. The chamfers are created.

Machine joggle

Select: LIMIT1 > | MACHINE | MACH JOG | SINGL RAD or DBL RAD

Create connecting arcs between two parallel lines

Select: LIMIT1 > | MACHINE | MACH JOG | SINGL RAD or DBL RAD | TRIM EL2 or NO TRIM

Prompt: REFERENCE LINE: SEL LN1

Single radius

Select: LIMIT1 > | MACHINE | MACH JOG | SINGL RAD | TRIM EL2 or NO TRIM

Prompt: REFERENCE LINE: SEL LN1

Upon selecting the SINGL RAD option, a single connecting curve tangent to the second selected line will be created. TRIM EL2 and NO TRIM work as described under the CORNER option above.

1. Select the first line in the region to be retained.
2. Select the second line parallel to the first in the region to be retained.
3. Indicate the region in which the connection is to be created, or select a point.
4. Enter a radius or click on YES to accept the suggested value. The connecting curve is created.

Double radius

Select: LIMIT1 > | MACHINE | MACH JOG | DBL RAD | TRIM EL2 or NO TRIM

Prompt: REFERENCE LINE: SEL LN1

Upon selecting the DBL RAD option, lines will be joined by two curves tangent to each line. TRIM EL2 and NO TRIM work as described under the CORNER option above.

1. Select the first line in the region to be retained.
2. Select a second line parallel to the first in the region to be retained.
3. Indicate the region in which the connecting arc is to be created, or select a point.
4. Enter values for RAD1 and RAD2 (the sum of which must be greater than or equal to the distance between the parallel lines), or click on YES to accept the suggested values.

Sheet joggle

Create connecting arcs between pairs of parallel lines or curves.

Select: LIMIT1 > MACHINE | SHT JOG | PARAL LN or CURVES

Create connecting arcs where connection length provides tangency points

Select: LIMIT1 > MACHINE | SHT JOG | PARAL LN | LNG TGT

Prompt: REFERENCE LINE: SEL 1st LN

1. Select the first line in the region to be retained.
2. Select the second line parallel to the first in the region to be retained.
3. Select two more lines parallel to each other at the same distance apart as the first two.
4. Indicate the region in which the connecting curves are to be created.
5. Enter values for inner radius and connection length, or click on YES to accept suggested values. Connecting curves are created.

When the LNG PT option is selected, the connecting length provides the points of intersection between connecting and selected lines.

Create connecting curves between pairs of parallel curves

Select: LIMIT1 > MACHINE | SHT JOG | CURVES

Prompt: REFERENCE: SEL 1ST LN/CRV

1. Select the first line or curve in the region to be retained.
2. Select the line or curve parallel to the first in the region to be retained.
3. Select two more lines or curves parallel to each other at the same distance apart as the first two.
4. Indicate the region in which the connecting curves are to be created, or select a point.
5. Enter values for inner radius and connection length, or click on YES to accept suggested values. Connect curves are created.

BREAK

Create breaks in DRAW elements.

Select: LIMIT1 > BREAK

Prompt: WSP MULTI-SEL SEL LN/CRV/CCV/CST/NRBC

The Tolerance panel provides the following options.

- STD. Break using default intersection projection tolerance.
- COARSE. Break using a different value of intersection projection tolerance.

To select a line or curve to be broken, you have three alternatives.

- Select a point. The break is made at the point if it is on a line or the projection point if it is not.
- Select a line. The break is made at the point of the actual or projected intersection.
- Select a curve. Proposed break points at intersections are displayed with symbols and numbers.

Next, enter the numbers of desired break points, or click on YES to break at all points.

To break a curve into elemental arcs, take the following steps.

1. Select a curve.
2. Select a temporary symbol, or click on YES to accept all break points.

CONCATEN

Concatenate element ends into single element.

Select: LIMIT1 > CONCATEN | CURVE or COMP CRV

Prompt: SEL LN/CRV/CST or SEL LN/CCV

1. Select a line or curve.
2. Select a second line or curve connected to the first. The concatenation is executed on selection of the second element. Tangent discontinuity is indicated by a double arrow symbol.
3. If automatic concatenation of additional elements is possible, click on YES to execute or NO to refuse.

EXTRAPOL

Extrapolate line or curve.

Select: LIMIT1 > EXTRAPOL

Prompt: SEL LN/CRV

1. Select a line or curve. Numbers are displayed at ends.

✓ *NOTE: Circle or part circle elements cannot be extrapolated. Use the RELIMIT | CLOSE option instead.*

2. Select the end to extrapolate by selecting or entering a number.
3. Enter a value for length or percentage, depending on element type. Alternatively, click on YES to accept suggested value.
4. Click on YES again as necessary to iterate the extrapolation.

✓ *NOTE: If the message SUPPRESS LIMIT CURVE BEFORE EXTRAPOLATION appears, use LIMIT1 > RELIMIT | CLOSE to relimit one or both ends of the element.*

APPR CRV

Approximate composite curve by general curve.

Select: LIMIT1 > APPR CRV

Prompt: SEL CCV

Select a composite curve. The approximation of the curve is executed upon selection.

LIMIT1 in 3D SPACE Mode

Chapters 6, 7

The LIMIT1 function in 3D SPACE mode is used to modify the limits of 3D SPACE elements.

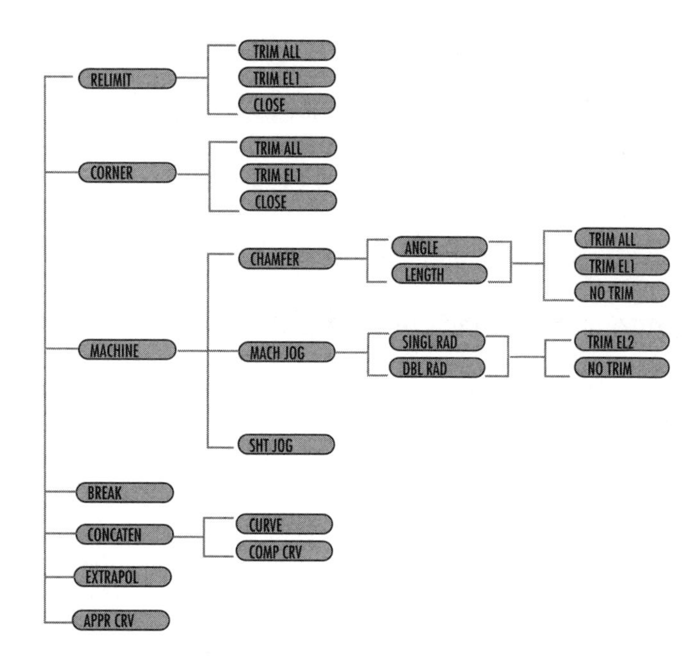

Option menu for LIMIT1 in 3D SPACE mode.

RELIMIT

Modify limits of DRAW elements.

Select: LIMIT1 > RELIMIT | TRIM ALL or TRIM EL1 or CLOSE

Trim or extend two elements simultaneously to intersection point

Select: LIMIT1 > RELIMIT | TRIM ALL

Prompt: SEL LN/CRV/NRBC

Proceed with one of the following alternatives.

- Select a line or curve. Indicate a point or select a point, line, or curve in the region to be retained. The line or curve is relimited to the point or projection point.
- Select a line or curve. Select a point, line, or curve. The line or curve is relimited to the point of actual or projected intersection.

Indicate in another quadrant for an alternative solution as necessary.

Relimit one end of DRAW element

Select: LIMIT1 > RELIMIT | TRIM EL1

Prompt: WSP MULTI-SEL SEL LN/CRV/CRV/NRBC / / IND REGION

1. Select line or curve elements individually or by multiselect in the region to be retained. If a multiselect option was used, click on YES to end selection.
2. Indicate a point, or select a line or a curve. If the first selection was through keyboard entry or multiselect, indicate the region to be retained. The first selected elements are relimited to the point of the actual or projected intersection with the second selected element.
3. If required, indicate in another quadrant for an alternative solution.

Restore default limits of DRAW elements

Select: LIMIT1 > RELIMIT | CLOSE

Prompt: SEL LN/CRV/NRBC

1. Select a line (LN) or curve (CRV). Numbers are displayed at the ends of the selected element.
2. Select a number if one end is to be modified or click on YES for both ends. A part circle or ellipse will close, and a line will change from segment to unlimited.

CORNER

Create circular arcs between SPACE elements.

Select: LIMIT1 > CORNER | TRIM ALL or TRIM EL1 or NO TRIM

Limit two SPACE elements with circular arc

Select: LIMIT1 > CORNER | TRIM ALL

Prompt: SEL 1st LN/CRV / / KEY RAD

1. Enter a new radius value if required. Select a line or curve on the side to be retained. (A new radius value can be entered at this stage as well.)
2. Select a line or curve. Click on YES to accept the implicit plane (which produces a 3D connecting curve), or select another plane. A concave connecting arc is created.
3. Click on YES to view the convex solution, if appropriate.
4. Click on YES again to create a line joining the tangent points.

5. If required, indicate in another region, and the corner will be recomputed.

Create circular arc joining two elements with only first element limited

Select: LIMIT1 > CORNER | TRIM EL1

Prompt: SEL 1st LN/CRV / / KEY RAD

1. Enter a new radius value if desired. Select a line or curve on the side to be retained. (A new radius value can be entered at this stage as well.)
2. Select a line or curve. Click on YES to accept the implicit plane (which produces a 3D connecting curve), or select another plane. A concave connecting arc is created.
3. If appropriate, click on YES to view the convex solution.
4. Click on YES again to create a line joining the tangent points.
5. If desired, indicate in another region and the corner will be recomputed.

Create circular arc joining two elements without relimiting

Select: LIMIT1 > CORNER | NO TRIM

Prompt: SEL 1st LN/CRV / / KEY RAD

1. Enter a new radius value if desired. Select a line or curve on the side to be retained. (A new radius value can be entered at this stage as well.)
2. Select a line or curve. A concave connecting arc is created.
3. Click on YES to view the convex solution.
4. Click on YES again to create a line joining the tangent points.
5. If required, indicate in another region and the corner will be recomputed.

MACHINE

Create chamfers and connecting curves.

Select: LIMIT1 > MACHINE | CHAMFER or MACH JOG or SHT JOG

Create chamfers

Select: LIMIT1 > MACHINE | CHAMFER | ANGLE or LENGTH | TRIM ALL or TRIM EL1 or NO TRIM

Create chamfer by specifying angle and one leg length

Select: LIMIT1 > MACHINE | CHAMFER | ANGLE | TRIM ALL or TRIM EL1 or NO TRIM

Prompt: SEL 1st LN/CRV

1. Select a line or curve in the region to be retained. Select a second line in the region to be retained.
2. If required, enter length and angle values. If length value only is entered, 45 degrees is assumed. Alternatively, click on YES to accepted suggested values. The chamfer is created.
3. TRIM ALL, TRIM EL1, and NO TRIM work as described in CORNER option entry. If desired, indicate in another region, and the corner will be recomputed.

Create chamfer by specifying two leg lengths

Select: LIMIT1 > MACHINE | CHAMFER | LENGTH | TRIM ALL or TRIM EL1 or NO TRIM

Prompt: SEL 1st LN/CRV

1. Select a line or curve in the region to be retained. Select a second line or curve in the region to be retained.
2. If desired, enter values for leg lengths LNG1 and LNG2. Alternatively, click on YES to accept values suggested. Chamfers are created.
3. If required, indicate in another region, and the corner will be recomputed.

TRIM ALL, TRIM EL1 and NO TRIM work as described under the CORNER option.

Create connecting arcs between two parallel lines

Select: LIMIT1 > | MACHINE | MACH JOG | SINGL RAD or DBL RAD | TRIM EL2 or NO TRIM

Prompt: REFERENCE LINE: SEL LN1

Upon selecting the SINGL RAD option, a single connecting curve tangent to the second selected line will be created.

When the DBL RAD option is selected, lines will be joined by two curves tangent to each line.

Create single connecting curve

Select: LIMIT1 > | MACHINE | MACH JOG | SINGL RAD | TRIM EL2 or NO TRIM

Prompt: REFERENCE LINE: SEL LN1

1. Select the first line in the region to be retained. Select the second line parallel to the first in the region to be retained.
2. Indicate the region in which the connection is to be created, or select a point.
3. Enter a radius or click on YES to accept value suggested. Connecting curve is created.

4. TRIM EL2 and NO TRIM work as described under CORNER option.

Join lines by two curves

Select: LIMIT1 > | MACHINE | MACH JOG | DBL RAD | TRIM EL2 or NO TRIM

Prompt: REFERENCE LINE: SEL LN1

1. Select the first line in the region to be retained. Select a second line parallel to the first in the region to be retained.
2. Indicate the region in which the connecting arc is to be created, or select a point.
3. Enter values for RAD1 and RAD2. (The sum of the two must be greater than or equal to the distance between the parallel lines.) Alternatively, click on YES to accept the values suggested.

TRIM EL2 and NO TRIM work as described under the CORNER option.

Create connecting arcs between pairs of parallel lines or curves

Select: LIMIT1 > MACHINE | SHT JOG

Prompt: REFERENCE LINE: SEL LN1

1. Select the first line in the region to be retained. Select the second line parallel to the first in the region to be retained.
2. Select two more lines parallel to each other at the same distance apart, and parallel to the first two.
3. Indicate the region in which the connecting curves are to be created.
4. Enter values for inner radius and connection length. Alternatively, click on YES to accept suggested values. Connecting curves are created.

BREAK

Create breaks in DRAW elements.

Select: LIMIT1 > BREAK

Prompt: WSP MULTI-SEL SEL LN/CRV/CCV/CST/NRBC

Upon selection of the BREAK option, a Tolerance panel is displayed offering the following options.

- STD. Break using default intersection projection tolerance.
- COARSE. Break using a different value of intersection projection tolerance.

To break a line or curve, take the following steps.

1. Select a point; the break is made at the point if it is on the line or at the projection point if it is not. Alternatively, if you select a line, the break is made at the point of actual or projected intersection. Finally, if you select a curve, the proposed break points at intersections are displayed with symbols and numbers.
2. Enter the numbers of desired break points, or click on YES to break at all points.

To break a curve into elemental arcs, take the following steps.

1. Select a curve.
2. Select a temporary symbol, or click on YES to accept all break points.

CONCATEN

Concatenate elements ends into a single element.

Select: LIMIT1 > CONCATEN | CURVE or COMP CRV

Prompt: SEL LN/CRV/CST or SEL LN/CCV

1. Select a line or curve.
2. Select a second line or curve connected to the first. The concatenation is executed on selection of the second element. Tangent discontinuity is indicated by a double arrow symbol.
3. If automatic concatenation of further elements is possible, click on YES to execute or NO to refuse.

EXTRAPOL

Extrapolate a line or a curve.

Select: LIMIT1 > EXTRAPOL

Prompt: SEL LN/CRV

1. Select a line or curve.

✓ **NOTE:** *Circle or part circle elements cannot be extrapolated. Use the RELIMIT | CLOSE option instead.*

1. Numbers are displayed at the ends. Select the end to extrapolate by selecting or entering a number.
2. Enter a value for length or percentage, depending on element type. Alternatively, click on YES to accept suggested value.
3. If desired, click on YES again to iterate the extrapolation.

✓ **NOTE:** *If the message SUPPRESS LIMIT CURVE BEFORE EXTRAPOLATION is encountered, use LIMIT1 > RELIMIT | CLOSE to relimit one or both ends of the element.*

APPR CRV

Approximate a composite curve by a general curve.

Select: LIMIT1 > APPR CRV

Prompt: SEL CCV

Select a composite curve. The approximation of the curve is performed upon selection.

LINE in DRAW Mode

Chapter 2

The LINE function is used to create and modify unlimited lines, line segments, and grids of lines.

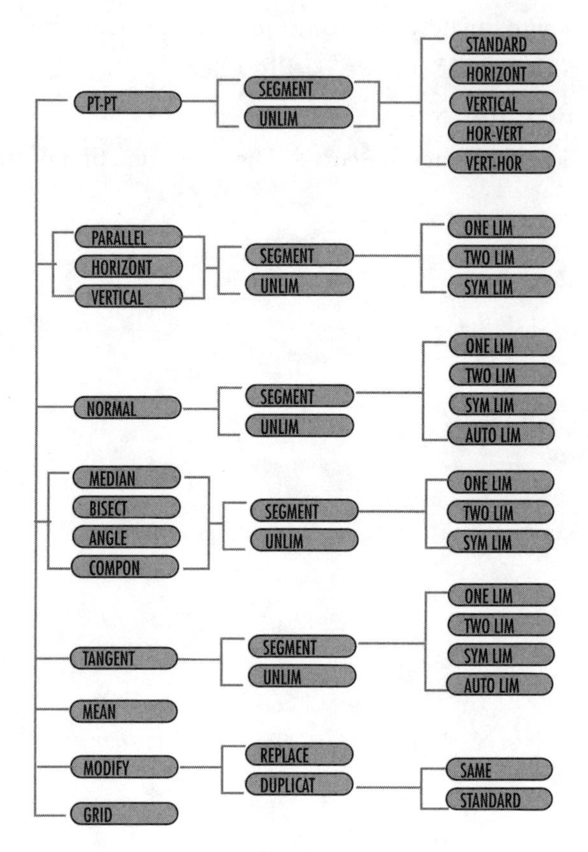

Option menu for LINE in DRAW mode.

Vicinity Selection

The vicinity selection facility is available for selecting the point(s) used in creating lines in the LINE function. The facility permits you to select a line or curve element to define the end point; the defined point is the end closest to wherever the line or curve is selected. The created end point serves as a temporary point used in creating a line. Vicinity selection is available when SEL PT <LN/CRV> appears in the message area.

Rubber Banding

The rubber banding facility is available when using the LINE function's PT-PT and MODIFY options. The facility is used to display the line to be created in the form of a rubber band. Rubber banding is switched on or off via the Display and Manipulation window. For details on this option and others accessed in the Display and Manipulation window, see Appendix G.

Line Limitation

You generally have the options of creating lines of unlimited length or line segments. Line segments can be defined by using the following suboptions: ONE LIM, TWO LIM, SYM LIM, and AUTO LIM.

When creating a line segment, the direction of the line to be created is displayed by a vector. (Reverse the vector by selecting it.) The direction of the vector is the positive direction. The length of the segment line can be defined using keyed in values or selecting limiting elements. Procedures for using the segment line options are discussed below.

ONE LIM

The start point of the line is defined by the vector origin, and the length of the line is defined by one of the following methods. (Recall that once you select the start point, you can reverse the vector direction by selecting it.)

- Click on YES to accept the displayed default value.
- Key in a value to the input information area.
- Select a limiting element among the types described below.
  - PT (Point). Define limitation as projection of the point onto the vector. With the BISECT and ANGLE options, the limitation is defined as follows: projection of segment line onto circle passing through the limitation point; center center is the origin point of segment line.
  - LND (Line). Define limitation as the intersection of the line with the vector direction. If the line is parallel to the vector, the line segment created will have the same length as the selected limiting line.
  - CRVD (Curve). Define limitation as the intersection of the curve with the vector direction. If more than one intersection with this curve and vector direction exists, the intersection point closest to the selected point will be taken.

TWO LIM

The line segment is defined by its two end points via one of several methods. (Recall that once you select the start point, you can reverse the vector direction by selecting it.)

Method 1

Define the limitations by algebraic values. Click on YES to accept the displayed default value(s), or key in a value(s) to the input information area.

Method 2

To define the first limitation element, use one of the following types.

- PT (Point). Define limitation as projection of the point onto the vector. With the BISECT and ANGLE options, the limitation is defined as follows: projection of the segment line onto a circle passing through the limitation point; the center is the origin point of the segment line.
- LND (Line). Define limitation as the intersection of the line with the vector direction. If the line is parallel to the vector, the line segment created will have the same length as the selected limiting line.
- PLN (Plane). Define limitation as the intersection of this plane with the vector direction.
- CRVD (Curve). Define limitation as the intersection of the curve with the vector direction. If more than one intersection with the curve and vector direction exists, the intersection closest to the selection point will be used.

An alternative to using one of the above types is to key in a value to the input information area. The first limitation will then be defined along the vector direction with the keyed value. Reverse the vector direction as necessary.

To define the second limitation, take one of the following alternatives: (a) Click on YES to accept the displayed default value. (b) Key in a value to the input information area. (c) Select a second limiting element of the following types.

- PT (Point). Define limitation as the projection of the point onto the vector. With the BISECT and ANGLE options, the limitation is defined as follows: projection of the segment line onto a circle passing through the limitation point; the center is the origin point of the segment line.
- LND (Line). Define limitation as the intersection of the line with the vector direction. If the line is parallel to the vector, the line segment created will have the same length as the selected limiting line.
- CRVD (Curve). Define limitation as the intersection of the curve with the vector direction. If more than one intersection with the curve and vector direction exists, the intersection closest to the selection point will be used.

SYM LIM

The origin of the vector symmetry is defined as the center of the symmetry, and the length of the line segment can be defined by one of the following methods.

- Click on YES to accept the displayed default value.
- Key in a value to the input information area.
- Select a limiting element of the following types.
    - PT (Point). Define limitation as the projection of the point onto the vector. With the BISECT and ANGLE options, the limitation is defined as follows: projection of the segment line onto a circle passing through the limitation point; the center is the origin point of the segment line.
    - LND (Line). Define limitation as the intersection of the line with the vector direction. If the line is parallel to the vector, the line segment created will have the same length as the selected limiting line.
    - CRVD (Curve). Define limitation as the intersection of the curve with the vector direction. If more than one intersection with the curve and vector direction exists, the intersection closest to the selection point will be used.

The created line segment will be symmetrical with respect to the origin of the vector symmetry.

AUTO LIM

In the AUTO LIM mode, the line segment created is automatically limited by the two selected elements. No vector direction is indicated.

PT-PT

Create lines between two selected points.

Select: LINE > PT-PT | UNLIM or SEGMENT

Create line between two points

Select: LINE > PT-PT | UNLIM or SEGMENT | STANDARD

Prompt: SEL 1ST PT <LN/CRV>

Successively select two point elements.

Create chain of lines

Select: LINE > PT-PT | UNLIM or SEGMENT | STANDARD

Prompt: SEL 1ST PT <LN/CRV>

1. Successively select two point elements. Click on YES to switch on the chain option.
2. Select further point elements as necessary. Click on YES to end the chain option.

Create unlimited horizontal or vertical line

Select: LINE > PT-PT | UNLIM | HORIZONT or VERTICAL

Prompt: SEL 1ST PT <LN/CRV>

Select a point element; the line created will pass through the selected point.

Create horizontal or vertical line segment

Select: LINE > PT-PT | SEGMENT | HORIZONT or VERTICAL

Prompt: SEL 1ST PT <LN/CRV>

Successively select two point elements to define the line segment. The first selected point will be the start point of the created line, and the second end point will be the projection of the second selected point onto the horizontal or vertical line passing through the first point.

✓ **NOTE:** *A chain of lines can also be created following the steps for PT-PT | SEGMENT | STANDARD.*

Create horizontal and vertical line passing through two points

Select: LINE > PT-PT | UNLIM or SEGMENT | HOR-VERT

Prompt: SEL 1ST PT <LN/CRV>

Successively select two point elements. The first selected point is on the horizontal line or line segment, and the second point is on the vertical line or line segment. In the case of line segments, the system creates the intersection of the two line segments to define the lengths of the segments.

✓ **NOTE:** *A chain of lines can also be created following the steps provided for PT-PT | SEGMENT | STANDARD.*

Create vertical and horizontal lines passing through two points

Select: LINE > PT-PT | UNLIM or SEGMENT | VERT-HOR

Prompt: SEL 1ST PT <LN/CRV>

Successively select two point elements. The first selected point is on the vertical line or line segment, and the second point is on the horizontal line or line segment. In the case of line segments, the system creates the intersection of the two line segments to define the lengths of the segments.

✓ **NOTE:** *A chain of lines can also be created following the steps for PT-PT | SEGMENT | STANDARD.*

TIP: *Remember that the vicinity selection facility can be used with all PT-PT options.*

PARALLEL

Create one or more parallel lines or line segments.

Select: LINE > PARALLEL | SEGMENT or UNLIM

Create parallel line segments

Select: LINE > PARALLEL | SEGMENT

Prompt: DIRECTION : SEL LN SEL PT/CRV

Create line segment parallel to line and passing through point

1. Select a line element, and then select a point element. The vector displayed indicates the direction, and will be parallel to the selected line. Its origin will be at the selected point.
2. Create the line segment using one of the line limitation procedures described at the beginning of this entry.

Create equidistant parallel line segment(s)

1. Select a line element.
2. Indicate the region on the side to which the parallel line segment(s) will be created.
3. Key in the distance that the created parallel line segment will be located from the first selected line. Next, the number of parallel lines desired can be keyed in (e.g., *20,3*, meaning three lines at a distance of 20 from the previous line). Alternatively, click on YES to accept the displayed default values.
4. Create the line segment(s) using one of the line limitation procedures described at the beginning of this entry.

Create chain set of equidistant parallel line segments

1. Follow the steps provided for creating equidistant parallel line segments.
2. Click on YES to switch on the chain option.
3. Key in a value for the parallel distance as well as the number of lines as necessary.
4. Create the line segment(s) using one of the line limitation procedures described at the beginning of this entry.
5. Click on YES to end the chain option.

Create line segment parallel to line and tangent to curve

1. Select a line element, and then select a curve element.
2. Create the line segment(s) using one of the line limitation procedures described at the beginning of this entry.

Create line segment equidistant between two parallel lines

1. Successively select two line elements.
2. Key in the number of desired line segments.
3. Create the line segment(s) using one of the line limitation procedures described at the beginning of this entry.

Create parallel unlimited lines

Select: LINE > PARALLEL | UNLIM

Prompt: DIRECTION : SEL LN SEL PT/CRV

Create unlimited line parallel to line and passing through point

Select a line element, and then select a point element. The vector displayed indicates the direction, and will be parallel to the selected line. Its origin will be at the selected point.

Create equidistant parallel unlimited line

1. Select a line element.
2. Indicate the region at the side to which the parallel line segment(s) will be created.
3. Key in the distance at which the created parallel line segment is to be located from the first selected line. The number of desired parallel lines can be keyed in as well (e.g., *20,3*, meaning three lines at a distance of 20 from the previous line). Alternatively, click on YES to accept the displayed default values.

Create chain set of equidistant unlimited parallel lines

1. Follow the steps provided for creating equidistant parallel unlimited lines.
2. Click on YES to switch on the chain option.
3. Key in a value for the parallel distance, as well as the number of lines as necessary.
4. Click on YES to end the chain option.

Create unlimited line parallel to line and tangent to curve

Select a line element, and then select a curve element.

Create unlimited line equidistant between two parallel lines

Successively select two line elements, and then key in the number of desired line segments.

HORIZONT or VERTICAL

Create horizontal or vertical lines.

Select: LINE > HORIZONT or VERTICAL | SEGMENT or UNLIM

Create horizontal or vertical line segments

Select: LINE > HORIZONT or VERTICAL | SEGMENT

Prompt: SEL PT/CRV // KEY DIST

Create horizontal or vertical line segment passing through point

1. Select a point type element.
2. Create the line segment using one of the line limitation procedures described at the beginning of this entry.

Create horizontal or vertical line segment tangent to curve

1. Select a curve type element.
2. Create the line segment using one of the line limitation procedures described at the beginning of this entry. The line segment created will be the nearest tangent to the selection position of the selected curve.

Create horizontal or vertical line segment at specific distance from origin

1. Key in a distance in the input information area to define the distance from the origin. If <Enter> is pressed with no value in the input information area, a line will be created passing through the origin.
2. Create the line segment using one of the line limitation procedures described at the beginning of this entry.

Create unlimited horizontal or vertical lines

Select: LINE > HORIZONT or VERTICAL | UNLIM

Prompt: SEL PT/CRV // KEY DIST

Create unlimited horizontal or vertical line passing through point

Select a point type element.

Create unlimited horizontal or vertical line tangent to curve

Select a curve type element. The line created will be the nearest tangent to the selection position of the selected curve.

Create unlimited horizontal or vertical line at specific distance from origin

Key in a distance to the input information area to define the distance from the origin. If <Enter> is pressed with no value in the input information area, a line will be created passing through the origin.

NORMAL

Create lines normal to selected elements.

Select: LINE > NORMAL | SEGMENT or UNLIM

Create normal line segments

Select: LINE > NORMAL | SEGMENT

Prompt: SEL 1ST PT <LN/CRV>

Create line segment normal to line and passing through point

1. Select a point element, and then select a line element.
2. Create the line segment using one of the line limitation procedures described at the beginning of this entry.

Create line segment normal to curve and passing through point

1. Select a point element, and then select a curve element. If there is more than one solution, the created normal segment will be the normal closest to the position the curve was selected.
2. Create the line segment using one of the line limitation procedures described at the beginning of this entry.

Create line segment normal to line and curve

1. Select a line element, and then select a curve element. If there is more than one solution, the created normal segment will be the normal closest to the position where the curve was selected.
2. Create the line segment using one of the line limitation procedures described at the beginning of this entry.

Create line segment normal to circle and curve

1. Select a curve element, and then select a circle element. If there is more than one solution, the created normal segment will be the normal closest to the position where the curve and circle was selected.

2. Create the line segment using one of the line limitation procedures described at the beginning of this entry.

Create unlimited normal line

Select: LINE > NORMAL | UNLIM

Prompt: SEL 1ST PT<LN/CRV>

Create unlimited line normal to line and passing through point

Select a point element, and then select a line element.

Create unlimited line normal to curve and passing through point

Select a point element, and then select a curve element. The created normal segment will be the normal closest to the position where the curve was selected.

Create unlimited line normal to line and curve

Select a line element, and then select a curve element. The created normal segment will be the normal closest to the position where the curve was selected.

Create unlimited line normal to circle and curve

1. Select a curve or circle element, and then select another circle element. The created line will pass through the center of the two circles, or through the center of the circle and normal to the curve.

MEDIAN

Create median lines.

Select: LINE > MEDIAN | SEGMENT or UNLIM

Create median segment of two points or line segment

Select: LINE > MEDIAN | SEGMENT

Prompt: SEL PT1 / LN

1. Successively select two point elements or a line element.
2. Create the line segment using one of the line limitation procedures described at the beginning of this entry.

Create unlimited median of two points or line segment

Select: LINE > MEDIAN | UNLIM

Prompt: SEL PT1 / LN

Successively select two point elements, or select a line element.

BISECT

Create bisector or set of equiangular lines between two angled lines. This option can also be used to create a set of equidistant lines between two parallel lines.

Select: LINE > BISECT | SEGMENT or UNLIM

Create segment bisector or set of equiangular segment lines between two angled lines

Select: LINE > BISECT | SEGMENT

Prompt: SEL 1ST LN

1. Successively select two line elements.
2. Key in the number of equiangular segment lines to be created, or click on YES to create the bisector.
3. Create the line segment using one of the line limitation procedures described at the beginning of this entry.

Create unlimited bisector or set of equiangular lines between two angled lines

Select: LINE > BISECT | UNLIM

Prompt: SEL 1ST LN

1. Successively select two line elements.
2. Key in the number of equiangular segment lines to be created, or click on YES to create the bisector.

ANGLE

Create angled lines or set of equiangular lines.

Select: LINE > ANGLE | SEGMENT or UNLIM

Create angled line segment or set of equiangular line segments passing through point

Select: LINE >ANGLE | SEGMENT

Prompt: REFERENCE : SEL LN SEL PT

1. Select a point element to define the point of origin. The default reference line is the current horizontal axis. If a different reference is desired, select a line element or select two points. (The line passing through the two points will be the reference line.)
2. Select the rotation vector to reverse the rotation angle as necessary.
3. Click on YES to accept the displayed default values, or key in the desired angle. If required, key in the number of line segments (max. 50) in the *xx,x* format (e.g., *20,3* means three lines at an angle of 20 degrees from the previous angled line). If only the angle value is keyed in, one line segment will be created.
4. If desired, select the angle vector to reverse its direction.
5. Create the line segment using one of the line limitation procedures described at the beginning of this entry.

Create unlimited angled line or set of equiangular lines passing through point

Select: LINE >ANGLE | UNLIM

Prompt: REFERENCE : SEL LN SEL PT

1. Select a point element to define the point of origin. The default reference line is the current horizontal axis. If a different reference is desired, select a line element or select two points (the line passing through the two points will be the reference line).
2. If required, select the rotation vector to reverse the rotation angle.
3. Click on YES to accept the displayed default values, or key in the desired angle. If required, key in the number of line segments (max. 50) in the following format *xx,x* (e.g., *20,3* means three lines at an angle of 20 degrees from the previous angled line). If only the angle value is keyed in, one line segment will be created.

TIP: *Remember that the vicinity selection facility can be used with all ANGLE options.*

COMPON

Create line based on its components. Components are the lengths of line projections onto each axis of the two-axis system.

Select: LINE > COMPON | SEGMENT or UNLIM

Creating line segments using components

Select: LINE > COMPON | SEGMENT

Prompt: SEL PT <LN/CRV> KEY A, B, C

Create line segment using Cartesian equation

1. Key in the values of coefficients A, B, and C in the following Cartesian equation: AX + BY + C = 0.
2. Create the line segment using one of the line limitation procedures described at the beginning of this entry.

Create line segment with specific components and passing through point

1. Select a point element.
2. Use one of the following alternatives: (a) Click on YES to accept the displayed default values. (b) Key in a value for the angle between the line segment and the current horizontal axis. (c) Key in values for the two components of the line direction vector. (If the same values, such as *20,20,* are keyed in, the line angle will be 45 degrees.)
3. Create the line segment using one of the line limitation procedures described at the beginning of this entry.

Create unlimited line using components

Select: LINE > COMPON | UNLIM

Prompt: SEL PT <LN/CRV> KEY A, B, C

Create unlimited line using Cartesian equation

Key in values of coefficients A, B, and C in the following Cartesian equation: AX + BY + C = 0.

Create unlimited line with specific components and passing through point

1. Select a point element.
2. Click on YES to accept the displayed default values, or key in a value for the angle between the line segment and the current horizontal axis. Alternatively, key in values for the two components of the line direction vector. (If the same values, such as *20,20,* are keyed in, the line angle will be 45 degrees.)

TIP: *Remember that the vicinity selection facility can be used with all COMPON options.*

TANGENT

Create tangent lines.

Select: LINE > TANGENT | SEGMENT or UNLIM

Create tangent line segments

Select: LINE > TANGENT | SEGMENT

Prompt: SEL PT / CRV1

Create line segment tangent to curve and passing through point

1. Select a point element.
2. Select a curve element. A direction vector will appear indicating a line tangent to the curve closest to the selection position, and which passes through and has its origin on the first selected point.
3. Create the line segment using one of the line limitation procedures described at the beginning of this entry.

Create line segment tangent to two curves

1. Successively select two curve elements. A direction vector will appear indicating a line tangent to the two curves closest to where they were selected.
2. Create the line segment using one of the line limitation procedures described at the beginning of this entry.

Create line segment tangent to curve with specific angle from reference line

1. Select a curve element.
2. Select a line element from which the angle is calculated. If no line is selected, the angle will be taken from the current horizontal axis.
3. If desired, select the rotation vector to reverse the angle.
4. Key in a value for the tangent angle. A tangent vector will appear closest to the position selected on the curve, and at the keyed angle relative to the current horizontal axis or the selected line. Alternatively, click on YES to accept the displayed default value.
5. Create the line segment using one of the line limitation procedures described at the beginning of this entry.

Create unlimited tangent lines

Select: LINE > TANGENT | UNLIM

Prompt: SEL PT / CRV1

Create unlimited line tangent to curve and passing through point

Select a point element, and then select a curve element. The line created will be a line tangent to the curve closest to the selection position, and which passes through, and has its origin on, the first selected point.

Create unlimited line tangent to two curves

Successively select two curve elements. The line created will be tangent to the two curves closest to where they were selected.

Create unlimited line tangent to curve with specific angle from reference line

1. Select a curve element.
2. Select a line element from which the angle is calculated. If no line is selected, the angle will be taken from the current horizontal axis.
3. If desired, select the rotation vector to reverse the angle.
4. Key in a value for the tangent angle, or click on YES to accept the displayed default value.

The tangent line created will be closest to the position selected on the curve, and at the keyed angle relative to the current horizontal axis or the selected line.

MEAN

Create mean line from two or more selected points.

Select: LINE > MEAN

Prompt: SEL IST PT <LN/CRV>

1. Successively select at least two points.
2. Click on YES to create the desired mean line. The line is created so that the sum of the squares of the distance between the points and the created line is minimized.

TIP: Remember that the vicinity selection facility can be used with the MEAN option.

MODIFY

Modify line geometry.

Select: LINE > MODIFY | REPLACE or DUPLICAT | STANDARD or SAME

Prompt: SEL LN

Upon modifying line geometry, you have the following choices.

- REPLACE. New line replaces old line. If dimensions or text are associated with the line, the dimensions will be recomputed.
- DUPLICAT. Selected line is duplicated. Suboptions are described below.
  - STANDARD. New line is created with the current settings as defined in the STANDARD function.
  - SAME. New line is created with the same attributes and so forth as the selected line.

Translate line segment

1. Select a line segment. Three symbols display on the line. Symbols 1 and 2 are at the end points and 3 at the midpoint.
2. Click on YES to initialize the translation process.
3. Select one of the symbols on the line segment to define the origin of the translation, and then select a point element to define the end point of the translation. Alternatively, key in values for the translation components.
4. If desired, click on YES to repeat the translation.

Translate unlimited line

1. Select a line segment.
2. Click on YES to initialize the translation process.
3. Select a point element to define the origin of the translation vector, or select a point element to define the end point of the translation. Alternatively, key in values for the translation components.
4. Click on YES to repeat the translation as necessary.

Rotate line

1. Select a line element to be rotated. If a line segment is selected, three symbols display on the line. Symbols 1 and 2 are at the end points and 3 at the midpoint.
2. Select a line element. The point of intersection of the two lines will be the center of rotation. Two alternatives follow: (a) Select a point or circle element. The center of rotation will be where the point or center of the circle is projected onto the line. (b) Select a symbol. If a line segment was selected in step 1, the center of rotation will be defined by the position of the symbol.
3. At this juncture, you again have three alternatives: (a) Key in a value to define the angle of rotation. (b) Select a line element. The created line will be parallel to this selected line. (c) Select a point element. The created line will be parallel to a line joining this point to the previously defined center of rotation.
4. Click on YES for an alternative solution as necessary.

Create symmetry around line

1. Select a line element. If a line segment is selected, three symbols display on the line. Symbols 1 and 2 are at the end points and 3 at the mid point.
2. Select a line element to define the axis of symmetry. Click on YES to create the symmetrical line.

Modify end point of line segment or stretch line segment

1. Select the line segment to be modified. Three symbols display on the line. Symbols 1 and 2 are at the end points and 3 at the midpoint.
2. Select a symbol, and then select a point element. The point corresponding to the previously selected symbol will move to this selected point.

GRID

Create line grid.

Select: LINE > GRID

Prompt: YES : UNSPEC   ORIGIN : SEL PT

Create orthogonal line grid parallel to axes of current axis system

Use diagonal point and number of lines in grid

1. Select a point element to define the grid reference. The four quadrants of the grid display.
2. Select a point element to define the limit of the grid diagonal to the previously selected reference point.
3. Key in values for the number of lines in each grid direction (e.g., *20,20* would produce 20 grid lines in each direction).

Define step and number of lines

1. Select a point element to define the grid reference. The four quadrants of the grid display.
2. Key in values for the step desired in both grid directions (e.g., *20,20* would produce 20 grid lines in each direction).
3. Key in values for the number of lines in each direction of the first quadrant.
4. Key in values for the number of lines in each direction of the third quadrant.

Define step and number of lines in first quadrant only

1. Select a point element to define the grid reference. The four quadrants of the grid display.
2. Key in values for the step desired in both grid directions(e.g., *20,20* would produce 20 grid lines in each direction).
3. Key in values for the number of lines in each direction of the first quadrant.
4. Click on YES to create the grid.

Define step and number of lines in third quadrant only

1. Select a point element to define the grid reference. The four quadrants of the grid display.
2. Key in values for the step desired in both grid directions (e.g., *20,20* would produce 20 grid lines in each direction).
3. Click on YES, and then key in values for the number of lines in each direction of the third quadrant to create the grid.

Define step and two limiting points

Use one of the following alternatives.

One

1. Select a point element to define the grid reference. The four quadrants of the grid display.
2. Key in values for the step desired in both grid directions(e.g., *20,20* would produce 20 grid lines in each direction).
3. Select a point element in the first or fourth quadrant.
4. Select a point element in the quadrant diagonal to the one selected in the previous step.

Two

1. Select a point element to define the grid reference. The four quadrants of the grid display.
2. Key in values for the step desired in both grid directions (e.g., *20,20* would produce 20 grid lines in each direction).
3. Select a point element in the first or fourth quadrant.
4. Click on YES to create the grid.

Three

1. Select a point element to define the grid reference. The four quadrants of the grid display.

2. Key in values for the step desired in both grid directions (e.g., *20,20* would produce a dimension of 20 between grid lines in each direction).
3. Click on YES, and then select a point element in the second or third quadrant.

Create unspecified orthogonal line grid or unspecified line grid

1. Click on YES to start the unspecified grid procedure.
2. Select a point element to define the grid reference.
3. Select a line element to define the reference line for the first direction of the grid.

The above steps are necessary to define the creation of all unspecified grids. In the next step, you define an unspecified orthogonal or nonorthogonal line grid.

4. Click on YES to allow the creation of an unspecified orthogonal line grid. Alternatively, select a line element to define the reference line for the second direction of the grid to create an unspecified line grid.
5. Complete the grid by using one of the following definition options.
   - Select the diagonal point and number of lines. Select a point element to define the limit of the grid diagonal to the previously selected reference point. Key in values for the number of lines in each grid direction, the grid will be created.
   - Key in values for the step desired in both directions. Key in values for the number of lines. Use one of the following three alternatives.
     - Key in values for the step desired in both grid directions. Key in values for the number of lines required in each direction of the first quadrant. Key in values for the number of lines desired in each direction of the third quadrant. The grid is created.
     - Key in values for the number of lines desired in each direction of the first quadrant. Click on YES. The number of lines in each direction of the third quadrant will be zero, and the grid is created.
     - Click on YES; the number of lines in each direction of the first quadrant will be zero. Key in values for the number of lines desired in each direction of the third quadrant. The grid is created.
   - Define the diagonal point and the two limiting points. Use one of the following three alternatives.
     - Key in values for the step desired in both grid directions. Select a point element in the first or fourth quadrant. Select a point element in the quadrant diagonal to the previously selected point. The grid will be created.
     - Key in values for the step required in both grid directions. Select a point element in the first or fourth quadrant. Click on YES; the number of lines in each

direction of the third or second quadrant (diagonal to the quadrant selected above) will be zero. The grid will be created.

- Key in values for the step desired in both grid directions. Click on YES; the number of lines in each direction of the first quadrant will be zero. Select a point element in the third or second quadrant. The grid will be created.

LINE in 2D SPACE Mode

Chapter 9

The LINE function is used to create and modify unlimited lines, line segments, and line grids.

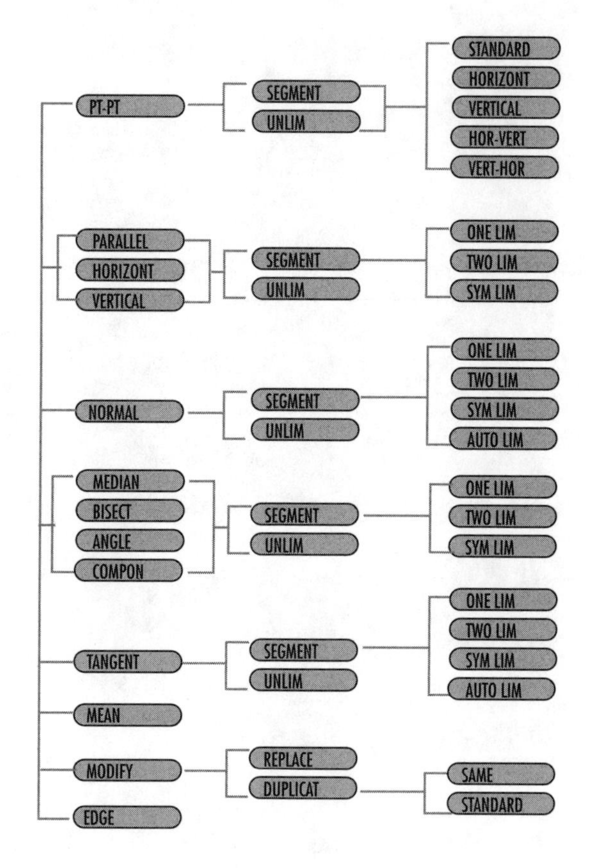

Option menu for LINE in 2D SPACE mode.

Vicinity Selection

The vicinity selection facility is available for selecting the point(s) used in creating lines in the LINE function. The facility permits you to select a line or curve element to define the end point; the defined point is the end closest to wherever the line or curve is selected. The created end point serves as a temporary point used in creating a line. Vicinity selection is available when SEL PT <LN/CRV> appears in the message area.

Rubber Banding

The rubber banding facility is available when using the LINE function's PT-PT and MODIFY options. The facility is used to display the line to be created in the form of a rubber band. Rubber banding is switched on or off via the Display and Manipulation window. For details on this option and others accessed in the Display and Manipulation window, see Appendix G.

Horizontal and Vertical

In 2D SPACE mode, the horizontal and vertical are defined with respect to the first and the second axes of the current axis system.

Line Limitation

When creating lines you generally have the options of creating lines of unlimited length or line segments. Line segments can be defined by using the following suboptions: ONE LIM, TWO LIM, SYM LIM, and AUTO LIM.

When creating segment lines, the direction of the line to be created is displayed by a vector. (Reverse the vector by selecting it.) The direction of the vector is the positive direction. The length of the segment line can be defined using keyed in values or selecting limiting elements. Procedures for using the segment line options are discussed below.

ONE LIM

The start point of the line is defined by the vector origin, and the length of the line is defined by one of the following methods. (Recall that once you select the start point, you can reverse the vector direction by selecting it.)

- Click on YES to accept the displayed default value.
- Key in a value to the input information area.
- Select a limiting element among the types described below.
    - PT (Point). Define limitation as projection of the point onto the vector. Upon using the BISECT and ANGLE options, the limitation is defined as the projection of the segment line onto a circle, which passes through the limitation point and whose center is the origin point of the segment line itself.
    - LN (Line). Define limitation as the intersection of the line with the vector direction. If the line is parallel to the vector, the line segment created will have the same length as the limiting line selected.

- CRV (Curve). Define limitation as the intersection of this curve with the vector direction. If there is more than one intersection between the curve and the vector direction, the intersection point closest to the selection point will be used.

TWO LIM

The line segment is defined by its two end points via one of several methods. (Recall that once you select the start point, you can reverse the vector direction by selecting it.)

Method 1

Define the limitations by algebraic values. Click on YES to accept the displayed default value(s), or key in a value(s) to the input information area.

Method 2

To define the first limitation element, use one of the following types.

- PT (Point). Define limitation as the projection of the point onto the vector. Upon selecting the BISECT and ANGLE options, the limitation is defined as the projection of the line segment onto a circle, which passes through the limitation point and whose center is the origin point of the segment line itself.
- LN (Line). Define limitation as the intersection of the line with the vector direction. If the line is parallel to the vector, the line segment created will have the same length as the limiting line selected.
- CRV (Curve). Define limitation as the intersection of this curve onto the vector direction. If there is more than one intersection between this curve and the vector direction, the intersection point closest to the selection point will be used.

An alternative to using one of the above types is to key in a value to the input information area. The first limitation will then be defined along the vector direction with the keyed value. Reverse the vector direction as necessary.

To define the second limitation, take one of the following alternatives: (a) Click on YES to accept the displayed default value. (b) Key in a value to the input information area. (c) Select a second limiting element of the following types.

- PT (Point). Define limitation as the projection of the point onto the vector. Upon selecting the BISECT and ANGLE options, the limitation is defined as the projection of the line segment onto a circle, which passes through the limitation point and whose center is the origin point of the segment line itself.
- LN (Line). Define limitation as the intersection of the line with the vector direction.

- CRV (Curve). Define limitation as the intersection of this curve onto the vector direction. If there is more than one intersection between this curve and the vector direction, the intersection point closest to the selection point will be used.

SYM LIM

The origin of the vector symmetry is defined as the center of the symmetry, and the length of the line segment can be defined by one of the following methods.

- Click on YES to accept the displayed default value.
- Key in a value to the input information area.
- Select a limiting element of the following types.
  - PT (Point). Define limitation as the projection of the point onto the vector. Upon selecting the BISECT and ANGLE options, the limitation is defined as the projection of the line segment onto a circle, which passes through the limitation point and whose center is the origin point of the segment line itself.
  - LN (Line). Define limitation as the intersection of the line with the vector direction. If the line is parallel to the vector, the line segment created will have the same length as the limiting line selected.
  - CRV (Curve). Define limitation as the intersection of this curve onto the vector direction. If there is more than one intersection between this curve and the vector direction, the intersection point closest to the selection point will be used.

The created line segment will be symmetrical with respect to the origin of the vector symmetry.

AUTO LIM

In the AUTO LIM mode, the line segment created is automatically limited by the two selected elements. No vector direction is indicated.

PT-PT

Create line between points.

Select: LINE > PT-PT | UNLIM or SEGMENT | STANDARD or HORIZONT or VERTICAL or HOR-VERT or VERT-HOR

Create line between two points

Select: LINE > PT-PT | UNLIM or SEGMENT | STANDARD

Prompt: SEL 1ST PT <LN/CRV>

Successively select two point elements.

Create chain of lines

Select: LINE > PT-PT | UNLIM or SEGMENT | STANDARD

Prompt: SEL 1ST PT <LN/CRV>

1. Successively select two point elements.
2. Click on YES to switch on the chain option.
3. Select additional point elements as necessary.
4. Click on YES to end the chain option.

Create unlimited horizontal or vertical line

Select: LINE > PT-PT | UNLIM | HORIZONT or VERTICAL

Prompt: SEL 1ST PT <LN/CRV>

Select a point element; the line created will pass through the selected point.

Create horizontal or vertical line segment

Select: LINE > PT-PT | SEGMENT | HORIZONT or VERTICAL

Prompt: SEL 1ST PT <LN/CRV>

Successively select two point elements to define the line segment. The first selected point will be the start point of the created line. The second end point will be the projection of the second selected point onto the horizontal or vertical line passing through the first point.

✓ ***NOTE:*** *A chain of lines can also be created following the steps for PT-PT | SEGMENT | STANDARD.*

Create horizontal line and vertical line passing through two points

Select: LINE > PT-PT | UNLIM or SEGMENT | HOR-VERT

Prompt: SEL 1ST PT <LN/CRV>

Successively select two point elements. The first selected point is on the horizontal line or line segment, and the second point is on the vertical line or line segment. In the case of line segments, an intersection of the two line segments to define the lengths of the segments is created.

✓ ***NOTE:*** *A chain of lines can also be created following the steps for PT-PT | SEGMENT | STANDARD.*

Create vertical line and horizontal line passing through two points

Select: LINE > PT-PT | UNLIM or SEGMENT | VERT-HOR

Prompt: SEL 1ST PT <LN/CRV>

Successively select two point elements. The first selected point is on the vertical line or line segment, and the second point is on the horizontal line or line segment. In the case of line segments, the system creates the intersection of the two line segments to define the segment lengths.

✓ **NOTE:** *A chain of lines can also be created following the steps for PT-PT | SEGMENT | STANDARD.*

➾ **TIP:** *Remember that the vicinity selection facility can be used with all PT-PT options.*

PARALLEL

Create one or more parallel lines or line segments.

Select: LINE > PARALLEL | SEGMENT or UNLIM

Create parallel line segments

Select: LINE > PARALLEL | SEGMENT | ONE LIM or TWO LIM or SYM LIM

Prompt: DIRECTION : SEL LN   SEL PT/CRV

Create line segment parallel to line and passing through point

1. Select a line element, and then select a point element. The vector displayed indicates the direction, and will be parallel to the selected line. Its origin will be at the selected point.
2. Create the line segment using one of the line limitation procedures described at the beginning of this entry.

Create equidistant parallel line segment(s)

1. Select a line element.
2. Indicate the region on the side to which the parallel line segment(s) will be created.
3. Key in the distance that the created parallel line segment will be located from the first selected line. Next, the number of desired parallel lines can be keyed in using the *xx,x* format (e.g., *20,3*, meaning three lines at a distance of 20 from the previous line). Alternatively, click on YES to accept the displayed default values.
4. Create the line segment(s) using one of the line limitation procedures described at the beginning of this entry.

Create chain set of equidistant parallel line segments

1. Follow the steps provided for creating equidistant parallel line segments.
2. Click on YES to switch on the chain option.
3. Key in a value for the parallel distance as well as the number of lines as necessary.
4. Create the line segment(s) using one of the line limitation procedures described at the beginning of this entry.
5. Click on YES to end the chain option.

Create line segment parallel to line and tangent to curve

1. Select a line element, and then select a curve element.
2. Create the line segment(s) using one of the line limitation procedures described at the beginning of this entry.

Create line segment equidistant between two parallel lines

1. Successively select two line elements.
2. Key in the number of desired line segments.
3. Create the line segment(s) using one of the line limitation procedures described at the beginning of this entry.

Create parallel unlimited lines

Select: LINE > PARALLEL | UNLIM

Prompt: DIRECTION : SEL LN SEL PT/CRV

Create unlimited line parallel to line and passing through point

Select a line element, and then select a point element. The vector displayed indicates the direction, and will be parallel to the selected line. Its origin will be at the selected point.

Create equidistant parallel unlimited lines

1. Select a line element.
2. Indicate the region at the side to which the parallel line segment(s) will be created.
3. Key in the distance at which the created parallel line segment is to be located from the first selected line. The number of desired parallel lines can be keyed in as well using the *xx,x* format (e.g., *20,3*, meaning three lines at a distance of 20 from the previous line). Alternatively, click on YES to accept the displayed default values.

Create chain set of equidistant unlimited parallel lines

1. Follow the steps provided for creating parallel equidistant unlimited lines.

2. Click on YES to switch on the chain option.
3. Key in a value for the parallel distance, as well as the number of lines as necessary.
4. Click on YES to end the chain option.

Create unlimited line parallel to line and tangent to curve

Select a line element, and then select a curve element.

Create unlimited line equidistant between two parallel lines

Successively select two line elements, and then key in the number of desired line segments.

HORIZONT or VERTICAL

Create horizontal or vertical lines.

Select: LINE > HORIZONT or VERTICAL | SEGMENT or UNLIM

Create horizontal or vertical line segments

Select: LINE > HORIZONT or VERTICAL | SEGMENT | ONE LIM or TWO LIM or SYM LIM

Prompt: SEL PT/CRV // KEY DIST

Create horizontal or vertical line segment passing through point

1. Select a point type element.
2. Create the line segment using one of the line limitation procedures described at the beginning of this entry.

Create horizontal or vertical line segment tangent to curve

1. Select a curve type element.
2. Create the line segment using one of the line limitation procedures described at the beginning of this entry. The line segment created will be the nearest tangent to the selection position of the selected curve.

Create horizontal or vertical line segment at specific distance from origin

1. Key in a distance in the input information area to define the distance from the origin. If <Enter> is pressed with no value in the input information area, a line will be created passing through the origin.
2. Create the line segment using one of the line limitation procedures described at the beginning of this entry.

Create unlimited horizontal or vertical lines

Select: LINE > HORIZONT or VERTICAL | UNLIM

Prompt: SEL PT/CRV // KEY DIST

Create unlimited horizontal or vertical line passing through point

Select a point type element.

Create unlimited horizontal or vertical line tangent to curve

Select a curve type element. The line created will be the nearest tangent to the selection position of the selected curve.

Create unlimited horizontal or vertical line at specific distance from origin

Key in a distance to the input information area to define the distance from the origin. If <Enter> is pressed with no value in the input information area, a line will be created passing through the origin.

NORMAL

Create lines normal to selected elements.

Select: LINE > NORMAL | SEGMENT or UNLIM

Create normal line segments

Select: LINE > NORMAL | SEGMENT | ONE LIM or TWO LIM or SYM LIM or AUTO LIM

Prompt: SEL PT / LN / CRV

Create line segment normal to line and passing through point

1. Select a point element, and then select a line element.
2. Create the line segment using one of the line limitation procedures described at the beginning of this entry.

Create line segment normal to curve and passing through point

1. Select a point element, and then select a curve element. If there is more than one solution, the created normal segment will be the normal closest to the position the curve was selected.
2. Create the line segment using one of the line limitation procedures described at the beginning of this entry.

Create line segment normal to line and curve

1. Select a line element, and then select a curve element. If there is more than one solution, the created normal segment will be the normal closest to the position where the curve was selected.
2. Create the line segment using one of the line limitation procedures described at the beginning of this entry.

Create line segment normal to circle and curve

1. Select a curve element, and then select a circle element. If there is more than one solution, the created normal segment will be the normal closest to the position where the curve and circle was selected.
2. Create the line segment using one of the line limitation procedures described at the beginning of this entry.

Create unlimited normal line

Select: LINE > NORMAL | UNLIM

Prompt: SEL PT / LN / CRV

Create unlimited line normal to line and passing through point

Select a point element, and then select a line element.

Create unlimited line normal to curve and passing through point

Select a point element, and then select a curve element. The created normal segment will be the normal closest to the position where the curve was selected.

Create unlimited line normal to line and curve

Select a line element, and then select a curve element. The created normal segment will be the normal closest to the position where the curve was selected.

Create unlimited line normal to circle and curve

1. Select a curve or circle element, and then select another circle element. The created line will pass through the center of the two circles, or through the center of the circle and normal to the curve.

MEDIAN

Create median lines.

Select: LINE > MEDIAN | SEGMENT or UNLIM

Create median segment of two points or line segment

Select: LINE > MEDIAN | SEGMENT | ONE LIM or TWO LIM or SYM LIM

Prompt: SEL PT1 / LN

1. Successively select two point elements or a line element.
2. Create the line segment using one of the line limitation procedures described at the beginning of this entry.

Create unlimited median of two points or line segment

Select: LINE > MEDIAN | UNLIM

Prompt: SEL PT1 / LN

Successively select two point elements, or select a line element.

BISECT

Create bisector or set of equiangular lines between two angled lines. This function can also be used to create a set of equidistant lines between two parallel lines.

Select: LINE > BISECT | SEGMENT or UNLIM

Create segment bisector or set of equiangular segment lines between two angled lines

Select: LINE > BISECT | SEGMENT | ONE LIM or TWO LIM or SYM LIM

Prompt: SEL 1ST LN

1. Successively select two line elements.
2. Key in the number of equiangular segment lines to be created, or click on YES to create the bisector.
3. Create the line segment using one of the line limitation procedures described at the beginning of this entry.

Create unlimited bisector or set of equiangular lines between two angled lines

Select: LINE > BISECT | UNLIM

Prompt: SEL 1ST LN

1. Successively select two line elements.
2. Key in the number of equiangular segment lines to be created, or click on YES to create the bisector.

ANGLE

Create angled lines or set of equiangular lines.

Select: LINE >ANGLE | SEGMENT or UNLIM

Create angled line segment or set of equiangular line segments passing through point

Select: LINE >ANGLE | SEGMENT | ONE LIM or TWO LIM or SYM LIM

Prompt: REFERENCE : SEL LN SEL PT

1. Select a point element to define the point of origin. The default reference line is the current horizontal axis. If a different reference is desired, select a line element or select two points. (The line passing through the two points will be the reference line.)
2. If required, select the rotation vector to reverse the rotation angle.
3. Click on YES to accept the displayed default values, or key in the desired angle. If required, key in the number of line segments (max. 50) in the *xx,x* format (e.g., *20,3* means three lines at an angle of 20 degrees from the previous angled line). If only the angle value is keyed in, one line segment will be created.
4. If desired, select the angle vector to reverse its direction.
5. Create the line segment using one of the line limitation procedures described at the beginning of this entry.

Create unlimited angled line or set of equiangular lines passing through point

Select: LINE >ANGLE | UNLIM

Prompt: REFERENCE : SEL LN SEL PT

1. Select a point element to define the point of origin. The default reference line is the current horizontal axis. If a different reference is desired, select a line element or select two points (the line passing through the two points will be the reference line).
2. Select the rotation vector to reverse the rotation angle as necessary.
3. Click on YES to accept the displayed default values, or key in the desired angle. If required, key in the number of line segments (max. 50) in the *xx,x* format (e.g., *20,3* means three lines at an angle of 20 degrees from the previous angled line). If only the angle value is keyed in, one line segment will be created.

TIP: *Remember that the vicinity selection facility can be used with all ANGLE options.*

COMPON

Create line based on its components. Components are the lengths of line projections onto each axis of the two-axis system.

Select: LINE > COMPON | SEGMENT or UNLIM

Create line segments using components

Select: LINE > COMPON | SEGMENT | ONE LIM or TWO LIM or SYM LIM

Prompt: SEL PT <LN/CRV> KEY IN A, B, C

Create line segment using Cartesian equation

1. Key in the values of coefficients A, B, and C in the following Cartesian equation: AX + BY + C = 0.
2. Create the line segment using one of the line limitation procedures described at the beginning of this entry.

Create line segment with specific components and passing through point

1. Select a point element.
2. Click on YES to accept the displayed default values. Alternatively, key in a value for the angle between the line segment and the current horizontal axis, or key in values for the two components of the line direction vector. (If the same values, such as *20,20*, are keyed in, the line angle will be 45 degrees.)
3. Create the line segment using one of the line limitation procedures described at the beginning of this entry.

Create unlimited line using components

Select: LINE > COMPON | UNLIM

Prompt: SEL PT <LN/CRV> KEY IN A, B, C

Create unlimited line using Cartesian equation

Key in values of coefficients A, B, and C in the following Cartesian equation: AX + BY + C = 0.

Create unlimited line with specific components and passing through point

1. Select a point element.
2. Click on YES to accept the displayed default values, or key in a value for the angle between the line segment and the current horizontal axis. Alternatively, key in val-

ues for the two components of the line direction vector. (If the same values, such as *20,20*, are keyed in, the line angle will be 45 degrees.)

TIP: *Remember that the vicinity selection facility can be used with all COMPON options.*

TANGENT

Create tangent lines.

Select: LINE > TANGENT | SEGMENT or UNLIM

Create tangent line segments

Select: LINE > TANGENT | SEGMENT | ONE LIM or TWO LIM or SYM LIM or AUTO LIM

Prompt: SEL PT / CRV1

Create line segment tangent to curve and passing through point

1. Select a point element.
2. Select a curve element. A direction vector will appear indicating a line tangent to the curve closest to the selection position, and which passes through and has its origin on the first selected point.
3. Create the line segment using one of the line limitation procedures described at the beginning of this entry.

Create line segment tangent to two curves

1. Successively select two curve elements. A direction vector will appear indicating a line tangent to the two curves closest to where they were selected.
2. Create the line segment using one of the line limitation procedures described at the beginning of this entry.

Create line segment tangent to curve with specific angle from reference line

1. Select a curve element.
2. Select a line element from which the angle is calculated. If no line is selected, the angle will be taken from the current horizontal axis.
3. Select the rotation vector to reverse the angle as necessary.
4. Key in a value for the tangent angle. A tangent vector will appear closest to the position selected on the curve, and at the keyed angle relative to the current horizontal

axis or the selected line. Alternatively, click on YES to accept the displayed default value.

5. Create the line segment using one of the line limitation procedures described at the beginning of this entry.

Create unlimited tangent lines

Select: LINE > TANGENT | UNLIM

Prompt: SEL PT / CRV1

Create unlimited line tangent to curve and passing through point

Select a point element, and then select a curve element. The line created will be a line tangent to the curve closest to the selection position, and which passes through and has its origin on the first selected point.

Create unlimited line tangent to two curves

Successively select two curve elements. The line created will be tangent to the two curves closest to where they were selected.

Create unlimited line tangent to curve with specific angle from reference line

1. Select a curve element.
2. Select a line element from which the angle is calculated. If no line is selected, the angle will be taken from the current horizontal axis.
3. Select the rotation vector to reverse the angle as necessary.
4. Key in a value for the tangent angle, or click on YES to accept the displayed default value.

The tangent line created will be closest to the position selected on the curve, and at the keyed angle relative to the current horizontal axis or the selected line.

MEAN

Create mean line from two or more selected points.

Select: LINE > MEAN

Prompt: SEL IST PT <LN/CRV>

1. Successively select at least two points.
2. Click on YES to create the desired mean line. The line is created so that the sum of the squares of the distance between the points and the created line is minimized.

➾ **TIP:** *Remember that the vicinity selection facility can be used with the MEAN option.*

MODIFY

Modify line geometry.

Select: LINE > MODIFY | REPLACE or DUPLICAT

Prompt: SEL LN

When modifying line geometry, you have the following choices.

- REPLACE. New line replaces old line. If dimensions or text are associated with the line, the dimensions will be recomputed.
- DUPLICAT. Selected line is duplicated. Suboptions are described below.
  - STANDARD. New line is created with the current settings as defined in the STANDARD function.
  - SAME. New line is created with the same attributes and so forth as the selected line.

Translate line segment

1. Select a line segment. Three symbols display on the line. Symbols 1 and 2 are at the end points and 3 at the midpoint.
2. Click on YES to initialize the translation process.
3. Select one of the symbols on the line segment to define the origin of the translation.
4. Select a point element to define the end point of the translation, or key in values for the translation components.
5. Click on YES to repeat the translation as necessary.

Translate unlimited line

1. Select a line segment.
2. Click on YES to initialize the translation process.
3. Select a point element to define the origin of the translation vector.
4. Select a point element to define the end point of the translation, or key in values for the translation components.
5. Click on YES to repeat the translation as necessary.

Rotate line

1. Select a line element to be rotated. If a line segment is selected, three symbols display on the line. Symbols 1 and 2 are at the end points and 3 at the midpoint.
2. Select a line element. The point of intersection of the two lines will be the center of rotation. Two alternatives follow: (a) Select a point or circle element. The center of rotation will be where the point or center of the circle is projected onto the line. (b)

Select a symbol. If a line segment was selected in step 1, the center of rotation will be defined by the position of the symbol.

3. At this juncture, you again have three alternatives: (a) Key in a value to define the angle of rotation. (b) Select a line element. The ˇeated line will be parallel to this selected line. (c) Select a point element. The created line will be parallel to a line joining this point to the previously defined center of rotation.
4. Click on YES for an alternative solution as necessary.

Symmetry around line

1. Select a line element. If a line segment is selected, three symbols display on the line. Symbols 1 and 2 are at the end points and 3 at the midpoint.
2. Select a line element to define the axis of symmetry.
3. Click on YES to create the symmetrical line.

Modify end point of line segment or stretch line segment

1. Select the line segment to be modified. Three symbols display on the line. Symbols 1 and 2 are at the end points and 3 at the midpoint.
2. Select a symbol, and then select a point element. The point corresponding to the previously selected symbol will move to this selected point.

EDGE

Create lines normal to current 2D plane passing through point.

Select: LINE > EDGE

Prompt: SEL PT <LN/CRV>

Select a point element. The created line will be seen in the current 2D plane, but does not belong to that plane. If you switch back to 3D mode and then return to the 2D plane used above, the line created will no longer be selectable.

TIP: *Remember that the vicinity selection facility can be used with the EDGE option.*

LINE in 3D SPACE Mode

Chapter 9

The LINE function is used to create and modify unlimited lines, line segments, and grids of lines.

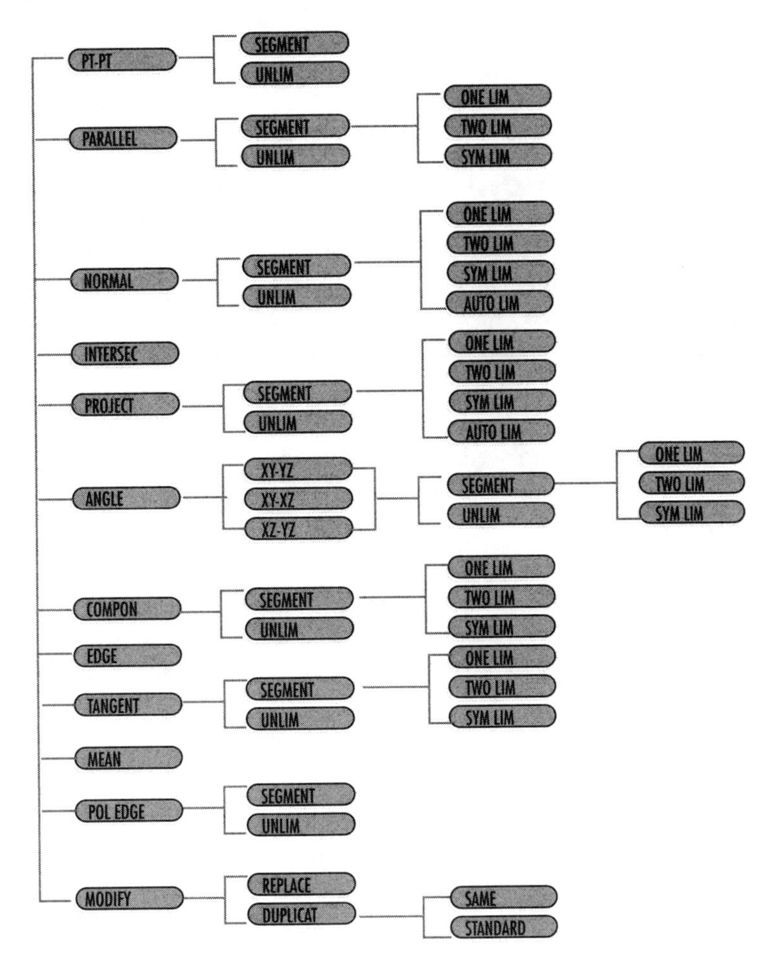

Option menu for LINE in 3D SPACE mode.

Vicinity Selection

The vicinity selection facility is available for selecting the point(s) used in creating lines in the LINE function. The facility permits you to select a line or curve element to define the end point; the defined point is the end closest to wherever the line or curve is selected.

The created end point serves as a temporary point used in creating a line. Vicinity selection is available when SEL PT <LN/CRV> appears in the message area.

Rubber Banding

The rubber banding facility is available when using the LINE function's PT-PT and MODIFY options. The facility is used to display the line to be created in the form of a rubber band. Rubber banding is switched on or off via the Display and Manipulation window. For details on this option and others accessed in the Display and Manipulation window, see Appendix G.

Line Limitation

When creating lines you generally have the options of creating lines of unlimited length or line segments. Line segments can be defined by using the following suboptions: ONE LIM, TWO LIM, SYM LIM, and AUTO LIM.

When creating segment lines, the direction of the line to be created is displayed by a vector. (Reverse the vector by selecting it.) The direction of the vector is the positive direction. The length of the segment line can be defined using keyed in values or selecting limiting elements. Procedures for using the segment line options are discussed below.

ONE LIM

The start point of the line is defined by the vector origin, and the length of the line is defined by one of the following methods. (Recall that once you select the start point, you can reverse the vector direction by selecting it.)

- Click on YES to accept the displayed default value.
- Key in a value to the input information area.
- Select a limiting element among the types described below.
  - PT (Point). Define limitation as projection of the point onto the vector.
  - LN (Line). Define limitation as the intersection of the line with the vector direction. If the line is parallel to the vector, the line segment created will have the same length as the selected limiting line.
  - PLN (Plane). Define limitation as intersection of plane with the vector direction.
  - SUR (Surface). Define limitation as intersection of surface with vector direction. If more than one intersection exists, the intersection nearest the selection point will be used.

TWO LIM

The line segment is defined by its two end points via one of several methods. (Recall that once you select the start point, you can reverse the vector direction by selecting it.)

Method 1

Define the limitations by algebraic values. Click on YES to accept the displayed default value(s), or key in a value(s) to the input information area.

Method 2

To define the first limitation element, use one of the following types.

- PT (Point). Define limitation as the projection of the point onto the vector.
- LN (Line). Define limitation as the intersection of the line with the vector direction. If the line is parallel to the vector, the line segment created will have the same length as the selected limiting line.
- PLN (Plane). Define limitation as the intersection of this plane with the vector direction.
- SUR (Surface). Define limitation as the intersection of the surface with the vector direction. If more than one intersection exists, the intersection nearest the selection point will be used.

An alternative to using one of the above types is to key in a value to the input information area. The first limitation will then be defined along the vector direction with the keyed value. Reverse the vector direction as necessary.

To define the second limitation, take one of the following alternatives: (a) Click on YES to accept the displayed default value. (b) Key in a value to the input information area. (c) Select a second limiting element of the following types.

- PT (Point). Define limitation as the projection of the point onto the vector.
- LN (Line). Define limitation as the intersection of the line with the vector direction.
- PLN (Plane). Define limitation as the intersection of this plane with the vector direction.
- SUR (Surface). Define limitation as the intersection of the surface with the vector direction. If more than one intersection exists, the intersection nearest the selection point will be used.

SYM LIM

The origin of the vector symmetry is defined as the center of the symmetry, and the length of the line segment can be defined by one of the following methods.

- Click on YES to accept the displayed default value.
- Key in a value to the input information area.
- Select a limiting element of the following types.
    - PT (Point). Define limitation as the projection of the point onto the vector.
    - LN (Line). Define limitation as the intersection of the line with the vector direction. If the line is parallel to the vector, the line segment created will have the same length as the limiting line selected.
    - PLN (Plane). Define limitation as the intersection of this plane with the vector direction.
    - SUR (Surface). Define limitation as the intersection of the surface with the vector direction. If more than one intersection exists, the intersection nearest the selection point will be used.

The created line segment will be symmetrical with respect to the origin of the vector symmetry.

AUTO LIM

In the AUTO LIM mode, the line segment created is automatically limited by the two selected elements. No vector direction is indicated.

PT-PT

Create lines between points.

Select: LINE > PT-PT | UNLIM or SEGMENT

Prompt: SEL 1ST PT <LN/CRV>

Create line between two points

Select: LINE > PT-PT | UNLIM or SEGMENT

Successively select two point elements.

Create line chain

Select: LINE > PT-PT | UNLIM or SEGMENT

1. Successively select two point elements. Click on YES to switch on the chain option.
2. Select additional point elements as necessary. Click on YES to end the chain option.

TIP: *Remember that the vicinity selection facility can be used with all PT-PT options.*

PARALLEL

Create parallel lines or line segments.

Select: LINE > PARALLEL | SEGMENT or UNLIM

Create parallel line segments

Select: LINE > PARALLEL | SEGMENT | ONE LIM or TWO LIM or SYM LIM or AUTO LIM

Prompt: ORIGIN : SEL PT DIRECTION : SEL LN

Create line segment parallel to another line and passing through point

1. Select a point element, and then select a line element. Alternatively, select a line element, and then select a point element.
2. Create the line segment using one of the line limitation procedures described at the beginning of this entry.

Create line segment parallel to intersection of two planes and passing through point

1. Select a point element, and then successively select two plane elements.
2. Create the line segment using one of the line limitation procedures described at the beginning of this entry. The line created will be parallel to a line on the intersection of the two planes.

Create line segment parallel to plane, normal to another line, and passing through point

1. Select a point element, and then a plane element and a line element.
2. Create the line segment using one of the line limitation procedures described at the beginning of this entry. The line created will be parallel to the intersection between the selected plane and a plane normal to the selected line.

Create parallel unlimited lines

Select: LINE > PARALLEL | UNLIM

Prompt: ORIGIN : SEL PT DIRECTION : SEL LN

Create unlimited line parallel to another line and passing through point

Select a point element, and then a line element. Alternatively, select a point element, and then a line element.

Create unlimited line parallel to intersection of two planes and passing through point

1. Select a point element.
2. Successively select two plane elements. The created line will be parallel to a line on the intersection of the two planes.

Create unlimited line parallel to plane, normal to another line, and passing through point

Select a point element, a plane element, and then a line element. The line created will be parallel to the intersection between the selected plane and a plane normal to the selected line.

TIP: *Remember that the vicinity selection facility can be used with all PARALLEL options.*

NORMAL

Create lines normal to selected elements.

Select: LINE > NORMAL | SEGMENT or UNLIM

Create normal line segments

Select: LINE > NORMAL | SEGMENT | ONE LIM or TWO LIM or SYM LIM or AUTO LIM

Prompt: SEL PT / LN1

Create line segment normal to plane, surface, or face, and passing through point

1. Select a point element.
2. Select a plane, surface, or face element.
3. Create the line segment using one of the line limitation procedures described at the beginning of this entry. The created line will be normal to the selected element.

Create line segment normal to two lines and passing through point

1. Select a point element, and then successively select two line elements.
2. Create the line segment using one of the line limitation procedures described at the beginning of this entry. The line will be created parallel to the normal common to both lines.

If the two selected lines are concurrent, the line created will be normal to the plane of the selected lines.

When the two selected lines are parallel, the line created will be parallel to the plane of the selected lines as well as normal to them.

Create segment line normal to line, parallel to plane, and passing through point

1. Select a point element, and then select a line element and a plane element.
2. The line segment can be created using one of the line limitation procedures described at the beginning of this entry. The created line will be parallel to the intersection between the selected plane and a plane normal to the selected line.

Create line segment normal to another line and passing through point

1. Select a point element, and then select a line element.
2. Click on YES to accept a line normal to the selected line and passing through the selected point.
3. The line segment can be created using one of the line limitation procedures described at the beginning of this entry. The line created will be normal to the selected line and passing through the selected point.

Create segment line on normal common to two lines

Successively select two line elements. The line segment can be created using one of the line limitation procedures described at the beginning of this entry. The line will be created on the normal common to both lines.

Create unlimited normal line

Select: LINE > NORMAL | UNLIM

Prompt: SEL PT / LN1

Create unlimited line normal to plane, surface, or face, and passing through point

Select a point element, and then select a plane, surface, or face element. The created line will be normal to the selected element.

Create unlimited line normal to two lines and passing through point

Select a point element, and then successively select two line elements. The line will be created parallel to the normal common to both lines. If the two selected lines are concurrent, the line created will be normal to the plane of the selected lines. When the two selected lines are parallel, the line created will be parallel to the plane of the selected lines as well as normal to them.

Create unlimited line normal to line, parallel to plane, and passing through point

Select a point element, and then select a line element and a plane element. The line created will be parallel to the intersection between the selected plane and a plane normal to the selected line.

Create unlimited line normal to another line and passing through point

1. Select a point element, and then select a line element.
2. Click on YES to accept. The line created will be normal to the selected line and passing through the selected point.

Create unlimited line on normal common to two lines

Successively select two line elements. The line will be created on the normal common to both lines.

INTERSEC

Create lines on the intersection between two planes.

Select: LINE > INTERSEC

Prompt: SEL PLN1

Successively select two plane elements.

PROJECT

Create lines projected onto plane.

Select: LINE > PROJECT | SEGMENT or UNLIM

Create line segment projected onto plane

Select: LINE > PROJECT | SEGMENT | ONE LIM or TWO LIM or SYM LIM or AUTO LIM

Prompt: SEL LN / PLN

1. Select a line element, and then select a plane element.
2. Create the line segment using one of the line limitation procedures described at the beginning of this entry.

Create unlimited line projected onto plane

Select: LINE > PROJECT | UNLIM

Prompt: SEL LN / PLN

Select a line element and then a plane element.

ANGLE

Create lines from two angles.

Select: LINE > ANGLE | XY-YZ or XY-XZ or XZ-YZ

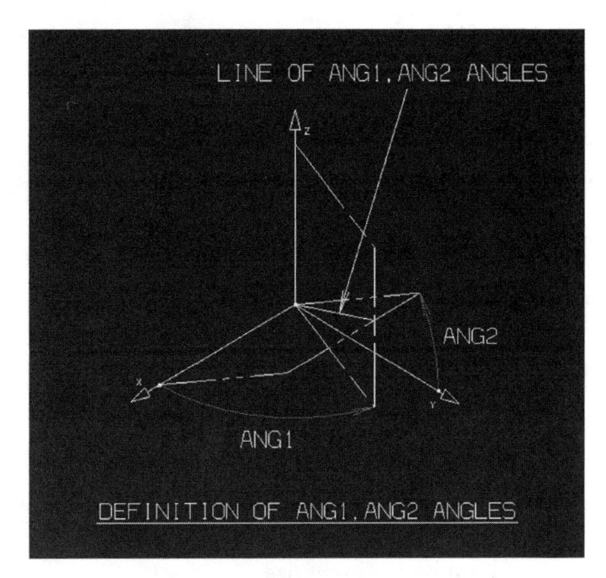

Definition of the two angles required for creating angled lines under ANGLE option.

Create line segment passing through point forming angle between projections onto two planes as specified on menu option selection and first axes of planes

Select: LINE > ANGLE | XY-YZ or XY-XZ or XZ-YZ | SEGMENT | ONE LIM or TWO LIM or SYM LIM

Prompt: SEL PT <LN/CRV> KEY ANG1, ANG2 // YES : STD

1. Select a point element.
2. Key in values for the angles with respect to the first axis of each plane as specified on menu option selection, or click on YES to accept the displayed default values.
3. Create the line segment using one of the line limitation procedures described at the beginning of this entry.

Create unlimited line passing through point forming angle between projections onto two planes as specified on menu option selection and first axes of planes

Select: LINE > ANGLE | XY-YZ or XY-XZ or XZ-YZ | UNLIM

Prompt: SEL PT <LN/CRV> KEY ANG1, ANG2 // YES : STD

1. Select a point element.
2. Key in values for the angles with respect to the first axis of each plane as specified on menu option selection, or click on YES to accept the displayed default values.

TIP: *Remember that the vicinity selection facility can be used with all ANGLE options.*

COMPON

Create line based on components. Components are the lengths of line projections onto each axis of the three-axis system.

Select: LINE > COMPON | SEGMENT or UNLIM

Create line segment passing through point with specific components

Select: LINE > COMPON | SEGMENT | ONE LIM or TWO LIM or SYM LIM

Prompt: SEL PT <LN/CRV>

1. Select a point element.
2. Key in values for the three components of the line, or click on YES to accept the displayed default values.
3. Create line segment using one of the line limitation procedures described at the beginning of this entry.

Create unlimited line passing through point with specified components

Select: LINE > COMPON | UNLIM

Prompt: SEL PT <LN/CRV>

Select a point element. Key in values for the three components of the line, or click on YES to accept the displayed default values.

TIP: *Remember that the vicinity selection facility can be used with all COMPON options.*

EDGE

Create line normal to plane of current window.

Select: LINE > EDGE

Prompt: SEL PT <LN/CRV>

Select a point element, or indicate a point. The line will be created passing through the selected point and normal to the plane of the current window. Consequently, the line will therefore be viewed from its end.

TIP: *Remember that the vicinity selection facility can be used with the EDGE option.*

TANGENT

Create tangent lines.

Select: LINE > TANGENT | SEGMENT or UNLIM

Create tangent segment lines

Select: LINE > TANGENT | SEGMENT | ONE LIM or TWO LIM or SYM LIM

Prompt: SEL CRV

Create line segment tangent to curve and passing through point on curve

Select a curve element, and then select a point element on the selected curve. The line segment can be created using one of the line limitation procedures described at the beginning of this entry. The created line will be tangential to the selected curve and pass through the selected point.

Create line segment tangent to curve end point

Select a curve element, and then select it again. The line segment can be created using one of the line limitation procedures described at the beginning of this entry. The created line will be tangential to the selected curve and passing through one of its end points.

Create unlimited tangent lines

Select: LINE > TANGENT | UNLIM

Prompt: SEL CRV

Create unlimited line tangent to curve and passing through point on curve

Select a curve element, and then select a point element on the selected curve. The created line will be tangential to the selected curve and pass through the selected point.

Create unlimited line tangent to curve end point

Select a curve element and then select the element again. The created line will be tangential to the selected curve and pass through one of its end points.

➾ ***TIP:*** *Remember that the vicinity selection facility can be used with all TANGENT options.*

MEAN

Create mean line from two or more selected points.

Select: LINE > MEAN

Prompt: SEL 1ST PT <LN/CRV>

Successively select at least two points. Click on YES to create the mean line. The line is created so that the sum of the squares of the distance between the points and the created line is minimized.

➾ ***TIP:*** *Remember that the vicinity selection facility can be used with the MEAN option.*

POL EDGE

Create lines superimposed on edges of solid or polyhedral surface.

Select: LINE > POL EDGE | UNLIM or SEGMENT

Prompt: SEL PIP / POL / SOL / STR EDGE

1. Select an edge of a solid or polyhedral surface. The created line will be a segment the same length as the selected edge or an unlimited line, depending on the menu option selected.
2. Click on YES to create lines on all edges of the selected solid or polyhedral surface. Lines will be created on all edges of the solid only if it is a polyhedral solid (i.e., created in the SOLIDM function).

MODIFY

Modify line geometry.

Select: LINE > MODIFY | REPLACE or DUPLICAT

Prompt: SEL LN

When modifying line geometry, you have the following choices.

- REPLACE. New line replaces old line. If dimensions or text are associated with the line, the dimensions will be recomputed.
- DUPLICAT. Selected line is duplicated. Suboptions are described below.
  - STANDARD. New line is created with the current settings as defined in the STANDARD function.
  - SAME. New line is created with the same attributes and so forth as the selected line.

Translate line segment

1. Select a line segment. Three symbols display on the line. Symbols 1 and 2 are at the end points and 3 at the midpoint.
2. Click on YES to initialize the translation process.
3. Select one of the symbols on the line segment to define the origin of the translation.
4. Select a point element to define the end point of the translation, or key in values for the translation components.
5. Click on YES to repeat the translation as necessary.

Translate unlimited line

1. Select a line segment.
2. Click on YES to initialize the translation process.
3. Select a point element to define the origin of the translation vector.
4. Select a point element to define the end point of the translation, or key in values for the translation components.
5. Click on YES to repeat the translation as necessary.

Create symmetry around plane

1. Select a line element. If a line segment is selected, three symbols display on the line. Symbols 1 and 2 are at the end points and 3 at the mid point.
2. Select a plane element to define the axis of symmetry and modify the selected line.

Modify end point of line segment or stretch line segment

1. Select the line segment to be modified. Three symbols display on the line. The symbols 1 and 2 are at the end points and 3 at the midpoint.
2. Select a symbol, and then select a point element. The point corresponding to the previously selected symbol will move to this selected point.

MARK UP in DRAW Mode

Chapter 3

The MARK UP function is used to create center lines, screw threads, and various arrows to annotate drawings.

Option menu for MARK UP in DRAW mode.

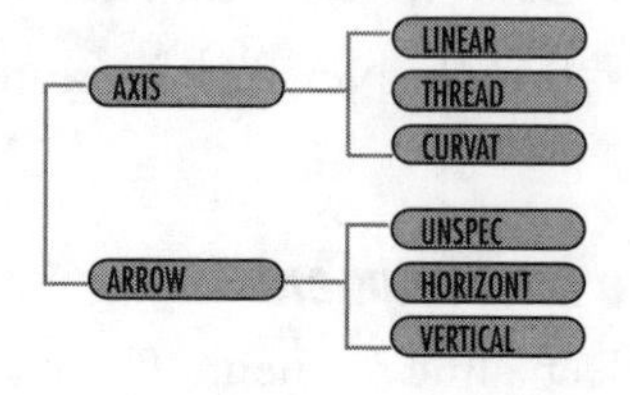

AXIS

The AXIS option is used to create center lines and screw thread representation. All lines created in this manner appear as dot-dash line with a thickness of 1 (0.1mm).

Illustration of options.

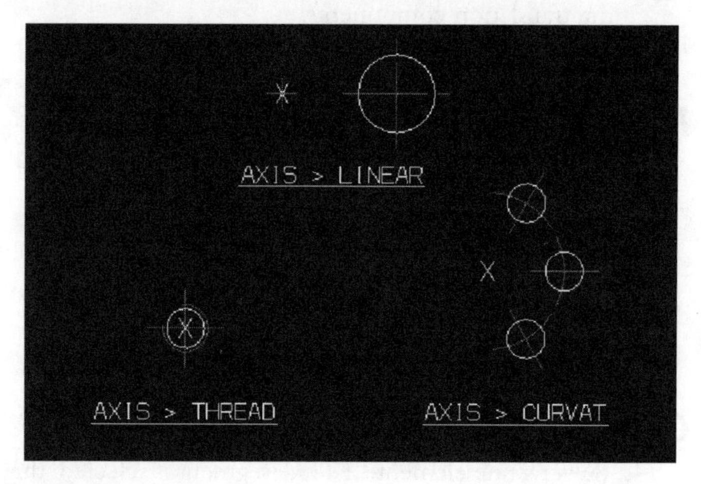

Create center lines through point or circle

Select: MARK UP > AXIS | LINEAR

Prompt: KEY AXIS OVERRUN VALUE SEL LN//SEL PT/CIR/CONIC/EDG1

The OVERRUN value is the length of the line outside the circle or from the point to the end of the line. The current OVERRUN value will display in the message area.

Create center line passing through selected point

1. If the current overrun value is incorrect, key in a value before proceeding.
2. Select a point; center lines are created on the selected point as shown in the preceding illustration.

Create center line passing through selected circle

1. If the current overrun value is incorrect, key in a value before proceeding.
2. Select a circle. Center lines are created on the selected circle as shown in the preceding illustration.

Create representation of screw thread

Select: MARK UP > AXIS | THREAD

Prompt: KEY THREAD OFFSET SEL LN//SEL CIR/EDG1

The OFFSET value is the distance between the circle you select and the broken concentric circle that you create. The current OFFSET value displays in the message area. If the OFFSET value is positive, the concentric circle will be larger than the selected circle. If the OFFSET value is negative, the concentric circle will be smaller.

1. If the current offset value is incorrect, key in a value before proceeding. If the value is correct, proceed to step 2.
2. Select a circle that represents the minor diameter of the thread; a broken concentric circle is created around the selected circle, as shown in the preceding illustration.

Create pitch circle center line

Select: MARK UP > AXIS | CURVAT

Prompt: SEL PT/CIR

1. To create a CURVAT center line, define a center point by selecting a point or circle (the center point of which becomes the center point of the line). The following prompt displays: KEY OVERRUN VALUE SEL PT/CIR // YES: END.
2. As in AXIS | LINEAR, the current OVERRUN value displays in the message area and can be changed by keying in a different value.
3. Select a circle or point. The pitch circle center lines are created as shown in the preceding illustration.

ARROW

Create arrows. In the ARROW PARAMETERS window, you can select parameters to control arrow size, end symbol, and orientation.

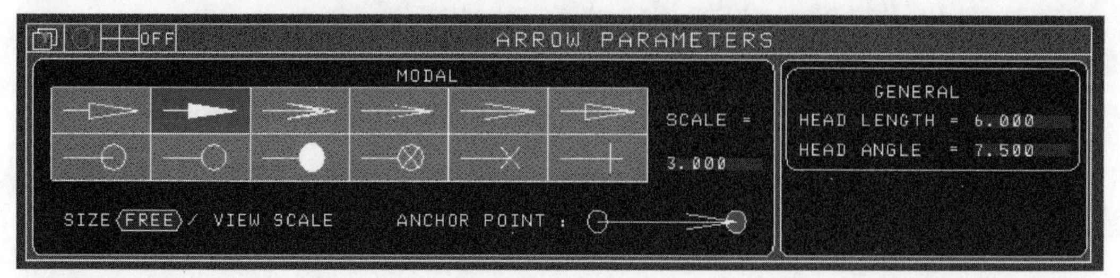

ARROW PARAMETERS window in MARK UP function.

Window options are described below.

- End symbols. Choose from arrows, circular, cross (X), or plus (+) symbols. Change the current (or highlighted) symbol by selecting another symbol.
- Symbol scale. Specify scale of end symbol.
- Symbol size. Choose size of arrow (i.e., free or linked to scale of view in which arrow is placed).
- Symbol anchor. Select symbol anchor point (i.e., symbol or opposite end).
- Head length and angle. Set arrowhead length and angle.

You can also create arrow lines that are exclusively horizontal or vertical, or at an angle.

✓ ***NOTE:*** *When length or angle is modified, existing arrows change to reflect the new values, and all subsequently created arrows will be based on the new values. All other options affect only the selected arrow.*

Create arrow

Select: MARK UP > ARROW | UNSPEC or HORIZONT or VERTICAL

Prompt: MODIFY:SEL ARROW   CREATE:SEL ANCHOR ELEM

1. Define the anchor position by selecting an element or indicating.
2. Define arrow length using one of the following alternatives: (a) Select an element. (b) Indicate its position. (c) Key in a length for the arrow line. (d) Click on the YES button to accept the standard length shown in the message area.

✓ **NOTE**: *Keying in a length or accepting the standard length in UNSPEC generates a horizontal arrow by default.*

Modify arrow

Select: MARK UP > ARROW | UNSPEC or HORIZONT or VERTICAL

Prompt: MODIFY:SEL ARROW   CREATE:SEL ANCHOR ELEM

Select the arrow. Options in the ARROW PARAMETERS window change as shown in the following illustration.

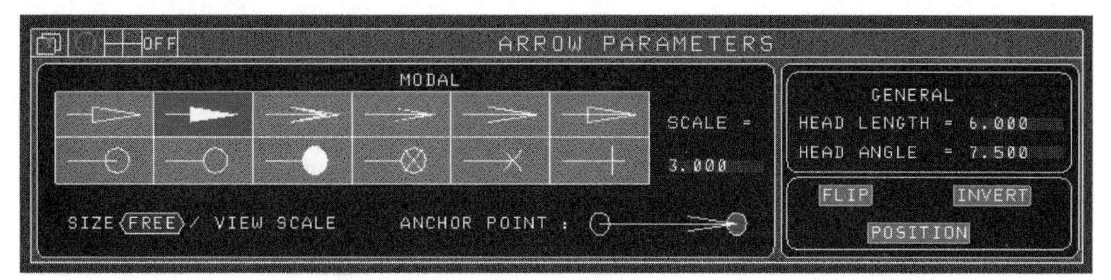

ARROW PARAMETERS window used for modifying arrows.

You can also modify the end symbol used, arrowhead size, symbol scale, and whether the arrow is free or linked to the view scale. Parameters in the dialog are described below.

- •FLIP. Flip the arrow by 180° to the anchor point.
- •INVERT. Move arrowhead to opposite end of line.
- •POSITION. Move end position of arrow but not anchor point.

Changing the three parameters described above in the window affects only the arrow you select. However, if you change the arrowhead length or angle, all arrows in the model will be resized.

MERGE in DRAW and SPACE Modes

Chapter 13

The MERGE function in DRAW and SPACE modes is used to copy elements from a sending model into a previously saved receiving model using the Save current model in tmp file button on the fixed menu.

Option menu for MERGE in DRAW and SPACE modes.

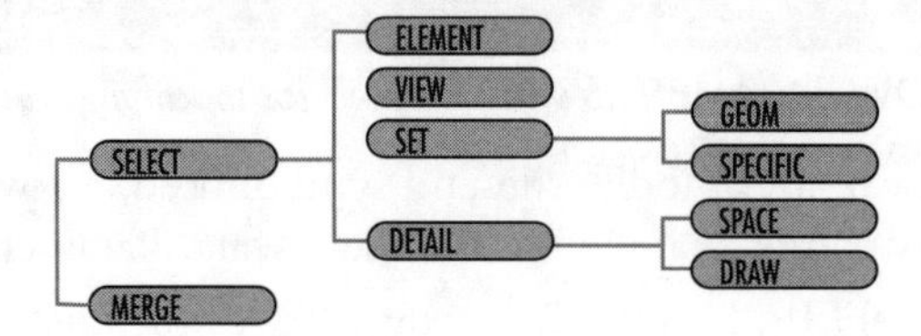

Before attempting a MERGE operation, the following steps must be taken.

1. The receiving model must be saved by clicking on the Save current model in tmp file button on the fixed menu.
2. Read the model which contains the items to be merged (or copied).
3. Select MERGE function.

SELECT

Select elements or details.

Select: MERGE > SELECT | ELEMENT or VIEW or SET or DETAIL

Select elements to be merged

Select: MERGE > SELECT | ELEMENT

Prompt: WSP MULTI SEL SEL ELEM / / YES: ELT

The DID YOU USE THE SV SWITCH message also displays.

1. Select elements to be merged individually or via multiselection, or click on YES to display a list of model elements (3D elements only) and select from the list. All elements logically linked to the selected elements will be selected.
2. Select the MERGE option from the menu, and then click on YES to complete the merge.

Select views to be merged

Select: MERGE > SELECT | VIEW

Prompt: SEL VIEW / / YES: LIST

1. Select a view from the sending model (which will be highlighted), or click on YES to display a list and select the required view from the list.
2. Select the MERGE option from the menu, and then click on YES to complete the merge.

Select sets to be merged

Select: MERGE > SELECT | SET | GEOM or SPECIFIC

Prompt: SEL SET / / YES: LIST

1. With GEOM selected, the associated geometric set is highlighted upon selection of an element. Alternatively, click on YES to display a list of sets and select from the list. With SPECIFIC selected, geometric elements required by another in the selected set will be duplicated.
2. When all sets required for merging have been selected, select the MERGE option from the menu, and then click on YES to complete the merge.

✓ **NOTE:** *If two sets have the same manual identifier, the identifier from the passive model will be modified. A dollar sign character ($) followed by an integer ranging from 1 to 9 will be placed before the original identifier.*

Select details to be merged

Select: MERGE > SELECT | DETAIL

Prompt: KEY STRING / / ENTER: LIST SEL DITTO / / YES: DISPLAY

1. Additional options display. Select a detail from the workspace, or click on YES to display a list of details and then select from the list.

✓ **NOTE:** *Additional options display only if details exist in the sending model.*

2. Select the MERGE option from the menu, and then click on YES to complete the merge.

✓ **NOTE:** *If two details have the same manual identifier, the identifier from the passive model will be modified. A dollar sign character ($) followed by an integer ranging from 1 to 9 will be placed before the original identifier.*

MERGE

Select: MERGE > MERGE

Prompt: YES: MERGE

The use of the MERGE option is described in previous option entries.

MODELS in DRAW and SPACE Modes

Chapter 15

The MODELS function is used in conjunction with the Motif CATIA File Interface to manage and modify models contained in a session or passively overlaid. The COPY option is also used to copy or merge elements from a passive overlaid model into the active model.

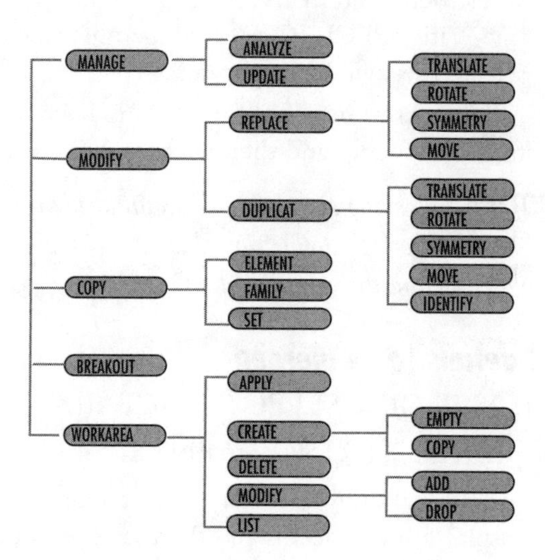

Option menu for MODELS in DRAW and SPACE modes.

MANAGE

Analyze and update models.

Select: MODELS > MANAGE | ANALYZE or UPDATE

Analyze models (active and passive)

Select: MODELS > MANAGE | ANALYZE

Prompt: MODEL: SEL ELEM   YES: SHOW ACTIVE MODEL

An Analysis window displays a list of all models in the current session by model number and name. The following information is also provided.

- Whether model is active or passive
- Whether model has swap capability

- Whether model was overlaid with DRAW on or off
- Standard color of model
- Pick or no pick
- Whether the model has been activated

Click on YES to highlight the active model in the work area and the list.

Check models for updating

Select: MODELS > MANAGE | UPDATE

Prompt: MODEL: SEL ELEM

1. An Update window is displayed listing all passive models in the session. Select from list individually, or click on Check all to determine whether models have been updated.
2. Click on YES to force update as necessary.
3. Select File in the Info column to display the Information panel (as in the Analysis option) for each model.

MODIFY

Transform model(s).

Select: MODELS > MODIFY | REPLACE or DUPLICAT | TRANSLATE or ROTATE or SYMMETRY or MOVE

Upon selection of the MODIFY option, the Model Selection window is displayed offering MONO-SEL or MULTI-SEL. Select MONO-SEL (default) to select a single model. Select MULTI-SEL to select more than one model.

Translate model(s)

Select: MODELS > MODIFY | REPLACE or DUPLICAT | TRANSLATE

Prompt: MODEL: SEL ELEM / / KEY STRING

1. Select the model(s) to be translated.
2. Define translation direction by selecting two points, a line, or entering coordinates.
3. Enter translation distance. Select the arrow to invert as necessary.
4. Click on YES to confirm translation.

Rotate model(s)

Select: MODELS > MODIFY | REPLACE or DUPLICAT | ROTATE

Prompt: MODEL: SEL ELEM / / KEY STRING

1. Select the model(s) to be rotated, and then define the rotation axis by selecting a line.
2. Define rotation angle by selecting points, lines, or planes, or entering an angle.
3. Select the arrow to invert as necessary.
4. Click on YES to confirm rotation.

Impose symmetry on model(s)

Select: MODELS > MODIFY | REPLACE or DUPLICAT | SYMMETRY

Prompt: MODEL: SEL ELEM / / KEY STRING

Select the model(s) for symmetry, and then define the plane of symmetry by selecting points, lines, or a plane. Click on YES to confirm the symmetry.

Move model(s)

Select: MODELS > MODIFY | REPLACE or DUPLICAT | MOVE

Prompt: MODEL: SEL ELEM / / KEY STRING

1. Select the model(s) to be moved.
2. Define the move by selecting the first axis and then the second axis.
3. Click on YES to confirm the move.

COPY

Copy components from passive to active model.

Select: MODELS > COPY | ELEMENT or FAMILY or SET

Copy elements

Select: MODELS > COPY | ELEMENT

Prompt: COPY: MSEL ELEM

1. In the passive model, select elements to be copied individually or by multiselect.
2. Click on YES to end selection, and click on YES again to confirm copy. Selected elements are copied into the active model.

Copy families

Select: MODELS > COPY | FAMILY

Prompt: COPY: MSEL ELEM

1. Select an element; all elements in the same family will be selected and highlighted.
2. Click on YES to end selection, and click on YES again to confirm copy. The selected family will be copied into the active model.

Copy sets

Select: MODELS > COPY | SET

Prompt: COPY: SEL SET

1. Select an element; all elements in the same family will be selected and highlighted.
2. Click on YES to end selection, and click on YES again to confirm copy. The selected sets will be copied into the active model.

BREAKOUT

Break out active model. If a duplicate occurrence of the active model has been created using the MODELS > MODIFY | DUPLICAT option, the active model can be given a new name and unlinked from duplicates.

Select: MODELS > BREAKOUT

Prompt: NEW MODEL: KEY NAME

Enter a new name for the active model. The BREAKOUT DONE message displays. The active model has been renamed and unlinked from duplicates.

WORKAREA

Create separate work areas in which different models can be active. Because you can only work on the active model, it provides a quick and easy method of working on various models in a single session by swapping work areas.

Select: MODELS > WORK AREA | APPLY or CREATE or DELETE or MODIFY

Apply active work area

Select: MODELS > WORK AREA | APPLY

Prompt: WORK AREA: KEY (STRING)

Enter a character string to select work area, or enter a blank field and select from the list. The selected work area becomes the active work area.

Create work area

Select: MODELS > WORK AREA | CREATE | EMPTY or COPY

Create empty work area

Select: MODELS > WORK AREA | CREATE | EMPTY

Prompt: WORK AREA: KEY NAME

1. Using 24 characters or less, enter a name for the new work area. The new work area is created.
2. Select models from the work area, or enter a blank field and select model(s) from the list to be added to the new work area.
3. Click on YES to end selection.

Create copy of existing work area

Select: MODELS > WORK AREA | CREATE | COPY

Prompt: WORK AREA: KEY STRING YES: ACTIVE WORK AREA

1. Enter a character string or blank field and select from the list.
2. Using 24 characters or less, enter a name for the work area. The work area is created and the WORK AREA COPIED message displays.

Delete work area

Select: MODELS > WORK AREA | DELETE

Prompt: Click on YES to confirm deletion.

Work area is deleted.

Modify work area

Select: MODELS > WORK AREA | MODIFY | ADD or DROP or LIST

Add models to work area

Select: MODELS > WORK AREA | MODIFY | ADD

Prompt: WORK AREA : KEY (STRING) YES: ACTIVE WORK AREA

1. Enter a string or blank field and select from the list. Models belonging to the selected work area are dimmed.
2. Select an element from a model to be added to the work area. Click on YES to select a new work area.

Drop models from work area

Select: MODELS > WORK AREA | MODIFY | DROP

Prompt: WORK AREA: KEY (STRING) YES: ACTIVE WORK AREA

1. Enter a string or a blank field and select from the list. Models not belonging to the selected work area are dimmed.

2. Select an element from a model to be dropped. Click on YES to select a new work area.

List models contained in work area

Select: MODELS > WORK AREA | LIST

Prompt: WORK AREA: KEY (STRING)   YES: ACTIVE WORK AREA

1. Enter a string or a blank field and select from the list.
2. The Model List window displays models contained in the selected work area.

PARAM3D in SPACE Mode

INSIDE CATIA

Chapter 17

The PARAM3D function is used to create, manage, and apply parameterized geometry to 3D elements by specifying relationships among elements and applying dimensions that can be modified later. The following elements can be parameterized: points, lines, circles, arcs, planes, conics, canonical surfaces, solid primitives, pipe elements, and composite curves.

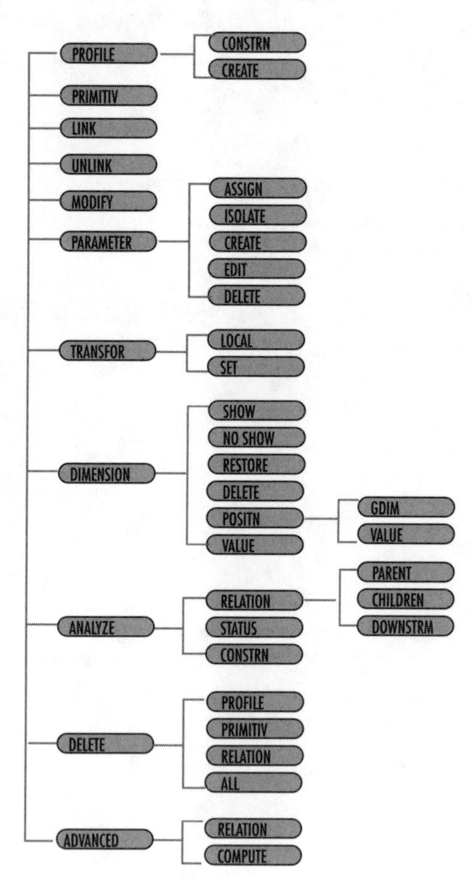

Option menu for PARAM3D in SPACE mode.

PROFILE

Define profile constraints and create profile.

Select: PARAM3D > PROFILE | CONSTRN or CREATE

Define constraints of profile used for solid contours and surfaces where profiles are parents of selected elements

Select: PARAM3D > PROFILE | CONSTRN

Prompt: SEL LN/CONIC/CRV/CCV / / SEL SOL-PRM

1. Select a solid primitive. The profile is highlighted, including inner domains. Alternatively, select an element of a created profile.

✓ **NOTE 1:** *The primitive must have a profile. Cuboids, cylinders, and spheres are not allowed because they are created without geometry.*

✓ **NOTE 2:** *The profile will be visualized and highlighted even if it has been previously placed in No Show. However, if the geometry is located in a hidden layer, a message to that effect will display.*

2. Select any additional elements to be included in the parameterized geometry.
3. Click on YES to complete the definition of the profile. The Parameter Definition panel displays. By clicking on AutoRel or the YES button, the profile will be automatically parameterized.

Parameter Definition window.

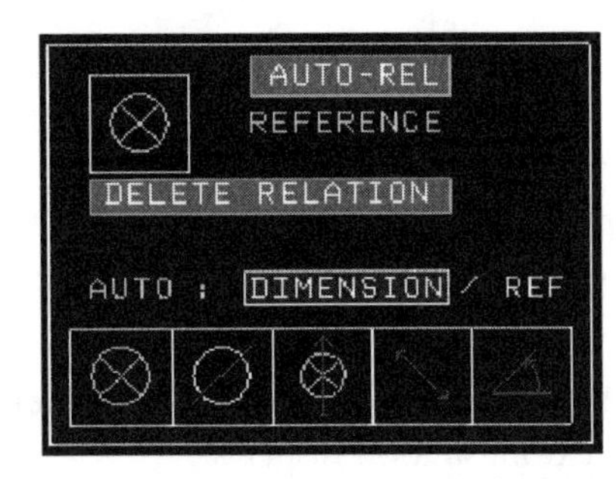

To manually assign relationships between profile geometry elements, take the following steps.

1. Click on the Reference button (shown at left) located at the top of the window. Clicking on the button results in a display at the top of the window.
2. Select profile elements which will become reference elements (i.e., fixed elements or start points).
3. Select one of the other relationship buttons described below, and then select profile elements to create relationships.

Radius or diameter of a circle or circular arc.

Offset between two points or between a point and a line or between two line supports.

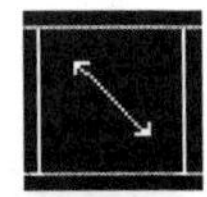

Reference vector.

Angle between two lines.

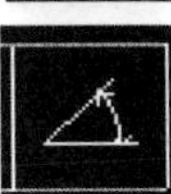

4. Upon selection of any of the above buttons, the window is expanded providing you the opportunity of giving the relationship a name if desired. At this point, if DRIVING is selected, the relationship dimension can be used to drive a modification. If MEASURE is selected, the relationship dimension cannot be modified directly, but only as a result of modifying driving parameters.
5. After making the requisite selections, click on YES to compute. Parametric dimensions showing the parameterized relationships display.

Define profile for parameterization

Select: PARAM3D > PROFILE | CREATE

Prompt: CONTOUR: MSELW LN/CONIC/CRV/CV CONTOUR: SEL PT

Select a closed contour or an unclosed contour element. Click on AutoSearch to close. The message PROFILE CREATED displays.

PRIMITIV

Parameterize solid primitives or wireframe elements, and transfer relationships among elements.

Select: PARAM3D > PRIMITIV

Prompt: SEL GDIM SEL SUR//SEL ELEM//SEL CRV/CCV

1. In the AutoRel window, select REFERENCE to automatically create reference elements and DIMENSION to automatically define dimensioned relationships.
2. Select an element. If the selected element is a spline or arc, the panel shown in the next illustration displays.

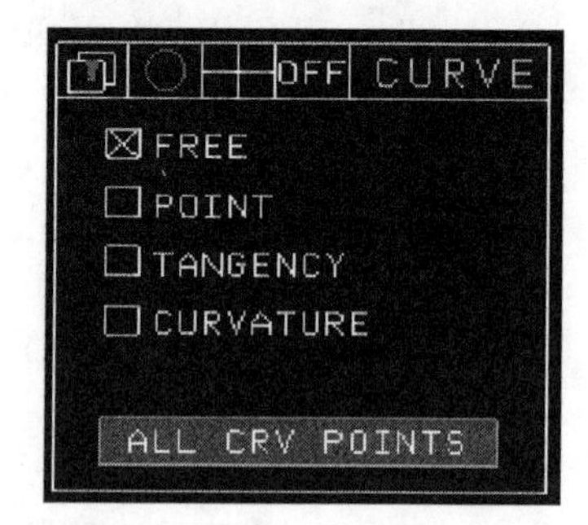

3. Select the type of constraints to be applied, and then select the points where references are to be placed. Alternatively, click on ALL CRV POINTS to place references on all points. Next, click on YES to compute the parameterization.
4. If the selected element is a solid or surface, the message RELATIONSHIP DEFINED displays and the parametric dimensions created.

✓ ***NOTE:*** *If the message INCOMPLETE SELECTION displays, reselect the primitive or click on YES to accept the current primitive.*

LINK

Replace reference with parametric relationship.

Select: PARAM3D > LINK

Prompt: REF TO BE REPLACED: SEL ELEM/GDIM   SEL FTR

1. Select a reference element or symbol.
2. Select another element of an appropriate type from which the first selected element will be dimensioned.
3. Repeat the above step until the definition is complete and the message LINK COMPLETE REFERENCE REPLACED displays, and parametric dimensions appear.

UNLINK

Replace relationship with reference element.

Select: PARAM3D > UNLINK

Prompt: SEL ELEM / / SEL GDIM/DIMN

Select a parametric dimension or parameterized element. The associated relationship is deleted and replaced with a reference.

MODIFY

Modify parameter values.

Select: PARAM3D > MODIFY

Prompt: SEL FTR / / SEL DIMN/GDIM

1. Select the parametric dimension to be modified from the work area or the Parameter window.
2. Enter a new value. Click on YES to execute modification.
3. Click on SOLID UPDATE as necessary to update the solid.
4. On selection of a parameter for modification, the following window displays. By selecting the type of relationship under modification, geometry can be used to specify a new value for the parameter.

OFF PARAMETER

VALUE	PARM IDENTIFIER
142.1285	INP6
44.94383	INP7
61.42111	INP8
94.72743	INP9
97.78783	INP10

PARAMETER

Manipulate parameters.

Select: PARAM3D > PARAMETER | ASSIGN or ISOLATE or CREATE or EDIT or DELETE

Assign parameter to relationships or parameters

Select: PARAM3D > PARAMETER | ASSIGN

Prompt: SEL DRIVING PARAMETER

The ASSIGN option enables you to simultaneously modify several parameters/relationships by assigning them to a driving parameter. A command bar displays containing the following choices.

- REL. Assign selected parameter to one or more relationships.
- PARM. Assign selected parameter to one or more parameters.
- CHANGE DRIVING PARM. End assignment process and execute update.

1. Select a parameter; the parameter will be highlighted.
2. Select other parameters to which the first selected parameter is to be assigned.
3. Select CHANGE DRIVING PARM. The assignment is executed.

Isolate parameter from relationship

Select: PARAM3D > PARAMETER | ISOLATE

Prompt: SEL DRIVEN RELATION : SEL GDIM/DIMN

The ISOLATE option is used to cancel an assignment (see previous entry). Only appropriate parameters are selectable. On selection of a parameter, the assignment is cancelled.

Define parameter to be included in algebraic expression

Select: PARAM3D > PARAMETER | CREATE

Prompt: SEL PARAMETER

1. Enter a name for the parameter to be created in Name field.
2. Enter a value for the parameter to be created in Function field, or select the EDIT option.

Edit parameter

Select: PARAM3D > PARAMETER | EDIT

Prompt: SEL PARAMETER

Select a parameter to be edited. Enter an algebraic expression in the Function field. Click on YES to validate.

Delete parameter

Select: PARAM3D > PARAMETER | DELETE

Prompt: SEL PARAMETER YES: DELETE ISOLATED PARAMETER

✓ **NOTE:** *After a parameter has been isolated it can be deleted with this option.*

Select isolated parameter. Parameter is deleted.

TRANSFOR

Apply transformations.

Select: PARAM3D > TRANSFOR | LOCAL or SET

Apply transformation to parameterized solid or profile

Select: PARAM3D > TRANSFOR | LOCAL or SET

Prompt: SEL PROFILE / / SEL SOL

✓ **NOTE:** *Only stored transformations can be used.*

1. In the window presenting stored transformations, select the desired transformation from the list.
2. Select REPLACE or DUPLICATE.
3. Select a solid or profile. Click on YES to apply the transformation.

Apply transformation to set with parametric data

Select: PARAM3D > TRANSFOR | SET

Prompt: SEL SET

1. Select transformation from window.
2. Select element. The entire set containing the element is selected.
3. Select REPLACE or DUPLICATE. Click on YES to apply.

DIMENSION

Manage parametric dimensions.

Select: PARAM3D > DIMENSION | SHOW or NO SHOW or RESTORE or DELETE or POSITN or VALUE

SHOW or NO SHOW

Select: PARAM3D > DIMENSION | SHOW or NO SHOW

Prompt: SEL ELEM SEL PRIM / / YES: CURRENT SET

1. Choose the desired method of solid selection in the panel shown in the next illustration. From left to right, select complete solid, primitive, or FSUR (functional surface or solid face).

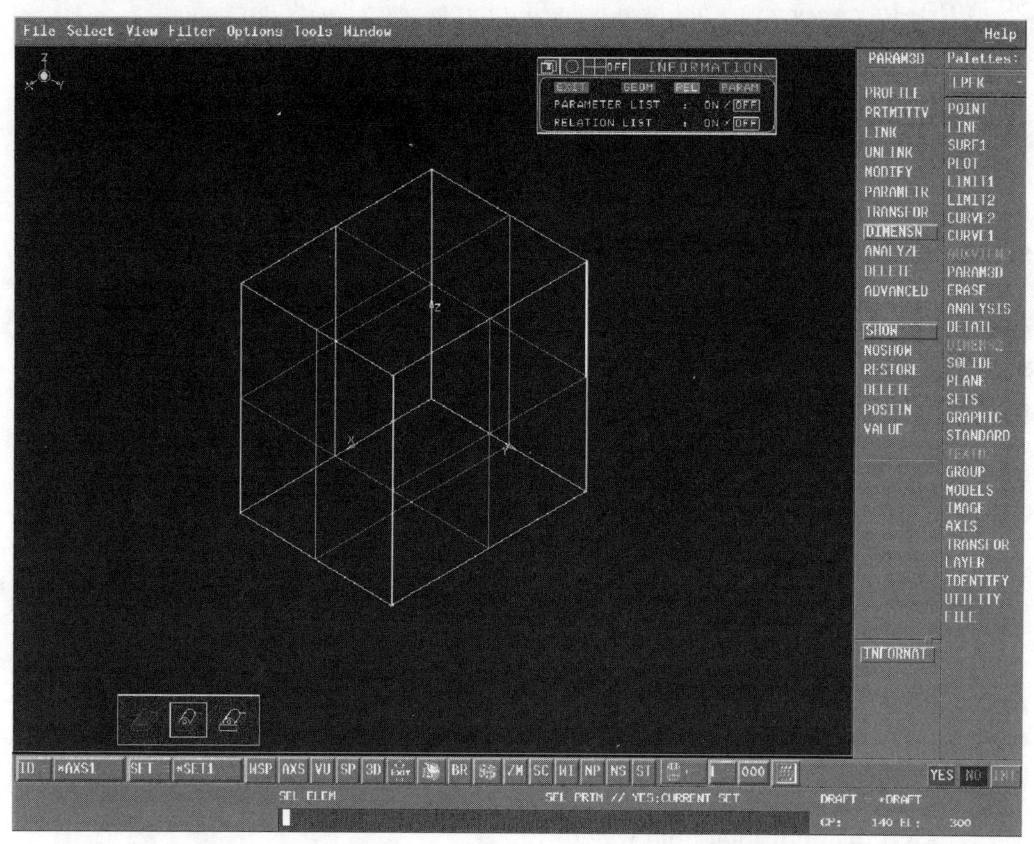

2. Select dimension to show or no show by displaying the parameter list and selecting parameter, or click on YES to show or no show all previously processed parameters.

Restore previously deleted dimension

Select: PARAM3D > DIMENSION | RESTORE

Prompt: SEL ELEM SEL PRIM / / YES: CURRENT SET

1. Take one of the following alternatives: (a) Select a parameterized element. (b) Select a solid primitive or FSUR. (c) Select a relationship from the panel. All associated parametric dimensions are restored.
2. Click on YES to restore all previously deleted dimensions.

Delete dimension

Select: PARAM3D > DIMENSION | DELETE

Prompt: SEL DIMN/GDIM / / SEL ELEM SEL PRIM / / YES: CURRENT SET

Select a dimension or parametric element from the work area or table, or click on YES to delete all dimension in the current set.

Reposition dimension

Select: PARAM3D > DIMENSION | POSITN | GDIM or VALUE

Reposition complete dimension

Select: PARAM3D > DIMENSION | POSITN | GDIM

Prompt: SEL GDIM

Select a dimension. Indicate a point to reposition the dimension.

Reposition dimension value

Select: PARAM3D > DIMENSION | POSITN | VALUE

Prompt: SEL GDIM / / KEY DX / / IND

Select the dimension value to be moved. Enter a value or indicate a point to reposition the value.

Modify display characteristics of dimensions

Select: PARAM3D > DIMENSION | VALUE

Prompt: MSELC GDIM

1. Change values in the GDIM Value window as desired, and then select dimensions to be modified.
2. If required, click on SAVE to save modified settings. Alternatively, click on RESET to return to default values.

ANALYZE

Analyze parametric and/or DIMENS2 type dimensions and constraints.

Select: PARAM3D > ANALYZE | RELATION or CONSTRN

Analyze dimension relations

Select: PARAM3D > ANALYZE | RELATION | PARENT or CHILDREN or DOWNSTRM

Analyze parents of parametric element

Select: PARAM3D > ANALYZE | RELATION | PARENT

Prompt: SEL REL / / SEL ELEM SEL PRIM / / SEL FTR / / SEL FTR

Select a dimension, wireframe element, or solid. The parents will be highlighted, and the Relationship window displayed.

Analyze children of parametric element

Select: PARAM3D > ANALYZE | RELATION | CHILDREN

Prompt: SEL REL / / SEL ELEM SEL PRIM / / SEL FTR / / SEL FTR

Select a dimension, wireframe element, or solid. The children will be highlighted, and the Relationship window displayed.

Analyze relationship path dependent on element

Select: PARAM3D > ANALYZE | RELATION | DOWNSTRM

Prompt: SEL ELEM / / SEL DIMN / GDIM

Select any element of a relationship or dimension. All associated elements in the same set are highlighted.

Analyze constraint status of parametric rule

Select: PARAM3D > ANALYZE | CONSTRN

A list of over- or underconstraints displays as appropriate.

DELETE

Delete selected parts of geometry.

Select: PARAM3D > DELETE | PROFILE or PRIMITIV or RELATION or ALL

Delete parameters of profile

Select: PARAM3D > DELETE | PROFILE

Prompt: SEL PROFILE / / SEL SOL-PRIM

In the Delete panel, you have the following choices.

- COMPUTE. Delete rules from relationship definition.
- PROFILE. Delete profile associated rules.
- RELATION. Delete relationship and associated parameterization.
- ALL. Delete profile and relationship.

Select profile or associated solid. Click on YES to delete.

Delete parameters of primitive or wireframe

Select: PARAM3D > DELETE | PRIMITIV

Prompt: SEL ELEM

Select a primitive or wireframe. Click on YES to delete all parameters associated with the selected element.

Delete relationship

Select: PARAM3D > DELETE | RELATION

Prompt: RELATION: SEL GDIM/DIMN/REL ID SEL PARAMETER

Display the Relationship panel. Select a relationship for deletion.

Delete all parametertized geometry

Select: PARAM3D > DELETE | ALL

Prompt: SEL ELEM / / YES: CURRENT SET

1. Click on YES to delete the current set, or select any element. Associated parametric set is highlighted.
2. Click on YES to delete.

ADVANCED

Define and compute relationships.

Select: PARAM3D > ADVANCED | RELATION or COMPUTE

Define valuated relationships and references

Select: PARAM3D > ADVANCED | RELATION

Prompt: SEL DIMN/GDIM SEL ELEM / / SEL BUTTON

On selection of the ADVANCED option, the Parametization Relationship panel displays. The ADVANCED | RELATION option is used to create relationships between elements which cannot be achieved using the LINK option (e.g., the position of an isolated feature in relation to the edges of a solid). Select the relevant button from the panel (as described under the PROFILE section above).

- If reference element button is selected, successively select elements to become reference elements.
- If reference vector button is selected, successively select elements to become reference vectors.
- If internal parameter button is selected, select an appropriate solid element. A relationship is created.
- If offset button is selected, select point, line, plane, or FSUR elements.
- If angle button is selected, successively select line, plane, or FSUR elements, or a point to define a new angle.

Define implicit relationships automatically

Select: PARAM3D > ADVANCED | COMPUTE

Prompt: SEL ELEM / / SEL SOL-PRIM

Select elements for parameterization. Click on YES to generate topological valuated relationships.

PATTERN in DRAW Mode

Chapter 7

The PATTERN function is used to define, apply, and manage hatching, dotting, coloring, and cell patterns.

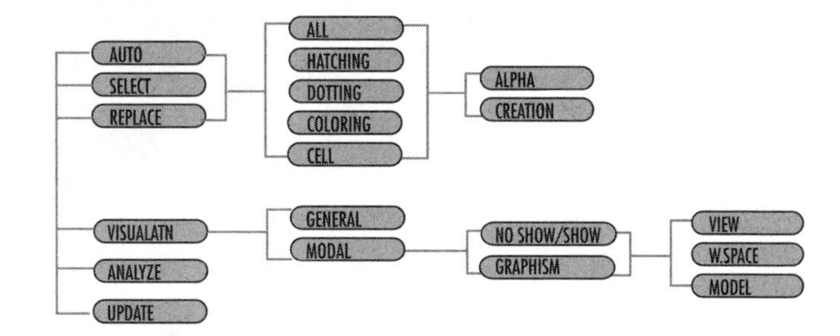

Option menu of PATTERN in DRAW mode.

A hatching pattern is a series of parallel equidistant lines defined by the following parameters.

- PITCH. Distance between the parallel lines.
- SLOPE ANGLE. Angle of the lines.
- OFFSET. Distance between one of the lines and the lower left corner of the origin.

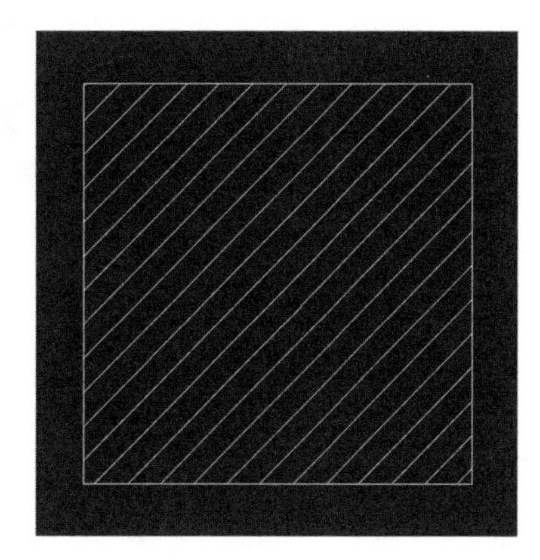

Hatching pattern.

A dotting pattern is a series of equidistant points arranged as a standard or staggered grid of points.

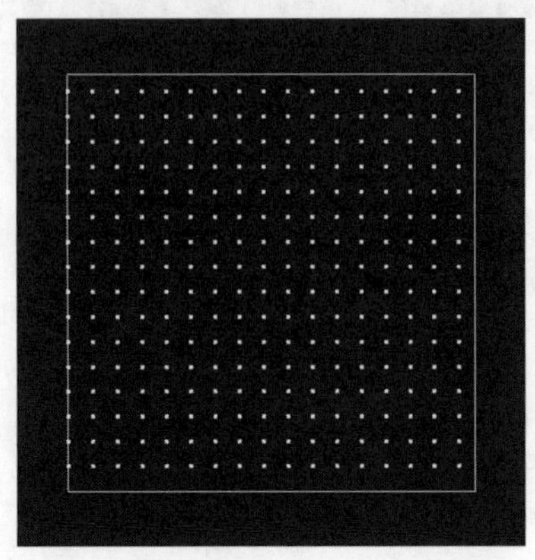

Dotting pattern.

To define the coloring pattern, select an element featuring the desired color or select the color number from the color table.

Coloring pattern.

To define the cell pattern, take the following steps.

- Select a DRAW detail containing points, lines, curves, point, or line grids.
- Define the size and slope of the cell.
- Define the repetition (direction and step) of the cell in the pattern.

A cell pattern displays by a substitute pattern defined at the time of creation in order to increase response times. The cell pattern will be shown correctly when a model is plotted.

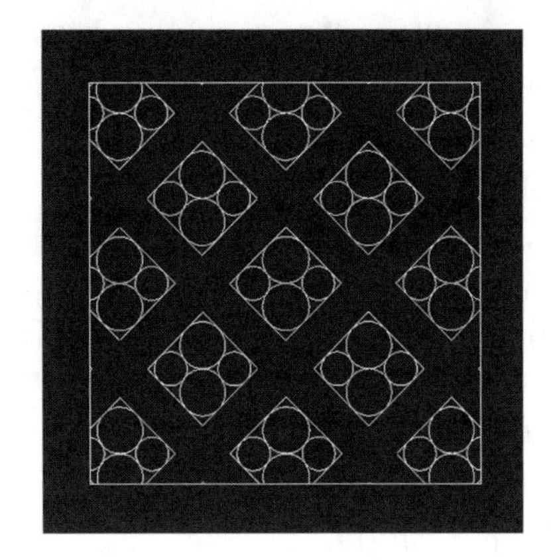

Cell pattern.

AUTO

Automatically place pattern on shape (SHAP) element or in closed area defined by elements in current DRAW view.

Select: PATTERN > AUTO | ALL or HATCHING or DOTTING or COLORING or CELL | ALPHA or CREATION

Prompt: SEL ELEM

Select pattern type

Select one of the four pattern types (HATCHING, DOTTING, COLORING, or CELL), or the ALL option to display available patterns for all types. Select display order. ALPHA displays the patterns alphabetically, and CREATION displays the patterns in the order created.

✓ **NOTE:** *As an alternative to displaying the lists of available patterns, use one of the following methods: (a) If the desired pattern has already been used in the model, it can be selected from the screen. (b) To accept the last used pattern, press <Enter>.*

Use one of the following alternatives: (a) Click on the YES button to display all samples of the available patterns. (A maximum of 30 can simultaneously display on the screen.) Click on the YES or NO button to scroll the available patterns. Select the desired pattern or identifier from the displayed patterns. (b) In the input information area, key in a character string, or leave blank and press <Enter>. Click on the YES or NO button to scroll the available patterns. Select the desired pattern identifier.

Apply pattern

All elements in the current view taken into account for applying the pattern are highlighted. A PATTERN dialog appears containing the following options.

- SAG. Accuracy of patterns on curves.
- BLANKING. Whether patterns are applied over text and dimensions.
- BLANKING OFFSET. Amount of clearance around text or dimensions when blanking is set to ON.

After setting dialog options, indicate in the area to which the selected pattern will be applied. If the area indicated is closed, the pattern will be applied taking into account the preset blanking. If the area indicated is not closed, the NO AREA FOUND message displays. Indicate a new area to apply the pattern, or click on the YES button to perform a pattern area research on all highlighted elements in the current view.

✓ **NOTE:** *Only solid lines of thickness greater than 1 are to automatically create contours for application of a pattern.*

Apply multiarea pattern

After indicating an area for pattern application, you could indicate an additional area. The pattern will be applied to the new area and added (via a Boolean operation) to the current pattern, thereby creating a multiarea pattern.

SELECT

Select a shape element or contours defined by lines, points, curves, text, or dimensions to apply a pattern.

Select: PATTERN > SELECT | ALL or HATCHING or DOTTING or COLORING or CELL | ALPHA or CREATION

Prompt: SEL ELEM

Select pattern type

Select one of the four pattern types (HATCHING, DOTTING, COLORING, or CELL), or the ALL option to display available patterns for all types. Select display order. ALPHA displays the patterns alphabetically, and CREATION displays the patterns in the order created.

✓ **NOTE:** *As an alternative to displaying the lists of available patterns, use one of the following methods: (a) If the desired pattern has already been used in the model, it can be selected from the screen. (b) To accept the last used pattern, press <Enter>.*

Use one of the following alternatives: (a) Click on the YES button to display all samples of the available patterns. (A maximum of 30 can simultaneously display on the screen.) Click on the YES or NO button to scroll the available patterns. Select the desired pattern or identifier from the displayed patterns. (b) In the input information area, key in a character string, or leave blank and press <Enter>. Click on the YES or NO button to scroll the available patterns. Select the desired pattern identifier.

Apply pattern

1. Select a shape (SHAP) element. If desired, select additional shape(s) to create a multi-area pattern.
2. Successively select at least three points to define a contour. Click on the YES button to close the contour defined by the three points.
3. Select a line or curve. If desired, click on the YES button to automatically search for elements joined to the first selected line or curve in order to create a closed contour. Alternatively, select additional line or curve elements until the contour is closed. Click on the YES button to accept the selected contour.

If text or dimensions appear in the selected contour and the pattern is not to be applied over them, select the required text and dimensions before clicking on the YES button to accept the closed contour.

Apply multiarea pattern

Upon applying a pattern as described above, take the following steps to create a multiarea pattern.

1. Click on the YES button, and select an additional area as necessary.
2. Click on YES again to end the selection. The pattern will be applied to the new area and will be added (result of a Boolean operation) to the current pattern, thereby creating a multiarea pattern.

REPLACE

Replace all references to pattern with different pattern.

Select: PATTERN > REPLACE | ALL or HATCHING or DOTTING or COLORING or CELL | ALPHA or CREATION

Prompt: SEL ELEM

1. Select the replacement pattern as described under the AUTO and SELECT options above.
2. Select the pattern to be replaced.
3. Click on the YES button to confirm the replacement.

VISUALTN

Visualize patterns and modify pattern visualization.

Select: PATTERN > VISUALTN | GENERAL or MODAL

Prompt: SEL ELEM

Define general pattern visualization

Select: PATTERN > VISUALTN | GENERAL

Prompt: YES : NO SHOW PATTERN

Prompt: YES : SHOW PATTERN

Click on the YES button to visualize/hide all the patterns already created. In the NO SHOW mode, no patterns will be visualized. In the SHOW mode, patterns will be visible or invisible according to how the selective display is defined by a shape. (See VISUALTN | MODAL | NO SHOW/SHOW.)

✓ **NOTE:** *Patterns placed in NO SHOW using PATTERN > VISULATN | GENERAL will be plotted.*

Modify local display mode of pattern

Select: PATTERN > VISUALTN | MODAL | NO_SHOW/SHOW or GRAPHISM

Transfer patterns into show or no show mode

Select: PATTERN > VISUALTN | MODAL | NO_SHOW or SHOW | VIEW or W.SPACE or MODEL

Prompt: SEL SHAP // YES : SWAP

To transfer patterns into the NO SHOW mode (from SHOW mode), select pattern(s) to be transferred.

To transfer patterns into the SHOW mode (from NO SHOW mode), click on the YES button to display patterns in NO SHOW mode and then select the pattern(s) to be transferred.

If all patterns are to be transferred, select the ALL PATTERNS command button.

Locally modify graphic characteristics of pattern

Select: PATTERN > VISUALTN | MODAL | GRAPHISM | VIEW or W.SPACE or MODEL

Prompt: SEL PATTERN

In the VISU PARM dialog, you can modify the following parameters by selecting the required item and entering the new value.

- REFERENCE. Pattern reference relative to the DRAFT or the shape ELEMENT.
- OFFSETX. Horizontal offset of pattern.
- OFFSETY. Vertical offset of pattern.
- ANGLE. Slope of pattern.
- SCALE. Scale of pattern.

After parameters are selected, the pattern will be modified. If you require all occurrences of a particular pattern identity, click on the YES button to change all patterns in the current view, workspace, or model.

ANALYZE

Analyze pattern.

Select: PATTERN > ANALYZE

Prompt: SEL PATTERN

Select pattern to be analyzed. A PATTERN ANALYSIS window appears containing the following information.

- Pattern type (hatching, dotting, coloring or cell)
- Creation order number of pattern
- Order number of substitution pattern
- Numerical values used for pattern creation

UPDATE

Update pattern when geometry and text used to create pattern are modified.

Select: PATTERN > UPDATE

Prompt: SEL ELEM UPDATE : SEL PATTERN

In the PATTERN window, you can set the following options.

- SAG. Accuracy of patterns on curves.
- BLANKING. Whether patterns are applied over text and dimensions.
- BLANKING OFFSET. Amount of clearance around text or dimensions when blanking is set to ON.

Select the pattern to be updated; the pattern will be modified according to the new conditions.

PLANE in SPACE Mode

Chapter 9

The PLANE function is used to create and modify planes. Planes are represented by square symbols of a standard size.

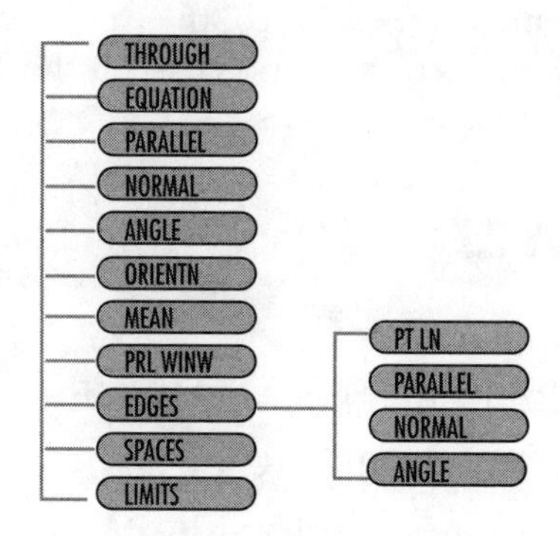

Option menu for PLANE in SPACE mode.

THROUGH

Create planes using points, lines, curves, surfaces, faces, and functional surfaces of solids or DRAW elements.

Select: PLANE > THROUGH

Prompt: SEL PT // SEL LN1 // SEL CRV/FAC DRAW_ELEM:SEL ELEM // SEL FSUR

As seen in the prompt line there are many options for creating through planes. Appearing below are some of the alternative elements that can be selected in order to create a through plane.

- Three PT (point) elements.
- A PT (point) element and a LN (line) element. The point must not lie on the line.
- A CRV (curve) element.
- A FAC (face) or SUR (surface) element.
- A planar FSUR (functional surface of a solid) element.
- Any DRAW element, as all DRAW elements lie on a plane defined by the view.

Once the plane has been defined, the plane graphics symbol should be positioned. Three alternate methods for accomplishing this task are described below.

- Indicate a point; the center of the symbol will lie on the indicated point.

- Select a point; the center of the symbol will lie on the selected point.
- Click on the YES button to automatically position the plane.

EQUATION

Create plane by defining respective equation.

Select: PLANE > EQUATION

Prompt: KEY PLANE EQUATION

Use one of the following alternatives for defining a plane equation.

- Key in the A, B, C, and D coefficients of the equation $AX + BY + CZ + D = 0$.
- Key in an equation in the form X or Y or Z = constant (e.g., $X = 0$ will provide a plane on the YZ axis).
- Key in the definition using the axis reference planes, that is, XY or XZ or YZ. These planes are the same as $Z = 0$, $Y = 0$, and $X = 0$.

Once the plane has been defined, the plane graphics symbol should be positioned. Three alternate methods for accomplishing this task are described below.

- Indicate a point; the center of the symbol will lie on the indicated point.
- Select a point; the center of the symbol will lie on the selected point.
- Click on the YES button to position the plane automatically.

PARALLEL

Create planes parallel to specified plane.

Select: PLANE > PARALLEL

Prompt: SEL PT1 SEL LN1 // SEL CRV/SUR // SEL>PLN

Parallel planes are normally created by selecting a point and defining the plane to which it will be parallel; the created plane will be centered on the first selected point. Alternate methods of creating the planes follow.

First, select a PT (point) element and then define the plane by one of the alternatives listed below.

- Select a PLN (plane) element.
- Select three PT (point) elements to define the plane.
- Select a PT (point) element and an LN (line) element.
- Select two LN (line) elements to define the plane.
- Select a CRV (curve) element.

- Select a PT (point) element and then a SUR (surface) element. In this instance, the created plane will be tangential to the surface to which the point is projected.

Parallel planes can also be created between two existing parallel planes by selecting the two planes and then keying in the number of required planes, or by clicking on the YES button to produce a single plane.

Once the plane has been defined, the plane graphics symbol should be positioned. Three alternate methods for accomplishing this task are described below.

- Indicate a point; the center of the symbol will lie on the indicated point.
- Select a point; the center of the symbol will lie on the selected point.
- Click on the YES button to automatically position the plane.

NORMAL

Create planes normal to selected elements.

Select: PLANE > NORMAL

Prompt: SEL PT1 // SEL CRV/LN

NORMAL planes can be created in several ways as described below.

- Plane passing through a point and normal to a line or curve.

1. Select a PT element; the created plane will lie on this point.
2. Select an LN or CRV element.
3. Alternatively, you could select an LN or CRV element before selecting the PT element.

- Plane passing through a point and normal to a line defined by two points.

1. Select a point; the created plane will lie on this point.
2. Select two additional points to define the line.

- Plane passing through a point and normal to two planes.

1. Select a point, the created plane will lie on this point.
2. Select two PLN (plane) elements.

- Plane passing through a line and normal to a plane.

1. Select a line; the created plane will lie on this line.
2. Select a PLN (plane) element.

- Plane passing through two points and normal to a plane.

1. Select two points.
2. Select a PLN (plane) element.

Once the plane has been defined, the plane graphics symbol should be positioned. Three alternate methods for accomplishing this task are described below.

- Indicate a point; the center of the symbol will lie on the indicated point.
- Select a point; the center of the symbol will lie on the selected point.
- Click on the YES button to position the plane automatically.

ANGLE

Create planes at angle to existing planes passing through line.

Select: PLANE > ANGLE

Prompt: SEL LN

1. Select a line.
2. Select a PLN (plane) element. If the line does not lie on the selected plane it will be projected onto it.
3. Key in an angle between the selected plane and the required plane, and if required, the required number of planes in the following manner: *45,2*. By default the number of planes will be 1. Alternatively, click on the YES button to accept the default values displayed in the message area.
4. If desired, you can now click the YES button to obtain the symmetrical plane or planes about the selected plane.

ORIENTN

Change normal orientation of selected plane.

Select: PLANE > ORIENTN

Prompt: SEL PLN

1. Select a PLN (plane) element.
2. Select the displayed vector and the direction will be reversed, or click on the YES button to reverse the direction of the vector.

✓ **NOTE:** *If the selected plane is topologically linked with a curve or a face its orientation cannot be modified. In this case, the following message will be displayed: ELEMENT ALREADY IN USE.*

MEAN

Create mean plane of at least three points.

Select: PLANE > MEAN

Prompt: SEL PT1

1. Select at least three points.
2. Click on the YES button to create the mean plane.
3. Indicate a point; the center of the symbol will lie on the indicated point. Alternatively, you can select a point (the center of the symbol will lie on the selected point), or click on the YES button to position the plane automatically.

The computation of the mean plane will be such that the sum of the squares of the distances between the points to the plane is minimal.

PRL WINW

Create plane parallel to current screen and passing through point.

Select: PLANE > PRL WINW

Prompt: SEL PT1

1. Select a point.
2. Indicate a window anywhere in the SPACE work area.
3. Indicate a point; the center of the symbol will lie on the indicated point. Alternatively, you can select a point (the center of the symbol will lie on the selected point), or click on the YES button to position the plane automatically.

EDGES

Create planes normal to the current screen window.

Select: PLANE > EDGES | PT/LN or PARALLEL or NORMAL or ANGLE

Create plane normal to current screen window passing through two points or line

Select: PLANE > EDGES | PT/LN

Prompt: SEL PT1 // SEL LN

1. Select two points or a line.
2. Indicate a window anywhere in the SPACE work area.
3. Indicate a point; the center of the symbol will lie on the indicated point. Alternatively, select a point (the center of the symbol will lie on the selected point), or click on the YES button to position the plane automatically.

Create plane normal to current screen window passing through point and parallel to line

Select: PLANE > EDGES | PARALLEL

Prompt: SEL LN // SEL PT1

1. Select a point.
2. Select a line, or select a line and then a point.
3. Indicate a window anywhere in the SPACE work area.
4. Indicate a point; the center of the symbol will lie on the indicated point. Alternatively, you can select a point (the center of the symbol will lie on the selected point), or click on the YES button to position the plane automatically.

Create plane normal to current screen window passing through point and normal to line

Select: PLANE > EDGES | NORMAL

Prompt: SEL LN

1. Select a line and a point.
2. Indicate a window anywhere in the SPACE work area.
3. Indicate a point; the center of the symbol will lie on the indicated point. Alternatively, you can select a point (the center of the symbol will lie on the selected point), or click on the YES button to position the plane automatically.

Create plane normal to current screen window passing through point and forming angle with line

Select: PLANE > EDGES | ANGLE

Prompt: SEL LN

1. Select a line and a point.
2. Indicate a window anywhere in the SPACE work area.
3. Key in the required angle.
4. Indicate a point; the center of the symbol will lie on the indicated point. Alternatively, you can select a point (the center of the symbol will lie on the selected point), or click on the YES button to automatically position the plane.

SPACES

Create equidistant planes.

Select: PLANE > SPACES

Prompt: LN CRV:SEL LN/CIR/ELL/HYP/PAR/CRV SEL PT1 // SEL PLN1

Create middle plane on line, curve, or between two points

Select a line or curve element. Click on the YES button to create the plane at the middle of the selected element. The created plane is perpendicular to the selected element at the point of intersection between the plane and the element.

Alternatively, select two points. Click on the YES button to create the plane at the midpoint between the two points. The created plane is perpendicular to the line passing through the two selected points.

Create equidistant planes along line or curve

1. Select a line or curve element or select two points to define a line.
2. For the point of origin, select a point. If the point does not lie on the previously selected element, its projection onto the selected element is used.
3. Key in the distance for the plane from the last selected point. You can also key in the number of planes required (e.g., *50,2*). If only the distance is keyed in, the default will be *1* plane.
4. If the plane is created the wrong side of the selected point click on the YES key to invert its position.

Create equidistant planes between two points along line or curve by imposing curvilinear distance between created planes

1. Select a line or curve element or select two points to define a line.
2. Click on the YES button to create a plane in the middle of the line or curve. Alternatively, you can select a point on the previously selected element, and then select a second point on the element. The two points define the origin and end points.
3. Key in the distance desired between planes.

The created planes will be distributed along the line or curve using the selected points as limits, so that the distance between them is as near as possible to the keyed value. These planes will be perpendicular to the element at their point of intersection.

LIMITS

Create plane parallel to current screen and passing through point.

Select: PLANE > LIMITS

Prompt: MSELW LN/CIR/ELL/HYP/PAR/CRV/CCV

Create limit planes on single arc curve or line segment

Select a line or curve element. The created planes will be perpendicular to the tangent vector at the ends of the curve or line.

Create limit planes on multi-arc curve

1. Select a line or curve element. The created planes will be perpendicular to the tangent vector at the ends of the curve or line.
2. Click on the YES button to create planes at all limits of the arc curves. The created planes will be perpendicular to the tangent vector at the ends of the curve end arcs.

PLOT in DRAW and SPACE Modes

INSIDE CATIA

Chapter 8

The PLOT function is used to create and manage plot sheets, capture windows, arrange plot windows on sheets, and preview sheets prior to plotting. A plot sheet contains the plot window(s) stored in a file or profile, and a plot window is the area of the screen that you wish to capture. Plotting arrangements vary considerably from site to site, but general principles are described below.

Option menu for PLOT in DRAW and SPACE modes.

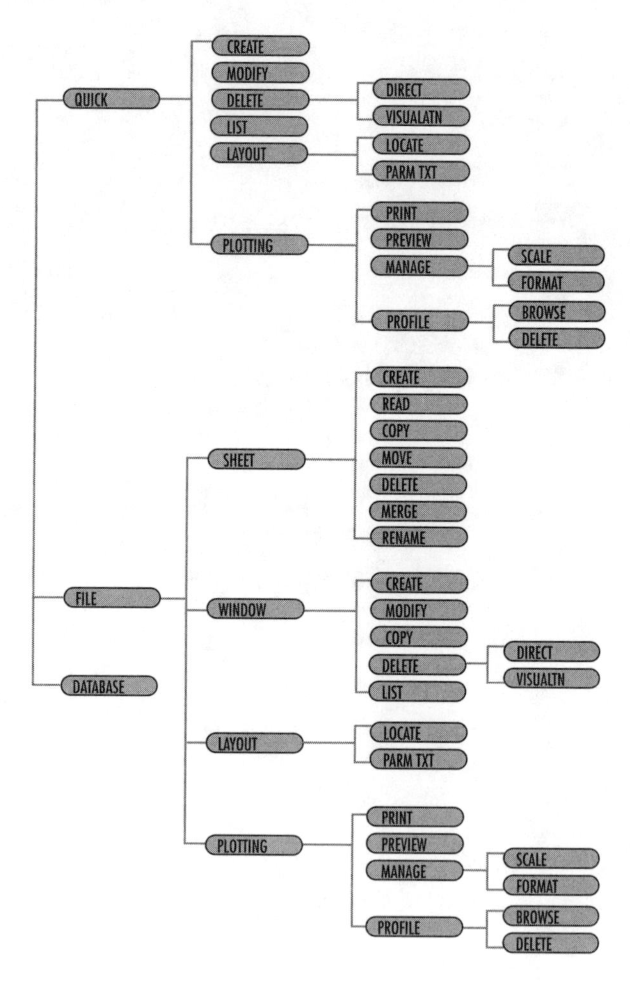

QUICK

Create and otherwise manipulate quick plot.

Select: PLOT > QUICK | CREATE or MODIFY or DELETE or LIST or LAYOUT or PLOTTING

Create quick plot without storage

Select: PLOT > QUICK | CREATE

Prompt: PT1: SEL> ELEM   INS PT1 / / YES: SIZE

The QUICK ENVIRONMENT DEFINED message displays, and the WINDOW INFORMATION dialog appears.

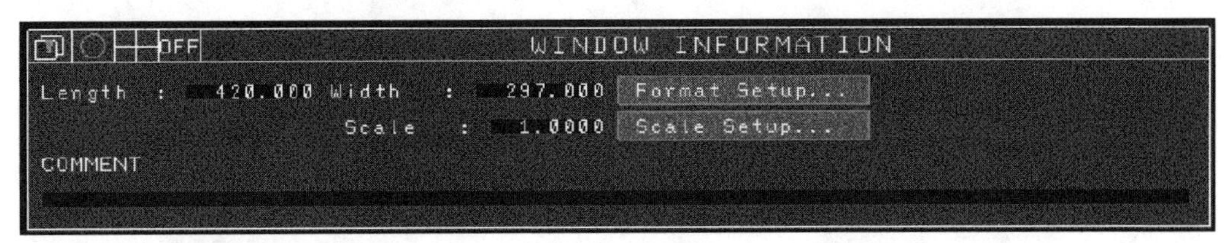

WINDOW INFORMATION dialog.

The above window is used to set the plot window size and scale. Follow one of the alternatives below.

- Enter values for length and width in the appropriate fields, or click on Format Set-up and select a format from the list. Enter a scale value, or click on Scale Set-up and select a scale from the list. Enter a comment.
- Select or indicate the points to define the corners of rectangle which will form the plot window. On selection or indication of the second corner point, the length and width values in the WINDOW INFORMATION dialog will change accordingly.
- Click on YES to accept settings shown in the dialog.

✓ ***NOTE:*** *Clicking on the BR symbol (bottom left of work area) works in the same way as the BR button on the fixed menu. The BR button is inactive when the PLOT function is selected.*

The WINDOW NUMBER # CREATED message displays along with the YES: NEW WINDOW prompt. If an additional plot window is desired, click on YES and then proceed as above.

Modify quick plot window

Select: PLOT > QUICK | MODIFY

Prompt: SEL ITEM

Upon selection of the MODIFY option, the LIST OF WINDOWS screen displays.

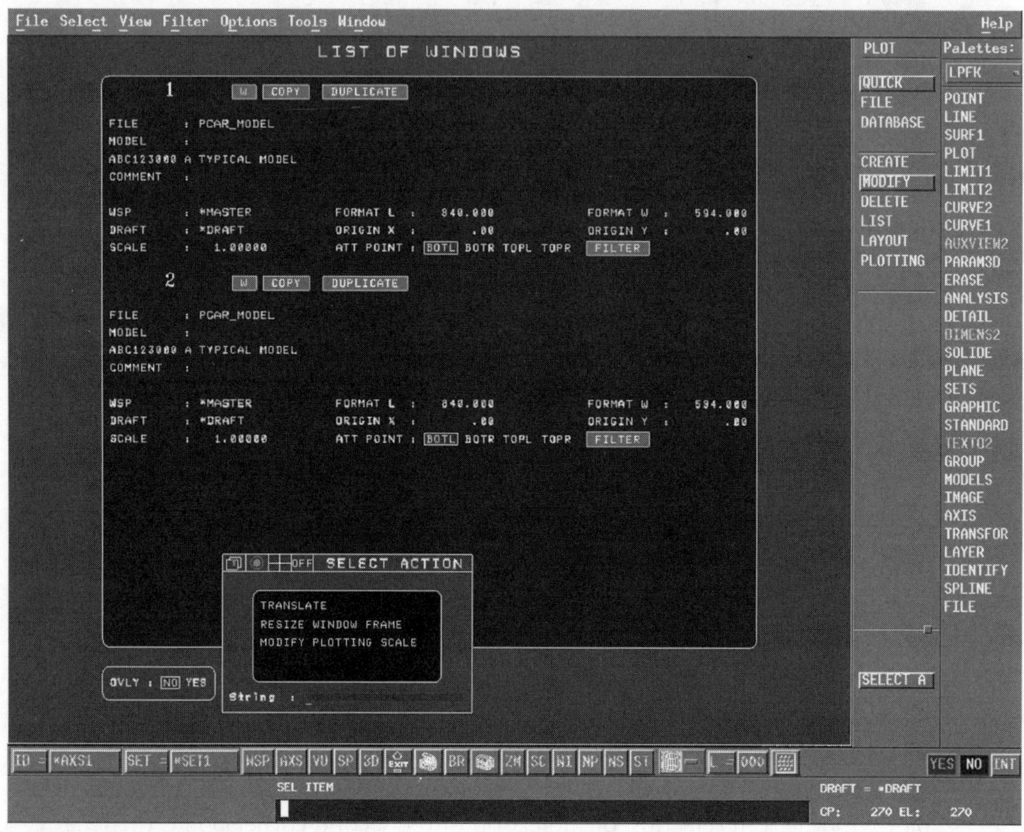

LIST OF WINDOWS dialog.

Click on the W button alongside the window number to be modified. In the SELECTION window, choose one of the following options.

- TRANSLATE. Move the plot window by indicating or selecting a point onto which the plot window center will be moved. Alternatively, select a frame element, and then indicate or select a point onto which the frame element will be moved, or enter coordinates *DX, DY*.
- RESIZE WINDOW FRAME. Return to position described under QUICK | CREATE where new values for window size can be entered or corner points indicated.
- MODIFY PLOTTING SCALE. Return to position described under QUICK | CREATE where new values for window scale can be entered or corner points indicated.

1. Click on the COPY button alongside the window number to which the characteristics of another window are to be copied.
2. Choose fields to copied from the displayed list, and click on YES to end the selection.
3. Click on the W button of the window you want to copy.
4. Click on the DUPLICATE button alongside the window number to be duplicated. A duplicate window is created.

✓ **NOTE:** *All blue fields in the LIST OF WINDOWS can be modified.*

5. Click on FILTER in the window where a filter is required. The following two options are available.
   - VISUALIZE ACTIVE FILTER. View details of the filter currently applied to the selected plot window. If passive models are overlaid and are to be included in the plot window, click on YES in the OVLY: window at bottom left corner of the screen.
   - APPLY FILTER WITHOUT DISPLAY.

Delete plot windows

Select: PLOT > QUICK | DELETE | DIRECT or VISUALTN

Prompt: YES: ALL

Select DIRECT to delete window without visualizing it first. Upon selection of VISUALTN, the window can be visualized before deletion.

1. Click on YES to delete all windows, or click on W for the window to be deleted.
2. Click on YES to confirm deletion.

✗ **WARNING:** *The NO option is not available after deleting a window.*

List plot windows for information only

Select: PLOT > QUICK | LIST

Prompt: SEL ITEM

The LIST OF WINDOWS screen as described under the MODIFY option displays, but for information only. No modification is allowed.

Lay out windows, define sheet format, preview plot, and manage parameterized texts

Select: PLOT > QUICK | LAYOUT | LOCATE or PARM TXT

Lay out windows, define sheet format, and preview plot

Select: PLOT > QUICK | LAYOUT | LOCATE

Prompt: SEL WINW / / SEL ELEM NEW ATT POINT: KEY CHAR / / INS PT

Upon selection of the LAYOUT | LOCATE option, the CURRENT SHEET LAYOUT screen displays.

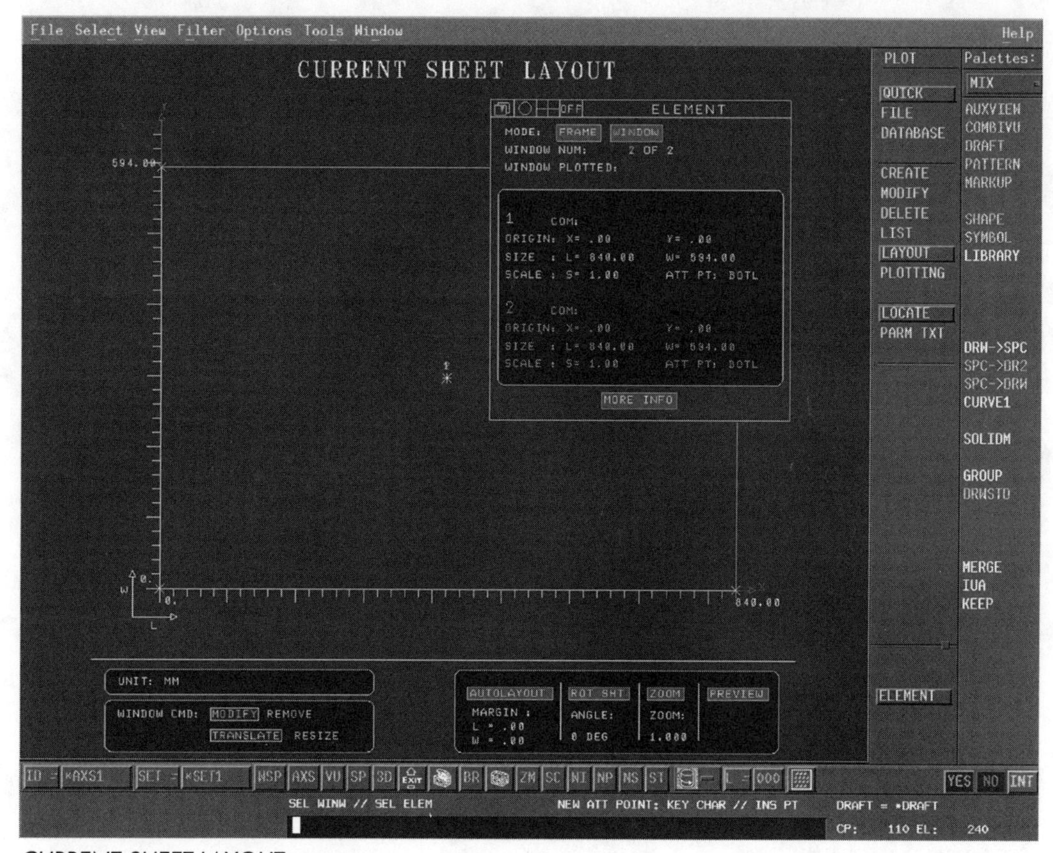

CURRENT SHEET LAYOUT screen.

Options available in the screen are described below.

- FRAME. Modify overall plot frame size or orientation. Upon selecting this option, the choices listed below are available at the bottom left of the screen.
  - MIN SIZE. Automatic sizing of plot sheet.
  - FORMAT. Select from list of standard formats.
- WINDOW. Modify window position.

- AUTO LAYOUT. Automatic arrangement of windows to fit plot sheet.
- ROT SHT. Rotate sheet by 90 degrees.
- ZOOM. Zoom screen view.
- PREVIEW. Preview display of exactly what you will see in the plot including true color. The Preview screen also allows further manipulation of displayed image, zoom, pan, and screen mapping.

- MORE INFO. Display detailed information about windows.

Create parameterized text to be printed on plot as comments

Select: PLOT > QUICK | LAYOUT | PARM TXT

Prompt: SEL ITEM

Check with your system administrator about the use of this option.

Execute plot, manage plot scale and format, and save and manage parameterized profiles

Select: PLOT > QUICK | PLOTTING | PRINT or PREVIEW or MANAGE or PROFILE

Print plot

Select: PLOT > QUICK | PLOTTING | PRINT

Prompt: SEL ITEM

Options in the PRINT ENVIRONMENT screen are described below.

- PROFILE. Plot profiles stored in a database. Check with your system administrator about availability and use of this option.
- PARAMETERS. Activates the SHEET PARAMETER window. Parameters in the dialog follow.
  - ENVIRONMENT
  - SCALING
  - VISUALIZATION
  - FRAMING
  - OPTIONS. Items viewed in plot, including SHEET information, SPACE elements, and DRAW elements.
- DEVICE. Select printing device.

- CONFIGURATION. Activates OUTPUT DEVICE PARAMETERS window. Set parameters pertaining to the plotting device, such as pen sorting and line thickness.
- EXECUTING PARAMETERS. Activates EXECUTION PARAMETERS window. Parameters include sheet rotation, black and white or color, and save profile.

When settings are complete, click on YES to execute plot. A PLOT SUBMISSION PANEL displays containing information on the plot, including whether it was successfully executed.

✓ ***NOTE:*** *Plotting arrangements at your site may be different from the above general principles. Check with your system administrator before plotting.*

Preview setup from PRINT option

Select: PLOT > QUICK | PLOTTING | PREVIEW

Select options described under PRINT. If execute is selected, a preview of the plot displays in exactly the same way as PLOT > QUICK | LAYOUT + Preview button.

Manage plotting scale and format

Select: PLOT > QUICK |PLOTTING | MANAGE | SCALE or FORMAT

This option is typically accessible by system administrators only.

Profile plots

Select: PLOT > QUICK |PLOTTING | PROFILE | BROWSE or DELETE

This option is typically accessible by system administrators only.

FILE

Manage plot sheets, windows, and layouts.

Select: PLOT > FILE | SHEET or WINDOW or LAYOUT or PLOTTING

Define current sheet

Select: PLOT > FILE | SHEET | CREATE or READ or COPY or MOVE or DELETE or MERGE or RENAME

Create plot sheet to be stored in sheet file

Select: PLOT > FILE | SHEET | CREATE

Prompt: SEL ELEM

Two windows display. The FILE INFORMATION window provides details of the sheet file, and the NEW CURRENT SHEET DEFINITION window, details on the current sheet.

1. Enter the file where the plot sheet is to be filed, or select the -> symbol at the right of the entry field to select the file from the list.
2. Enter the name of the sheet.

Read sheet from sheet file

Select: PLOT > FILE | SHEET | READ

Prompt: SEL ITEM

1. Upon selection of this option, the two windows described above remain displayed. In addition, the CURRENT SHEET window displays. Enter the file name in the FILE entry field, or click on -> and select from the list.
2. Enter the name of sheet to be read in the MEMBER entry field, or click on -> and select from the list.

Copy plot sheet from one sheet file to another

Select: PLOT > FILE | SHEET | COPY

Prompt: SEL ITEM

1. The COPY SHEET MEMBER window displays. Enter the name of the sending file in the FROM FILE entry field, or click on -> and select from the list.
2. Enter the name of the receiving file in the TO FILE entry field, or click on -> and select from the list.
3. Enter the name of the member to be copied in the MEMBER entry field, or click on -> and select from the list.
4. Enter a new name for the copied member in the NEW MEMBER NAME entry field, or click on -> and select from the list.
5. Click on YES to execute copy.

Move plot sheet from one sheet file to another

Select: PLOT > FILE | SHEET | MOVE

Prompt: SEL ITEM

MOVE SHEET MEMBER window displays. The steps to move a sheet from one sheet file to another are identical to those described for the COPY option above.

Delete sheet from sheet file

Select: PLOT > FILE | SHEET | DELETE

Prompt: SEL ITEM

1. The FILE INFORMATION and DELETE SHEET MEMBER windows display. Enter the name of the sheet file containing the sheet(s) to be deleted in the FILE entry field, or click on -> and select from the list.
2. Enter the name of the sheet to be deleted in the MEMBER entry field, or click on -> and select from the list.
3. Click on YES to confirm deletion.

Merge contents of two plot sheets

Select: PLOT > FILE | SHEET | MERGE

Prompt: SEL ITEM

1. The MERGE SHEET MEMBERS window displays. Enter the name of the sheet file containing the sheet for merging in the FILE entry field, or click on -> and select from the list.
2. Enter the name of the first sheet in the FIRST MEMBER entry field, or click on -> and select from the list.
3. Enter the name of the second sheet in the TO MERGE TO THE SECOND MEMBER entry field, or click on -> and select from the list.
4. Click on NAME OF THE FIRST MEMBER to use the name of the first selected sheet as the name for the merged sheet, or click on NEW NAME to enter a new name.

Rename sheet in sheet file

Select: PLOT > FILE | SHEET | RENAME

Prompt: SEL ITEM

1. The FILE INFORMATION and RENAME SHEET MEMBER windows display. Enter the name of the file containing the sheet to be renamed in the FILE entry field, or click on -> and select from the list.
2. Enter the name of the sheet to be renamed in the MEMBER entry field, or click on -> and select from the list.
3. Enter a new name for the sheet in the NEW SHEET MEMBER NAME entry field.

Manipulate windows in stored plot

Select: PLOT > FILE | WINDOW | CREATE or MODIFY or COPY or DELETE or LIST

Create window in stored plot

Select: PLOT > FILE | WINDOW | CREATE

Prompt: PT1: SEL > ELEM INS PT1 / / YES: SIZE

1. The WINDOW INFORMATION dialog displays. You have three alternatives to set the plot window size and scale, and add a comment as necessary.
   - Enter values for length and width in the appropriate fields, or click on Format Set-up and select a format from the list. Enter a scale value, or click on Scale Set-up and select a scale from the list. Enter a comment.
   - Select or indicate the points to define the corners of the rectangle that will form the plot window. On selection or indication of second corner point, the length and width values in the WINDOW INFORMATION window will change accordingly.
   - Click on YES to accept settings shown in the window.

✓ **NOTE:** *Clicking on the BR symbol (bottom left of work area) works in the same way as the BR button on the fixed menu. The BR button is inactive while the PLOT function is selected.*

2. The WINDOW NUMBER # CREATED message and YES: NEW WINDOW prompt display. If an additional plot window is required, click on YES and then proceed with the previous step.

Modify windows in stored plot

Select: PLOT > FILE | WINDOW | MODIFY

Prompt: SEL ITEM

Upon selection of the MODIFY option, the LIST OF WINDOWS screen displays. Click on the W button alongside the window number to be modified. In the SELECTION window, choose one of the following options.

- TRANSLATE. Move the plot window by indicating or selecting a point onto which the plot window center will be moved. Alternatively, select a frame element, and then indicate or select a point onto which the frame element will be moved, or enter coordinates *DX, DY*.
- RESIZE WINDOW FRAME. Return to position described under FILE | WINDOW | CREATE where new values for window size can be entered or corner points indicated.
- MODIFY PLOTTING SCALE. Return to position described under FILE | WINDOW | CREATE where new values for window scale can be entered or corner points indicated.

1. Click on the COPY button alongside the window number to which the characteristics of another window are to be copied.
2. Choose fields to be copied from the displayed list, and click on YES to end the selection.
3. Click on the W button of the window you want to copy.
4. Click on the DUPLICATE button alongside the window number to be duplicated. A duplicate window is created.

✓ **NOTE:** *All blue fields in the LIST OF WINDOWS can be modified.*

5. Click on FILTER in the window where a filter is required. The following two options are available.
   - VISUALIZE ACTIVE FILTER. View details of the filter currently applied to selected plot window. If passive models are overlaid and are to be included in the plot window, click on YES in the OVLY window at bottom left corner of the screen.
   - APPLY FILTER WITHOUT DISPLAY.

Copy plot windows from one plot sheet to another

Select: PLOT > FILE | WINDOW | COPY

Prompt: SEL ITEM

1. The SELECTION OF SOURCE SHEET window displays. Enter the name of the file containing the source sheet in the FILE entry field, or click on -> and select from the list.
2. Enter the name of the sheet containing the window(s) to be copied in the MEMBER entry field, or click on -> and select from the list.
3. Click on YES to display the list of windows in the source sheet.
4. Select W for each window to be copied, or click on YES to copy all windows.
5. Click on YES to confirm. The WINDOW NO ## COPIED or WINDOWS COPIED message displays.

✓ **NOTE:** *If copied windows are to be overlaid in the plot, verify that the OVLY switch at the bottom left of the LIST OF WINDOWS screen is set to YES.*

Delete windows from stored sheet

Select: PLOT > FILE | WINDOW | DELETE | DIRECT or VISUALATN

Prompt: WINDOW COMB: KEY TEXT / / YES: ALL

With DIRECT selected, the window will be deleted without visualizing it first. Upon selecting VISUALATN, the window is visualized before deletion.

1. Select W for windows to be deleted.
2. Click on YES to delete, or click on YES to delete all windows. The WINDOWS DELETED message displays.

List plot windows for information only

Select: PLOT > FILE | WINDOW | LIST

Prompt: SEL ELEM

The LIST OF WINDOWS screen displays. Modifications cannot be executed.

Lay out windows, define sheet format, preview plot, and manage parameterized text in stored plot sheet

Select: PLOT > FILE | LAYOUT | LOCATE or PARM TXT

Lay out windows, define sheet format, and preview plot

Select: PLOT > FILE | LAYOUT | LOCATE

Prompt: SEL WINW / / SEL ELEM NEW ATT POINTS KEY CHAR / / INS pt

Upon selection of the LAYOUT | LOCATE option, the CURRENT SHEET LAYOUT screen displays. The following options are available in the windows comprising the screen.

- FRAME. Modify overall plot frame size or orientation. Upon selecting this option, the choices listed below are available at the bottom left of the screen.
  - MIN SIZE. Automatic sizing of plot sheet.
  - FORMAT. Select from list of standard formats.
- WINDOW. Modify window position.
  - AUTO LAYOUT. Automatic arrangement of windows to fit plot sheet.
  - ROT SHT. Rotate sheet by 90 degrees.
  - ZOOM. Zoom screen view.
  - PREVIEW. Preview display of exactly what you will see in the plot including true color. The Preview screen also allows further manipulation of displayed image, zoom, pan, and screen mapping.
- MORE INFO. Display detailed information about windows.

Create parameterized texts to be printed on plot

Select: PLOT > FILE | LAYOUT | PARM TXT

Prompt: SEL ITEM

With this option, you can plot parameterized DRAW text as comments on the plot sheet. Check with your system administrator about the use of this option.

Execute plot, manage plot scale and format, and save and manage parameterized profiles

Select: PLOT > FILE | PLOTTING | PRINT or PREVIEW or MANAGE or PROFILE

The options available in FILE | PLOTTING are identical to those described under the QUICK | PLOTTING option above.

DATABASE

Select: PLOT > DATABASE | SHEET or WINDOW or LAYOUT or PLOTTING

Check with your system administrator about the use of this option.

POINT in DRAW Mode

Chapter 2

The POINT function is used to create points, grids of points, and tangency or limit points.

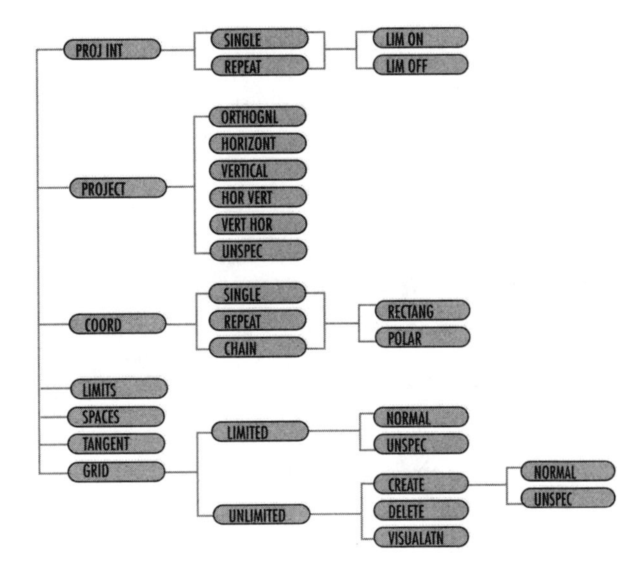

Option menu for POINT in DRAW mode.

VICINITY SELECTION

Vicinity selection is also available for selecting the point(s) for use in creating points. This facility allows you to select a line or curve element to define its end point at the end closest to where the line or curve is selected. The end point created will be a temporary point used in creating a point.

PROJ INT

Create points by intersecting elements or projecting a point onto an element.

Select: POINT > PROJ INT | SINGLE or REPEAT | LIM ON or LIM OFF

Upon creating points using PROJ INT, points can be created depending on whether the intersection or projection is within the limits of the selected elements. Use the LIM ON option to create points only by intersection or projection when either is within limits of selected elements. Select the LIM OFF option when you wish for points to be created by intersection or projection when either is outside the limits of selected elements.

Create single point

Select: POINT > PROJ INT | SINGLE | LIM ON or LIM OFF

Prompt: 1ST : MSELW ELEM // SEL VU

Create point by indication

Indicate a position for the point.

Create point midway between two points

Successively select two point elements.

Create point by projecting point onto element

1. Select a point element.
2. Select a line or curve element. If a curve is selected, the point closest to the projection point will be created.
3. If desired, click on the YES button to create all projection points.

Create point(s) on intersection of two elements

1. Successively select two line or curve elements. If there is more than one intersection between the selected elements, all intersection points will be created.
2. If required, where no intersection exists between the selected elements, click on the YES button to create two points with the least distance between the two selected elements.

Create several points

Select: POINT > PROJ INT | REPEAT | LIM ON or LIM OFF

Prompt: 1ST : MSELW ELEM // SEL VU

To create several points, the first selected element will be used as the reference element. The procedure for creating several points is the same as for single points; the first selected element is used as the reference.

To select a new reference element, click on the YES button to end the procedure and select a new reference element.

TIP: *Remember that elements can be selected using multiselect options. (See Appendices A and B for further information.) The vicinity selection facility can be used with all PROJ INT options.*

PROJECT

Create points by projecting points onto an element.

Select: POINT > PROJECT | ORTHOGNL or HORIZONT or VERTICAL or HOR VERT or VERT HOR or UNSPEC

Create point via orthogonal projection of point

Select: POINT > PROJECT | ORTHOGNL

Prompt: MSELW PTD // SEL VU

1. Select the point element to be projected.
2. Select another point element. The created point will be at the midpoint of the segment which joins the two selected points. Alternatively, select a line or curve element. The created point will be the orthogonal projection of the selected point onto the selected element. A maximum of 10 points will be created.

Create point via horizontal projection of point

Select: POINT > PROJECT | HORIZONT

Prompt: MSELW PTD // SEL VU

1. Select the point element to be projected. A horizontal vector displays.
2. Select another point element. The created point will be where the point is projected orthogonally onto the horizontal vector which passes through the first selected point. Alternatively, select a line or curve element. The created point will be where the point is projected horizontally onto the selected element.

Create point by vertical projection of point

Select: POINT > PROJECT | VERTICAL

Prompt: MSELW PTD // SEL VU

1. Select the point element to be projected. A vertical vector displays.
2. Select another point element. The created point will be where the point is projected orthogonally onto the vertical vector which passes through the first selected point. Alternatively, select a line or curve element. The created point will be where the point is projected vertically onto the selected element.

Create point via horizontal and vertical projection of point

Select: POINT > PROJECT | HOR VERT

Prompt: MSELW PTD // SEL VU

Successively select two point elements. A point will be created at the intersection of the horizontal vector which passes through the first selected point, and the vertical vector which passes through the second selected point.

Alternatively, select two line or curve elements, or a line and a curve element. The end points closest to the selection position of the lines or curves will be used to create a point at the intersection of the horizontal vector which passes through the first selected point, and the vertical vector which passes through the second selected point.

Create point via vertical and horizontal projection of point

Select: POINT > PROJECT | VERT HOR

Prompt: MSELW PTD // SEL VU

Successively select two point elements. A point will be created at the intersection of the vertical vector which passes through the first selected point, and the horizontal vector which passes through the second selected point. Alternatively, select two line or curve elements, or one line and one curve element. The end points closest to the selection position of the lines or curves will be used to create a point at the intersection of the vertical vector which passes through the first selected point, and the horizontal vector which passes through the second selected point.

Create point via projection of point in specific direction

Select: POINT > PROJECT | UNSPEC

Prompt: MSELW PTD // SEL VU

1. Select the point element to be projected.
2. Choose one of the following alternatives: (a) Select a line element to define the projection direction. (b) Key in a value for the angle of projection. This angle will be referenced to the current horizontal axis. (c) Click on the YES button to accept the displayed default values.
3. Select a point element. The point created will be on a line which passes through the first selected point and projected orthogonally from the second selected point. Alternatively, select a line or curve element. The created point will be where the first selected point is projected at the specified angle onto the line or curve.

COORD

Select: POINT > COORD | SINGLE or REPEAT or CHAIN | RECTANG or POLAR

Create points by keying in rectangular or polar coordinates. When creating points using the COORD option you have two choices. Upon selecting RECTANG, the point is defined by the X and Y directions. When selecting POLAR, the point is defined by its radius and angle.

Create single point

Select: POINT > COORD | SINGLE | RECTANG

Prompt: REFERENCE POINT : MSELW PTD SEL VU // KEY X,Y // IND POINT

Select: POINT > COORD | SINGLE | POLAR

Prompt: REFERENCE POINT : MSELW PTD SEL VU // KEY RAD, ANG // IND POINT

1. Choose one of the following alternatives: (a) Key in values for the point coordinates. The created point will be referenced to the current axis. (b) Indicate a position for the point. (c) Select a point element to serve as the reference point for the created point.
2. Key in values for the point coordinates. The created point references the previously selected point. Alternatively, click on the YES button to accept the displayed default values.

Create set of repeated points

Select: POINT > COORD | REPEAT | RECTANG or POLAR

Prompt: SEL VU // SEL PTD

1. Select a point element to serve as the reference point for the created set of points. Alternatively, indicate a position for the point to serve as the reference point for the created set of points.
2. Choose one of the following alternatives: (a) Key in values for the point coordinates. The created point will reference the previously selected point. Only one point will be created. (b) Click on the YES button to accept the displayed default values. Only one point will be created. (c) Key in the values for the coordinates of the first point to be created with reference to the first selected point and the number of points to be created. The points will be created evenly spaced on a line passing through the first selected point, the reference point, and the first point created.

Create chain set of points

Select: POINT > COORD | CHAIN | RECTANG or POLAR

Prompt: SEL VU // REFERENCE POINT : SEL PTD

1. Select a point element to serve as the reference point for the first created point.

2. Key in values for the point coordinates to be created. The first created point will reference the previously selected point. Additional coordinates can be keyed in; the points will then reference the previously created point. Alternatively, click on the YES button to accept the displayed default values. The YES button may be clicked as many times as required; the first created point will reference the previously selected point. Additional points will then reference the previously created point.

TIP: *Remember that the vicinity selection facility can be used with all COORD options.*

LIMITS

Create points on the limits of elements.

Select: POINT > LIMITS

Prompt: MSELW ELEM // SEL VU

Create line segment limit points

Select a line element. The two end points will be created.

Create curve limit points

Select a curve element. The two end points of the arc formed by the curve will be created. If the curve is closed, the center and open points will be created.

Create circle limit points

Select a circle element. The center point and the two end points of an arc of the circle will be created. If the circle is closed, the open point will be created.

Create ellipse limit points

Select an ellipse element. The center point, foci, and end points of an ellipse arc will be created. If the ellipse is closed, the center point, foci, and open point will be created.

Create parabola limit points

Select a parabola element. The focus, center point, and two end points will be created.

Create hyperbola limit points

Select a hyperbola element. The focus, center point, and two end points will be created.

Create axis limit point

Select a axis element. A point on the origin will be created.

Create contour limit points

1. Select a shape or pattern element. The end points of the selected contour edge will be created.
2. If desired, click on the YES button to create end points of all edges for the selected element.

Create limit point text

Select a text element. The anchor point of the text will be created.

TIP: *Remember that elements can be selected using the multiselect options. (See Appendices A and B for further information.)*

SPACES

Create equidistant points on selected element.

Select: POINT > SPACES

Prompt: SEL VU // LN : SEL ELEM   SEL CIRD // ORIGIN : SEL PT

Create equidistant points on line with specified distance

1. Select a point element to define the origin of the spaces, and then select a line element to define the direction of the spaces. Alternatively, select a line element to define the direction of the spaces, and then select a point element to define the origin of the spaces.
2. Key in a value for the distance. If required, key in the number of points (e.g., *20,3*, meaning three points at a distance of 20). If no value is entered for the number of points, the default is 1.
3. If desired, click on the YES button to reverse the position of the point in relation to the origin.

Create equidistant points on line with specified number of points

1. Select a line element to define the direction for the spaces.
2. Key in a value for the number of points required.
3. Click on the YES button to create end points, or select two points to define the start and end points.

Create equidistant points on circle using start point and angular dimension(s)

1. Select a circle element.
2. Select a point element to define the origin point.

3. Key in values for the angle of the radius vectors. If desired, key in the number of points.
4. Click on the YES button to reverse the direction of the created points as necessary.

Create midpoint on circle

Select a circle element. Click on the YES button to create the midpoint.

Create equidistant points on circle using start and end points and number of points

1. Select a circle element.
2. Key in values for the number of points required.
3. Click on the YES button to create the last point.

Create equidistant points on circle using start and end points and curvilinear dimension

1. Select a circle element.
2. Select a point element to define the origin point.
3. Select a point element to define the end point.
4. Key in a value for the curvilinear distance created. Points will be created between the two selected points at a distance as close as possible to the keyed value.

Create equidistant points on circle using start point and curvilinear dimension(s)

1. Select a circle element.
2. Select a point element to define the origin point.
3. Click on the YES button to select a curvilinear distance.
4. Key in a value for the curvilinear distance. If required, key in the number of points (e.g., *20,3*, meaning three points at a distance of 20). If no value is entered for the number of points, the default is 1.

TANGENT

Create points of tangency on curve.

Select: POINT > TANGENT

Prompt: SEL VU // CURVE : SEL ELEM SEL LND // SEL PTD

Select a curve element, and then select a point or line element. Alternatively, select a point or line element, and then select the curve element. The points of tangency between the point or line and the curve will be created, and the points of tangency and tangent directions temporarily display.

TIP: *Remember that elements can be selected using the multiselect options when selecting the second element. (See Appendices A and B for further information.)*

GRID

Create and manage point grids.

Select: POINT > GRID | LIMITED or UNLIM

Create limited point grid

Select: POINT > GRID | LIMITED | NORMAL or UNSPEC

Create orthogonal point grid parallel to current axis system

Select: POINT > GRID | LIMITED | NORMAL

Prompt: SEL VU // ORIGIN : SEL PTD

Use diagonal point and number points in grid

1. Select a point element to define the grid reference point. The four quadrants of the grid display.
2. Select a point element to define the limit of the grid diagonal to the previously selected reference point.
3. Key in values for the number points in each grid direction.
4. If required, key in values for the step required of the secondary grid. The secondary points will be represented by an x on the point grid. Alternatively, click on the YES button to create the grid.

Define step and number of points

1. Select a point element to define the grid reference point. The four quadrants of the grid display.
2. Key in values for the step required in both grid directions.
3. Key in values for the number of points in each direction of the first quadrant. Alternatively, click on the YES button.
4. Key in values for the number of points in each direction of the third quadrant. Alternatively, click on the YES button to create the grid.

5. If desired, key in values for the step required of the secondary grid. The secondary points will be represented by an x on the point grid. Alternatively, click on the YES to create the grid.

Define diagonal point and two end points

1. Select a point element to define the grid reference point. The four quadrants of the grid display.
2. Key in values for the step required in both grid directions.
3. Select a point element in the first or fourth quadrant. Alternatively, click on the YES button.
4. Select a point element in the quadrant diagonal to the previously selected point. Alternatively, select a point element in the second or third quadrant, or click on the YES button.
5. If desired, key in values for the step required of the secondary grid. The secondary points will be represented by an x on the point grid. Alternatively, click on the YES button to create the grid.

Create unspecified orthogonal or unspecified point grid

Select: POINT > GRID | LIMITED | UNSPEC

Prompt: SEL VU // ORIGIN : SEL PTD

The following steps are necessary to define the creation of all unspecified grids.

1. Select a point element to define the grid reference point.
2. Select a line element to define the first direction of the grid.

The next step allows you to define an unspecified orthogonal or nonorthogonal point grid.

Click on the YES button to create an unspecified orthogonal grid of points. Alternatively, select a line element to define the reference line for the second direction of the grid to allow the creation of an unspecified point grid.

To complete the grid definition, choose one of the following three alternatives and proceed with respective steps.

- Select the diagonal point and the number of points.

1. Select a point element to define the limit of the grid diagonal to the previously selected reference point.
2. Key in values for the number of points in each grid direction.
3. If desired, key in values for the step required of the secondary grid. The secondary points will be represented by an x on the point grid. Alternatively, click on the YES button to create grid.

- Define the step and the number of points.

1. Key in values for the step required in both grid directions, or click on the YES button to accept the displayed default values.
2. Key in values for the number of points required in each direction of the first quadrant. Alternatively, click on the YES button: the number of points in each direction of the first quadrant will be zero.
3. Key in values for the number of points required in each direction of the third quadrant. Alternatively, click on the YES button. The number of points in each direction of the third quadrant will be zero.
4. If desired, key in values for the step required of the secondary grid. The secondary points will be represented by an x on the point grid. Alternatively, click on the YES button to create the grid.

- Define the diagonal point and the two end points.

1. Key in values for the step required in both grid directions, or click on the YES button to accept the displayed default values.
2. Select a point element in the first or fourth quadrant. Alternatively, click on the YES button; the number of lines in each direction of the first quadrant will be zero.
3. Select a point element in the quadrant diagonal to the previously selected point. Alternatively, click on the YES button. The number of lines in each direction of third or second quadrant (diagonal to the quadrant selected in step 2) will be zero. A second alternative is to select a point element in the second or third quadrant.
4. If desired, key in values for the step required of the secondary grid. The secondary points will be represented by an x on the point grid. Alternatively, click on the YES button to create grid.

Create, delete, or visualize infinite point grid

Select: POINT > GRID | UNLIMITED | CREATE or DELETE or VISUALTN

Create infinite point grid

Select: POINT > GRID | UNLIMITED | CREATE | NORMAL or UNSPEC

Infinite point grids are represented by parallel lines in the direction of the grid. The points created are at the intersections of the parallel lines. An infinite point grid is not a geometric element. It is better described as a snap grid used as a graphical aid. There can only be one infinite point grid in a workspace.

Create normal infinite point grid

Select: POINT > GRID | UNLIMITED | CREATE | NORMAL

Prompt: SEL VU // ORIGIN : SEL PTD

1. Select a point element to define the reference point of the grid. The four quadrants of the grid display.
2. Key in values for the step required in both grid directions.
3. If desired, key in values for the step required of the secondary grid. The secondary points will be represented by an x on the point grid. Alternatively, click on the YES button to create grid.

Create unspecified point grid

Select: POINT > GRID | UNLIMITED | CREATE | UNSPEC

Prompt: SEL VU // ORIGIN : SEL PTD

1. Select a point element to define the grid reference point. The four quadrants of the grid display.
2. Select a line element to define the first direction of the grid.
3. Select a line element to define the second direction of the grid, or click on the YES button; the second direction will be normal to the first line selected.
4. Key in values for the step required in both directions of the grid.
5. If desired, key in values for the step required of the secondary grid. The secondary points will be represented by an x on the point grid. Alternatively, click on the YES button to create grid.

Delete infinite point grid

Select: POINT > GRID | UNLIM | DELETE

Prompt: SEL VU // YES : DELETE

Click on the YES button to delete the infinite point grid in the current workspace. There is only one infinite point grid in any workspace.

✓ **NOTE:** *The ERASE function cannot be used to delete an infinite point grid.*

Manage display of infinite grid

Select: POINT > GRID | UNLIM | VISUALTN

Prompt: SEL VU // YES : SWAP

Click on the YES button to change the current display of the infinite grid. If the grid is in SHOW mode it will be swapped to the NO SHOW mode and vice versa.

✓ **NOTE:** *The ERASE | SHOW or NO SHOW function cannot be used to manage the display of an infinite point grid.*

POINT in 2D SPACE Mode

Chapter 9

The POINT function is used to create points, and tangency or limit points.

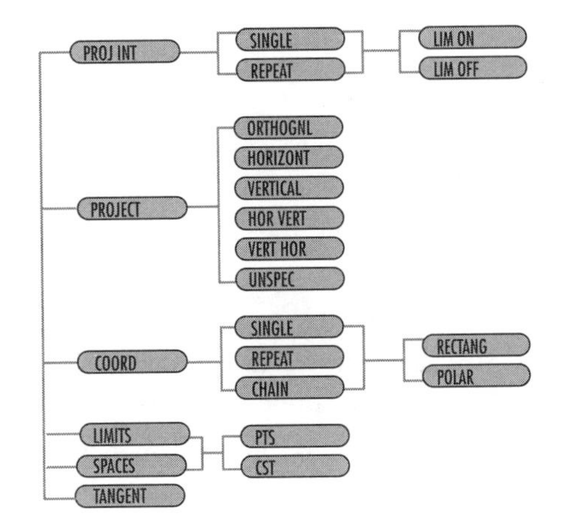

Option menu for POINT in 2D SPACE mode.

VICINITY SELECTION

Vicinity selection is also available for selecting the point(s) for use in creating points. This facility allows you to select a line or curve element to define its end point at the end closest to where the line or curve is selected. The end point created will be a temporary point used in creating a point.

PROJ INT

Create points by intersecting elements or projecting point onto element.

Select: POINT > PROJ INT | SINGLE or REPEAT | LIM ON or LIM OFF

Upon creating points using PROJ INT, points can be created depending on whether the intersection or projection is within the limits of the selected elements. Use the LIM ON option to create points only by intersection or projection when either is within limits of selected elements. Select the LIM OFF option when you wish for points to be created by intersection or projection when either is outside the limits of selected elements.

Create single point

Select: POINT > PROJ INT | SINGLE | LIM ON or LIM OFF

Prompt: 1ST : MSELW ELEM > ELEM

Create point by indication

Indicate a position for the point.

Create point midway between two points

Successively select two point elements.

Create point by projecting point onto element

1. Select a point element.
2. Select a line or curve element. If a curve is selected, the point closest to the projection point will be created.
3. If required, click on the YES button to create all projection points.

Create point(s) on intersection of two elements

1. Successively select two line or curve elements. If there is more than one intersection between the selected elements, all intersection points will be created.
2. If required, where no intersection exists between the selected elements, click on the YES button to create two points with the least distance between the two selected elements.

Create several points

Select: POINT > PROJ INT | REPEAT | LIM ON or LIM OFF

Prompt: 1ST : MSELW ELEM > ELEM

To create several points, the first selected element will be used as the reference element. The procedure for creating several points is the same as for single points; the first selected element is used as the reference.

To select a new reference element, click on the YES button to end the procedure and select a new reference element.

TIP: *Remember that elements can be selected using multiselect options. (See Appendices A and B for further information.) The vicinity selection facility can be used with all PROJ INT options.*

PROJECT

Create points by projecting points onto element.

Select: POINT > PROJECT | ORTHOGNL or HORIZONT or VERTICAL or HOR VERT or VERT HOR or UNSPEC

In SPACE 2D mode, the horizontal and vertical vectors are defined by the first and second axes of the current axis system.

Create point via orthogonal projection of point

Select: POINT > PROJECT | ORTHOGNL

Prompt: MSELW PT

1. Select the point element to be projected.
2. Select another point element. The created point will be at the midpoint of the segment which joins the two selected points. Alternatively, select a line or curve element. The created point will be the orthogonal projection of the selected point onto the selected element. A maximum of 10 points can be created.

Create point via horizontal projection of point

Select: POINT > PROJECT | HORIZONT

Prompt: MSELW PT

1. Select the point element to be projected. A horizontal vector displays.
2. Select another point element. The created point will be where the point is projected orthogonally onto the horizontal vector which passes through the first selected point. Alternatively, select a line or curve element. The created point will be where the point is projected horizontally onto the selected element.

Create point by vertical projection of point

Select: POINT > PROJECT | VERTICAL

Prompt: MSELW PT

1. Select the point element to be projected. A vertical vector displays.
2. Select another point element. The created point will be where the point is projected orthogonally onto the vertical vector which passes through the first selected point. Alternatively, select a line or curve element. The created point will be where the point is projected vertically onto the selected element.

Create point via horizontal and vertical projection of point

Select: POINT > PROJECT | HOR VERT

Prompt: MSELW PT

Successively select two point elements. A point will be created at the intersection of the horizontal vector which passes through the first selected point, and the vertical vector which passes through the second selected point.

Alternatively, select two line or curve elements, or a line and a curve element. The end points closest to the selection position of the lines or curves will be used to create a point at the intersection of the horizontal vector which passes through the first selected point, and the vertical vector which passes through the second selected point.

Create point via vertical and horizontal projection of point

Select: POINT > PROJECT | VERT HOR

Prompt: MSELW PT

Successively select two point elements. A point will be created at the intersection of the vertical vector which passes through the first selected point, and the horizontal vector which passes through the second selected point. Alternatively, select two line or curve elements, or one line and one curve element. The end points closest to the selection position of the lines or curves will be used to create a point at the intersection of the vertical vector which passes through the first selected point, and the horizontal vector which passes through the second selected point.

Create point via projection of point in specific direction

Select: POINT > PROJECT | UNSPEC

Prompt: MSELW PT

1. Select the point element to be projected.
2. Choose one of the following alternatives: (a) Select a line element to define the projection direction. (b) Key in a value for the angle of projection. This angle will be referenced to the current horizontal axis. (c) Click on the YES button to accept the displayed default values.
3. Select a point element. The point created will be on a line which passes through the first selected point and projected orthogonally from the second selected point. Alternatively, select a line or curve element. The created point will be where the first selected point is projected at the specified angle onto the line or curve.

COORD

Select: POINT > COORD | SINGLE or REPEAT or CHAIN | RECTANG or POLAR

Create points by keying in rectangular or polar coordinates. When creating points using the COORD option you have two choices. Upon selecting RECTANG, the point is defined by the X and Y directions. When selecting POLAR, the point is defined by its radius and angle.

Create single point

Select: POINT > COORD | SINGLE | RECTANG

Prompt: REFERENCE POINT : MSELW PT KEY X,Y // IND POINT

Select: POINT > COORD | SINGLE | POLAR

Prompt: REFERENCE POINT : MSELW PT KEY RAD, ANG // IND POINT

1. Choose one of the following alternatives: (a) Key in values for the point coordinates. The created point will be referenced to the current axis. (b) Indicate a position for the point. (c) Select a point element to serve as the reference point for the created point.
2. Key in values for the point coordinates. The created point references the previously selected point. Alternatively, click on the YES button to accept the displayed default values.

Create set of repeated points

Select: POINT > COORD | REPEAT | RECTANG or POLAR

Prompt: SEL PT

1. Select a point element to serve as the reference point for the created set of points. Alternatively, indicate a position for the point to serve as the reference point for the created set of points.
2. Key in values for the coordinates of the first point to be created with reference to the first selected point. If required, key in the number of points to be created. The points will be created evenly spaced on a line passing through the first selected point, the reference point, and the first point created. Alternatively, click on the YES button as many times as necessary to accept the values of the relative coordinates and the number of points. The last created point in each sequence is taken as the reference.

Create chain set of points

Select: POINT > COORD | CHAIN | RECTANG or POLAR

Prompt: REFERENCE POINT : SEL PT

1. Select a point element to serve as the reference point for the first created point.
2. Key in values for the point coordinates to be created. The first created point will reference the previously selected point. Additional coordinates can be keyed in; the points will then reference the previously created point. Alternatively, click on the YES button to accept the displayed default values. The YES button may be clicked

as many times as required; the first created point will reference the previously selected point. Additional points will then reference the previously created point.

TIP: *Remember that the vicinity selection facility can be used with all COORD options.*

LIMITS

Create points on limits of elements and element constraint on end point curve arcs.

Select: POINT > LIMITS | PTS or CST

Create points on limits of elements

Select: POINT > LIMITS | PTS

Prompt: MSELW ELEM

Create line segment limit points

Select a line element. The two end points will be created.

Create curve limit points

1. Select a curve element. The two end points of the arc formed by the curve will be created. If the curve is closed, only the open point will be created.
2. Click on the YES button to create the end points of the basic arcs as necessary.

Create circle limit points

Select a circle element. The center point and the two end points of an arc of the circle will be created. If the circle is closed, the center and open points will be created.

Create ellipse limit points

Select an ellipse element. The center point, foci, and end points of an ellipse arc will be created. If the ellipse is closed, the center point, foci, and open point will be created.

Create parabola limit points

Select a parabola element. The focus, center point, and two end points will be created.

Create hyperbola limit points

Select a hyperbola element. The focus, center point, and two end points will be created.

Create axis limit point

Select an axis element. A point on the origin will be created.

Create element constraint on curve arc end points

Select: POINT > LIMITS | CST

Prompt: MSELW CCV / CRV

Select a curve.

TIP: *Remember that elements can be selected using the multiselect options. (See Appendices A and B for further information.)*

SPACES

Create equidistant points on selected element and with constraints.

Select: POINT > SPACES | PTS or CST

Create equidistant points

Select: POINT > SPACES | PTS

Prompt: CURVE : SEL ELEM // ORIGIN : SEL PT

Create equidistant points from specified origin at specified distance for specified number of points

1. Select a SPACE element.
2. Select a point element to define the origin. If the selected point is not on the previously selected element, then the origin will be the orthogonal projection of the point onto the element.
3. Key in a value for the distance. If required, key in the number of points (e.g., *20,3* meaning three points at a distance of 20). If no value is entered for the number of points, the default is 1.
4. Click on the YES button to reverse the direction of the created points as necessary.

Create equidistant points parallel to line

1. Select a point element to define the origin.
2. Select a line element.
3. Key in a value for the distance. If required, key in the number of points (e.g., *20,3* meaning three points at a distance of 20). If no value is entered for the number of points, the default is 1.
4. Click on the YES button to reverse the direction of the created points as necessary.

Create equidistant points between two points separated by specified curvilinear length

1. Select a line, curve, or circle element.
2. Select two point elements to define the origin and end points. If the selected points are not on the previously selected element, then the origin and end points will be the orthogonal projection of the points onto the element.
3. Key in a value for the curvilinear distance created. Points will be created between the two selected points at the closest possible distance to the keyed value.

Create equidistant points between two points by specified number of points

1. Select a line, curve, or circle element.
2. Key in a value for the number of points required.
3. Select two point elements from the previously selected element to define the origin and end points. Points will be created between the origin and end points. Alternatively, click on the YES button to use the end points of the line or curve as the origin and end points. Points will be created between the origin and end points.

Create midpoint of element

1. Select a line, curve, or circle element.
2. Click on the YES button to create the midpoint.

Create equidistant points

Select: POINT > SPACES | CST

Prompt: CURVE : SEL ELEM // ORIGIN : SEL PT

Create equidistant points from specified origin at specified distance for specified number of points

1. Select a SPACE element.
2. Select a point element to define the origin. If the selected point is not on the previously selected element, then the origin will be the orthogonal projection of the point onto the element.
3. Key in a value for the distance, and then key in the number of points (e.g., *20,3* meaning three points at a distance of 20).
4. Click on the YES button to reverse the direction of the created CST as necessary.

Create equidistant points between two points separated by specified curvilinear length

1. Select a line, curve, or circle element.
2. Select two point elements on the previously selected element to define the origin and end points of the CST element.
3. Key in a value for the curvilinear distance created. The points of the CST element will be created between the two selected points at the closest possible distance to the keyed value.

Create equidistant points between two points by specified number of points

1. Select a line, curve, or circle element.
2. Key in a value for the number of points required.
3. Select two point elements from the previously selected element to define the origin and end points. If the selected points are not on the previously selected element, then the origin and end points will be the orthogonal projection of the points onto the element. CST points will be created between the origin and end points. Alternatively, click on the YES button to use the end points of the line or curve as the origin and end points. CST points will be created between the origin and end points.

Create constraint at midpoint of element

1. Select a line, curve, or circle element.
2. Click on the YES button to create the constraint at the midpoint.

TANGENT

Create points of tangency on curve.

Select: POINT > TANGENT

Prompt: CURVE : SEL ELEM // SEL LN

Select a curve element (the selected curve must be planar), and then select a line element. Alternatively, select the line element, and then the (planar) curve element. The points of tangency between the line and the curve will be created, and the points of tangency and tangent directions temporarily display.

TIP: *Remember that elements can be selected using the multiselect options when selecting the second element. (See Appendices A and B for further information on multiselect.)*

POINT in 3D SPACE Mode

Chapter 9

The POINT function is used to create points, and tangency or limit points.

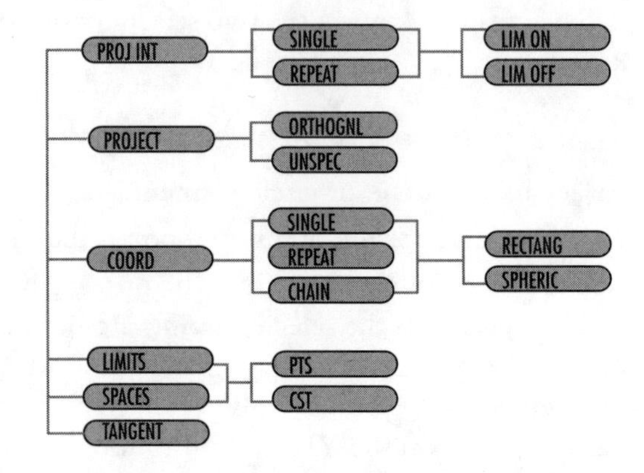

Option menu for POINT in 3D SPACE mode.

VICINITY SELECTION

Vicinity selection is also available for selecting the point(s) for use in creating points. This facility allows you to select a line or curve element to define its end point at the end closest to where the line or curve is selected. The end point created will be a temporary point used in creating a point.

PROJ INT

Create points by intersecting elements or projecting a point onto an element.

Select: POINT > PROJ INT | SINGLE or REPEAT | LIM ON or LIM OFF

Upon creating points using PROJ INT, points can be created depending on whether the intersection or projection is within the limits of the selected elements. Use the LIM ON option to create points only by intersection or projection when either is within limits of selected elements. Select the LIM OFF option when you wish for points to be created by intersection or projection when either is outside the limits of selected elements.

Create single point

Select: POINT > PROJ INT | SINGLE | LIM ON or LIM OFF

Prompt: 1ST : MSELW ELEM > ELEM

Create point midway between two points

Successively select two point elements.

Create point by projecting point onto element

1. Select a point element.
2. Select a SPACE element. If a curve, surface, face, or skin element is selected, the point closest to the projection point will be created.
3. If required, click on the YES button to create all the projection points. A maximum of 10 points can be created.

Create point(s) on intersection of two elements

1. Successively select two line or curve elements. If there is more than one intersection between the selected elements, all intersection points will be created.
2. If desired, where no intersection exists between the selected elements, click on the YES button to create two points with the least distance between the two selected elements.

Create several points

Select: POINT > PROJ INT | REPEAT | LIM ON or LIM OFF

Prompt: 1ST : MSELW ELEM > ELEM

1. To create several points, the first selected element will be used as the reference element. The procedure for creating several points is the same as for single points; the first selected element is used as the reference.
2. To select a new reference element, click on the YES button to end the procedure and select a new reference element.

TIP: *Remember that elements can be selected using multiselect options. (See Appendices A and B for further information.) The vicinity selection facility can be used with all PROJ INT options.*

PROJECT

Create points by projecting points onto element.

Select: POINT > PROJECT | ORTHOGNL or UNSPEC

Create point via orthogonal projection of point

Select: POINT > PROJECT | ORTHOGNL

Prompt: MSELW PT

1. Select the point element to be projected.
2. Select another point element. The created point will be at the midpoint of the segment which joins the two selected points. Alternatively, select a SPACE element. The created point will be the orthogonal projection of the selected point onto the selected element. A maximum of 10 points will be created.

Create point via projection of point in specified direction

Select: POINT > PROJECT | UNSPEC

Prompt: MSELW PT

1. Select the point element to be projected.
2. Select a line element to define the projection direction.
3. Select a point element. The point created will be on a line that passes through the first selected point and projected orthogonally from the second selected point. Alternatively, select a SPACE element. The created point will be where the first selected point is projected at the selected angle onto the element.

COORD

Select: POINT > COORD | SINGLE or REPEAT or CHAIN | RECTANG or SPHERIC

Create points by keying in rectangular or spheric coordinates. When creating points using the COORD option, you have two suboptions. Upon selecting RECTANG, the point is defined by the X and Y directions. Select SPHERIC to define the point by its radius and angle.

Create single point

Select: POINT > COORD | SINGLE | RECTANG

Prompt: REFERENCE POINT : MSELW PT KEY X,Y, Z

Select: POINT > COORD | SINGLE | SPHERIC

Prompt: REFERENCE POINT : MSELW PT KEY RAD, TETA, PSI

1. Use one of the following alternatives: (a) Key in values for point coordinates. The created point will be referenced to the current axis. (b) Indicate a position for the point. (c) Select a point element to serve as the reference point for the created point.

2. Key in values for point coordinates. The created point references the previously selected point. Alternatively, click on the YES button to accept the displayed default values.

Create set of repeated points

Select: POINT > COORD | REPEAT | RECTANG or SPHERIC

Prompt: SEL PT

1. Select a point element to serve as the reference point for the created set of points.
2. Key in the values for the coordinates of the first point to be created with reference to the first selected point. If required, key in the number of points to be created. The points will be created evenly spaced on a line passing through the first selected point (reference point) and the first point created. Alternatively, click on the YES button as many times as necessary to accept the values of the relative coordinates and number of points. The last created point in each sequence is taken as the reference.

Create chain set of points

Select: POINT > COORD | CHAIN | RECTANG or SPHERIC

Prompt: REFERENCE POINT : SEL PT

1. Select a point element to serve as the reference point for the first created point.
2. Key in values for the coordinates of the points to be created. The first created point will reference the previously selected point. Additional coordinates can be keyed in; the points will then reference the previously created point. Alternatively, click on the YES button to accept the displayed default values. The YES button may be clicked as many times as necessary. The first created point will reference the previously selected point. Additional points will then reference the previously created point.

TIP: *Remember that the vicinity selection facility can be used with all COORD options.*

LIMITS

Create points on element limits and element constraint on curve arc end points.

Select: POINT > LIMITS | PTS or CST

Create points on limits of selected elements

Select: POINT > LIMITS | PTS

Prompt: MSELW ELEM

Create limit points of line segment

Select a line element. The two end points will be created.

Create limit points of curve

1. Select a curve element. The two end points of the arc formed by the curve will be created. If the curve is closed, only the open point will be created.
2. Click on the YES button to create the end points of the basic arcs as necessary.

Create limit points of circle

Select a circle element. The center point and the two end points of an arc of the circle will be created. If the circle is closed, the center and open points will be created.

Create limit points of ellipse

Select an ellipse element. The center point, foci, and end points of an arc will be created. If the ellipse is closed, the center, foci, and open points will be created.

Create limit points of parabola

Select a parabola element. The focus, center point, and the two end points will be created.

Create limit points of hyperbola

Select a hyperbola element. The focus, center point, and the two end points will be created.

Create limit point of axis

Select an axis element. A point on the origin will be created.

Create limit point of face

1. Select a face element.
2. Click on the YES button to create the points on the corners of the selected face. Alternatively, key in a value for the distance the points will be created normal to each corner of the selected face.
3. If required, click on the YES button to reverse the direction of the points.

Create limit point of plane

Select a plane element. A point on the center of the plane symbol will be created.

Create limit point of constraint

Select a CST element. Definition points on which the constraint element rely will be re-created.

Create limit point of solid mock up or polyhedral surface

1. Select a solid mock up (SOLM) or polyhedral surface (POL) element. The end points of the selected edge will be created.
2. In the event a solid mock up was selected, click on the YES button to create the end points of all polyhedral edges as necessary.

Create limit point of exact solid

1. Select an edge or surface of the solid. Points will be created at the limit of the selected element.
2. If desired, click on the YES button to create the end points of the solid.

Create element constraint on curve arc end points

Select: POINT > LIMITS | CST

Prompt: MSELW CCV / CRV

Select a curve.

➾ ***TIP:*** *Remember that elements can be selected using the multiselect options. (See Appendices A and B for further information.)*

SPACES

Create equidistant points on selected element and with constraints.

Select: POINT > SPACES | PTS or CST

Create equidistant points

Select: POINT > SPACES | PTS

Prompt: CURVE : SEL ELEM SEL SUR / PLN / FAC // ORIGIN : SEL PT

Create equidistant points from specified origin at specified distance for specified number of points

1. Select a SPACE element.

2. Select a point element to define the origin. If the selected point is not on the previously selected element, then the origin will be the orthogonal projection of the point onto the element.
3. Key in a value for the distance. If desired, key in the number of points (e.g., *20,3* meaning three points at a distance of 20). If no value is entered for the number of points, the default is 1.
4. Click on the YES button to reverse the direction of the created points as necessary.

Create equidistant points parallel to line

1. Select a point element to define the origin.
2. Select a line element.
3. Key in a value for the distance. If required, key in the number of points (e.g., *20,3* meaning three points at a distance of 20). If no value is entered for the number of points, the default is 1.
4. Click on the YES button to reverse the direction of the created points as necessary.

Create equidistant points between two points separated by specified curvilinear length

1. Select a line, curve, or circle element.
2. Select two point elements to define the origin and end points. If the selected points are not on the previously selected element, then the origin and end points will be the orthogonal projection of the points onto the element.
3. Key in a value for the curvilinear distance created. Points will be created between the two selected points at the closest possible distance to the keyed value.

Create equidistant points between two points by specified number of points

1. Select a line, curve, or circle element.
2. Key in a value for the number of points required.
3. Select two point elements from the previously selected element to define the origin and end points. Points will be created between the origin and end points. Alternatively, click on the YES button to use the end points of the line or curve as the origin and end points. Points will be created between the origin and end points.

Create midpoint of element

1. Select a line, curve, or circle element.
2. Click on the YES button to create the midpoint.

Create equidistant points

Select: POINT > SPACES | CST

Prompt: CURVE : SEL ELEM SEL SUR / PLN / FAC // ORIGIN : SEL PT

Create equidistant points from specified origin at specified distance for specified number of points

1. Select a SPACE element.
2. Select a point element to define the origin. If the selected point is not on the previously selected element, then the origin will be the orthogonal projection of the point onto the element.
3. Key in a value for the distance, and then key in the number of points (e.g., *20,3* meaning three points at a distance of 20).
4. Click on the YES button to reverse the direction of the created CST as necessary.

Create equidistant points between two points separated by specified curvilinear length

1. Select a line, curve, or circle element.
2. Select two point elements on the previously selected element to define the origin and end points of the CST element.
3. Key in a value for the curvilinear distance created. The points of the CST element will be created between the two selected points at the closest possible distance to the keyed value.

Create equidistant points between two points by specified number of points

1. Select a line, curve, or circle element.
2. Key in a value for the number of points required.
3. Select two point elements from the previously selected element to define the origin and end points. If the selected points are not on the previously selected element, then the origin and end points will be the orthogonal projection of the points onto the element. CST points will be created between the origin and end points. Alternatively, click on the YES button to use the end points of the line or curve as the origin and end points. CST points will be created between the origin and end points.

Create constraint at midpoint of element

1. Select a line, curve, or circle element.

2. Click on the YES button to create the constraint at the midpoint.

TANGENT

Create points of tangency on curve.

Select: POINT > TANGENT

Prompt: CURVE : SEL ELEM // SEL LN

Select a curve element (the selected curve must be planar), and then select a line element. Alternatively, select the line element and then the (planar) curve element. The points of tangency between the line and the curve will be created, and the points of tangency and tangent directions temporarily display.

➡ ***TIP:*** *Remember that elements can be selected using the multiselect options when selecting the second element. (See Appendices A and B for further information on multiselect.)*

SETS in DRAW and SPACE Modes

Chapters 5, 13

The SETS function in DRAW and SPACE mode is used to create and manage sets.

Option menu for SETS in DRAW and SPACE modes.

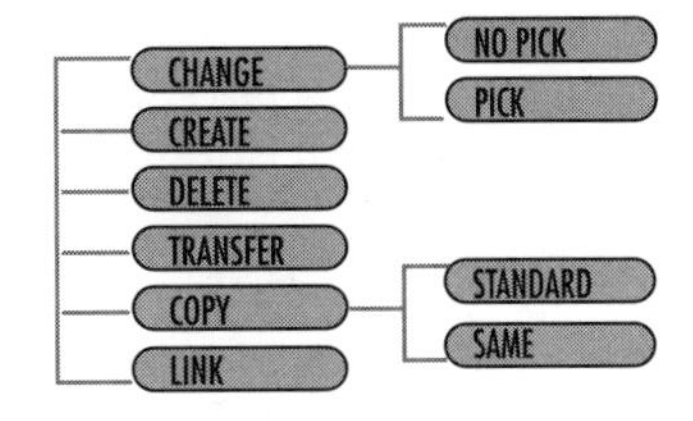

CHANGE

Change current set.

Select: SETS > CHANGE | PICK or NO PICK

Prompt: SEL SET / / YES: LIST

The NO PICK option makes the current nonselectable, whereas the PICK option makes it selectable.

1. Click on YES to display the list of sets in the model. Select a set from the list.
2. Click on YES to confirm. The selected set becomes the current set.

CREATE

Create sets.

Select: SETS > CREATE

Prompt: SET ID: KEY TEXT / / YES: AUTO ID

Enter an name for the new set or click on YES to accept the name suggested. The new set is created and becomes the current set.

DELETE

Delete sets.

Select: SETS > DELETE

Prompt: SEL SET / / YES: LIST

1. Click on YES to display the list of sets in the model. Select a set from the list.
2. Click on YES to confirm. The selected set is deleted.

✓ **NOTE:** *If the selected elements are referenced by notes or dimensions, a message to that effect will be displayed. The current set cannot be deleted.*

TRANSFER

Move elements into current set.

Select: SETS > TRANSFER

Prompt: MSELW ELEM

Select elements individually or by multiselect. Click on YES to confirm transfer. Selected elements are transferred into the current set.

COPY

Copy elements into current set.

Select: SETS > COPY | STANDARD or SAME

Prompt: MSELW ELEM

Upon selecting the STANDARD option, selected elements will be copied into the current set on the current layer, with the current standard display characteristics. With the SAME option selected, elements selected will be copied into the current set on the same layer and with the same display characteristics as the selected elements.

Select elements individually or by multiselect. Upon selection the elements will be copied into the current set.

LINK

Link set with current set.

Select: SETS > LINK

Prompt: SEL SET / / YES: LIST

Click on YES to display the list of sets in the model. Select the set to be linked from the list or in the work area. Click on YES to confirm the link operation.

SHAPE in DRAW Mode

Chapter 6

The SHAPE function is used in DRAW mode only to create and modify open and closed contours. The shapes can be closed or open polygonal contours, squares, rectangles, or multidomain shapes. Patterns can be applied to the shapes, and the shapes can also be used to perform section analysis.

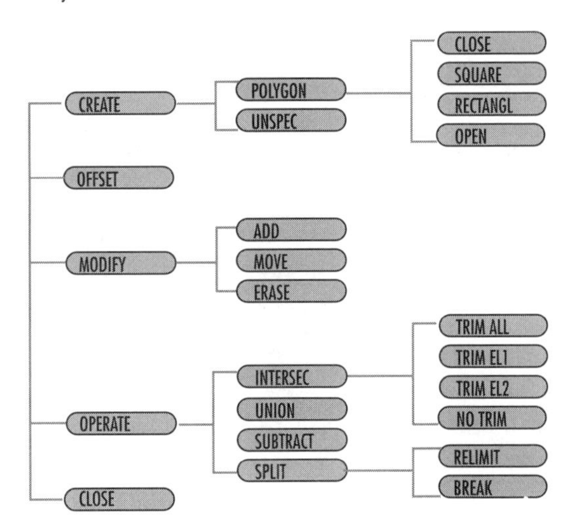

Option menu for SETS in DRAW mode.

CREATE

Create polygonal domains.

Select: SHAPE > CREATE | POLYGON | CLOSE

Create closed polygonal domains

Select: SHAPE > CREATE | POLYGON | CLOSE or SQUARE or RECTANGL or OPEN

Prompt: SEL 1ST PT/LN

✓ ***NOTE:*** *During the process of creating closed domains, rubber banding shows the contour being closed and is linked to the cursor position.*

1. Select at least three points to define the vertices of a desired closed domain, and at least three lines to define the sides of a desired closed domain.
2. Click on YES to end contour definition.

3. Click on YES to define further contour(s) inside the first as necessary.
4. Select points or lines as before to define inner domain. Click on YES to end definition.
5. If required, click on YES to display an identifier at the first vertex or select a point for the identifier.
6. Enter a name of up to 70 characters as necessary. Click on YES to end.

Create square domain

Select: SETS > CREATE | POLYGON | SQUARE

Prompt: SEL PT/LN

1. Select a point (lower left vertex of the square).
2. Enter a value for the length of the side. The resulting lines will be parallel to the current axis. Alternatively, click on YES to accept the suggested value.
3. Select a line segment. Click on YES. The selected line forms the diagonal of the square domain. Alternatively, select or indicate two points. The line joining the two points forms the diagonal of the square domain.
4. Select a line segment. The selected line defines the side of the square domain.
5. Indicate the desired half plane.
6. Select a point. The point defines the center of the square domain.
7. Select a line. The side of the square domain will have the same length as the selected line and the two parallel sides.
8. Select a point as necessary. The identifier will be displayed at this point. Alternatively, click on YES and the identifier will be displayed at one of the square vertices.
9. Enter a name of up to 70 characters as necessary.
10. Select a point or click on YES to position the name.

Create rectangular domain

Select: SETS > CREATE | POLYGON | RECTANGL

Prompt: SEL PT/LN / / SEL TXT

- To create a domain defined by a text or note, select the text (TXTD) element. The rectangular domain boxes the text.
- To create a domain by a vertex and two dimensions, select a point (the lower left vertex of the rectangle). Enter a value for length and width. Alternatively, click on YES to accept values offered. The rectangle will be parallel to the axis.
- To create a domain defined by a side and a point on the opposite side, select a line segment. The line defines one side of the rectangle. Select a point. The side parallel to the first side will pass through the point.

- To create a domain defined by a side and a dimension, select a line segment. The line defines one side of the rectangle. Enter a length for the second side. Indicate the half plane. Alternatively, click on YES to accept the suggested value.
- To create a domain defined by two opposite vertices, select two points. The line joining the two points defines the diagonal of the rectangle. Select or indicate a point to position the identifier. Alternatively, click on YES to display the identifier at the first selected point.

Create open polygonal contours

Select: SHAPE > CREATE | POLYGON | OPEN

Prompt: SEL 1ST PT/LN

Use one of the following alternatives.

- Select at least three points to define vertices of contour, and at least three lines to define the vectors of the contour. Click on YES to complete the contour definition.
- Select at least four lines. Click on YES to complete contour. Select a line to position the identifier.
- Click on YES to position the identifier at the first vertex.

Create unspecified closed shapes

Select: SHAPE > CREATE | UNSPEC

Prompt: KEY SAG SEL LN/CRV

1. If desired, enter a sag value, or continue accepting the suggested value. Select a line or curve. Click on YES to search automatically for a closed domain.
2. Select additional lines and curves to complete the domain as necessary. Click on YES to accept the closing.
3. Click on YES to create an inner domain as necessary. Select lines and curves to complete the inner domain. Click on YES to complete the creation.
4. Select a point to position the identifier, or click on YES to display the identifier at the closing point.

OFFSET

Create shape parallel to another shape.

Select: SHAPE > OFFSET

Prompt: SEL SHAPE

1. Select the reference shape.
2. Select a point to obtain the offset value, or indicate the region in which the offset is to be created.
3. Enter length and number of offsets.
4. Indicate the region in which offsets are to be created.
5. Click on YES to accept the suggested values.

MODIFY

Modify geometry of shape.

Select: SHAPE > MODIFY | MOVE or ADD or ERASE

Modify position of shape vertices

Select: SHAPE > MODIFY | MOVE

Prompt: SEL SHAPE

1. Select the shape. The shape is dimmed and a number displays at the vertices.
2. Select or enter a number.
3. Select a point. The selected vertex will be moved to the new point.
4. Click on YES to display the modified shape.

✓ **NOTE:** *If the shape contains a pattern, the modification will be applied to the pattern.*

Add vertex to shape

Select: SHAPE > MODIFY | ADD

Prompt: SEL SHAPE

1. Select a shape. The shape is dimmed and a number displays at each vertex.
2. Select or enter a number of a point after which the new point will be inserted.
3. Select a point. The new vertex is inserted at the point.
4. Click on YES to redisplay shape.

Delete vertex from shape

Select: SHAPE > MODIFY | ERASE

Prompt: SEL SHAP

1. Select a shape. The shape is dimmed and a number displays at each vertex.
2. Select or enter a number of the vertex to be deleted. The vertex is deleted if topology allows.

3. Click on YES to redisplay the shape.

OPERATE

Execute various operations on shapes.

Select: SHAPE > OPERATE | INTERSEC or UNION or SUBTRACT or SPLIT

Intersection of two closed shapes

Select: SHAPE > OPERATE | INTERSEC | TRIM ALL or TRIM EL1 or TRIM EL2 or NO TRIM

Prompt: SEL SHAP

- TRIM ALL. First selected element replaced by result of operation while the second is deleted.
- TRIM EL1. First element replaced by result of operation while the second remains unaltered.
- TRIM EL2. First element selected not modified, while second is erased after the operation.
- NO TRIM. First element selected not modified, while second is also unaltered.

Select a closed shape and then select a second closed shape, text, or dimension. The resulting shape is the intersection of the selected shapes.

Union of two closed shapes

Select: SHAPE > OPERATE | UNION

Prompt: SEL SHAP

Select a closed shape. Select a second closed shape, text, or dimension. The resulting shape is the union of the selected shapes.

Subtract closed shape from another

Select: SHAPE > OPERATE | SUBTRACT | TRIM ALL or TRIM EL1 or TRIM EL2 or NO TRIM

Prompt: SEL SHAP

Select a closed shape. Select a second closed shape or text or dimension. The result is a subtraction of the two shapes depending on the trim option selected.

Limit or break shape using line

Select: SHAPE > OPERATE | SPLIT

Prompt: SEL SHAP

Select a closed shape in the area to be retained. Select a line.

CLOSE

Close open shape.

Select: SHAPE > CLOSE

Prompt: SEL OPEN SHAPE

Select shape to be closed.

✓ **NOTE:** *This option can be used only on open shapes.*

SOLIDE in SPACE Mode

Chapter 11

The SOLIDE function is used to create and modify exact solids.

Option menu for SOLIDE in SPACE mode.

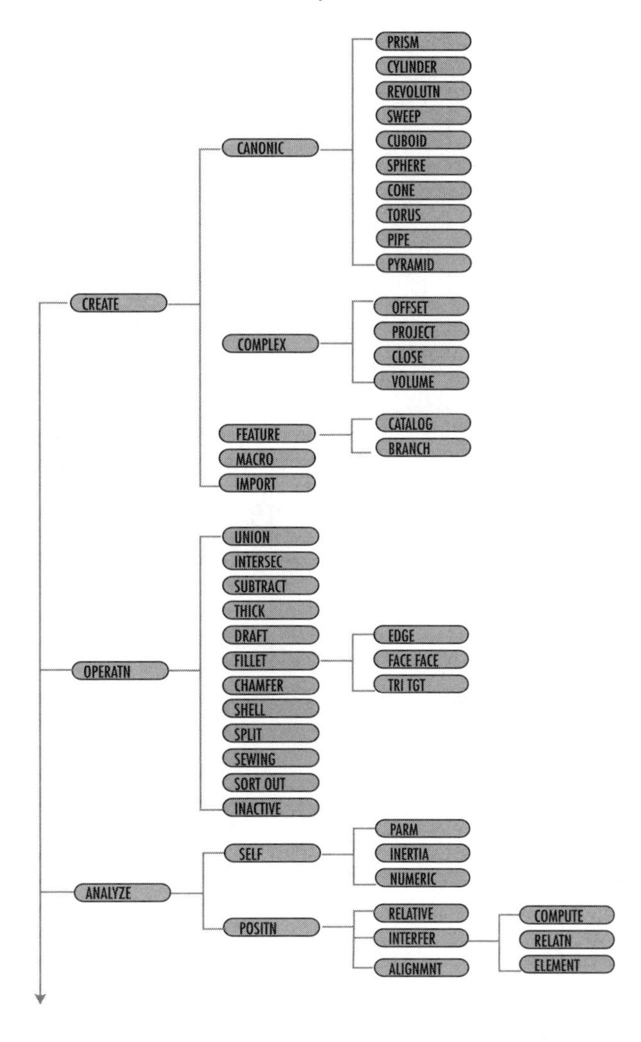

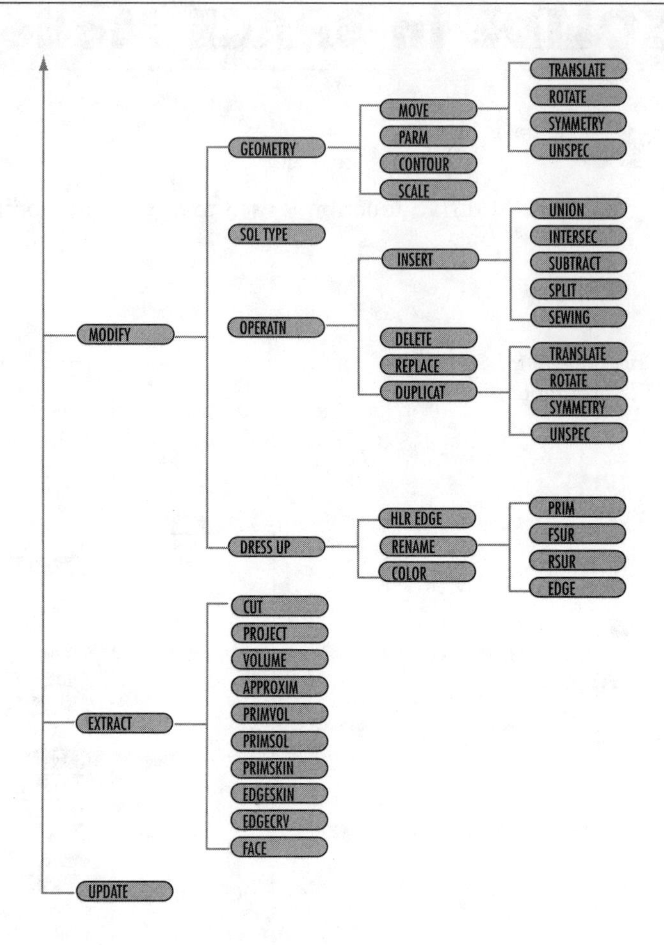

CREATE

Create exact solids and macroprimitives.

Select: CREATE | CANONIC or COMPLEX or FEATURE or MACRO or IMPORT

Create canonical solids

A canonical solid is defined by geometric values and/or a single parametric construction element known as a contour.

Select: SOLIDE > CREATE | CANONIC | PRISM or CYLINDER or REVOLUTION or SWEEP or CUBOID or SPHERE or CONE or TORUS or PIPE or PYRAMID

Create prism

Select: SOLIDE > CREATE | CANONIC | PRISM

Prompt: CONTOUR: MSELW LN/CONIC/CRV/CCV CONTOUR: SEL PT/FAC

In the Prism window, you will see the buttons illustrated below.

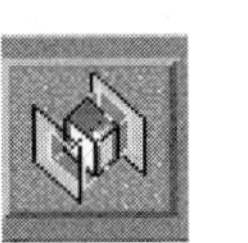

1  2   3 4

✓ **NOTE:** *As you place the cursor over each button (icon), a message describing the button's function appears in the message area.*

- Create a prism with no associative limits (button 1).
- Create an "until from to" solid using the functional surfaces (FSURs) of the current solid as prism limits (button 2).
- Create an "until from to" solid with drafts and fillets (button 3).

✓ **NOTE:** *When the Figure button is depressed during the above selections, a window providing a graphical prompt displays.*

- If the Display Dynamic Sketcher is available, the sketcher displays after clicking on button 4 and elements are selected. (See Frameviewer online documentation for a Dynamic Sketcher tutorial.)

In the Manage window, you will see the buttons illustrated below.

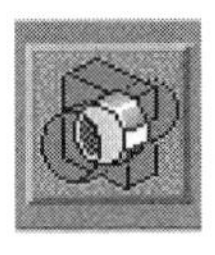

5 6 7 8

- Create a prism which is a new and separate feature (button 5).
- Create a prism to be added to the current solid (button 6).
- Create a prism which is the result of the intersection between the new prism and the current solid (button 7).
- Create a prism that will be subtracted from the current solid (button 8).

The following buttons are also available in the Manage window.

9

10

- With the CURRENT button, you can select a new solid to become the current solid.
- Switch the Part Editor window on or off (button 9). The Part Editor provides access to the CSG tree of the current solid.
- Change the solid selection method from edge to face or vice versa (button 10).

After completing selections in the windows, take the following steps.

1. Select an element of the contour geometry. Continue selecting elements until a closed contour is achieved.
2. Select a second contour element that is not adjacent to the first. (If possible, the contour will close.)
3. Click on the AutoSearch button.
4. Select additional geometry as necessary to define an inner domain, and then click on YES to end contour selection.
5. To change the extrusion direction, select one of the red arrows and select a line to define the new direction.
6. Define the limits of the prism by modifying the OFF1 and OFF2 values in the Prism window, or by entering new values in the input area.
7. For an "until from to" solid, select faces (FSURs) of the current solid.
8. Click on YES to create the prism.

Create cylinder

Select: SOLIDE > CREATE | CANONIC | CYLINDER

Prompt: REF PT: MSELW PT DIR: SEL LN/CIR/PLN

The options in the Cylinder window and Manage window are the same as described above for Prism, with the exception of the limit values in the Cylinder window.

Once you have made selections in the windows, take the following steps.

1. Select a point, or multiselect several points if more than one solid is to be created.
2. Select a line or plane to determine the direction of the cylinder, and then select the arrow to invert as necessary.
3. Define the limits in the Cylinder window work area or by entering values in the input area.

4. Click on YES to create the cylinder(s).

Create revolution

Select: SOLIDE > CREATE | CANONIC | REVOLUTION

Prompt: SEL REF PT / / AXIS: SEL LN/CIR/PLN

Parameters are entered in the Revolution window. Options in the Manage window are the same as described above for the PRISM option.

Once you have made selections in the windows, take the following steps.

1. Define the revolution axis by using one of the following alternatives: (a) Select a line. (b) Select a plane, and then a line. A revolution axis through the center of the plane and norman to the plane is created. (c) Select a point and then a line. A revolution axis parallel to the selected line passing through the point is created. (d) Select a curve, and then a point. A revolution axis normal to the curve passing through the point is created.
2. Define the contour by using one of the following alternatives: (a) Select an element of the contour geometry, and continue selecting elements until a closed contour is achieved. Select a second contour element that is not adjacent to the first. (If possible, the contour will close.) Click on the AutoSearch button. (b) Select a face.
3. Click on YES to end contour selection. Define the limits in the Revolution window work area or by entering new values in the input area.
4. Click on YES to create the revolution solid.

Create sweep

Select: SOLIDE > CREATE | CANONIC | SWEEP

Prompt:

✓ ***NOTE:*** *The methodology for creating the swept solid will vary depending on whether the CTR (center) curve and/or contour are open or closed and whether a pulling direction is specified. A CTR curve and contour cannot be open simultaneously.*

The Sweep window displays the Figure and Sketcher buttons. Manage window options are the same as described above for the Prism window.

Take the following steps to create a closed contour and closed CTR curve.

1. Successively select line or curve elements, or select the first element and then click on AutoSearch to select the complete CTR curve. If the CTR curve is planar, it can be discontinuous in tangency. In 3D mode the curve must be continuous in tangency.

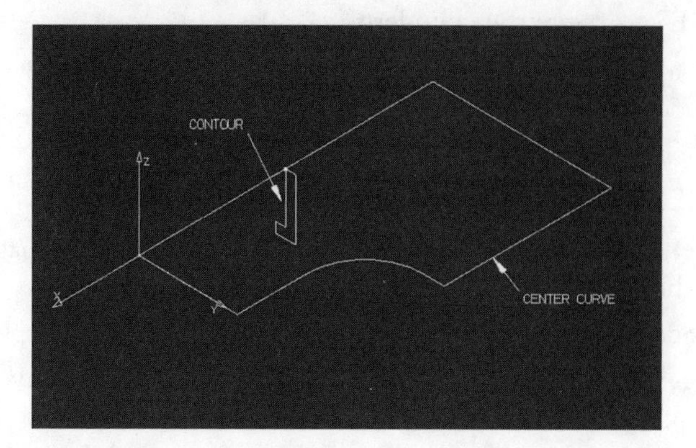

1. Click on YES to accept the CTR curve and the highlighted reference plane, or select an alternative reference plane. Select an element of the contour, and then click on AutoSearch.

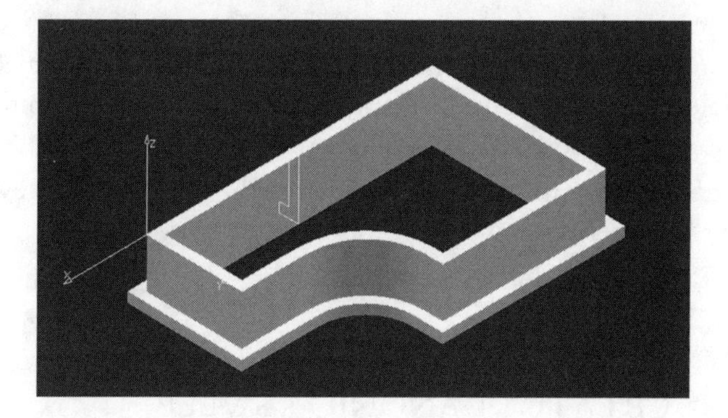

2. Duplications of the contour display along the curve providing a preview of the sweep solid to be created.
3. Enter an angle value as necessary to reorient the contour to the CTR curve, and/or D1 D2 displacement values to shift the contour in relation to the CTR curve.
4. Click on YES to create the solid.

Take the following steps to create an open contour with a closed planar CTR curve.

1. Successively select line or curve elements, or select the first element and then click on AutoSearch to select the complete CTR curve.

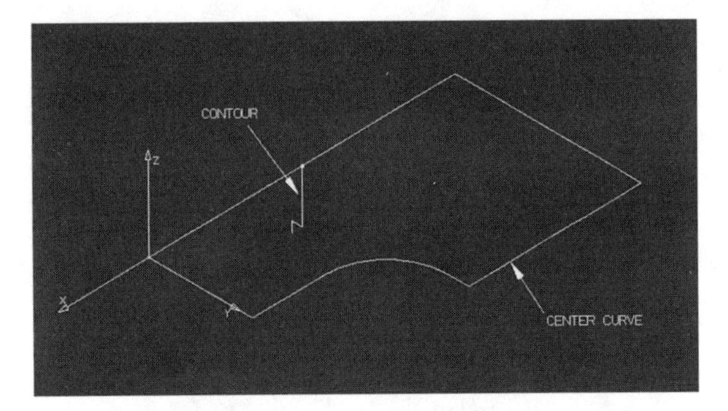

2. Click on YES to accept the CTR curve and the highlighted reference plane or select an alternative reference plane. Select an element of the contour then click on Auto-Search.
3. Click on YES to end contour definition.

4. Duplications of the contour display along the curve providing a preview of the sweep solid to be created.
5. Enter an angle value as necessary to reorient the contour to the CTR curve, and/or D1 D2 displacement values to shift the contour in relation to the CTR curve.
6. Click on YES to create the solid.

Take the following steps to create an open CTR curve (without pulling direction).

1. Successively select line or curve elements, or select the first element then click on AutoSearch to select the complete CTR curve.

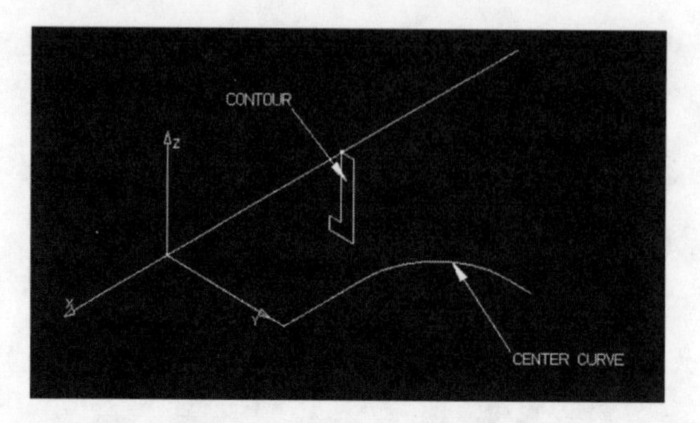

2. Click on YES to accept the CTR curve and the highlighted reference plane, or select an alternative reference plane.
3. Select an element of the contour and then click on AutoSearch. Duplications of the contour display along the curve providing a preview of the sweep solid to be created.

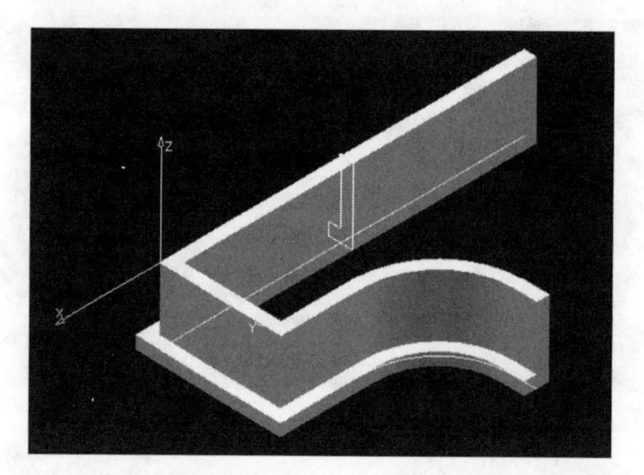

4. Enter an angle value as necessary to reorient the contour to the CTR curve, and/or D1 D2 displacement values to shift the contour in relation to the CTR curve.
5. Click on YES to create the solid.

Take the following steps to create a closed contour with an open CTR curve (with pulling direction).

1. Successively select line or curve elements, or select the first element and then click on AutoSearch to select the complete CTR curve.

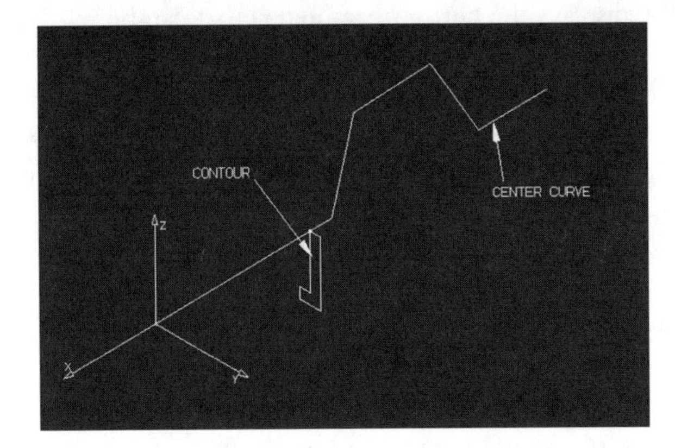

2. Click on Pulling Direction, and then select a line in the direction in which the contour is pulled.
3. Click on YES to accept the CTR curve and the highlighted reference plane, or select an alternative reference plane. Select an element of the contour and then click on AutoSearch. Duplications of the contour display along the curve providing a preview of the sweep solid to be created.

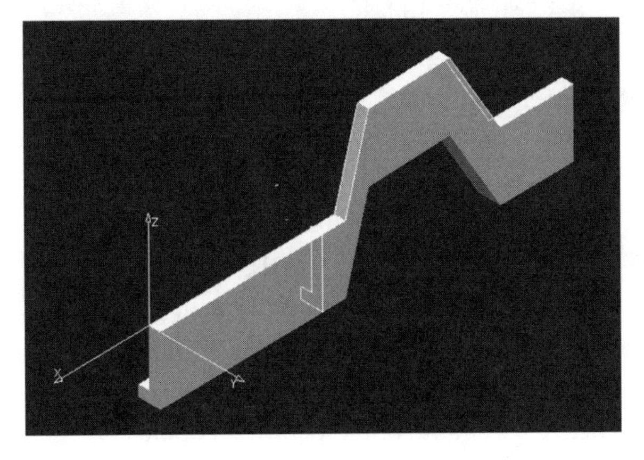

4. Enter an angle value as necessary to reorient the contour to the CTR curve, and/or D1 D2 displacement values to shift the contour in relation to the CTR curve.
5. Click on YES to create the solid.

Create cuboid

Select: SOLIDE > CREATE | CANONIC | CUBOID

Prompt: REF PT: MSELW PT / / DIR: SEL LN/PLN

The Cuboid window displays the current parameters that can be modified. The options in the Manage window are the same described above for the Prism window.

1. To define the anchor point and direction of the cuboid, make one of the following selections: (a) a point, and then a line or a point; (b) a point or two lines; (c) a plane, point, or line. A wireframe preview of the cuboid displays with parameters as in the Cuboid window.
2. Enter new values as necessary for LX, LY, and LZ in the input area or in the Cuboid window.
3. Click on YES to create the solid.

Create sphere

Select: SOLIDE > CREATE | CANONIC | SPHERE

Prompt: REF PT: MSELW PT DIR: SEL LN/CIR/PLN

The Sphere window displays the current parameters, which can be modified. The options in the Manage window are the same as described above for the Prism window.

1. Select a point, a line and a point, or two lines to determine the reference or anchor point of the sphere. A wireframe preview displays containing the sphere with parameters as in the Sphere window.
2. Enter a new value for R1 as necessary in the input area or the Sphere window.
3. Click on YES to create the solid.

Create cone

Select: SOLIDE > CREATE | CANONIC | CONE

Prompt: REF PT: MSELW PT DIR: SEL LN/CIR/PLN

The Cone window displays the current parameters, which can be modified. The options in the Manage window are the same as described above for the Prism window.

1. Select a point, a line and a point, or two lines to determine the reference or anchor point of the cone. A wireframe preview of the cone displays containing parameters as in the Cone window.
2. Enter new values as necessary for R1, R2, H1, and H2 in the input area or the Cone window.
3. Click on YES to create the solid.

Create torus

Select: SOLIDE > CREATE | CANONIC | TORUS

Prompt: REF PT: MSELW PT DIR: SEL LN/CIR/PLN

The Torus window displays the current parameters, which can be modified. The options in the Manage window are the same as described above for the Prism window.

1. Select a point, a line and a point, or two lines to determine the reference or anchor point of the torus. A wireframe preview of the torus with parameters as in the Torus window displays.
2. Enter new values as necessary for R1 and R2 in the input area or in the Torus window.
3. Click on YES to create the solid.

Create pipe

Select: SOLIDE > CREATE | CANONIC | PIPE

Prompt: CONTOUR: MSELW LN/CONIC/CRV/CCV CONTOUR: SEL PT/FAC

The Pipe window displays the current parameters which can be modified. The options in the Manage window are the same as described above for the Prism window.

1. Select a line or curve element. Click on AutoSearch to complete the contour as necessary.
2. Click on YES to end selection. A wireframe preview of the pipe displays containing parameters as in the Pipe window.
3. Enter a new value as necessary for R1 in the input area or Pipe window.
4. Click on YES to create the solid.

Create pyramid

Select: SOLIDE > CREATE | CANONIC | PYRAMID

Prompt: SEL: TOP VERTEX

The Pyramid window displays the Sketcher and Figure buttons as well as current parameters which can be modified. The options in the Manage window are the same as described above for the Prism window.

1. Select a point to be the top vertex of the pyramid.
2. Select an element of the bas contour, and click on AutoSearch to complete the contour.
3. Click on YES to create the solid.

Create solids via diverse element types

Create solid from surface (SUR), face (FAC), skin (SKI), or volume (VOL) type elements.

Select: SOLIDE > CREATE | COMPLEX | OFFSET or PROJECT or CLOSE or VOLUME

Create offset solid

Select: SOLIDE > CREATE | COMPLEX | OFFSET

Prompt: SEL SKD/SUR/FAC/SKI

1. The Offset window displays the current parameters, which can be modified. Options in the Manage window are the same as described above for the Prism window.
2. Select a surface, skin, or nonplanar face type element. Arrows and parameters display in the work area providing a preview of the solid to be created.
3. Enter values for OFF1 and OFF2 in the entry area or modify the current parameters in the Offset window as necessary.
4. Click on YES to create the solid.

Create solid by projecting surfaces or nonplanar faces onto plane

Select: SOLIDE > CREATE | COMPLEX | PROJECT

Prompt: SEL SKD/SUR/FAC/SKI

1. Options in the Manage window are the same as described above for the Prism window. Select a surface, skin, or nonplanar face element.
2. Select a plane. Define the projection direction by selecting a line or two points or a plane as necessary.
3. Click on YES to create the solid.

Create solid by closing surface or nonplanar face

Select: SOLIDE > CREATE | COMPLEX | CLOSE

Prompt: SEL SKD/SUR/FAC/SKI

Select a surface, skin, or non planar face element. Click on YES to create the solid.

Create solid from volume

Select: SOLIDE > CREATE | COMPLEX | VOL

Prompt: SEL VOL

Select a volume type element. Click on YES to create the solid.

Create features for library storage

Whether you are able to use this option will depend on how your system is set up.

Select: SOLIDE > CREATE | FEATURE | CATALOG

Frequently used features, such as bosses, ribs, pockets, and so forth can be created and stored with parameterized geometry in a library, and then used by numerous users to build solids. In this way, company design standards can be established and maintained.

Features are created using the SOLIDE function by taking the following steps.

- Create a detail workspace.
- Define geometry or solid.
- Parameterize geometry or solid.

Upon using the FTR CLASS and PART functions, the features definition attributes are defined, and the feature can be stored in a library and used in a mode similar to a library detail.

Create or delete feature using branch

A feature can also be created or deleted using the above option by selecting a branch from the CSG tree. The feature can then be modified using one of the following options: SOLIDE > MODIFY | OPERATN | INSERT or SOLIDE > MODIFY | OPERATN | DELETE.

Create macroprimitive solids

Select: SOLIDE > CREATE | MACRO

Prompt: SEL DIT

A macroprimitive is a solid created from a ditto. This process makes it possible to use numerous solids based on the same detail because features are controlled by current features in the detail workspace, and additional features are carried out in the master workspace. If a change is made to the detail at the root of the macroprimitive, it will affect all details and macroprimitives using the affected detail.

Select a ditto. The ditto is transformed into a macroprimitive solid.

Import previously published passive solids into model

Select: SOLIDE > CREATE | IMPORT

Prompt: SEL PASSIVE SOLID

Contact your system administrator for advice on the availability and use of this option.

OPERATN

Perform operations on solids. Operations are defined as topological modifications to solids.

Select: SOLIDE > OPERATN | UNION or INTERSECT or SUBTRACT or THICK or DRAFT or FILLET or CHAMFER or SHELL or SPLIT or SEWING or SORT OUT or INACTIVE

Union operations

Select: SOLIDE > OPERATN | UNION

Prompt: 1st OPERAND: MSELW SOL/DIT/VOL

In the Mode window you can choose to trim all elements during the intersection operation, trim the first selected element (duplicate the second), or duplicate all elements without trimming.

1. Select the first solid (ditto or volume).
2. Select the second solid (ditto or volume). The new solid resulting from the union displays. If a ditto is selected, it will be automatically converted to a macroprimitive.

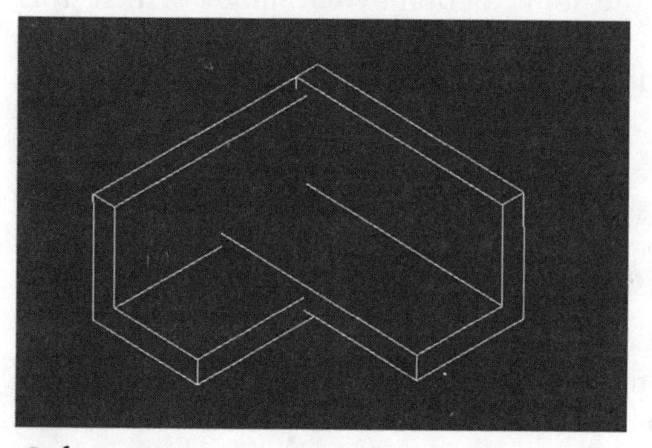

Before union.

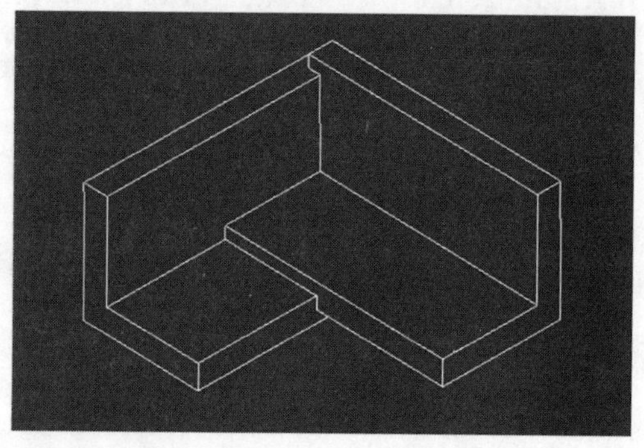

After union.

Intersection operations

Select: SOLIDE > OPERATN | INTERSEC

Prompt: 1st OPERAND: MSELW SOL/DIT/VOL

In the Mode window, you can choose to trim all elements during the intersection operation, trim the first selected element (duplicate the second), or duplicate all elements without trimming.

1. Select the first solid (ditto or volume).
2. Select the second solid (ditto or volume). The new solid resulting from the intersection volume of the selected solids displays. If a ditto is selected, it will be automatically converted to a macroprimitive.

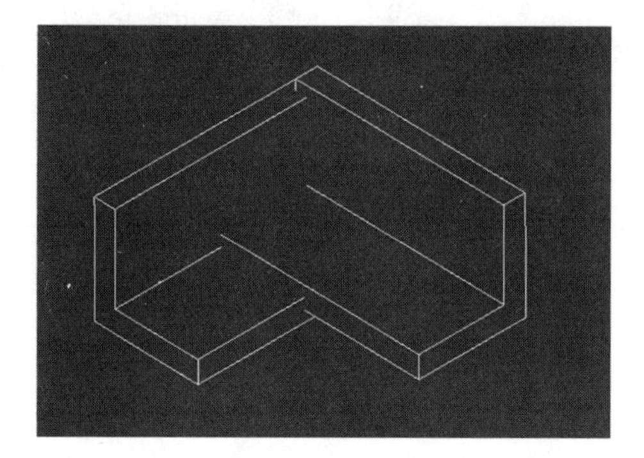

Before intersection.

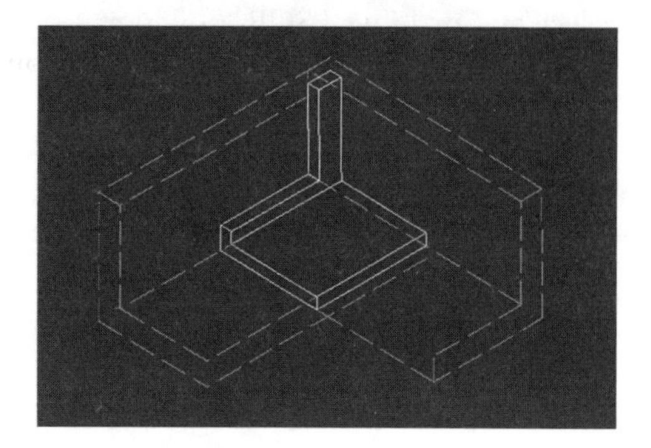

After intersection.

Subtraction operations

Select: SOLIDE > OPERATN | SUBTRACT

Prompt: 1st OPERAND: MSELW SOL/DIT/VOL

In the Mode window, you can choose to trim all elements during the subtraction operation, trim the first selected element (duplicate the second), or duplicate all elements without trimming.

1. Select the first solid (ditto or volume).
2. Select the second solid (ditto or volume). The new solid resulting from the volume remaining after the subtraction displays. If a ditto is selected, it will be automatically converted to a macroprimitive.

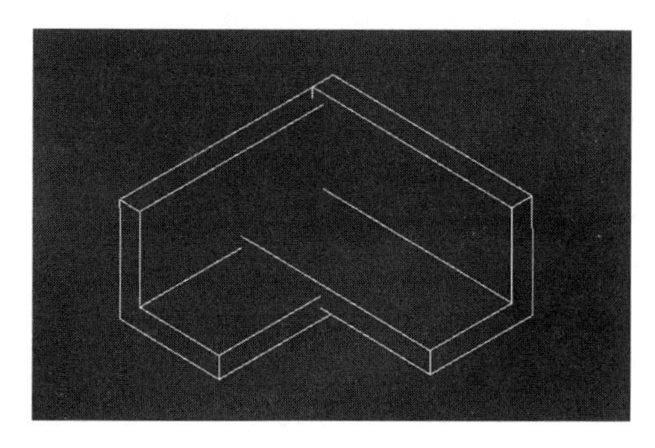

Before subtraction.

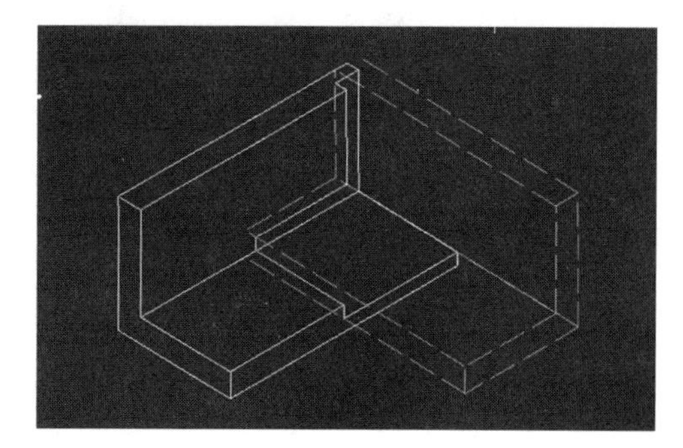

After subtraction.

Modify solid thickness

Select: SOLIDE > OPERATN | THICK

Prompt: ADD THICKNESS: SEL FSUR KEY THICKNESS

Select one or more FSUR (functional surface) elements of a solid. Enter a value for the thickness to be added. Click on YES to complete the operation.

Draft operation on solid

Create angled faces with a pulling direction.

Select: SOLIDE > OPERATN | DRAFT

Prompt: PULL DIR: SEL VECT ADD DRAFT: SEL FSUR / / KEY ANGLE

1. A default pulling direction represented by a green arrow displays. If a different direction is desired, select the arrow and then select a line.
2. Enter a draft angle (i.e., the angle between the pulling direction and the draft face).
3. Click on the Parting Element button in the Draft mode window as necessary.
4. Select a plane, solid face (FSUR) face surface, or skin. A green plane displays representing the plane at which the solid will be cut in two for drafting purposes.
5. Click on the Switch pulling direction button in the DRAFT mode window to draft in both directions from the parting plane as necessary.
6. If desired, define the neutral plane. (If there is only one pulling direction, the first selected surface of the solid defines the neutral plane.) If there are two pulling directions, the first neutral plane is created as described here, and the second is created symmetrically to the first and normal to the second pulling direction.
7. Select the solid faces (FSUR) to be drafted.
8. Modify draft angle as necessary.
9. Click on YES to perform draft operation.

Create fillets on solids

Create internal or external rounded corners of constant or variable radius on solids which are tangent to the surfaces being joined.

Select: SOLIDE > OPERATN | FILLET | EDGE or FACE FACE or TRI TGT

Fillet edge

Select: SOLIDE > OPERATN | FILLET | EDGE

Prompt: SEL REDG/VERTEX/RSUR / / KEY RAD

The Fillet Mode window provides the following options.

- Propagation. With Propagation set to Auto, the filleting process continues beyond the selected edge until an edge which is discontinuous in tangency is encountered.

With Propagation set to Manual, the filleting process continues beyond the selected edge only when it cannot do otherwise.

- Rolling Edge. With Rolling Edge selected, the fillet radius is "rolled" around another edge.

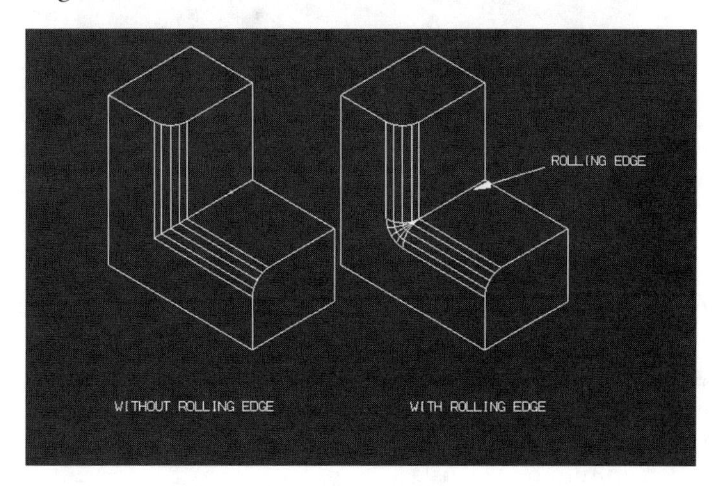

- Variable Fillet. The Linear and Imposed Tangency options are available upon clicking on the Switch Variable Fillet button (shown at left).

After selecting the edge to be filleted, various radii can be imposed at selected points.

1. Select an edge, vertex, or face (FSUR).
2. Click on YES to compute the fillet.

Create fillet radius between adjacent faces

Select: SOLIDE > OPERATN | FILLET | FACE FACE

Prompt: SEL> RSUR1 : FIRST FILLET SUPPORT

1. Select the first face (FSUR).
2. Select the second face (FSUR)
3. Enter a radius value or click on YES to accept the standard radius. The fillet operation is computed.

Create fillet between three faces by removing third selected face

Select: SOLIDE > OPERATN | FILLET | TRI TGT

Prompt: SEL> RSUR1 : FIRST FILLET SUPPORT

Successively select three faces as shown in the next illustration. A full fillet radius is created.

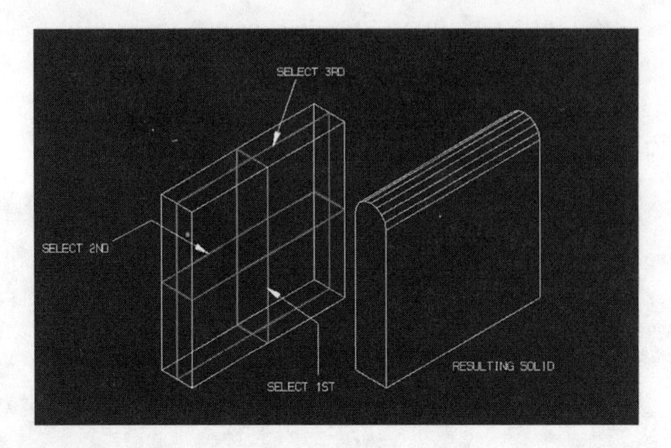

Create chamfered edges

Select: SOLIDE > OPERATN | CHAMFER

Prompt: SEL REDG/RSUR / / KEY (PARM1,PARM2)

The Chamfer Mode window provides the following options.

- Propagation at Auto or Manual.
- Parm Mode L1, L2 to specify both leg lengths or L1, A1 to specify one leg length and an angle.
- Reset to apply values to all edges.

1. Select an edge or a face.
2. Enter new parameter values as necessary.
3. Click on YES to compute.

Shell operation

Select: SOLIDE > OPERATN | SHELL

Prompt: SEL SOL

1. Select the solid on which the shell operation is to be performed.
2. Select face(s) to be removed (e.g., open ends) as necessary.
3. Enter offset values.
4. Click on YES to compute the operation.

Split operation

Select: SOLIDE > OPERATE | SPLIT

Prompt: SPLIT ELEM: SEL SKD/PLN/SUR/FAC/SKI

The Mode window provides the following options.

- Both Sides results in two solids.
- One Side results in one solid.
- Trim results in no duplication of solids.
- No Trim results in duplication of solids.

1. Select the splitting element (plane, surface, face, or skin).
2. Select the arrow to invert as necessary.
3. Select the solid to be split.
4. If one side was selected, click on YES as desired to change the side to be retained.

Sewing operation

Create depressions or bulges on solids using surfaces "sewn" onto solid.

Select: SOLIDE > OPERATN | SEWING

Prompt: SEWING ELEM: SEL SKD/SUR/FAC/SKI

The Mode window provides two options: Trim (no duplication of solids) and No Trim (duplication of solids).

1. Select the sewing element (surface, face, or skin).
2. Select the solid onto which the sewing operation is to be performed. The resulting element and the No of Bodies = # message display.
3. Click on YES to change side to be kept as necessary.

Sort out operation

Remove unwanted parts of a multibodied solid (e.g., a solid comprised of unconnected elements).

Select: SOLIDE > OPERATN | SORT OUT

Prompt: SEL SOL

1. Select the multibodied solid to be sorted out. Highlighted cubes display enclosing each separate part of the solid.
2. Select the cubes to be removed. The cubes are dimmed as you select them.
3. Click on YES to compute.

Inactivate solid features

Hide solid features no longer needed for visualization in the current session, or as part of the process of creating several different versions of a feature before making a final choice.

Select: SOLIDE > OPERATN | INACTIVE

Prompt: SEL BRANCH

In the Inactive window, you can select the features to be inactivated (All, Fillets, or Chamfers).

Select the branch to be inactivated on the solid or from the CSG tree via displaying the Part Editor. Click on YES to modify.

Reselect the feature in the CSG tree to reactivate it as necessary.

ANALYZE

Analyze solid data and positioning.

Select: SOLIDE > ANALYZE | SELF or POSITN

Analyze solid data

Select: SOLIDE> ANALYZE | SELF | PARM or INERTIA or NUMERIC

Analyze and create solid primitive parameters

Select: SOLIDE > ANALYZE | SELF | PARM

Prompt: SEL > SOL / / YES: ANALYZE

1. Activate the alphanumeric window by using <Alt>+<+> or selecting the Alpha wind on option in the Display and Manipulation window.
2. Select a solid to be analyzed. The parameters of the primitive solid display.

Analyze mechanical data

Select: SOLIDE > ANALYZE | SELF | INERTIA

Prompt: SEL > FEATURE

1. Activate the alphanumeric window by using <Alt>+<+> or selecting the Alpha wind on option in the Display and Manipulation window.
2. Select a solid. Enter a density value as necessary. The following values display: volume, mass, wetted area, center of gravity coordinates, main inertia axes, and moments of inertia.
3. Click on Keep Elements to create geometry at the center of gravity and axes of inertia.

Analyze modeling data

Select: SOLIDE > ANALYZE | SELF | NUMERIC

Prompt: SEL > FEATURE / / SEL PARM SEL SYMB / / YES: CREATE ALL PARAMS

1. Activate the alphanumeric window by using <Alt>+<+> or selecting the Alpha wind on option in the Display and Manipulation window.
2. Select a solid. The following values display: number of vertices, number of edges, number of faces, number of connex parts, memory size, and number of primitives.

Analyze position of solids

Select: SOLIDE > ANALYZE | POSITN | RELATIVE or INTERFER

Analyze relative position of two solids

Select: SOLIDE > ANALYZE | POSITN | RELATIVE

Prompt: SEL > ELEM1

Activate the alphanumeric window by using <Alt>+<+> or selecting the Alpha wind on option in the Display and Manipulation window.

The Analyze window provides the following options.

- Create Elements, Parameters or Intersections. Upon selection, these items will be stored.
- Background, Pick or No Pick. Unanalyzed items are retained in No Pick. Reset to return to normal display mode.
- Computational Sag. Amount the visualized curved solid deviates from true curve for the analysis.

Analysis of collision or clearance is provided in the alphanumeric window.

Analyze relative position of group of solids

Select: SOLIDE > ANALYZE | POSITN | INTERFER | COMPUTE or RELATN or ELEMENT

Prompt: SEL ELEM1

1. With the COMPUTE suboption selected, multiselect a group of solids for analysis. Click on YES to end selection.
2. Click on YES to complete. The ## (number of) INTERFERENCES DETECTED message displays.
3. Select the RELATN option. An Interference list window displays listing each solid and the solids they interfere with.
4. Select from the table. The clash analysis displays.
5. Select the ELEMENT option. In the First Operand window, select solids from the list.
6. In the Second Operand window, a list of clashes with the first selected solid displays. Select from the list. The clash analysis displays.

MODIFY

Modify geometry, operations, and display of solids.

Select: SOLIDE > MODIFY | GEOMETRY or OPERATN or DRESS UP

✓ **NOTE:** *Whenever the instruction to update a solid is given, either select the UPDATE option from the SOLIDE menu, or click on the Update button (shown at left) in the Part Editor window.*

Modify geometry of solid

Modify geometry via diverse methods.

Select: SOLIDE > MODIFY | GEOMETRY | MOVE or CONTOUR or SCALE or PARM

Perform transformation, rotation, symmetry, or stored transformation on solid geometry

Select: SOLIDE > MODIFY | GEOMETRY | MOVE | TRANSLATE or ROTATE or SYMMETRY or UNSPEC

Transformation

Select: SOLIDE> MODIFY | GEOMETRY | MOVE | TRANSLATE

Prompt: SEL BRANCH / / YES: CURRENT

1. Select solid element to be modified. The Translation window displays.

✓ **NOTE:** *If the INCOMPLETE SELECTION message displays, reselect the feature to be moved.*

2. In the Translation window, select Length to input a value for the translation or Measure to use existing geometry to define the translation.
3. Select Family mode to automatically select logically linked elements of the selected solid branch, or Element to select only the solid.
4. Define the Translation by using one of the following methods: Select a line, and then enter a value in the Translation window; select two points; enter coordinates; or select a plane, and then enter a length, or select a point or a line.
5. Select the arrow to invert as necessary. If desired, enter the number of times the translation is to be repeated in the input area.

✗ **WARNING:** *The translation length must be entered in the translation window, not in the input area.*

Click on YES to apply translation. Update the solid as necessary as described in the Note at the beginning of the section on the MODIFY option.

Rotation

Select: SOLIDE > MODIFY | GEOMETRY | MOVE | ROTATE

Prompt: SEL BRANCH

1. Select solid element to be modified.

✓ **NOTE:** *If the INCOMPLETE SELECTION message displays, reselect the feature to be moved.*

2. In the Rotation window, select Angle to input an angle value for the rotation, or Measure to use existing geometry to define the rotation.
3. Select Family mode to automatically select logically linked elements of the selected solid branch, or Element to select only the solid.
4. Define the rotation by one of the following methods: (a) Select a line or two points, and then enter an angle value in the Rotation window to select planes. (b) Select a plane after which the rotation axis passes through the center of the plane normal to the plane.
5. Select the arrow to invert as necessary.
6. Click on YES to apply the rotation. Update the solid as necessary as described in the Note at the beginning of the section on the MODIFY option.

Symmetry

Select: SOLIDE > MODIFY | GEOMETRY | MOVE | SYMMETRY

Prompt: SEL BRANCH / / YES: MODIFY

1. Select the solid element to be modified. The Mode window displays.
2. Select Family mode to automatically select logically linked elements of selected solid branch, or Element to select only the solid.
3. Define the symmetry by one of the following methods: (a) Select a plane. (b) Select two lines. (c) Select three points or a line and a point. (d) Select two points or a line to define axial symmetry.
4. Click on YES to apply the symmetry. Update the solid as necessary as described in the Note at the beginning of the section on the MODIFY option.

Stored transformations

Select: SOLIDE > MODIFY | GEOMETRY | MOVE | UNSPEC

Prompt: SEL BRANCH / / YES: APPLY

1. Select the solid element to be modified.

2. In the Transformation window, select the desired transformation.
3. Click on YES to apply the transformation.
4. Update the solid as necessary as described in the Note at the beginning of the section on the MODIFY option.

Modify primitives with contours

✓ **NOTE:** *Only primitives with contours (prisms, revolutions, and pyramids, but NOT cuboids, cylinders or spheres) can be modified using the CONTOUR option.*

Select: SOLIDE > MODIFY | GEOMETRY | CONTOUR

Prompt: SEL FEATURE

1. Select the feature to be modified.
2. Select other elements as necessary in the plane of the contour to be included in the parameterization (e.g., center lines).
3. Click on YES to end contour selection. The Parameter panel contains the following five relationship selection icons: Reference, Radius/Diameter, Vector reference, Offset, and Angle. The panel also contains the Auto Rel button which can be used to automatically parameterize the contour, and the Delete Relation button to delete automatic or manual relations.

✓ **NOTE:** *For further information on the parameter panel and relation icons, see the PARAM3D function entry in this book.*

1. Select the Reference icon in the Parameter panel, and then select an element which will become a reference element (i.e., fixed element or start point).
2. To create valuated relationships, select an icon from the five available. Make further selections from the panel as necessary.
3. Select contour elements to be linked with the selected relationship. Continue until all desired relationships are parameterized, or click on Auto Rel to compute parameterization automatically. Click on YES to compute.
4. Any of the listed parameters can now be modified by reselecting them and entering new values. Click on YES to end.
5. Update the solid as necessary as described in the Note at the beginning of the section on the MODIFY option.

Apply scaling

Select: SOLIDE > MODIFY | GEOMETRY | SCALE

Prompt: SEL BRANCH

1. Select the feature to be modified, and select a point from which the scaling is to be applied.
2. Enter a scaling ratio. Scaling is automatically applied.
3. Update the solid as necessary as described in the Note at the beginning of the section on the MODIFY option.

✓ **NOTE:** *If the solid outline appears to be dotted, or if the background in the Part Editor window is green, updating is required.*

Modify parameters of solid primitive

Select: SOLIDE > MODIFY | GEOMETRY | PARM

Prompt: SEL FEATURE / / YES: CURRENT FEATURE

1. From the solid or the CSG tree, select the feature to be modified, or click on YES to accept the current primitive. The parameters display in the work area and in an appropriate window (depending on primitive type).
2. If necessary, enter new values for parameters, or where planes display, select plane to be moved and then select new plane.

3. Click on YES to modify. If necessary, select the UPDATE option from the menu and click on YES to update, or click on the Update button (see left) in the Part Editor window.

✓ **NOTE:** *Under the PARM option, an old contour can be swapped for a new contour of suitable solids.*

4. Select the feature to be modified. Click on the CONT button in the Swap Parent window.
5. Select an element of the new contour. Continue until the YES Modify message displays.
6. Click on YES to modify. Update the solid as necessary as described in the Note at the beginning of the section on the MODIFY option.

The extrusion direction of the solid can also be changed by taking the following steps.

1. Select the feature to be modified, and select an element of the contour.
2. Select a line or plane. Click on YES to modify. The solid is recreated using the new extrusion direction and the original limits.
3. Update the solid as necessary as described in the Note at the beginning of the section on the MODIFY option.

Change B-REP type of solid

Select: SOLIDE > MODIFY | SOL TYPE

Prompt: MSELC SOL

1. After the Option window displays, select Smart to create smart solid as necessary.
2. Select an SOLM, approximate, or mockup solid. The message SOLID TYPE: APPROX ---------> EXACT displays, and the solid changes from approximate to exact.

✗ **WARNING:** *This operation cannot be performed if the selected solid is isolated.*

3. Update the solid as necessary as described in the Note at the beginning of the section on the MODIFY option.

Modify history of solid and duplicate branches of solid

Under the OPERATN option, previously created operations can be modified in various ways and branches of solids can be duplicated via diverse methods.

Select: SOLIDE > MODIFY | OPERATN | INSERT or DELETE or REPLACE or DUPLICAT

Insert operation into history of solid

Select: SOLIDE > MODIFY | OPERATN | INSERT

1. From the CSG tree in the Part Editor window for the solid containing the feature, select the branch to be inserted, or click on YES to select the current branch.
2. Select a solid or ditto in which the branch is to be inserted.
3. Update the solid as necessary as described in the Note at the beginning of the section on the MODIFY option.

Delete branch from history of solid

Select: SOLIDE > MODIFY | OPERATN | DELETE

Prompt: SOLIDE > SEL BRANCH / / YES: DELETE

1. In the Delete option window, select whether the deleted branch is to be stored after deletion.
2. Select a branch from the solid or from the CSG tree. Click on YES to delete.
3. Update the solid as necessary as described in the Note at the beginning of the section on the MODIFY option.

Replace branch in history of solid

Select: SOLIDE > MODIFY | OPERATN | REPLACE

Prompt: SEL BRANCH

Select a branch of the current solid or ditto as the replacement feature. Select a point to position the feature as necessary. Click on YES to replace.

Duplicate branch of solid

Select: SOLIDE > MODIFY | OPERATN | DUPLICAT | TRANSLAT or ROTATE or SYMMETRY or UNSPEC

Duplicate branch of solid by translation

Select: SOLIDE > MODIFY | OPERATN | DUPLICAT | TRANSLAT

Prompt: SEL BRANCH

1. Select the branch to be modified from the solid or CSG tree.
2. In the Translation window, select Length to input a value for the translation or Measure to use existing geometry to define the translation.
3. Select Family mode to automatically select logically linked elements of the selected solid branch, or Element to select only the solid.
4. Define the translation by one of the following methods: (a) Select a line and then enter a value in the Translation window. (b) Select two points. (c) Enter coordinates. (d) Select a plane and then enter a length or select a point or line.
5. Select the arrow to invert as necessary.
6. Enter the number of times the translation is to be repeated in the input area as necessary.

✗ **WARNING:** *The translation length must be entered in the Translation window, not in the input area.*

7. Click on YES to apply the translation.
8. Update the solid as necessary as described in the Note at the beginning of the section on the MODIFY option.

Duplicate branch of solid by rotation

Select: SOLIDE > MODIFY | OPERATN | DUPLICAT | ROTATE

Prompt: SEL BRANCH

1. In the Rotation window, select Angle to input an angle value for the rotation, or Measure to use existing geometry to define the rotation.
2. Select Family mode to automatically select logically linked elements of the selected solid branch, or Element to select only the solid.
3. Define the rotation by one of the following methods: (a) Select a line or two points, and then enter an angle value in the Rotation window while selecting planes. (b)

Select a plane, after which the rotation axis passes through the center of the plane normal to the plane.

4. Select the arrow to invert as necessary.
5. Click on YES to apply rotation. Update the solid as necessary as described in the Note at the beginning of the section on the MODIFY option.

Duplicate branch of solid by symmetry

Select: SOLIDE > MODIFY | OPERATN | DUPLICAT | SYMMETRY

Prompt: SEL BRANCH / / YES: CURRENT

1. Select the solid element to be modified.
2. Define the symmetry using one of the following methods: (a) Select a plane. (b) Select two lines. (c) Select three points or a line and a point. (d) Select two points or a line to define axial symmetry.
3. Click on YES to apply the symmetry. Update the solid as necessary as described in the Note at the beginning of the section on the MODIFY option.

Duplicate branch of solid using stored translation

Select: SOLIDE > MODIFY | OPERATN | DUPLICAT | UNSPEC

Prompt: SEL BRANCH KEY NUMBER / / YES: APPLY

Select the branch to be modified, and then select the transformation in the Transformation window. Click on YES to apply. Update the solid as necessary as described in the Note at the beginning of the section on the MODIFY option.

Modify solid display

Modify solid display elements.

Select: SOLIDE > MODIFY | DRESS UP | HLR EDGE or RENAME or COLOR

Remove hidden lines

Select: SOLIDE > MODIFY | DRESS UP | HLR EDGE

Prompt: SEL EDGE / / YES: CHANGE FEATURE VISU

A window displays in which you choose the features to be modified. Select an edge. Click on YES to switch between the sharp edge and HLR (hidden lines removed) modes.

Modify identifier of solid subelement

Select: SOLIDE > MODIFY | DRESS UP | RENAME | PRIM or FSUR or RSUR or EDGE

Prompt: SEL PRIM or FSUR or RSUR or EDGE

Select a primitive, FSUR, RSUR, or edge. Enter a new identifier for the element (up to 70 characters).

Modify color of primitive

Select: SOLIDE > MODIFY | DRESS UP | COLOR

Prompt: SEL FEATURE / / YES: CURRENT

To display the color selector, select an element. Select a color or enter a number. Click on YES. The COLOR CHANGED message displays.

EXTRACT

Create geometry and faces for solids.

Select: SOLIDE > EXTRACT | CUT or PROJECT or VOLUME or APPROXIM or PRIMVOL or PRIMSOL or PRIMSKIN or EDGESKIN or EDGECRV or FACE

Create geometry and faces by intersecting solids with planes

Select: SOLIDE > EXTRACT | CUT

Prompt: MSELW PLN

Select intersecting plane(s), and then select a solid. Geometry representing the cut is created.

Create geometry by projecting solids onto plane

Select: SOLIDE > EXTRACT | PROJECT

Prompt: MSELW PLN

1. Select the projection plane(s). An arrow indicating the projection direction displays. Select to invert as necessary.
2. Select a solid. Geometry is created representing the orthographic projection of the solid onto the plane(s).

Extract volumes from solids

Select: SOLIDE > EXTRACT | VOLUME

Prompt: SEL SOL

Select a solid. Volumes, faces, surfaces, lines, and curves are created. (Surfaces are created in the No Show area.)

Create SOLM (approximate solid) from exact solid

Select: SOLIDE > EXTRACT | APPROXIM

Prompt: SEL SOL/VOL

The Analyze window displays. Enter a value as necessary for computational sag (deviation of solid facet to true curved geometry) and maximum facet size. Select an exact solid. An isolated approximate (SOLM) is created.

✓ **NOTE:** *An isolated solid lacks history.*

Extract primitive subelements from solid

Select: SOLIDE > EXTRACT | PRIMVOL

Prompt: SEL> FEATURE

Select a primitive subelement. Volume, faces, surfaces, lines, and curves are created. (Surfaces are created in the No Show area.)

Duplicate solid primitive

Select: SOLIDE > EXTRACT | PRIMSOL

Prompt: SEL BRANCH

Select a primitive. A duplicate primitive is created. Update the solid as described in the Note at the beginning of the MODIFY section.

Extract a skin from primitive subelement

Select: SOLIDE > EXTRACT | PRIMSKIN

Prompt: SEL FEATURE

Select a primitive subelement. A skin, faces, surfaces, lines, and curves are created. (Surfaces are created in the No Show area.)

Extract skins from primitives forming solid edge

Select: SOLIDE > EXTRACT | EDGESKIN

Prompt: SEL > EDGE

Select the edges of a solid. Click on YES to create elements adjacent to selected element. Skins, faces, surfaces, lines, and curves are created. (Surfaces are created in the No Show area.)

Extract edge geometry from solid

Select: SOLIDE > EXTRACT | EDGECRV

Prompt: SEL> EDGE

Select the edge of a solid. Geometry representing the edge is created.

Extract faces from solids

Select: SOLIDE > EXTRACT | FACE

Prompt: SOLID FACE: SEL RSUR

Select a solid face (FSUR). A face is created.

UPDATE

Update B-REP (boundary representation) of solid.

Select: SOLIDE > UPDATE

Prompt: MSELW / / YES: FORCE UPDATE

The Update window displays.

1. Select Local Solid Visualization to view local solids as the update is executed.
2. Select Information to display information about the update throughout process.
3. Select Inactive errors and Continue to inactivate errors encountered and continue the update process.
4. Select the solid(s) to be updated. The message ALREADY UPDATED will be displayed if appropriate. Upon clicking to confirm the update, the UPDATE DONE message displays.

PART EDITOR

The Part Editor window is switched on and off by clicking on the Part Editor button (shown at left) in the Manage window. The Part Editor window displays the CSG (constructive solid geometry) of the current solid, which graphically defines the history of the solid in the form of a tree or a list. (Click on View to choose.) Additional buttons are described below.

Smart Solid. Provides access to the local solid scanner which can be used like a VCR to play through the history of a solid. You can also stop and access an individual local solid that is also part of the current solid's history.

Part Editor window.

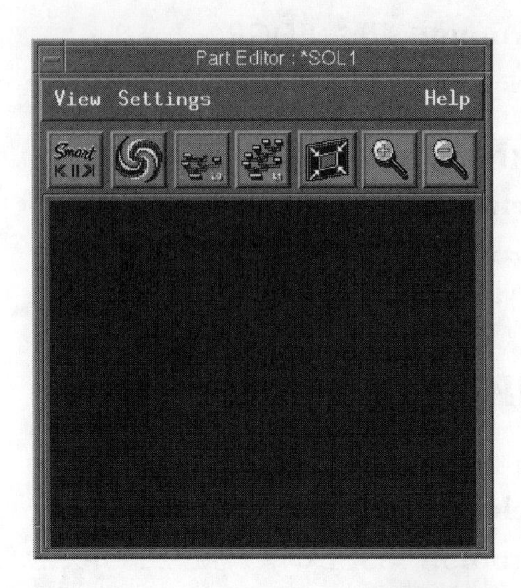

Update. Works in the same way as the UPDATE option in the SOLIDE menu.

Reframes CSG tree in window.

Displays simplified view of CSG tree.

Zoom in.

Displays a detailed view of CSG tree.

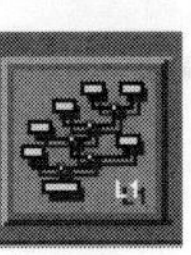

Zoom out.

You can drag, zoom, and pan in the CSG tree window using mouse button 2. Next, you can drag while pressing button 2 and moving the cursor horizontally. To zoom and pan, press button 2 while moving the cursor vertically.

In addition, by placing the cursor over a branch or feature in the CSG tree and pressing on button 3, the following actions can be carried out.

- Rename or change feature identifier.
- Reorder or move feature or branch to another position in CSG tree.
- Change color of feature.
- Display Parent Management window.
- Delete feature or branch.

- Break or mimic MODIFY | OPERATE | DELETE plus Keep Branch from SOLIDE menu.
- Inactivate or mimic OPERAT | INACTIVE from SOLIDE menu.

By placing the cursor over the background in the CSG window and pressing button 3, additional options are available.

- Make solid Smart or Unsmart.
- Rename or change identifier of current solid.
- Change color of current solid.
- Display Parent Management window.
- Delete current solid.
- Collapse the CSG structure.

For more information on solids, see the "Part Design User Guide" available in Frame-viewer.

SOLIDM in SPACE Mode

Chapter 11

The SOLIDM function is used to create and modify mock up (or approximate) solids.

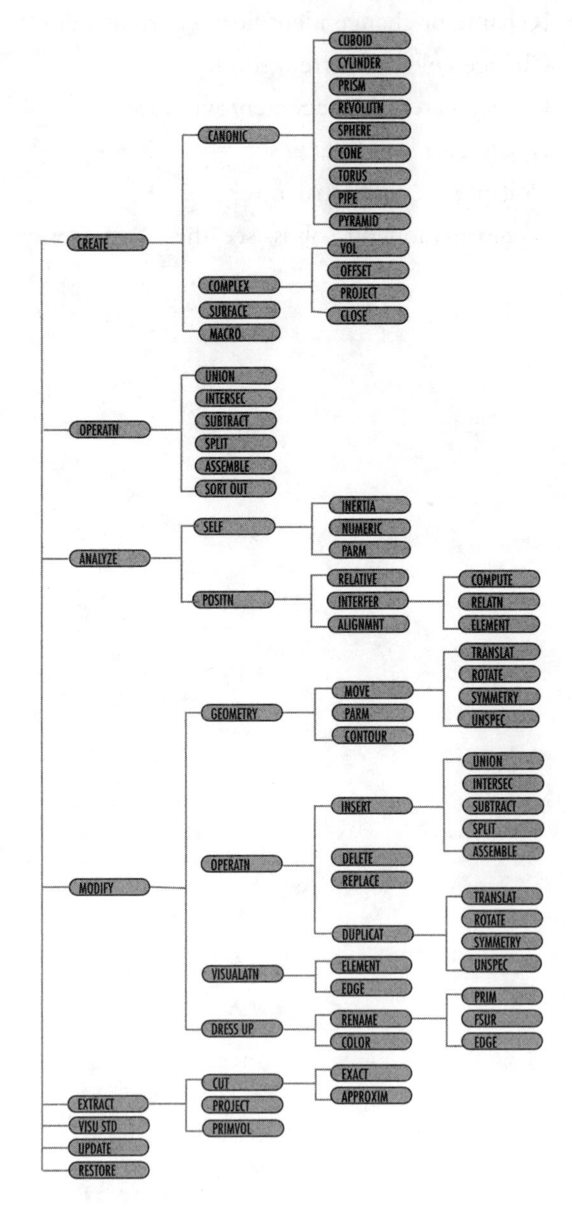

Option menu for SOLIDM in SPACE mode.

CREATE

Select: CREATE > CANONIC or COMPLEX or SURFACE or MACRO

Create canonical solids

Select: SOLIDM > CREATE | CANONIC | CUBOID or CYLINDER or PRISM or REVOLUTN or SPHERE or CONE or TORUS or PIPE or PYRAMID

When a canonic option is selected, a parameter window appropriate to the selected option displays. A small circle appears at the center of each dimension in the window. Upon selecting the circle, geometry can be used as a means to modify the parameter.

A canonical solid is defined by geometric values and/or a single parametric construction element known as a "contour."

Create cuboid

Select: SOLIDM > CREATE | CANONIC | CUBOID

Prompt: REF PT: MSELW PT / / DIR: SEL LN/PLN

1. In the Cuboid window, modify current parameters as necessary.
2. The Management window provides access to CSG representation. You have the option of displaying the solid history as a tree structure or a list. Use the Current Solid option to select a new solid as the current one. With the Solid Access option, choose the method of selecting a solid (face or edge).
3. Select a point, and then a line or a point.
4. Select a point or two lines.
5. Select a plane, point, or line to define the anchor point and direction of the cuboid.
6. Enter new values for LX, LY, and LZ in the input area or in the Cuboid window as necessary.
7. If desired, select a line to change the direction in which the cuboid will be created.
8. Click on YES to create the solid.

Create cylinder

Select: SOLIDM > CREATE | CANONIC | CYLINDER

Prompt: REF PT: MSELW PT DIR: SEL LN/CIR/PLN

1. In the Cylinder window, modify current parameters as necessary.
2. Options in the Management window are the same as described under the CREATE | CANONIC | CUBOID option. Make desired selections.

3. In the Discretization window, set the line application (number of facet lines per quadrant) for the large and small radii.
4. Select a point, or multiselect several points if more than one solid is to be created.
5. Select a line or plane to determine the direction of the cylinder, and then select the arrow to invert if desired.
6. Define the limits in the Cylinder window work area or enter values in the input area.
7. Click on YES to create the cylinder(s).

Create prism solid

Select: SOLIDM > CREATE | CANONIC | PRISM

Prompt: CONTOUR: MSELW LN/CONIC/CRV/CCV CONTOUR: SEL PT/FAC

1. In the Prism window, modify current parameters as necessary.
2. Options in the Management window are the same as those described under the CREATE | CANONIC | CUBOID option. Make desired selections.
3. In the Discretization window, set the sag value (deviation of facet line from true curve geometry).
4. Select an element of the contour geometry, and then choose one of the following alternatives: (a) Continue selecting elements until a closed contour is achieved. (b) Select a second contour element not adjacent to the first (the contour will close if possible). (c) Click on the AutoSearch button.
5. If desired, select additional geometry to define an inner domain, and then click on YES to end contour selection.
6. If required, select the small circle on the ELEVATION DIRECTION arrow in the PRISM window, and then select a line to change the extrusion direction to an oblique angle.
7. Click on YES to create the solid.

Create revolution

Select: SOLIDM > CREATE | CANONIC | REVOLUTION

Prompt: SEL REF PT / / AXIS: SEL LN/CIR/PLN

1. In the Revolution window, modify current parameters as necessary.
2. Options in the Management window are the same as described under the CREATE | CANONIC | CUBOID option. Make desired selections.
3. In the Discretization window, set the sag (deviation of facet from true curve geometry), and the line approximation (number of facets per quadrant).

4. Define the revolution axis by using one of the following alternatives: (a) Select a line. (b) Select a plane, and then a line. A revolution axis through the center of the plane normal to the plane is created. (c) Select a point, and then a line. A revolution axis parallel to the selected line passing through the point is created. (d) Select a curve, and then a point. A revolution axis normal to the curve passing through the point is created.
5. Select an element of the contour geometry, and then use one of the following alternatives: (a) Continue selecting elements until a closed contour is achieved. (b) Select a second contour element not adjacent to the first (if possible, the contour will close). (c) Click on the AutoSearch button.
6. Click on YES to end the contour.
7. Click on YES to create the revolution solid.

Create sphere

Select: SOLIDM > CREATE | CANONIC | SPHERE

Prompt: REF PT: MSELW PT   DIR: SEL LN/CIR/PLN

1. In the Sphere window, modify current parameters as necessary.
2. Options in the Management window are the same as described under the CREATE | CANONIC | CUBOID option. Make desired selections.
3. In the Discretization window, set the line application (number of facet lines per quadrant) for the large and small radii.
4. Select a point, line, and a point or two lines to determine the reference or anchor point of the sphere.
5. If desired, enter new values for R1 and R2 in the input area or in the Sphere window.
6. Click on YES to create the solid.

Create cone

Select: SOLIDM > CREATE | CANONIC | CONE

Prompt: REF PT: MSELW PT   DIR: SEL LN/CIR/PLN

1. In the Cone window, modify current parameters as necessary.
2. Options in the Management window are the same as described above under the CREATE | CANONIC | CUBOID option. Make desired selections.
3. In the Discretization window, enter the value for line application (number of facet lines per quadrant).
4. Select a point, line, and point or two lines to determine the reference or anchor point of the cone.

5. If desired, enter new values for R1, R2, H1, and H2 in the input area or Cone window.
6. Click on YES to create the solid.

Create torus

Select: SOLIDM > CREATE | CANONIC | TORUS

Prompt: REF PT: MSELW PT DIR: SEL LN/CIR/PLN

1. In the Torus window, modify current parameters as necessary.
2. Options in the Management window are the same as described under the CREATE | CANONIC | CUBOID option. Make desired selections.
3. In the Discretization window, enter values for line application (number of facet lines per quadrant) for large and small radii.
4. Select a point, line, and point or two lines to determine the reference or anchor point of the torus.
5. Select a line to define the direction of the torus.
6. If desired, enter new values for R1 and R2 in the input area or Torus window.
7. Click on YES to create the solid.

Create pipe

Select: SOLIDM > CREATE | CANONIC | PIPE

Prompt: CONTOUR: MSELW LN/CONIC/CRV/CCV CONTOUR: SEL PT/FAC

1. In the Pipe window, modify current parameters as necessary.
2. Options in the Management window are the same as described above under the CREATE | CANONIC | CUBOID option. Make desired selections.
3. In the Discretization window, enter values for contour sag (deviation of facet from true curve geometry), and radius line application (number of facets per quadrant).
4. Select a line or curve element. If desired, click on AutoSearch to complete the contour.
5. Click on YES to end selection.
6. If desired, enter a new value for R1 in the input area or Pipe window.
7. Click on YES to create the solid.

Create pyramid

Select: SOLIDM > CREATE | CANONIC | PYRAMID

Prompt: SEL: TOP VERTEX

1. In the Pyramid window, modify current parameters as necessary.
2. Options in the Management window are the same as described above under the CREATE | CANONIC | CUBOID option. Make desired selections.
3. In the Discretization window, set the sag (deviation of facet from true curve geometry).
4. Select a point to serve as the top vertex of the pyramid.
5. Select an element of the bas contour, and then click on AutoSearch to complete contour.
6. Click on YES to create the solid.

Create solids from surface, face, skin or volume type elements

Select: SOLIDM > CREATE | COMPLEX | VOLUME or OFFSET or PROJECT or CLOSE

Create solid from volume

Select: SOLIDM > CREATE | COMPLEX | VOL

Prompt: SEL VOL

1. Options in the Management window are the same as described above under the CREATE | CANONIC | CUBOID option. Make desired selections.
2. In the Discretization window, enter a value for the volume faces NCT (number of points used to approximate edge).
3. Select a volume type element. Click on YES to create the solid.

Create offset solid

Select: SOLIDM > CREATE | COMPLEX | OFFSET

Prompt: SEL POL/SUR/FAC

1. In the Offset window, modify current parameters as necessary.
2. Options in the Management window are the same as described above under the CREATE | CANONIC | CUBOID option. Make desired selections.
3. In the Discretization window, enter values for surface NCU (number of segments used to approximate edge in U direction), and NCV (number of segments used to approximate edge in V direction).
4. Select a surface, polyhedron, or face type element. Arrows and parameters display in the work area providing a preview of the solid to be created.
5. If desired, enter values for OFF1 and OFF2 in the entry area, or modify current parameters in the Offset window.
6. Click on YES to create the solid.

Create solid by projecting surfaces or nonplanar faces onto plane

Select: SOLIDM > CREATE | COMPLEX | PROJECT

Prompt: SEL POL/SUR/FAC/

1. Options in the Management window are the same as described above under the CREATE | CANONIC | CUBOID option. Make desired selections.
2. In the Discretization window, enter values for surface NCU (number of segments used to approximate edge in U direction), and NCV (number of segments used to approximate edge in V direction).
3. Select a surface, polyhedron, or face element.
4. Select a plane.
5. If desired, define the projection direction by selecting one of the following: a line or two points, or a plane.
6. Click on YES to create the solid.

Create solid by closing surface or nonplanar face

Select: SOLIDM > CREATE | COMPLEX | CLOSE

Prompt: SEL POL/SUR/FAC

1. Options in the Management window are the same as described above under the CREATE | CANONIC | CUBOID option. Make desired selections.
2. In the Discretization window, enter values for surface NCU (number of segments used to approximate edge in U direction), and NCV (number of segments used to approximate edge in V direction).
3. Select a surface, polyhedron, or face element. Click on YES to create the solid.

Create polyhedral surface

Select: SOLIDM > CREATE | SURFACE

Prompt: SEL SUR/FAC

1. Options in the Management window are the same as described above under the CREATE | CANONIC | CUBOID option. Make desired selections.
2. In the Discretization window, enter values for the surface NCU (number of segments used to approximate edge), and planar face NCT (number of points used to approximate edge).
3. Select a surface or face. Click on YES to create the polyhedron.

OPERATN

Perform operations on solids. Operations are defined as topological modifications to solids.

Select: SOLIDM > OPERATN | UNION or INTERSEC or SUBTRACT or SPLIT or ASSEMBLE or SORT OUT

Union operation

Select: SOLIDM > OPERATN | UNION

Prompt: 1st OPERAND: MSELW SOL/DIT/VOL

1. In the window, select among the following options: (a) trim all elements during the union operation, (b) trim the first selected element (duplicate the second), or (c) duplicate all elements without trimming.
2. Select the first solid (ditto or volume).
3. Select the second solid (ditto or volume). The new solid resulting from the union displays. If a ditto is selected, it will be automatically converted to a macroprimitive.

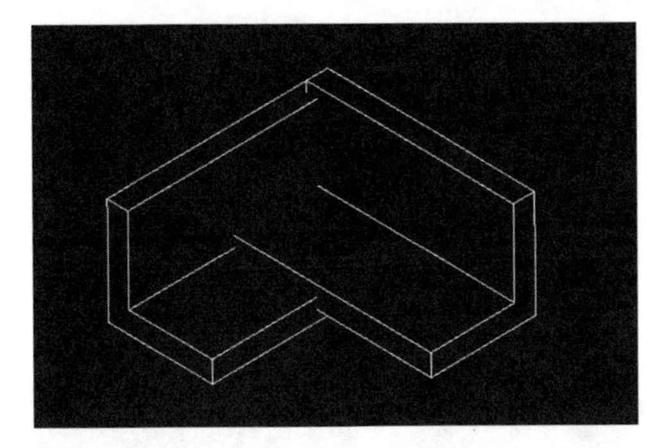

Before union.

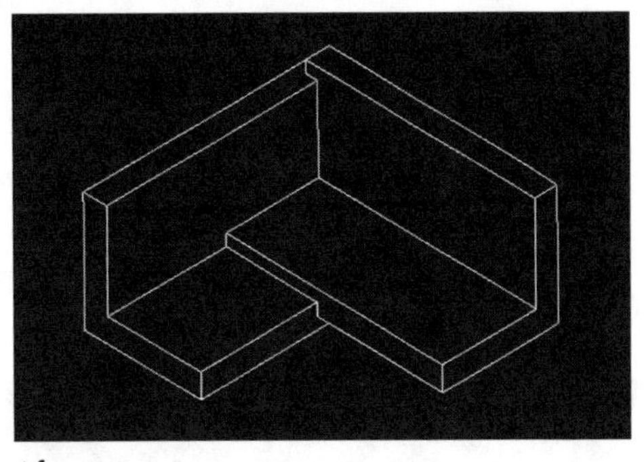

After union.

Intersection operation

Select: SOLIDM > OPERATN | INTERSEC

Prompt: 1st OPERAND: MSELW SOL/DIT/VOL

1. In the window, make a selection among the following options: (a) trim all elements during the intersection operation, (b) trim the first selected element (duplicate the second), (c) or duplicate all elements without trimming.
2. Select the first solid (ditto or volume).
3. Select the second solid (ditto or volume). The new solid resulting from the intersection volume of the selected solids displays. If a ditto is selected, it will be automatically converted to a macroprimitive.

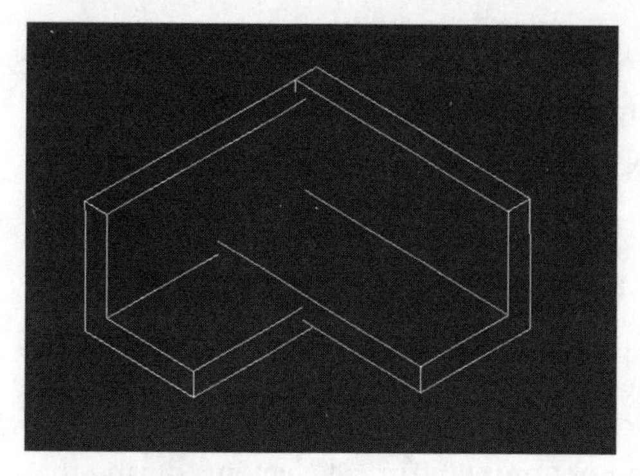

Before intersection.

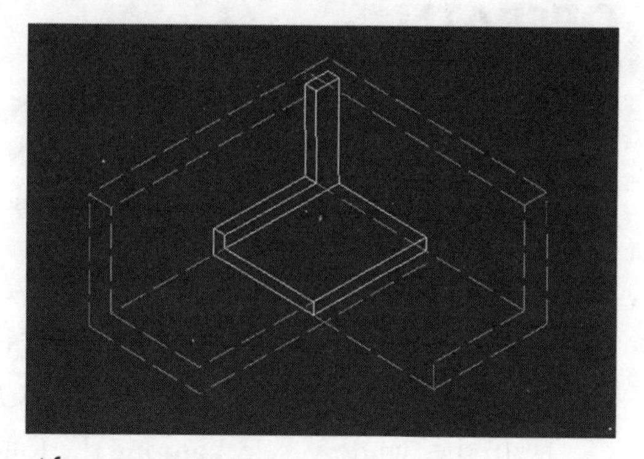

After intersection.

Subtraction operation

Select: SOLIDM > OPERATN | SUBTRACT

Prompt: 1st OPERAND: MSELW SOL/DIT/VOL

1. In the window, make desired selection among the following options: (a) trim all elements during the subtraction operation, (b) trim the first selected element (duplicate the second), or (c) duplicate all elements without trimming.
2. Select the first solid (ditto or volume).
3. Select the second solid (ditto or volume). The new solid results from the volume remaining after the subtraction displays. If a ditto is selected, it will be automatically converted to a macroprimitive.

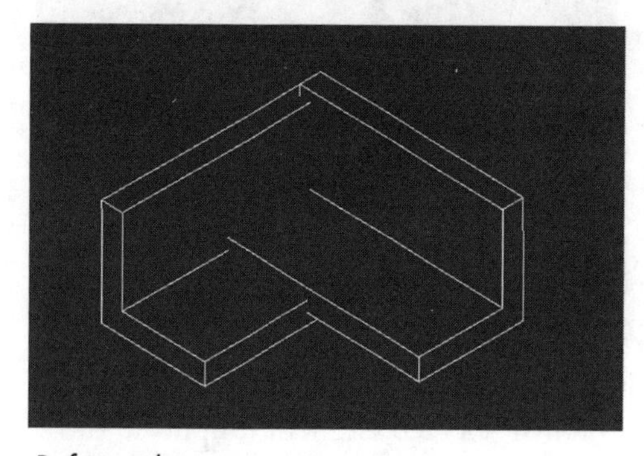

Before subtraction.

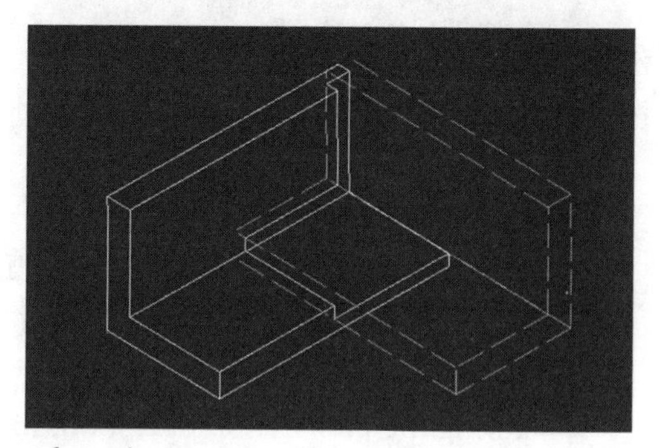

After subtraction.

Split operation

Select: SOLIDM > OPERATN | SPLIT

Prompt: SPLIT ELEM: SEL PLN/POL/SUR/FAC

1. In the window make necessary selections among the following options.
   - Both Sides. Results in two solids.
   - One Side. Results in one solid.
   - Trim. No duplication of solids.
   - No Trim. Solids are duplicated.
2. Select the splitting element (plane, surface, or face).
3. Select arrow to invert as necessary.
4. Select the solid to be split.
5. If one side was selected, click on YES. Change side to be retained as necessary.

Assemble operation

Select: SOLIDM > OPERATN | ASSEMBLE

Prompt: SEL POL1 displays.

1. Options in the Management window are the same as described above under the CREATE | CANONIC | CUBOID option. Make desired selections.
2. Select a polyhedron, and then select a second polyhedron. If selected polyhedrons form an open element, a polyhedron is created. If selected polyhedrons form a closed shape, a solid is created.

Sort out operation

Remove unwanted parts of multibodied solid (i.e., solid consisting of unconnected elements).

Select: SOLIDM > OPERATN | SORT OUT

Prompt: SEL SOL displays.

1. Options in the Management window are the same as described above under the CREATE | CANONIC | CUBOID option. Make desired selections.
2. If desired, select OUTSIDE or INSIDE.
3. Select the multibodied solid to be sorted out. Highlighted cubes are displayed enclosing each separate part of the solid.
4. Select the cubes to be removed; the cubes are dimmed upon selection.
5. Click on YES to compute.

ANALYZE

Analyze solid data and positioning.

Select: SOLIDM > ANALYZE | SELF or POSITN

Analyze solid data and parameters

Select: SOLIDM > ANALYZE | SELF | INERTIA or NUMERIC or PARM

Analyze solid mechanical data

Select: SOLIDM > ANALYZE | SELF | INERTIA

Prompt: SEL > SOL/POL/PIP/STR

1. The above prompt expands to include the following options. Make desired selections.
   - ELEMENT DENSITY. Enter a value as necessary.
   - DENSITY CUBE. Select YES to display density cube used in inertia calculation.
   - RETAIN ELEM. Select YES if temporary elements (e.g., inertia axes are to be retained).
2. Activate the alphanumeric window with <Alt> + <+> or select the Alpha wind on option in the Display and Manipulation window.
3. Select a solid.
4. If desired, enter a density value. The following values are displayed: volume, mass, wetted area, coordinates of the center of gravity, main inertia axes, and moments of inertia.
5. Click on Retain Elements to create geometry at the center of gravity and axes of inertia.

Analyze solid modeling data

Select: SOLIDM > ANALYZE | SELF | NUMERIC

Prompt: SEL > SOL/POL displays.

1. Options in the Management window are the same as described above under the CREATE | CANONIC | CUBOID option. Make desired selections.
2. Activate the alphanumeric window with <Alt> + <+> or select the Alpha wind on option in the Display and Manipulation window.
3. Select a solid. The following values display: number of vertices, number of edges, number of facets, memory size, and number of primitives.

Analyze and create solid primitive parameters

Select: SOLIDM > ANALYZE | SELF | PARM

Prompt: SEL > PRIM displays.

1. Options in the Management window are the same as described above under the CREATE | CANONIC | CUBOID option. Make desired selections.
2. Activate the alphanumeric window with <Alt> + <+> or select the Alpha wind on option in the Display and Manipulation window.
3. Select a solid to be analyzed. The parameters of the primitive solid are displayed.

Analyze position of solids

Select: SOLIDM > ANALYZE | POSITN | RELATIVE or INTERFER or ALIGNMNT

Analyze relative position of two solids

Select: SOLIDM > ANALYZE | POSITN | RELATIVE

Prompt: SEL> ELEM1

1. Activate the alphanumeric window with <Alt> + <+> or select the Alpha wind on option in the Display and Manipulation window.
2. In the Management window, make desired selections to the following options.
   - Create Elements, Parameters or Intersections. Upon selection these items will be retained.
   - Background, Pick or No Pick. Items other than those under analysis will be located in No PICK. Reset to return to normal display mode.
   - Computational Sag. Amount by which the visualized curved solid deviates from the true curve for the analysis.

Analysis of any collision or clearance is provided in the alphanumeric window.

Analyze relative position of group of solids

Select: SOLIDM > ANALYSIS | POSITN | INTERFER | COMPUTE or RELATN or ELEMENT

Prompt: MSELW > ELEM

1. The above prompt expands to incorporate the following options. Make desired selections.
   - CLEARANCE. Enter a value below which contact is assumed.
   - COMPUTATIONAL SAG. Enter a value for deviation of facets from true curve geometry.
   - RETAIN CONTACTS. Select if contact geometry is to be retained.

2. Options in the Management window are the same as described above under the CREATE | CANONIC | CUBOID option. Make desired changes.
3. With the COMPUTE option selected, multiselect a group of solids for analysis. Click on YES to end selection, and then click on YES to complete. The ## (number of) INTERFERENCES DETECTED message displays.
4. Select the RELATN option. In the Interference list of each solid and respective interferences, select from the list. The clash analysis displays.
5. Select the ELEMENT option. In the First Operand window, select a solid from the list. In the Second Operand window, a list of clashes with the first selected solid displays. Select from the list; the clash analysis displays.

Analyze tenon and mortise alignment

Select: SOLIDM > ANALYZE | POSITN | ALIGNMNT

Prompt: MSELW SOL is expanded to include FEATURE: CYLINDER ALL

1. Select the first group of elements, and then select the second group of elements. Click on YES to end. The PINS or HOLES ARE DETECTED message displays as appropriate.
2. Click on YES to compute. One of the following messages displays: ALL AXIS MATCHED or ALL AXIS UNMATCHED.
3. Click on YES to retain misalignment axis.

MODIFY

Modify solid geometry, operations, visualization, and dress up.

Select: SOLIDM > MODIFY | GEOMETRY or OPERATN or VISUALTN or DRESS UP

✓ **NOTE:** *Whenever the instruction to update a solid is given, select the UPDATE option from the SOLIDM menu.*

Modify solid geometry

Select: SOLIDM > MODIFY | GEOMETRY | MOVE or PARM or CONTOUR

GEOMETRY expands to include CSG SCANNING + or -.

Transform solid geometry

Select: SOLIDM > MODIFY | GEOMETRY | MOVE | TRANSLAT or ROTATE or SYMMETRY or UNSPEC

Perform translation on solid geometry

Select: SOLIDM > MODIFY | GEOMETRY | MOVE | TRANSLAT

Prompt: SEL BRANCH / / YES: CURRENT

1. In the window select ELEMENT to transform only the solid, or select FAMILY to transform solid and logically linked elements.
2. Select the solid element to be modified.

✓ **NOTE:** *If the INCOMPLETE SELECTION message displays, reselect the feature to be moved.*

3. In the Translation window, select Length to input a value for the translation.
4. Define the translation via one of the following alternatives: (a) Select a line, and then enter a value in the Translation window. (b) Select two points. (c) Enter coordinates. (d) Select a plane, and then enter a length or select a point or a line.
5. If desired, select the arrow to invert.

✗ **WARNING:** *The translation length must be entered in the Translation window, not in the input area.*

6. Click on YES to apply the translation.
7. If desired, update solid as described at the start of the MODIFY entry.

Perform rotation on solid geometry

Select: SOLIDM > MODIFY | GEOMETRY | MOVE | ROTATE

Prompt: SEL BRANCH // YES: CURRENT

1. Select solid element to be modified.

✓ **NOTE:** *If INCOMPLETE SELECTION message displays, reselect the feature to be moved.*

2. In the Rotation window, select Angle to input an angle value for the rotation. Select Family mode to automatically select logically linked elements of the selected solid branch, or Element to select only the solid.
3. Define the rotation by selecting a line or two points, and then entering an angle value in the Rotation window. Alternatively, select a plane; the rotation axis passes through the center of the plane normal to the plane.
4. Select arrow to invert as necessary.
5. Click on YES to apply rotation.
6. If desired, update solid as described at the start of the MODIFY entry.

Perform symmetry on solid geometry

Select: SOLIDM > MODIFY | MOVE | SYMMETRY

Prompt: SEL BRANCH / / YES: CURRENT

1. Define the symmetry by using one of the following alternatives: (a) Select a plane. Select two lines. (c) Select three points or a line and a point. (d) Select two points or a line.
2. Click on YES to apply the symmetry. If desired, update solid as described at the start of the MODIFY entry.

Perform stored transformations on solid geometry

Select: SOLIDM > MODIFY | GEOMETRY | MOVE | UNSPEC

Prompt: SEL BRANCH / / YES: APPLY

1. In the Transformation window, select the solid element to be modified, and then select the desired transformation.
2. Click on YES to apply the transformation. If desired, update the solid as described at the start of the MODIFY entry.

Modify parameters of solid primitive

Select: SOLIDM > MODIFY | GEOMETRY | PARM

Prompt: SEL PRIM / / YES: CURRENT FEATURE

1. From the solid or CSG tree, select the feature to be modified, or click on YES to accept the current primitive. The parameters are displayed in the work area and in an appropriate window (depending on primitive type).
2. If desired, enter new values for parameters. Alternatively, where planes are displayed, select the plane to be moved and then select a new plane.
3. Update the solid by selecting the UPDATE option from the menu as necessary.
4. Under the PARM option, an old contour can be swapped for a new contour of suitable solids.
5. Select the primitive to be modified. Click on the CONT symbol in the work area.
6. Select an element of the new contour. Continue until the YES Modify message displays. Click on YES to modify.
7. Update the solid as described at the start of the MODIFY entry.
8. The extrusion direction of the solid can also be changed as follows: (a) Select the feature to be modified. (b) Select an element of the contour. (c) Select a line or a plane.
9. Click on YES to modify. The solid is recreated using the new extrusion direction and the original limits.
10. If desired, update the solid as described at the start of the MODIFY entry.

Modify primitives with contours

✓ **NOTE:** *Only primitives with contours (e.g., prisms, revolutions, and pyramids, but NOT cuboids, cylinders, and spheres) can be modified with the CONTOUR option.*

Select: SOLIDM > MODIFY | GEOMETRY | CONTOUR

Prompt: SEL PRIM

1. Select the primitive to be modified.
2. If desired, select other elements in the plane of the contour to be included in the parameterization (e.g., center lines).
3. Click on YES to end contour selection.
4. The Parameter panel displays the following five relationship selection icons: Reference, Radius/Diameter, Vector reference, Offset, and Angle. In addition, the panel contains the Auto Rel button (automatically parameterize the contour); Delete Relation button (delete automatic or manual relations). Select the Reference icon in the Parameter panel, and then select an element which will become a reference element (i.e., fixed element or start point).

✓ **NOTE:** *For further information on the Parameter panel and relation icons, see the PARAM3D function entry in this book.*

5. To create valuated relationships, select an icon from the available five in the Parameter panel. Make additional selections from the displayed choices as necessary.
6. Select contour elements to be linked with the selected relationship. Continue until all desired relationships are parameterized. Alternatively, click on Auto Rel to automatically compute parametization, and then click on YES to compute.
7. All listed parameters can now be modified by reselecting them and entering new values. Click on YES to end.
8. If desired, update solid as described at the start of the MODIFY entry.

Modify history of solid

Select: SOLIDM > MODIFY | OPERATN | INSERT or DELETE or REPLACE or DUPLICAT

Insert operation into history of solid

Select: SOLIDM > MODIFY | OPERATN | INSERT | UNION or INTERSEC or SUBTRACT or SPLIT or SEWING or ASSEMBLE

Prompt: SEL BRANCH // YES: CURRENT

1. From the CSG tree for the solid containing the feature, select the branch to be inserted. Alternatively, click on YES to select the current branch.
2. Select a solid or ditto in which the branch is to be inserted.
3. If desired, update the solid as described at the start of the MODIFY entry.

Delete branch from history of solid

Select: SOLIDM > MODIFY | OPERATN | DELETE

Prompt: SEL BRANCH / / YES: DELETE

1. In the Delete option window, select whether the deleted branch is to be retained after deletion.
2. Select a branch from the solid or CSG tree. Click on YES to delete.
3. If desired, update the solid as described at the start of the MODIFY entry.

Replace branch in history of solid

Select: SOLIDM > MODIFY | OPERATN | REPLACE

Prompt: SEL BRANCH // YES: CURRENT

1. Select a branch of the current solid or ditto as the replacement feature.
2. If desired, select a point to position the feature. Click on YES to replace.

Duplicate branch of solid

Select: SOLIDM > MODIFY | OPERATN | DUPLICAT | TRANSLAT or ROTATE or SYMMETRY or UNSPEC

Duplicate branch of solid by translation

Select: SOLIDM > MODIFY | OPERATN | DUPLICAT | TRANSLAT

Prompt: SEL BRANCH // YES: CURRENT

1. Select the branch to be modified from the solid or CSG tree.
2. In the Translation window, select Length to input a value for the translation.
3. Define the translation by using one of the following alternatives: (a) select a line, and then enter a value in the Translation window; (b) select two points; (c) enter coordinates; (d) select a plane, and then enter a length, or (e) select a point or a line.
4. Select the arrow to invert as necessary.
5. If desired, enter the number of times the translation is to be repeated in the input area.

✗ **WARNING:** *The translation length must be entered in the translation window, not in the input area.*

6. Click on YES to apply the translation.
7. If desired, update the solid as described at the start of the MODIFY entry.

Duplicate branch of solid by rotation

Select: SOLIDM > MODIFY | OPERATN | DUPLICAT | ROTATE

Prompt: SEL BRANCH

1. The Rotation window displays. In the window, select Angle to input an angle value for the rotation.
2. Define the rotation by using one of the following alternatives: (a) select a line or two points, and then enter an angle value in the Rotation window; (b) select planes; (c) select a plane (rotation axis passes through the center of the plane normal to the plane).
3. Select arrow to invert as necessary.
4. Click on YES to apply rotation.
5. If desired, update the solid as described at the start of the MODIFY entry.

Duplicate branch of solid by symmetry

Select: SOLIDM > MODIFY | OPERATN | DUPLICAT | SYMMETRY

Prompt: SEL BRANCH / / YES: CURRENT

1. Select the solid element to be modified.
2. Define the symmetry by selecting a plane according to one of the following alternatives: (a) select two lines; (b) select three points or a line and a point; (c) select two points or a line.
3. Click on YES to apply the symmetry. Update the solid as described at the start of the MODIFY entry as necessary.

Duplicate branch of solid using stored transformation

Select: SOLIDM > MODIFY | OPERATN | DUPLICAT | UNSPEC

Prompt: SEL BRANCH KEY NUMBER / / YES: APPLY

In the Transformation window, select the desired transformation. Select the branch to be modified from the CSG tree or the solid. Click on YES to apply. Update the solid as described at the start of the MODIFY entry as necessary.

Apply and modify display attributes of solids

Select: SOLIDM > MODIFY | VISUALTN | ELEMENT or EDGE

Apply and modify display attributes of elements

Select: SOLIDM > MODIFY | VISUALTN | ELEMENT

Prompt: MSELW SOL/POL

Options in the PROCESS OPTIONS window are described below.

- CHOOSE PARAMETERS. Display the SPECIAL PARAMETERS window where visualization parameters can be selected.
- RESET TO STANDARD. Reset to standard settings.
- SAME ELEMENT. Use the same visualization attributes as a selected reference element.
- RESTORE. Restore previously altered attributes to standard.

Apply and modify display attributes of edges

Select: SOLIDM > MODIFY | VISUALTN | EDGE

Prompt: SEL EDGE

Select an edge of a solid. Click on YES to modify the visualization.

Modify solid display

Select: SOLIDM > MODIFY | DRESS UP | RENAME or COLOR

Modify identifier of solid subelement

Select: SOLIDM > MODIFY | DRESS UP | RENAME | PRIM or FSUR or EDGE

Prompt: SEL PRIM or FSUR or EDGE // YES: CURRENT

1. Select a primitive, FSUR , RSUR, or edge.
2. Enter a new identifier for the element (up to 70 characters).

Modify color of primitive

Select: SOLIDM > MODIFY | DRESS UP | COLOR

Prompt: SEL PRIM / / YES: CURRENT

1. Select an element.
2. Use one of the following alternatives: (a) Select an element of the desired color. (b) Select a color. (c) Enter a number.
3. Click on YES. The COLOR CHANGED message displays.

EXTRACT

Create solid geometry and faces.

Select: SOLIDM > EXTRACT | CUT or PROJECT or PRIMVOL

Create geometry and faces by intersecting solids with planes

Select: SOLIDM > EXTRACT | CUT | EXACT or APPROXIM

Prompt: MSELW PLN

Select intersecting plane(s), and then select a solid. Geometry representing the cut is created (as exact geometry with the EXACT option selected, and as facets when the APPROXIM option is selected).

Create geometry by projecting solids onto plane

Select: SOLIDM > EXTRACT | PROJECT

Prompt: MSELW PLN

1. Select the projection plane(s). An arrow indicating the projection direction displays.
2. Select arrow to invert as necessary.
3. Select a solid. Geometry is created representing the orthographic projection of the solid onto the plane(s).

Extract primitive subelements from solid

Select: SOLIDE > EXTRACT | PRIMVOL

Prompt: SEL > PRIM

Select a primitive subelement. Volume, faces, surfaces, lines, and curves are created. (Surfaces are created in the No Show area.)

VISU STD

Set display standards.

Select: SOLIDM > VISU STD

Prompt: REF ELEM: SEL SOL/POL

1. Select desired values in the HLR Parameters window. Select a solid or polyhedron which becomes the reference element. Make further selections from the window as necessary.
2. The MODIFICATION BR SWITCH REQUIRED message displays. Click on the BR button on the fixed menu.

UPDATE

Update solid boundary representation (B-REP).

Select: SOLIDM > UPDATE

Prompt: MSELW / / YES: FORCE UPDATE

1. The Update window displays. Select VISU ON or OFF to view or not view the local solid as the update is executed. Select INFO ON or OFF to display or not display information about the update during the process.
2. Select Inactive errors and Continue to inactivate errors encountered and continue update process, respectively.
3. Select the solid(s) to be updated. The ALREADY UPDATED message displays as appropriate.
4. Click on YES to confirm update. The UPDATE DONE message displays.

RESTORE

Restore solid to state of last update.

Select: SOLIDM > RESTORE

Prompt: SEL SOL/POL

1. Select a solid or polyhedron. Alternatively, click on YES for the current solid.
2. Click on YES to confirm restoration. One of the following messages displays: RESTORE DONE, ALREADY UPDATED, or RESTORE NOT ALLOWED.

SPC->DR2 in DRAW Mode

Chapter 14

The SPC->DR2 function is used to create geometry in DRAW mode by cutting SPACE mode solids with the use of a plane or projection of SPACE mode solids, volumes, or sets of faces onto a plane. Use this function in conjunction with the AUX-VIEW function to create views featuring cross sections of 3D solids, volumes, or faces.

Option menu for SPC->DR2 in DRAW mode.

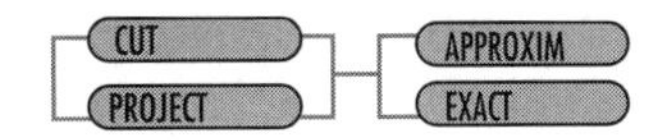

CUT

Cut SPACE mode solids and project the elements into DRAW mode views.

Select: SPC->DR2 > CUT | APPROXIM or EXACT

Create shapes of cut polyhedral form of solid

Select: SPC->DR2 > CUT | APPROXIM

Prompt: SEL VIEW

1. Select a view.
2. Click on the YES button to accept the displayed cutting depth, or key in a value for the cutting depth.
3. Select SPACE element(s) (solids or pipes) to be cut. Multiselection can be used. (See Appendices A and B for further information.)

Create exact form of cut solid

Select: SPC->DR2 > CUT | EXACT

Prompt: SEL VIEW

1. Select a view.
2. Click on the YES button to accept the displayed cutting depth, or key in a value for the cutting depth.
3. Select a SPACE element(s) (solids or pipes) to be cut. Multiselection can be used. (See Appendices A and B for further information.)

PROJECT

Project SPACE mode solids, volumes, or faces into DRAW mode views.

Select: SPC->DRW > PROJECT | APPROXIM or EXACT

Orthogonally project edges of solids onto one or more DRAW views

Select: SPC->DR2 >PROJECT | APPROXIM

Prompt: SEL VIEW // YES : ALL VIEWS

1. Click on the YES button to accept all views in which to project the solids, or select a view(s), and click on the YES button to end view selection.
2. In the Hidden Lines Mode window, make selections to control the DRAW elements. Selections are listed below. An image showing selection results appears in the window.
   - CONTOUR
     - REMOVE
     - DIMMED | SOLID or DOTTED or DASHED or DOT-DASH
     - NORMAL | SOLID or DOTTED or DASHED or DOT-DASH
   - WIREFRAME
     - REMOVE
     - DIMMED | SOLID or DOTTED or DASHED or DOT-DASH
     - NORMAL | SOLID or DOTTED or DASHED or DOT-DASH
3. Select SPACE element(s) to be projected into the selected view(s). Multiselection can be used. (See Appendices A and B for further information.)

✓ **NOTE:** *To obtain the projection of a solid contained within a detail, use the following multiselection: *SOL>.*

Orthogonally project one or more complex volumes, faces, or exact solids onto DRAW views

Select: SPC->DR2 > PROJECT | EXACT

Prompt: SEL VIEW // YES : ALL VIEWS

1. Click on the YES button to accept all views in which to project the solids, or select a view(s), and click on the YES button to end view selection.
2. In the Hidden Lines Mode window, make selections to control the DRAW elements. Selections are listed below. An image showing selection results appears in the window.

- CONTOUR
  - REMOVE
  - DIMMED | SOLID or DOTTED or DASHED or DOT-DASH
  - NORMAL | SOLID or DOTTED or DASHED or DOT-DASH

3. Select SPACE element(s) to be projected into the selected view(s). Multiselection can be used. (See Appendices A and B for further information.)

✓ **NOTE:** *To obtain the projection of a solid contained within a detail, use the following multiselection: *SOL>.*

SPC->DRW in DRAW Mode

Chapter 14

The SPC->DRW function is a DRAW mode function used to create geometry in DRAW mode by cutting SPACE mode wireframe and surface elements. It is also used to create geometry in DRAW mode by projecting SPACE mode wireframe and surface elements. To create views featuring views and cross sections of 3D objects, use this function in conjunction with AUXVIEW.

Option menu for SPC->DRW in DRAW mode.

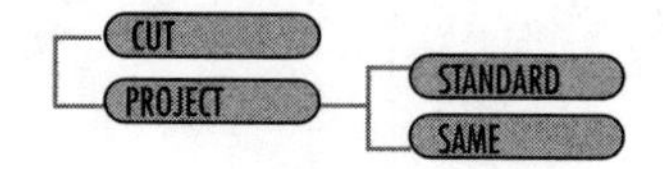

CUT

Cut SPACE mode wireframe and surface elements and project the elements into DRAW mode views.

Select: SPC->DRW > CUT

Prompt: SEL VIEW // YES : ALL VIEWS

1. Click on the YES button to accept all views. In this case, each view defines a cutting plane with a depth of zero from the respective creation plane. Alternatively, select a view, and click on the YES button to accept the displayed cutting depth, or key in a value for the cutting depth.
2. Select a SPACE element(s) to be cut. Multiselection can be used. (See Appendices A and B for further information.)
3. If desired, key in a value for a new depth of the cutting plane.
4. Click on the YES button to end selection of the SPACE elements to be cut.

✓ ***NOTE:*** *All DRAW elements created are placed on the current layer and contain graphical attributes as set in the STANDARD function.*

PROJECT

Project SPACE mode wireframe and surface elements into DRAW mode views.

Select: SPC->DRW > PROJECT | STANDARD or SAME

Prompt: SEL VIEW // YES : ALL VIEWS

1. Click on the YES button to accept all views in which to project the 3D elements, or select a view(s), and click on the YES button to end view selection.
2. Select a SPACE element(s) to be projected into the selected view(s). Multiselection can be used. (See Appendices A and B for further information.)
3. Click on the YES button to end selection of SPACE elements to be projected.

✓ **NOTE:** *The graphical attributes of the DRAW elements created are controlled by the STANDARD or SAME option selection. Under the STANDARD option, the DRAW elements are created in the current layer with graphical attributes as set in the STANDARD function. With the SAME option selected, the DRAW elements are created with the same graphical attributes and layer as the cut SPACE elements.*

STANDARD in DRAW and SPACE Modes

The STANDARD function in DRAW and SPACE modes is used to set the graphic attributes and geometric parameters of DRAW and SPACE elements. Some of the options available in the STANDARD function can also be accessed by clicking on the ST button on the fixed menu.

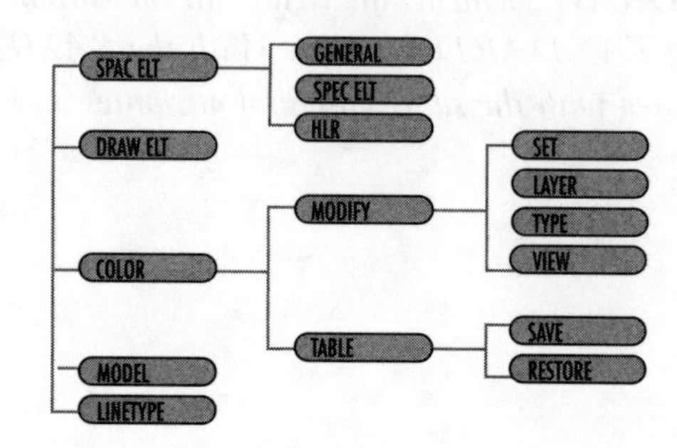

Option menu for STANDARD in DRAW and SPACE modes.

SPACE ELT

Define graphic attributes of SPACE elements.

Select: STANDARD > SPAC ELT | GENERAL or SPEC ELT or HLR

Define general standards

Select: STANDARD > SPAC ELT | GENERAL

Prompt: SEL ITEM / / YES: DISPLAY

In the SPACE ELEMENT GRAPHIC STANDARDS screen, standards can be set for the following SPACE elements: points, lines, curves, thickness, forced color, planes, and implicit points.

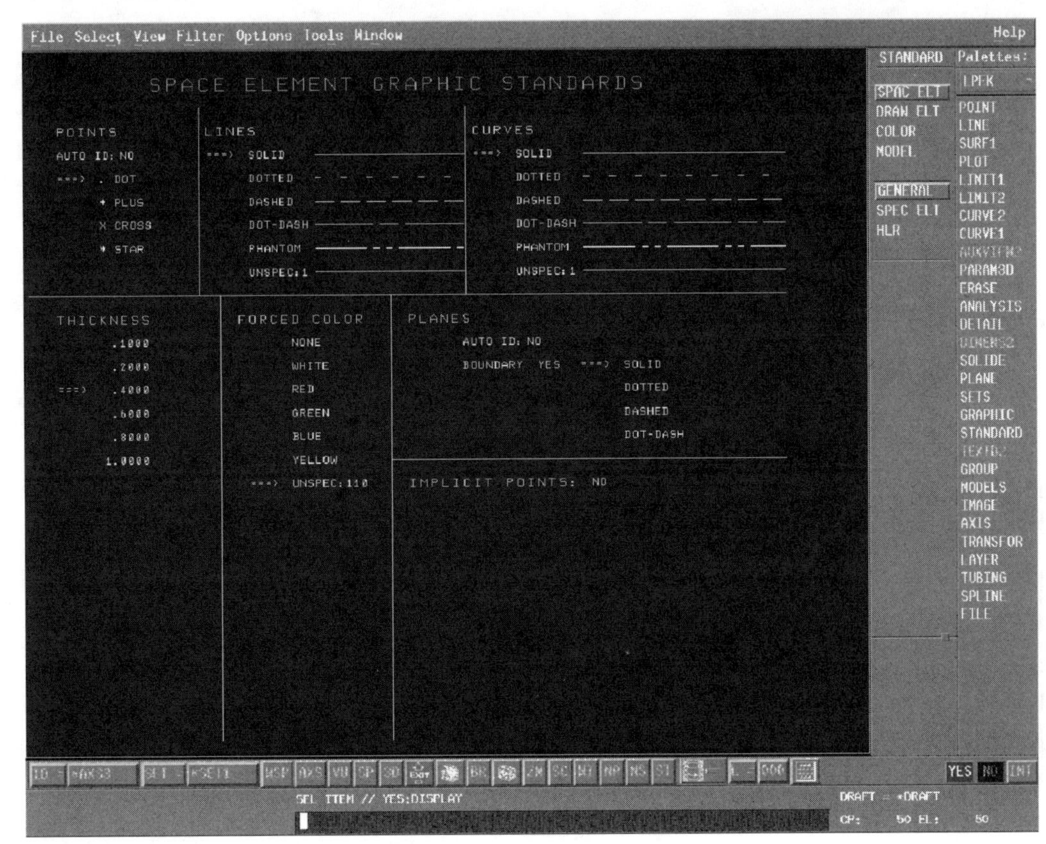

SPACE ELEMENT GRAPHIC STANDARDS screen.

Define standard settings for special elements

Select: STANDARD > SPAC ELT | SPEC ELT

Prompt: SEL FUNCTION

Select a special element from the list appearing in the ELEMENT selection window.

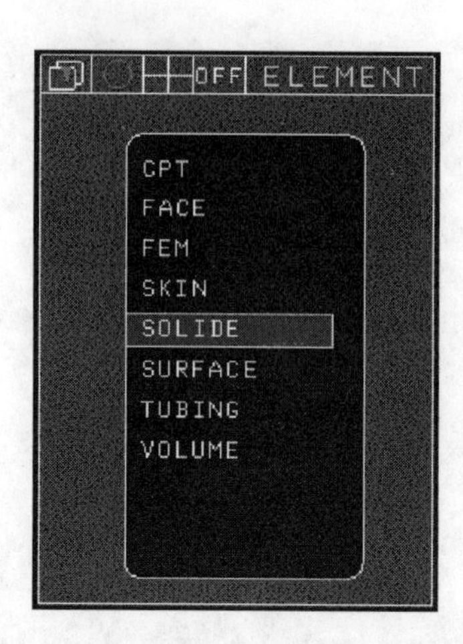

ELEMENT selection window.

Depending on the selection, additional windows are displayed (see SOLIDE example below) in which more graphic attributes are set.

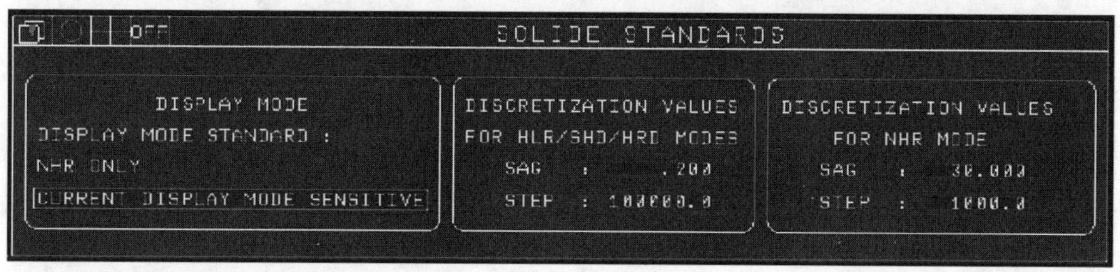

SOLIDE STANDARDS window.

Modify hidden line parameters

Select: STANDARD > SPAC ELT | HLR

Prompt: REF ELEM: SEL ELEM

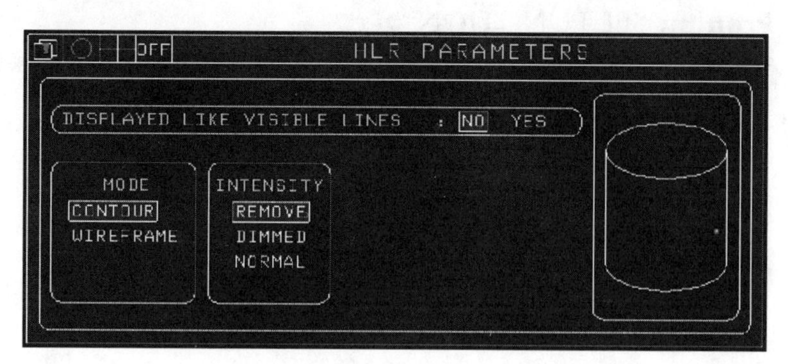

HLR PARAMETERS window.

In the HLR PARAMETERS window, solids can be displayed as CONTOUR or WIRE-FRAME by selecting in the MODE panel. Set the visualization standard and make hidden lines visible or invisible in the INTENSITY panel.

DRAW ELT

Define graphic attributes of DRAW elements.

Select: STANDARD > DRAW ELT

Prompt: SEL ITEM / / YES: DISPLAY

In the DRAW ELEMENTS GRAPHIC STANDARDS screen, set standards for the following elements: points, lines, curves and shapes, grid mode, implicit points, and polar or rectangular coordinates.

COLOR

Define or modify color of sets, layers, types, or views, and save or restore color table.

Select: STANDARD > COLOR | MODIFY or TABLE

Define or modify color of set

Select: STANDARD > COLOR | MODIFY | SET

Prompt: YES: LIST SEL SET

1. Select or enter an element in the set to be modified. Alternatively, click on YES for the list of available sets, and select from the list.
2. Enter a number for the desired color.

Define or modify color of layer

Select: STANDARD > COLOR | MODIFY | LAYER

Prompt: YES: LIST KEY COLOR / / SEL ELEM

1. Select or enter an element on the layer to be modified. Alternatively, click on YES for the list of layers and select from list.
2. Enter a number for the desired color.

Select: STANDARD > COLOR | MODIFY | TYPE

Prompt: YES: LIST SEL ELEM

1. Select or enter an element of the type to be modified. Alternatively, click on YES for a list and select from the list.
2. Enter a number for the desired color.

Select: STANDARD > COLOR | MODIFY | VIEW

Prompt: YES: LIST SEL VIEW / / SEL SPC ELEM

1. Select a view, or click on YES for a list of views; select from the list.
2. Enter a number for the desired color.

Save or restore color table

Select: STANDARD > COLOR | TABLE | SAVE or RESTORE

Save color table

Select: STANDARD > COLOR | TABLE | SAVE

Prompt: YES: CONFIRM

The color table is modified using the COL option in the Display and Manipulation window (see Appendix G). Click on YES to save color table.

Restore color table

Select: STANDARD > COLOR | TABLE | RESTORE

Prompt: YES: RESTORE

Click on YES to restore the standard color table.

MODEL

Define geometric standards.

Select: STANDARD > MODEL

Prompt: SEL ITEM

Page 1 of the GEOMETRIC STANDARDS screen is activated.

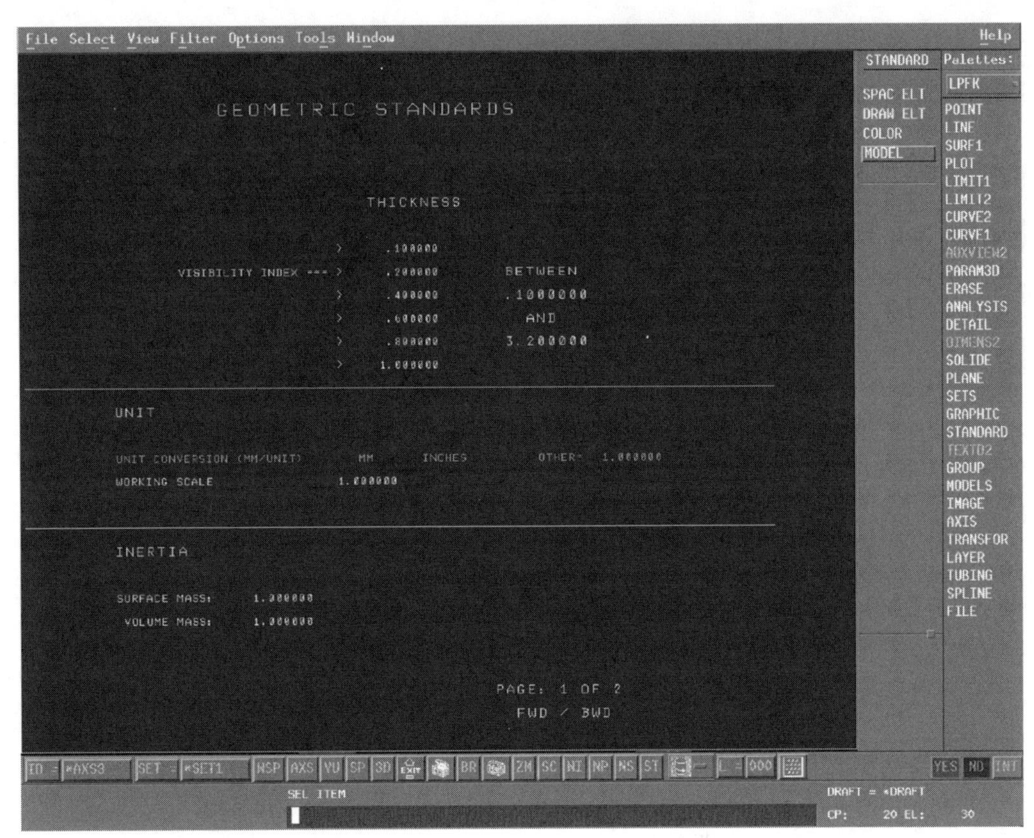

GEOMETRIC STANDARDS screen (page 1).

Standards defined in the screen are described below.

- THICKNESS. Select thickness value to be modified and enter value between .1mm and 3.2mm.
- VISIBILITY INDEX. Select angle bracket symbol (>) next to a particular thickness. If the visibility index is greater than the current thickness index and the latter also points to the first value in the list of thicknesses, then the element is displayed in dimmed mode but is still selectable.
- UNIT. Select desired units (mm, inches, or other). This selection must be made before any elements are created in the model. Once elements are created, the current selection is nonselectable and cannot be changed.
- WORKING SCALE. Select the value to be modified; enter a new value as necessary.
- INERTIA. Enter values for density.

Select Page 2. Define model tolerances in this screen.

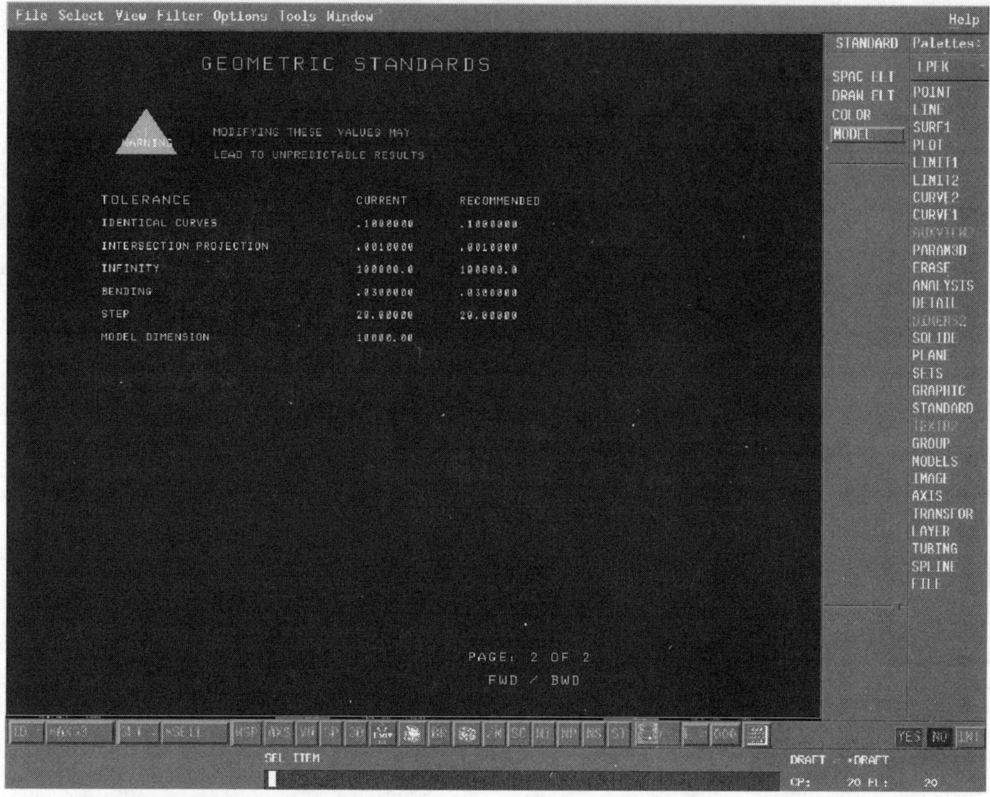

GEOMETRIC STANDARDS screen (page 2).

- IDENTICAL CURVES. Enter value from which the system will consider two curves as being identical.
- INTERSECTION PROJECTION. Enter value from which the system will consider two points being one and the same.
- INFINITY. Enter value from which the system will consider data infinite.
- BENDING. Enter value from which the system approximates curves using short straight lines.
- STEP. Enter value that defines the maximum line length used as a facet to approximate a curve.

LINETYPE

Typically accessible only by the system administrator, this option is used to set available unspecified line types.

Select: STANDARD > LINETYPE

SYMBOL in DRAW Mode

INSIDE CATIA

Chapter 7

The SYMBOL function is used only in DRAW mode to create and manage symbols. Symbols are similar to details, except that a symbol consists of a single element that symbolizes an assembly of geometric elements. A symbol is advantageous in that it uses less storage space than a detail. However, the geometric elements that comprise the symbol can be accessed only through the SYMBOL function.

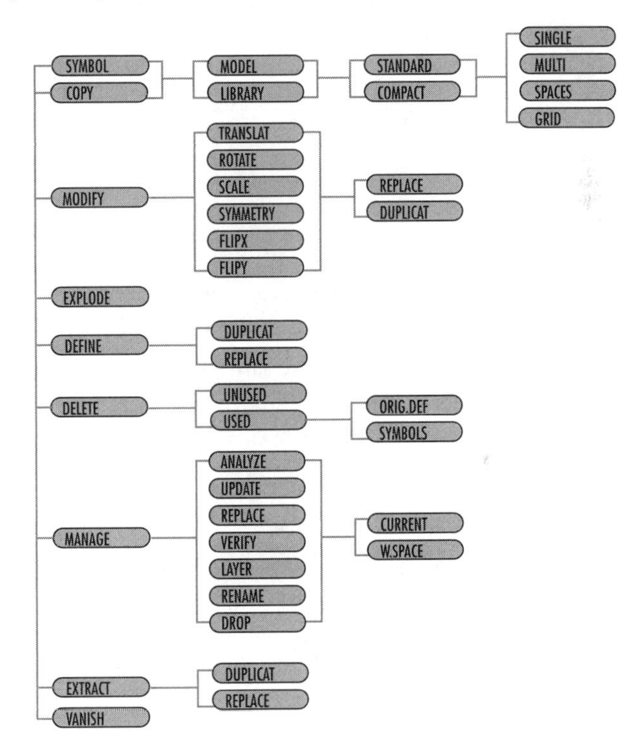

Option menu for SYMBOL in DRAW mode.

SYMBOL

Place existing symbol(s) in model layer or library.

Select: SYMBOL > SYMBOL | MODEL or LIBRARY | STANDARD or COMPACT | SINGLE or MULTI or SPACES or GRID

Place symbol(s) in model

Select: SYMBOL > SYMBOL | MODEL | STANDARD or COMPACT | SINGLE or MULTI or SPACES or GRID

Place single symbol in model

Select: SYMBOL > SYMBOL | MODEL | STANDARD or COMPACT | SINGLE

Prompt: KEY STRING / / ENTER LIST SEL SYMBOL / / YES: DISPLAY

✓ **NOTE:** *Upon selection of STANDARD, symbol occurrences are placed on the layer on which they were created. With COMPACT selected, symbol occurrences are placed on the current layer.*

1. Proceed with one of the following alternatives: (a) Click on YES to display available symbols, and then select the desired symbol. (b) Enter a blank field to display the list of available symbols, and then select the desired symbol. (c) Enter a string (up to 30 characters) to display the list of available symbols containing the string, and then select the desired symbol.
2. To place the symbol occurrence, begin by entering a new scale value as necessary.
3. Select or indicate a point, or select two lines which intersect (the intersect point will be selected) to position the symbol.
4. Enter an orientation angle if desired.
5. Click on YES to end orientation. You can now repeat steps 2 through 4 to place additional occurrences of the same symbol or click on YES to select a new one.

Place multiple symbols in model

Select: SYMBOL > SYMBOL | MODEL | STANDARD or COMPACT | MULTI

Prompt: KEY STRING / / ENTER LIST SEL SYMBOL / / YES: DISPLAY

1. Proceed with one of the following alternatives: (a) Click on YES to display available symbols, and then select the desired symbol. (b) Enter a blank field to display the list of available symbols, and then select the desired symbol. (c) Enter a string (up to 30 characters) to display the list of available symbols containing the string, and then select the desired symbol.
2. To position the symbol occurrence, enter a new scale value if desired, and then select or indicate a series of points or pairs of intersecting lines.
3. Enter an orientation angle if desired.
4. Click on YES to end orientation. You can now repeat the above or click on YES to select a new symbol.

Place spaced symbols in model

Select: SYMBOL > SYMBOL | MODEL | STANDARD or COMPACT | SPACES

Prompt: KEY STRING / / ENTER LIST SEL SYMBOL / / YES: DISPLAY

1. Proceed with one of the following alternatives: (a) Click on YES to display available symbols, and then select the desired symbol. (b) Enter a blank field to display the list of available symbols, and then select the desired symbol. (c) Enter a string (up to 30 characters) to display the list of available symbols containing the string, and then select the desired symbol.
2. Enter a new scale value if desired.
3. Select two points, and then enter the number of desired equally spaced symbol occurrences.
4. Select a line or curve, and then select two points. Proceed as above or key in the desired number of equally spaced symbol occurrences along the selected element.
5. Click on YES, and then enter an approximate distance between symbol occurrences.
6. Enter an orientation angle if desired. You can now repeat the above or click on YES to select a new symbol.

Place symbols in grid

Select: SYMBOL > SYMBOL | MODEL | STANDARD or COMPACT | GRID

Prompt: KEY STRING / / ENTER LIST SEL SYMBOL / / YES: DISPLAY

1. Proceed with one of the following alternatives: (a) Click on YES to display available symbols, and then select the desired symbol. (b) Enter a blank field to display the list of available symbols, and then select the desired symbol. (c) Enter a string (up to 30 characters) to display the list of available symbols containing the string, and then select the desired symbol.
2. Enter a new scale value if desired, and then select a point type grid. Symbol occurrences will be placed on all points.
3. Enter an orientation angle as necessary. You can now repeat the above or click on YES to select a new symbol.

Place library symbols

Select: SYMBOL > SYMBOL | LIBRARY | STANDARD or COMPACT | SINGLE or MULTI or SPACES or GRID

Prompt: KEY STRING / / ENTER:LIST SEL SYMBOL YES: DISPLAY

1. On selection of the LIBRARY option, a list of available symbols in the last used family displays. If a family has not been used, enter a blank field to display the list of available library families.
2. Click on NO to select an alternative library, or select the desired family.

3. From the displayed list, select the desired symbol.
4. Click on YES to confirm or NO to return to the list.
5. Position symbol occurrence using the methods described in the MODEL options above.

✓ **NOTE:** *All MODEL options are also available under LIBRARY.*

COPY

Create copies of symbols in element form. With the COPY option, the symbol elements are copied into the workspace and the link with the symbol is broken.

Select: SYMBOL > COPY | MODEL or LIBRARY | STANDARD or COMPACT | SINGLE or MULTI or SPACES or GRID

Prompt: KEY STRING / / ENTER:LIST SEL SYMBOL YES: DISPLAY

✓ **NOTE:** *Methods of copy selection, placement, and orientation are identical to those described above for symbols.*

MODIFY

Modify position of symbol occurrences or copies.

Select: SYMBOL > MODIFY | TRANSLATE or ROTATE or SCALE or SYMMETRY or FLIPX or FLIPY | REPLACE or DUPLICAT

Upon selection of REPLACE, the original symbol occurrence or copy will be replaced on execution of the modification. With DUPLICAT selected, the original symbol occurrence or copy will be retained.

✓ **NOTE 1:** *Copies can be modified immediately following placement only. Copies are handled as elements thereafter.*

✓ **NOTE 2:** *If a symbol occurrence is incorrectly selected, click on YES to switch to a new one.*

Translate symbol

Select: SYMBOL > MODIFY | TRANSLATE | REPLACE or DUPLICAT

Prompt: SEL SYMBOL

1. Select the symbol occurrence to be translated.
2. Define the translation by using one of the following alternatives: (a) Select two points. (b) Select two parallel lines. (c) Enter X, Y coordinates. (d) Reselect a line

and then a point or vice versa. (e) Select two nonparallel lines. (f) Select a line to define the direction, and then enter a length.

3. Click on YES to select a new symbol occurrence.

Rotate symbol

Select: SYMBOL > MODIFY | ROTATE | REPLACE or DUPLICAT

Prompt: SEL SYMBOL

1. Select the symbol occurrence to be translated.
2. Define the rotation according to one of the following alternatives: (a) Select two points. (b) Select a point and then enter an angle. (c) Select two lines or select a line, and then enter an angle.
3. Click on YES to select a new symbol occurrence.

Modify symbol scale

Select: SYMBOL > MODIFY | SCALE | REPLACE or DUPLICAT

Prompt: SEL SYMBOL

1. Select the symbol occurrence to be translated.
2. Define the scaling by using one of the following alternatives: (a) Enter a scale value. (b) Select two lines (LND1 and LND2) where the current scale is multiplied by the quotient of the two lengths. (c) Select two circles (CIRD1 and CIRD2) where the current scale is multiplied by the quotient of the two radii.
3. Click on YES to select a new symbol occurrence.

✓ **NOTE:** *Text items are not affected by scaling.*

Modify symbol symmetry

Select: SYMBOL > MODIFY | SYMMETRY | REPLACE or DUPLICAT

Prompt: SEL SYMBOL

Select the symbol occurrence. To define the symmetry, select a point or a line, and then click on YES to select a new symbol occurrence.

Symmetry around X or Y axis

Select: SYMBOL > MODIFY | FLIPX or FLIPY | REPLACE or DUPLICAT

Prompt: SEL SYMBOL

The selected symbol occurrence or copy will be copied or replaced by symmetry around the X or Y axis.

EXPLODE

Transform symbol occurrence into copy.

Select: SYMBOL > EXPLODE

Prompt: SEL SYMBOL

1. Select the symbol to be exploded. If the symbol occurrence was created as a compact, the copied elements will be placed on the layer of the symbol occurrence. If the symbol occurrence was created as a standard, the copied elements will be placed on the layer on which the original symbol was created.
2. Click on the YES button to confirm the explode. The symbol occurrence will be copied into the current workspace.

DEFINE

Define (create) new symbol.

Select: SYMBOL > DEFINE | DUPLICAT or REPLACE

Prompt: 16 CHARACTERS FOR SYMBOL ID   KEY NEW SYMBOL NAME

In order to be able to create (define) a new symbol, you must be in a detail workspace. To switch to a detail workspace, use the WSP button or the DETAIL > CHANGE function and option. Elements that are visible when the detail workspace has a LAYER FILTER ALL applied will become the defined symbol. The NO SHOW option can be used to hide undesired elements.

1. To define a symbol, you have two options. Upon selecting DUPLICAT, the symbol is defined by the elements in the detail workspace, and the detail workspace will be retained. Consequently, the symbol cannot have the same name as the detail workspace. With REPLACE selected, the symbol is defined by the elements in the detail workspace, and the detail will then be deleted. Therefore, the symbol can use the same name as the defining detail workspace.
2. Enter a name for the new symbol.

✓ ***NOTE:*** *The new name must have between 1 and 16 characters and must not be the same as any existing detail or symbol, except in the case of REPLACE as mentioned above.*

3. Enter a comment of up to 48 characters, or click on YES if no comment is desired.
4. Click on YES to confirm the definition of the symbol.

DELETE

Delete symbol occurrence and/or original definition.

Select: SYMBOL > DELETE | UNUSED or USED

The lists of symbols available for deletion can be obtained in the same manner as stated in the SYMBOL > MODEL or LIBRARY options.

Delete unused symbol

Select: SYMBOL > DELETE | UNUSED

Prompt: KEY STRING YES: DISPLAY

Select the symbol to be deleted, and then click on YES to confirm.

Delete all occurrences of symbol and original definition

Select: SYMBOL > DELETE | USED | ORIG.DEF

Prompt: KEY STRING SEL SYMBOL // YES: DISPLAY

Select the symbol to be deleted, and then click on YES to confirm.

Delete all occurrences of symbol but retain original definition

Select: SYMBOL > DELETE | USED | SYMBOLS

Prompt: KEY STRING SEL SYMBOL // YES: DISPLAY

Select the symbol to be deleted, and then click on YES to confirm.

MANAGE

Manage symbols.

Select: SYMBOL > MANAGE | ANALYZE or UPDATE or REPLACE or VERIFY or LAYER or RENAME or DROP

Analyze details or workspaces

Select: SYMBOL > MANAGE | ANALYZE

Prompt: KEY STRING / / ENTER: LIST SEL SYMBOL / / YES: DISPLAY

1. The lists and displays of symbols available for analysis can be obtained in the same manner as stated in the SYMBOL > MODEL or LIBRARY options.

2. Select a symbol to be analyzed. The analysis results are displayed in the alphanumeric window. A symbol description includes identifier, type (model or library), and number of occurrences and/or unused.
3. Reselect a symbol occurrence to be analyzed. The analysis results are displayed in the alphanumeric window. Symbol occurrence descriptions include workspace identifier, set and layer on which symbol occurs, origin, scale, and angle of insertion.

Update external (library) symbol regarding current library

Select: SYMBOL > MANAGE | UPDATE

Prompt: KEY STRING / / ENTER: LIST SELSYMBOL / / YES: DISPLAY

The lists and displays of symbols available for updating can be obtained in the same manner as stated in the SYMBOL > MODEL or LIBRARY options. Dates and times of the current and external symbol are displayed for comparison.

1. Select a symbol to be updated.
2. Click on YES to confirm the choice of this version.
3. Click on YES to confirm the version of the symbol from the library to be retrieved.

Replace symbol with another symbol

Select: SYMBOL > MANAGE | REPLACE

Prompt: SEL SYMBOL TO REPLACE

1. Select the symbol occurrence to be replaced.
2. Select the replacement symbol. The lists and displays of symbols available for replacement can be obtained in the same manner as stated in the SYMBOL > MODEL or LIBRARY options.
3. Click on YES to confirm the replacement of the first selected symbol occurrence by the second selected symbol.
4. If desired, click on YES to replace all occurrences of the first selected symbol with the second selected symbol.
5. Click on YES to confirm replacement.

Verify nature of symbols

Select: SYMBOL > MANAGE | VERIFY | CURRENT or W.SPACE

Prompt: SEL SYMBOL / / YES: ALL SYMBOLS

With the CURRENT option selected, verification takes place in the current set. Upon selecting W.SPACE, verification takes place in the current workspace.

1. Select a symbol occurrence; all occurrences of the selected symbol will be highlighted. Alternatively, click on YES to display all symbol occurrences.
2. Select a symbol. All symbols at the same scale will be highlighted and the current scale shown in the message area.
3. Click on NO, and all symbols at a different scale will be highlighted.

Transform standard symbol occurrences into compact symbol occurrences and vice versa

Select: SYMBOL > MANAGE | LAYER | CURRENT or W.SPACE

Prompt: SEL SYMBOL TO TRANSFORM

With the CURRENT option selected, transformation takes place in the current set. Upon selecting W.SPACE, transformation takes place in the current workspace.

1. Select a symbol occurrence.
2. Click on YES. If the symbol occurrence was compact, it will be transformed into a standard symbol occurrence and vice versa.
3. If desired, click on YES to display all occurrences of the previously selected symbol occurrence.
4. Click on YES to transform all compact symbol occurrences into standard symbol occurrences and vice versa.

Rename symbols (or modify symbol comment)

Select: SYMBOL > MANAGE | RENAME

Prompt: KEY STRING / /ENTER: LIST SEL SYMBOL / / YES: DISPLAY

Select the symbol to be renamed. The lists and displays of symbols available for renaming can be obtained in the same manner as stated in the SYMBOL > MODEL or LIBRARY options.

1. Key in the new symbol name, or click on YES if the name is to remain unchanged.
2. Key in the new symbol comment, or click on YES if the comment is to remain unchanged.

Break link between externally stored symbol and respective library

Select: SYMBOL > MANAGE | DROP

Prompt: KEY STRING / /ENTER: LIST SEL SYMBOL / / YES: DISPLAY

Select a symbol to be dropped. The lists and displays of symbols available for renaming can be obtained in the same manner as stated in the SYMBOL > MODEL or LIBRARY options. Click on the YES button to confirm the symbol to be dropped.

✓ **NOTE:** *On the right side of the list of available symbols, MODEL or LIBRARY is shown providing the status of each symbol.*

EXTRACT

Copy geometric element from symbol occurrence and superimpose on corresponding element.

Select: SYMBOL > EXTRACT | DUPLICAT or REPLACE

Prompt: SEL SYMBOL

With the DUPLICAT option selected, the geometric element will be copied without modifying the symbol occurrence. Upon selecting REPLACE, the geometric element will be copied and the original element will be hidden in the symbol occurrence.

1. Select a symbol occurrence.
2. Select the element to be extracted. The created element will be placed on the current layer. The created element can then be modified using other functions depending on the type of element extracted. For example, if a text element was extracted it can be modified using TEXTD2 modification options.

VANISH

Hide geometric element from symbol occurrence.

Select: SYMBOL > VANISH

Prompt: SEL SYMBOL

Select a symbol occurrence, and then select the element to be hidden.

TEXT in DRAW Mode

The TEXT function is used to create text nodes controlled and linked to points, lines, structures, and shapes used to display attributes, identifiers, and data resulting from calculation programs.

✓ **NOTE:** *Availability of the TEXT function in DRAW mode will depend your system setup.*

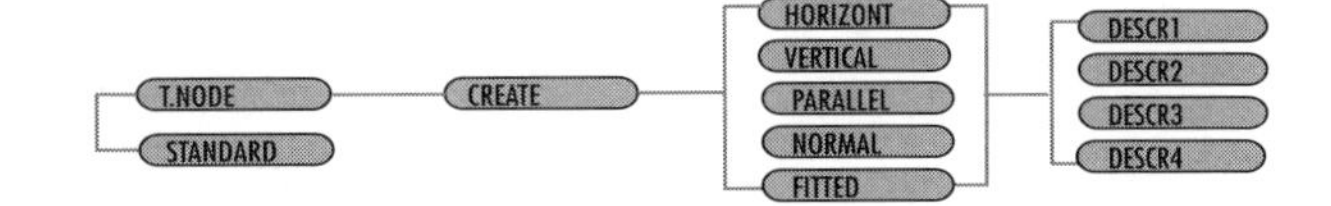

Option menu for TEXT in DRAW mode.

T.NODE

Create text nodes. A text node contains text that enables the visualization of values of attributes, element identifiers, user program results, and preformatted texts contained in a text file. Because the creation of text nodes requires the creation of the abovementioned items, and the creation of said items is beyond the scope of this book, text nodes are not covered in detail.

Horizontal or vertical text node creation require the following selections.

Select: TEXT > T.NODE | CREATE | HORIZONT or VERTICAL | DESCR1 or DESCR2 or DESCR3 or DESCR4

Prompt: POSITION T.NODE : SEL PR / LN / CIR

Selections for the creation of parallel or normal text nodes follow.

Select: TEXT > T.NODE | CREATE | PARALLEL or NORMAL | DESCR1 or DESCR2 or DESCR3 or DESCR4

Prompt: ORIENTATION : KEY ANG // SEL TXT / LN

✓ **NOTE:** *The TEXT > T.NODE | CREATE | FITTED option is not available.*

Upon selection of one of the above options, a TEXT-NODE dialog displays as shown in the next illustration.

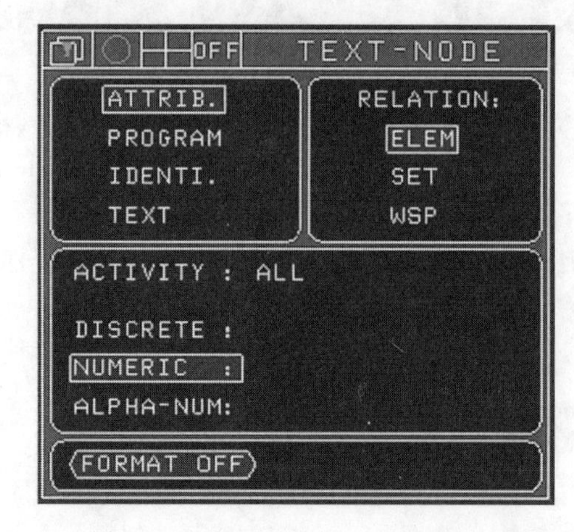

TEXT-NODE dialog window.

The above dialog is used to define the text nodes to be created, and contains the characteristics listed below.

- ATTRIB. Attribute.
- PROGRAM. Calculation program output.
- IDENTI. CATIA identifier.
- TEXT. Text from file or keyboard entry.
- ELEM. Displayed data related to element.
- SET. Displayed data related to set.
- WSP. Displayed data related to workspace.
- Define attribute type when ATTRIB option is selected.
- Define display format (used for placing optional text before or after text node).

After defining parameters to be used for the text node, you can proceed with creating the node.

1. To define the position of the text node, select a point, circle, or infinite line element.
2. To apply the text node and related attributes, select an element.

STANDARD

Define parameters used to create and display text created in TEXT function.

Select: TEXT > STANDARD

Prompt: SEL PARM // KEY NEW VALUE YES : END

Upon selection of the option, the Text Standards parameter window displays as shown in the next illustration.

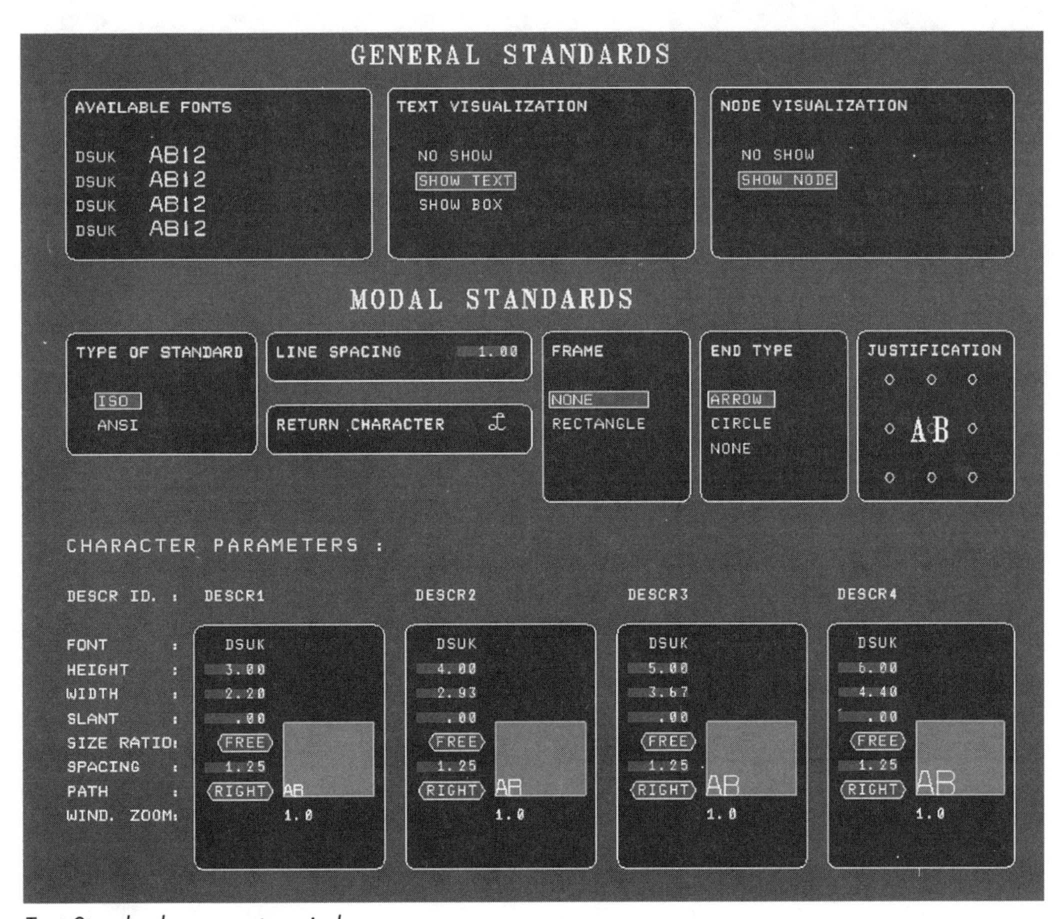

Text Standards parameter window.

Define text parameters used in the TEXT function from this window. Parameters are described below.

- AVAILABLE FONTS. Defines fonts used in four text options (DESCR 1 to 4).
- TEXT VISUALIZATION. Define visualisation of text.
- NODE VISUALIZATION. Define visualisation of nodes.
- TYPE OF STANDARD. Select between ISO or ANSI standards.
- LINE SPACING. Define spacing between lines of text.
- RETURN CHARACTER. Define character used to end line.
- END TYPE. Define end symbol.
- JUSTIFICATION. Define justification.

- CHARACTER PARAMETERS. Define parameters for character standards (DESCR 1 to 4). Available parameters are listed below.
  - Height of text
  - Width of text
  - Slant angle of text
  - Size ratio of text
  - Spacing of text characters
  - Path of text (right to left or left to right)

Upon changing any of the above values, click on the YES button to end selection.

TEXT in SPACE Mode

Chapter 16

The TEXT function in SPACE mode is used to create and modify text associated with SPACE elements.

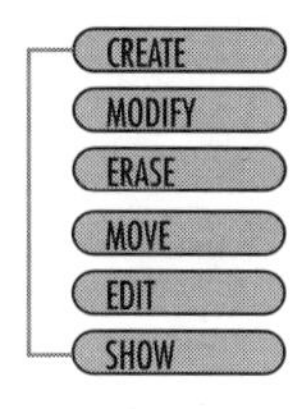

Option menu for TEXT in SPACE mode.

CREATE

Create text associated with SPACE elements.

Select: TEXT > CREATE

Prompt: MSELW ELEM // SEL ELEM

1. Select a SPACE geometric element. The TEXT DEFINITION window displays containing parameters of the text to be created.

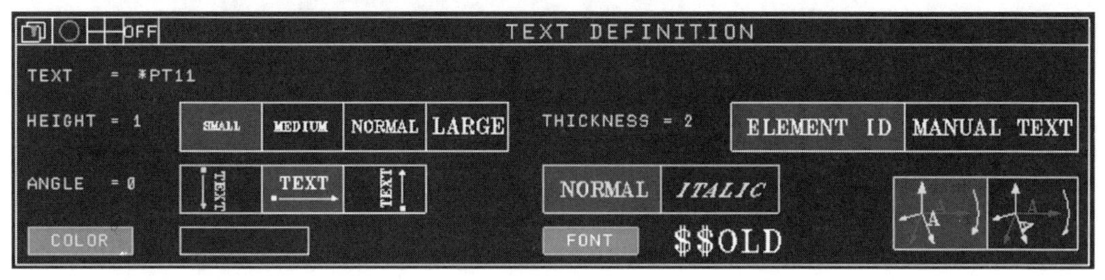

TEXT DEFINITION window.

Change the following parameters as necessary.

- TEXT. Key in text upon selection of MANUAL TEXT.
- HEIGHT. Text height. Also allows the selection of predefined text values.
- ANGLE. Angle of text insertion. Also allows selection of predefined insertion values.
- COLOR. Color of text.
- THICKNESS. Thickness of text. Applicable only when model is plotted.

- NORMAL/ITALIC. Upright or angled text.
- FONT. Font style.
- FIXED. Whether text is fixed with respect to display plane.

2. Position the created text by taking one of the following alternatives: (a) Select a point element. (b) Press <Enter>. The text will be positioned at the center of the previously selected element. (c) Indicate a position.
3. Click on the YES button to validate text position.

MODIFY

Modify parameters of SPACE text.

Select: TEXT > MODIFY

Prompt: MSELW ELEM // SEL ELEM

Select SPACE text or element with which the text is associated. Upon selection, the TEXT MODIFICATION dialog displays.

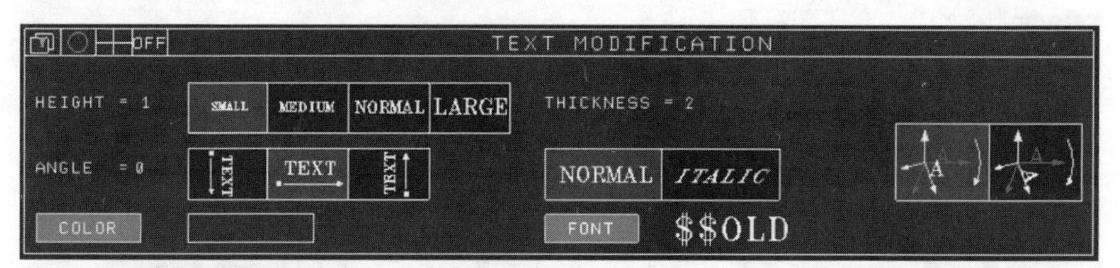

TEXT MODIFICATION dialog.

Change parameters as necessary.

- HEIGHT. Text height. Also allows the selection of predefined text values.
- ANGLE. Angle of text insertion. Also allows the selection of predefined insertion values.
- COLOR. Color of text.
- THICKNESS. Thickness of text. Applicable only when the model is plotted.
- NORMAL/ITALIC. Upright or angled text.
- FONT. Font style.
- FIXED. Whether to fix text with respect to display plane.

ERASE

Delete SPACE text.

Select: TEXT > ERASE

Prompt: MSELW ELEM // SEL ELEM

Select SPACE text or element to which the text is associated.

MOVE

Move SPACE text.

Select: TEXT > MOVE

Prompt: SEL TEXT

1. Select SPACE text.
2. If the selected text was previously positioned by indicating or selecting a position, press <Enter> to reposition the text on the center of the element.
3. Select or indicate a new position for the text.
4. Click on the YES button to end the selection.

EDIT

Edit SPACE text.

Select: TEXT > EDIT

Prompt: MSELW ELEM // SEL ELEM

Select a SPACE text or element to which the text is associated. Upon selection, the TEXT EDIT dialog displays.

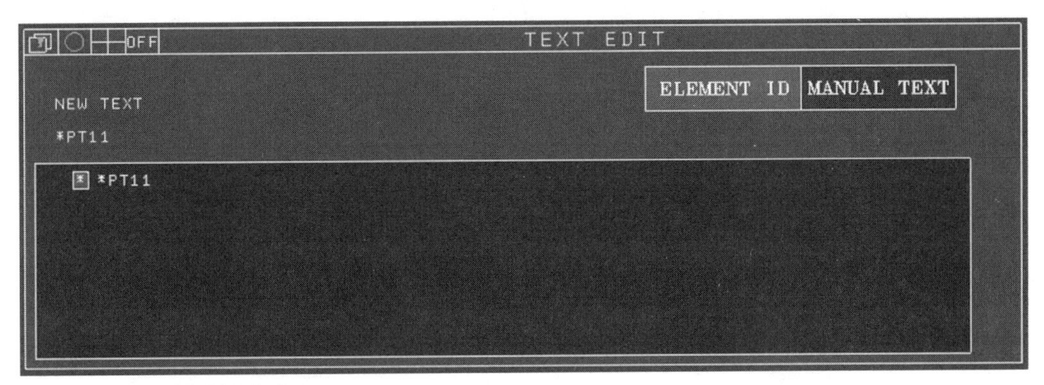

TEXT MODIFICATION dialog.

The SPACE text can then be edited from the displayed window. Text linked to the selected text or element can also be edited.

SHOW

Modify visualization (show/no show) of SPACE text.

Select: TEXT > SHOW

Prompt: MSELW ELEM // SEL ELEM

Select a SPACE text or element associated with the text. Upon selection, the TEXT VISUALISATION dialog displays.

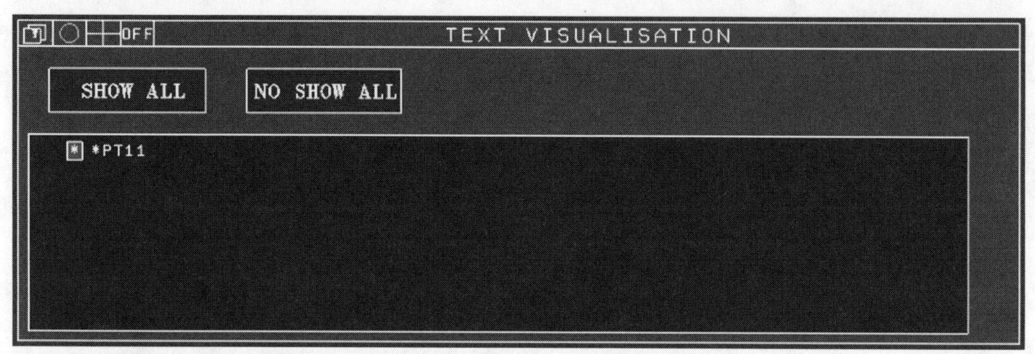

TEXT VISUALISATION dialog.

Visualization of the selected SPACE text and text associated with the SPACE element can be placed in SHOW or NO SHOW mode. Click on the YES button to end selection.

TEXTD2 in DRAW Mode

Chapter 7

The TEXTD2 function is used to create and modify text, geometric tolerances, datum features, datum targets, balloons, and roughness (machining) symbols in DRAW mode.

Option menu for TEXTD2 in DRAW mode.

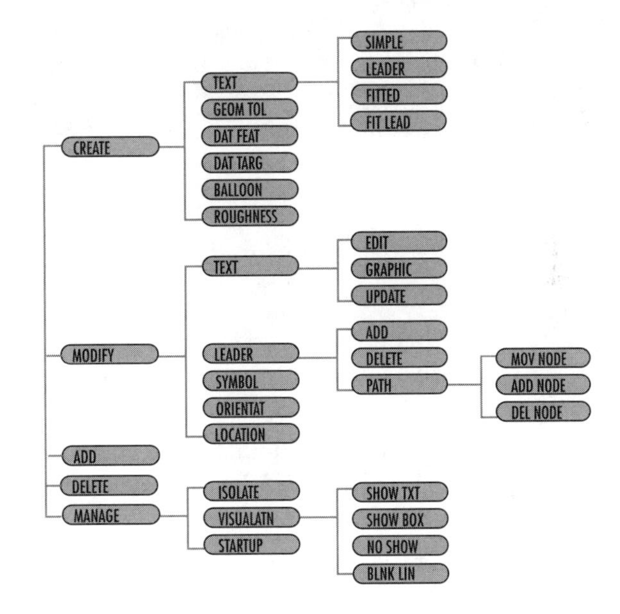

CREATE

Create text, geometric tolerances, datum features, datum targets, balloons, and roughness symbols.

Select: TEXTD2 > CREATE | TEXT or GEOM TOL or DAT FEAT or DAT TARG or BALLOON or ROUGHNESS

When the CREATE option is selected, the Management window appears. To expand the window, select the MORE button at the top right of the window. To return the window to condensed size, select the LESS button at the top right. This window is used to define the presentation of dimensions. Buttons (icons) are described below.

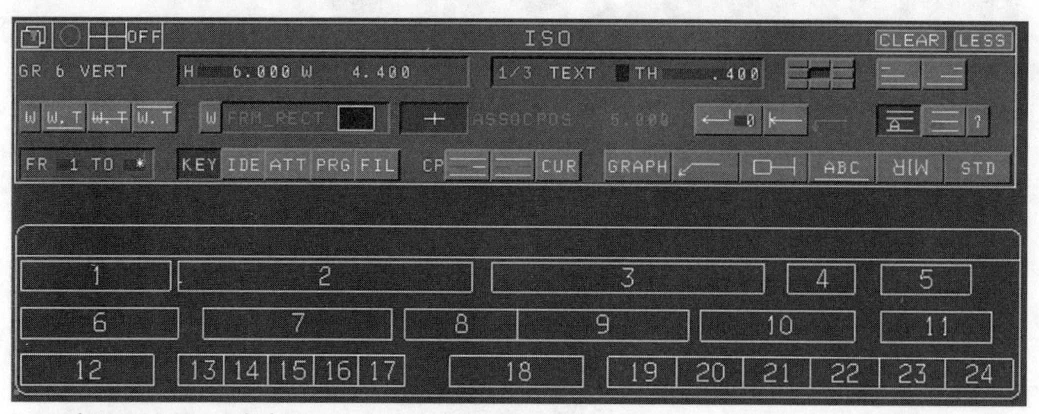

Typical TEXTD2 Expanded Management window.

✓ **NOTE:** *Your TEXTD2 Expanded Management window may differ from the illustration. The window configuration is set in the DRWSTD function, and window modification may be available only to your system administrator.*

- Button 1. Graphism name. Defines font style. Upon selecting the letters in white, a list of available text fonts appears. Selecting a font from the list makes it the current font.
- Button 2. Height and Width. Defines height and width of selected font. The width option may not be selectable, depending on the character graphism defined in the DRWSTD function.
- Button 3. Thickness and Color. Defines thickness and color for the text, score, frame, leader, and symbol. The TH box refers to the thickness of the various elements of text; in this case the text will be 0.4mm thick. If you select the second number, the first number and the word next to the numbers will change to allow you to change the thickness of the various text elements. If you select the word next to the numbers, in this case TEXT, a window appears in which you can define the color of various text elements.
- Button 4. Anchor point. Defines anchor point for text relative to the selected insertion point.
- Button 5. Justification. Defines text justification. The left button activates left justified text, and the right activates right justified text. If neither button is selected, the text will be centered.
- Button 6. Score. Defines position of score. The score can be under, through, or over the text. You also have a choice of whether the score is applied to all text being entered or to a piece of same (subtext). The W button refers to whole text and the S button to subtext. Selecting W will switch to S, and vice versa.

- Button 7. Framing. Defines framing (e.g., box, circle, arrow, or any other frame defined in DRWSTD). You have the option of applying the frame to whole text or subtext. To change the frame in use, select the frame description text (the case shown is FRM_RECT), and a list of available frames displays.
- Button 8. Associative text. Link text to a particular element.
- Button 9. Associative text positioning. Defines distance of text from the element to which text is linked. You can change this dimension only if the Associative text button is selected.
- Button 10. Edit. When using multiple lines of text, defines whether a new line of text starts after, below, or at the beginning of the existing text.
- Button 11. Text grid. Defines creation of subscript and superscript text.
- Button 12. Format. Defines length of text entered.
- Button 13. Key text. Key in desired text or copy from existing text.
- Button 14. Identification. Allows CATIA identification of an element to be used for text purposes.
- Button 15. Attribute. Use attributes of a selected element for text purposes.
- Button 16. Program. Create text from results of external calculation program.
- Button 17. File. Import text from external file.
- Button 18. Copy mode. Copy existing text.
- Button 19. Graphic parameter. Activates another window from which the user can alter the graphic display of text via the following options.
  - DISPLAY FOR EACH ANNOTATION. Defines how annotation displays.
  - SCALE. Defines whether text scale is linked to current view.
  - WRITING RULE. Defines the range in which annotation can be oriented in relation to the axis system.
- Button 20. Leader parameter. Activates window from which the user can set the text-leader relationships via the following options.
  - LEADER TEXT POSITIONING AND DISTANCES. Defines position of text relative to leader.
  - GAP. Defines position of text when text is inserted into leader.
- Button 21. Datum parameter. Selection of this button activates another window from which you can change the datum symbol.
- Button 22. Score parameter. Selection of this button activates another window from which you can set text-score relationships.

- Button 23. Mirroring parameter. Activates a window from which you can create mirrored text.
- Button 24. Standard parameter. Activates another window from which you can alter the text standard.

The following illustration shows one of the additional parameter windows, Button 22 or the Score parameter window. This window is used to define the position of text relative to score lines. The window may not match the next illustration, depending on how the DRWSTD function is set by your system administrator.

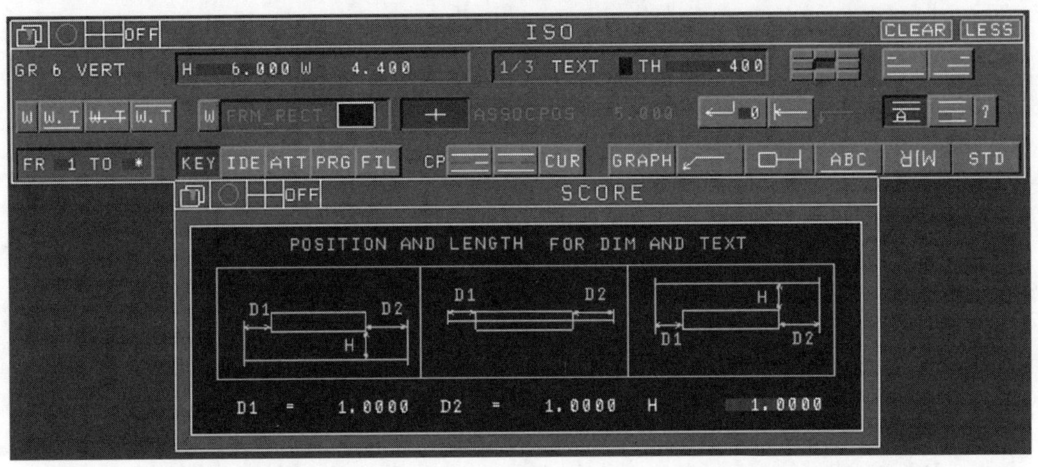

Typical text parameter window.

The small window at the bottom left of the screen when you select the TEXTD2 function is used to define text orientation. Buttons (icons) are described below.

TEXTD2 orientation options window.

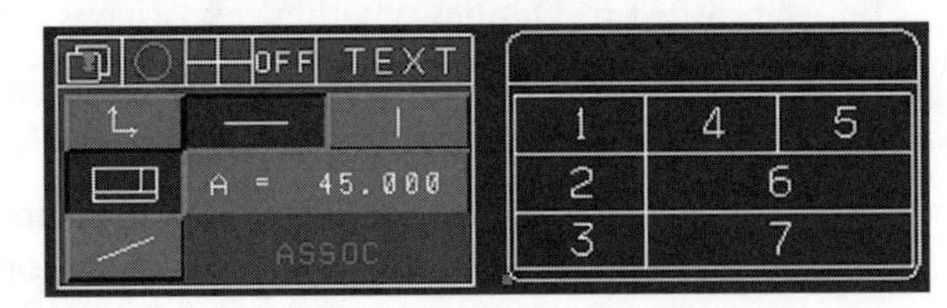

- Button 1. Position text relative to current view.
- Button 2. Position text relative to the screen.
- Button 3. Position text relative to a selected element.
- Button 4. Position text horizontal relative to current view or screen, depending on selection of Button 1 or 2.
- Button 5. Position text vertical relative to current view or screen, depending on selection of Button 1 or 2.

- Button 6. The number entered here defines the angle of created text if button 3 is selected. If zero is entered, the text created will be parallel to the selected element.
- Button 7. Text created is associated with a reference element.

✓ **NOTE:** *As mentioned previously, not all parameter options shown will be available to you. In addition, some of the options within the parameters may not be available. Availability of parameters is determined by the system administrator.*

Before creating annotations, you must select desired settings in the main window, including style of text (graphism), height of text, thickness, and color of text elements, position of the location anchor, scoring, and framing. In addition, you must also select desired settings in the orientation window, such as whether the text is relative to the current view, a selected element, or the screen.

Create simple, leader, fitted, and fitted leader text

Select: TEXTD2 > CREATE | TEXT | SIMPLE or LEADER or FITTED or FIT LEAD

Create simple text

Select: TEXTD2 > CREATE | TEXT | SIMPLE

Prompt: TEXT_POSITION : SEL > ELEM // IND POS

Create single line of simple text

1. Set the windows as desired.
2. Indicate or select a position to define the text anchor point.
3. Key in the desired text to the KEY TEXT window and press <Enter>. Alternatively, select an existing piece of text; this selected text will be copied onto the new position previously selected.
4. If you wish to change the position of the newly created text, indicate or select a new position.
5. Click on the YES button to end text creation. Alternatively, enter additional text in the KEY TEXT window to be positioned on the same line as the first piece of text. The new text may have different window settings from the first piece of text (e.g., different height, color).
6. If you wish to change the position of the newly created text, indicate or select a new position.
7. Click on the YES button to end text creation.

Create multiple lines of simple text

1. Set the windows as desired.
2. Indicate or select a position to define the text anchor point.
3. Key in the desired text to the KEY TEXT window and press <Enter>. Alternatively, select an existing piece of text; this selected text will be copied onto the new position previously selected.
4. If you wish to change the position of the newly created text, indicate or select a new position.
5. In the main window select the return button (Button 10 at right). Enter additional text in the KEY TEXT window and press <Enter>, or select existing text. The new text may have different window settings compared to the first piece of text (e.g., different height, color). This step may be repeated to create as many lines of text as desired.
6. If you wish to change the position of the newly created text, indicate or select a new position.
7. Click on the YES button to end text creation.

Create single line of simple text associated with selected element

1. Set the windows as desired.
2. In the main window, select the associative text button (Button 8 at right). If desired, change the dimension to position the associative text (Button 9).

3. Select the element to which the text will be associated.
4. If desired, select the vector to reverse direction for text position, or indicate a different position for the text.
5. Click on the YES button to accept the displayed position.
6. Key in the desired text to the KEY TEXT window and press <Enter>. Alternatively, select an existing piece of text; this text will be copied onto the new position previously selected.
7. Click on the YES button to end text creation.

✓ **NOTE 1:** *Associated text can also be created with additional text of different attributes. (See single line simple text creation.) Next, associated text can be created with multiple lines of text. (See multiple line simple text creation.)*

✓ **NOTE 2:** *If the element to which the text is associated is moved, the text will also move.*

Create single line of simple text with superscript or subscript

An example of text with a superscript is 4 HOLES Ø20.$^{(0.01)}$

1. Set the windows as desired.
2. Indicate or select a position to define the text anchor point.
3. Key in the desired text (e.g., 4 HOLES Ø20) to the KEY TEXT window and press <Enter>.
4. Select the icon at the right from the Button 11 group in the main window. A grid matching the next illustration displays.

5. Change height of the text in the main window as necessary.
6. Select a grid line on which to place the new text; the cursor position will move to the selected insertion line.
7. Key in the desired text [e.g., *(0.01)*] to the KEY TEXT window and press <Enter>.
8. Click on the YES button to end text creation.

Create leader text

Select: TEXTD2 > CREATE | TEXT | LEADER

Prompt: TEXT_POSITION : SEL > ELEM // IND POS

In the Leader Management window, you can determine how the leader will be created. The following illustration depicts the window, and buttons are described below.

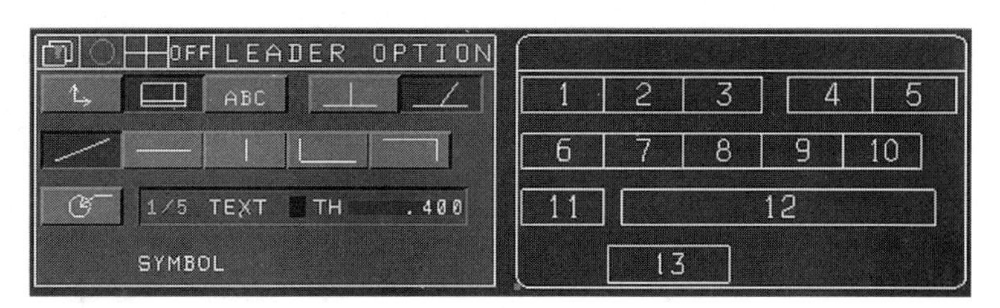

Leader Management window.

- Button 1. If leader options shown in buttons 7 to 10 are chosen with this button active, the leader line horizontal or vertical will reference the current view.

- Button 2. If leader options shown in buttons 7 to 10 are chosen with this button active, the leader line horizontal or vertical will reference the screen.
- Button 3. If leader options shown in buttons 7 to 10 are chosen with this button active, the leader line horizontal or vertical will reference the current text.
- Button 4. Leader lines will always be normal to the element selected for the anchor point of the end point.
- Button 5. Leader lines will always be at a straight angle between the start point and the element selected for the anchor point of the end point.
- Button 6. Leader line will be straight between the two points selected or indicated for the anchor points regardless of how buttons 1 to 3 are set.
- Button 7. Leader lines will always be horizontal with reference to buttons 1 to 3.
- Button 8. Leader lines will always be vertical with reference to buttons 1 to 3.
- Button 9. Leader lines will always be horizontal and then vertical with reference to buttons 1 to 3.
- Button 10. Leader lines will always be vertical and then horizontal with reference to buttons 1 to 3.
- Button 11. When the element selected to position the leader is a cone or cylinder, this button will determine whether the element edge or element axis is used.
- Button 12. These selections are used to change the thickness and color of the leader. The TH box refers to the thickness of the various leader elements. If you select the second number, the first number and the word next to the numbers will change to allow you to modify the thickness of the various text elements. If you select the word next to the numbers, in this case TEXT, a window appears that allows you to define the color of various leader elements.
- Button 13. Define end symbol. The end symbol can be set differently for indicated versus selected positions. Selecting the symbol button will activate an additional management window containing options to define symbols used for indicated and selected positions.

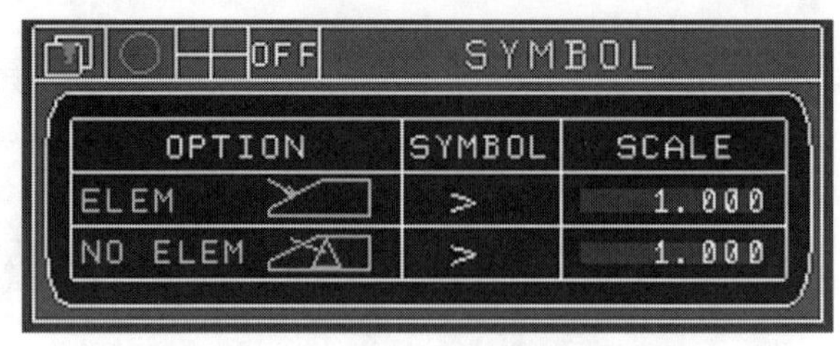

Symbol window.

If either of the symbols shown in the above window is selected, another window is activated in which available symbol options display. These options are defined in the DRW-STD function and are controlled by your system administrator.

From the window at right you can select the end symbol.

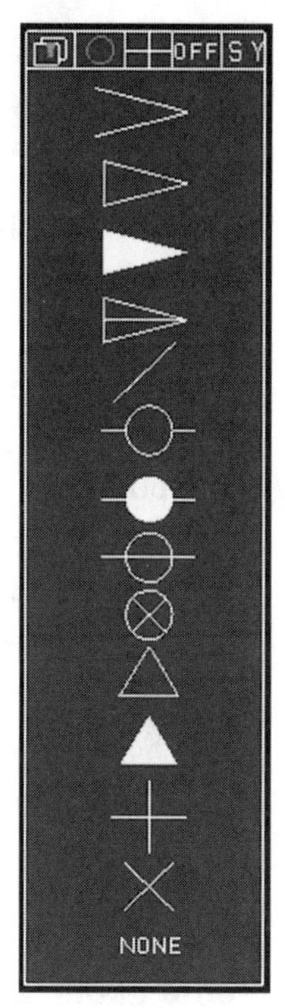

Symbol Selection window.

Create single line of leader text

1. Set the windows as desired.
2. Indicate or select a position to define the text anchor point.
3. Key in the desired text to the KEY TEXT window and press <Enter>. Alternatively, select an existing piece of text; this selected text will be copied onto the new position previously selected.
4. If you wish to change the position of the newly created text, indicate or select a new position.
5. The text created displays with an underscore and three symbols: 1 and 2 are the end points and 3 is the midpoint. Select the symbol from which you wish the leader to be attached.
6. In the LEADER OPTION windows set the options for leader creation.
7. Select or indicate a position for the end of the leader.
8. If the position of the leader end is indicated, additional indications can be made to define the path of the leader. Select an element to position the leader end point, or click on the YES button to end leader creation.

✓ **NOTE:** *Leader text can also be created with additional text of different attributes. (See single line simple text creation.) Leader text can also be created with multiple lines of text. (See multiple line simple text creation.)*

Create fitted text

Select: TEXTD2 > CREATE | TEXT | FITTED

Prompt: FIRST_POINT : SEL > ELEM

1. Set the windows as desired.
2. Select two points; the created text will be fit between these two points.
3. Key in the desired text to the KEY TEXT window and press <Enter>. Alternatively, select an existing piece of text; this selected text will be copied.
4. If you wish to change the position of the newly created text, indicate or select a new position.
5. Click on the YES button to end text creation.

Create fitted leader text

Select: TEXTD2 > CREATE | TEXT | FIT LEAD

Prompt: FIRST_POINT : SEL > ELEM

1. Set the windows as desired.
2. Select two points; the created text will be fit between these two points.
3. Key in the desired text to the KEY TEXT window and press <Enter>. Alternatively, select an existing piece of text; the selected text will be copied.
4. If you wish to change the position of the newly created text, indicate or select a new position.
5. The text created displays with an underscore and three symbols: 1 and 2 are the end points and 3 is the midpoint. Select the symbol from which you wish the leader to be attached.
6. From the LEADER OPTION windows, set the options for the leader creation.
7. Select or indicate a position for the end of the leader.
8. If the position of the leader end was indicated, additional indications can be made to define the path of the leader. Select an element to position the leader end point, or click on the YES button to end the leader creation.

Create geometric tolerances

Select: TEXTD2 > CREATE | GEOM TOL

Prompt: TEXT_POSITION : SEL > ELEM // IND POS

Upon creation of a geometric tolerance, the GEOM TOL window appears. The following illustration describes window options. The first group of buttons (icons) is comprised of available geometric tolerance symbols. The second group of buttons consists of available symbols for use with geometric tolerancing.

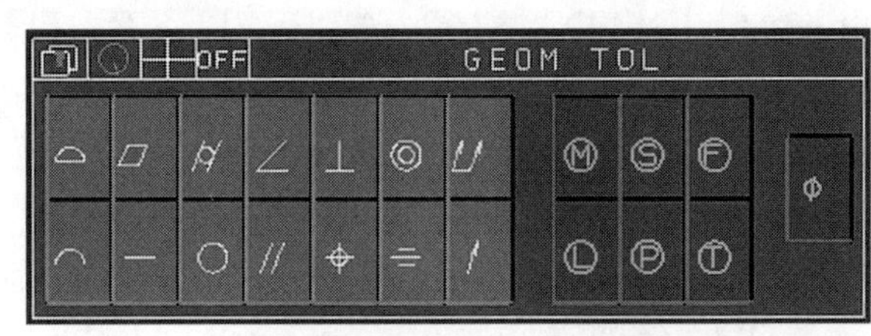

GEOM TOL window.

To create the geometric tolerance shown in the next illustration, take the following steps.

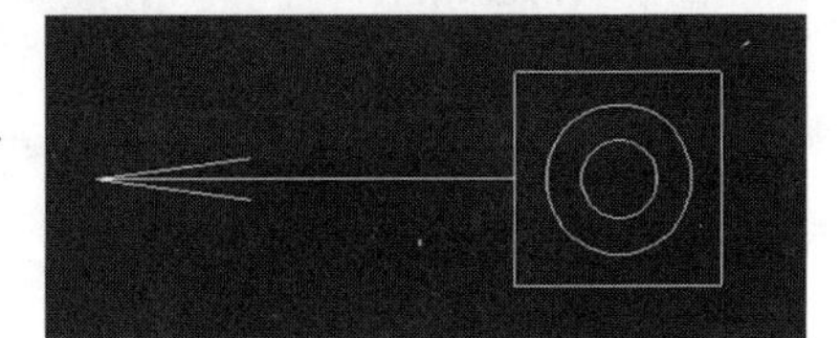

1. Set the windows as desired.
2. Indicate or select a position to define the geometric tolerance anchor point.

3. In the GEOM TOL window select the desired symbol.
4. When the symbol is created, the numbers appearing around the frame are used to define the location point for leaders. Select a symbol number; the Leader window appears as described in the LEADER option section. Set desired leader parameters. If no leader is desired, click on the YES button to end selection.
5. Select or indicate a position for the end of the leader.
6. If the position of the leader end is indicated, additional indications can be made to define the path of the leader. Select an element to position the leader end point, or click on the YES button to end the leader creation.
7. Indicate or select an additional point to reposition the geometric tolerance frame, or click on the YES button to end selection.

To create the geometric tolerance shown in the next illustration, take the following steps.

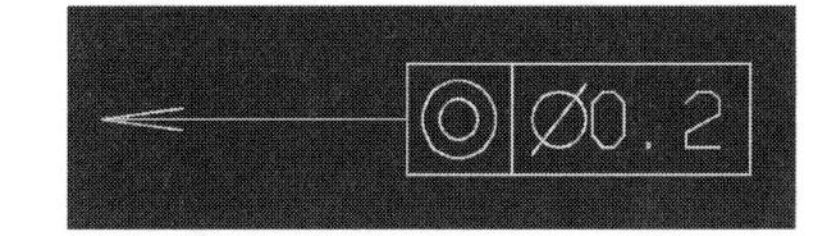

1. Set the windows as desired.
2. Indicate or select a position to define the geometric tolerance anchor point.
3. In the GEOM TOL window select the desired symbol.
4. Key in the desired text to the KEY TEXT window and press <Enter>.
5. Click on the YES button to close the geometric tolerance frame.
6. When the symbol is created, the numbers appearing around the frame are used to define the location point for leaders. Select a symbol number; the Leader window appears as described in the LEADER option section. Set desired leader parameters. If no leader is desired, click on the YES button to end selection.
7. Select or indicate a position for the end of the leader.
8. If the position of the leader end is indicated, additional indications can be made to define the path of the leader. Select an element to position the leader end point, or click on the YES button to end the leader creation.
9. Indicate or select an additional point to reposition the geometric tolerance frame, or click on the YES button to end selection.

To create the geometric tolerance shown in the next illustration, take the following steps.

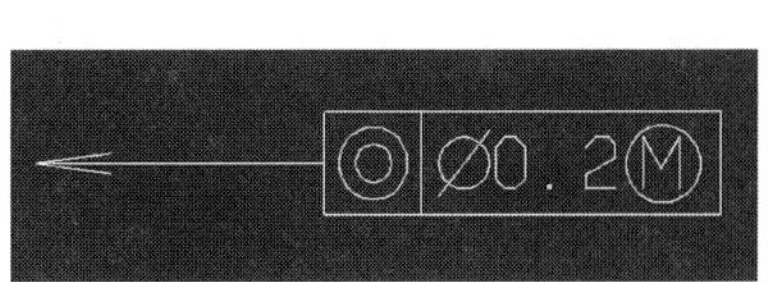

1. Set the windows as desired.
2. Indicate or select a position to define the geometric tolerance anchor point.
3. In the GEOM TOL window select the desired symbol.
4. Key in the desired text to the KEY TEXT window and press <Enter>.
5. Select an additional symbol from the GEOM TOL window.

6. Click on the YES button to close the geometric tolerance frame.
7. When the symbol is created, a series of numbers appear around the frame. These numbers are used to define the location point for leaders. Select a symbol number; the Leader window appears as described in the LEADER option section. Set desired leader parameters. If no leader is desired, click on the YES button to end selection.
8. Select or indicate a position for the end of the leader.
9. If the position of the leader end is indicated, additional indications can be made to define the path of the leader. Select an element to position the leader end point, or click on the YES button to end the leader creation.
10. Indicate or select an additional point to reposition the geometric tolerance frame, or click on the YES button to end selection.

To create the geometric tolerance shown in the next illustration, take the following steps.

1. Set the windows as desired.
2. Indicate or select a position to define the geometric tolerance anchor point.
3. In the GEOM TOL window select the desired symbol.
4. Key in the desired text to the KEY TEXT window and press <Enter>.
5. Click on the YES button to close the geometric tolerance frame.
6. Key in the desired text to the KEY TEXT window and press <Enter>.
7. When the symbol is created, a series of numbers appear around the frame. These numbers are used to define the location point for leaders. Select a symbol number; the Leader window appears as described in the LEADER option section. Set desired leader parameters. If no leader is desired, click on the YES button to end selection.
8. Select or indicate a position for the end of the leader.
9. If the position of the leader end is indicated, additional indications can be made to define the path of the leader. Select an element to position the leader end point, or click on the YES button to end the leader creation.
10. Indicate or select an additional point to reposition the geometric tolerance frame, or click on the YES button to end selection.

Create datum feature

Select: TEXTD2 > CREATE | DAT FEAT

Prompt: TEXT_POSITION : SEL > ELEM // IND POS

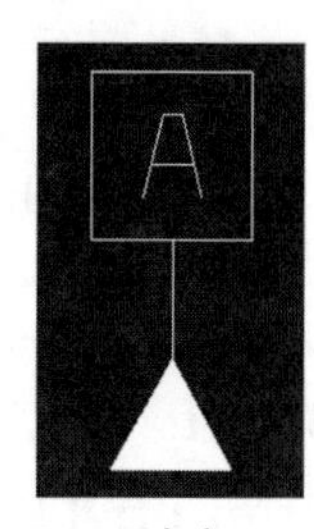

Typical datum feature.

1. Set the windows as desired.
2. Indicate or select a position to define the datum feature anchor point.
3. Key in the desired text for the upper half to the KEY TEXT window and press <Enter>. Alternatively, if no text is desired, click on the YES button.
4. Key in the desired text for the upper half to the KEY TEXT window and press <Enter>. Alternatively, if no text is desired, click on the YES button.
5. When the datum feature is created, a series of numbers appear around the frame. These numbers are used to define the location point for leaders. Select a symbol number; the Leader window appears as described in the LEADER option section. Set desired leader parameters.
6. Select or indicate a position for the end of the leader
7. If the position of the leader end was indicated, additional indications can be made to define the path of the leader. Select an element to position the leader end point, or click on the YES button to end the leader creation.

Create datum target

Select: TEXTD2 > CREATE | DAT TARG

Prompt: TEXT_POSITION : SEL > ELEM // IND POS

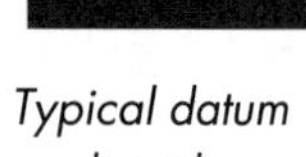

Typical datum target.

1. Set the windows as desired.
2. Indicate or select a position to define the datum target anchor point.
3. Key in the desired text for the lower half to the KEY TEXT window and press <Enter>.
4. Key in the desired text for the upper half to the KEY TEXT window and press <Enter>.
5. When the datum target is created, a series of numbers appear around the frame. These numbers are used to define the location point for leaders. Select a symbol number; the Leader window appears as described in the LEADER option section. Set desired leader parameters.
6. Select or indicate a position for the end of the leader
7. If the position of the leader end was indicated, further indications can be made to define the path of the leader. Use one of the following alternatives: (a) Select an element to position the leader end point or click on the YES button to end the leader creation. (b) Click on the YES button after keying in the lower text to create the

datum target with no text in the upper half and no leader. (c) Click on the YES button after keying in the upper text to create the datum target with text in both halves but no leader.

Create balloons

Select: TEXTD2 > CREATE | BALLOON

Prompt: TEXT_POSITION : SEL > ELEM // IND POS

Create single balloon

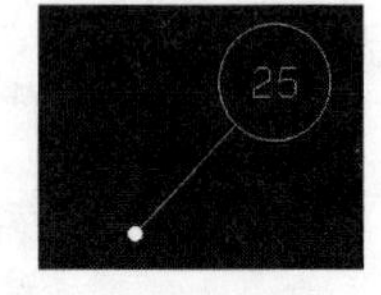

Typical balloon feature.

1. Set the windows as desired.
2. Indicate or select a position to define the balloon anchor point.
3. Key in the desired text to the KEY TEXT window and press <Enter>.
4. When the datum target is created, a series of numbers appear around the frame. These numbers are used to define the location point for leaders. Select a symbol number; the Leader window appears as described in the LEADER option section. Set desired leader parameters. Alternatively, click on the YES button to create the balloon without a leader.
5. Select or indicate a position for the end of the leader.
6. If the position of the leader end is indicated, additional indications can be made to define the path of the leader.
7. Select an element to position the leader end point, or click on the YES button to end the leader creation.

Create multiple balloons

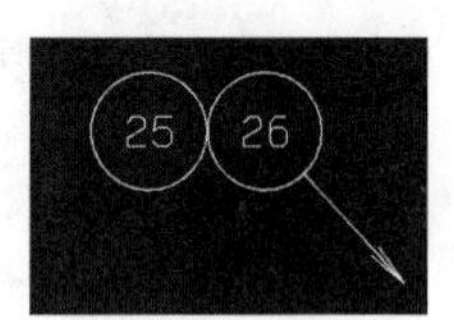

Typical balloon feature.

1. Set the windows as desired.
2. Indicate or select a position to define the balloon anchor point.
3. Key in the desired text to the KEY TEXT window and press <Enter>.
4. Key in a value for the angle at which the second balloon will be relative to the first balloon. Alternatively, select a reference line; the balloons will lie on a line parallel to the selected line.
5. If desired, select the vector to reverse the direction.
6. Key in the desired text for the second balloon in the KEY TEXT window and press <Enter>.

7. Continue keying in text for subsequent balloons as desired.
8. When the balloon(s) is created, a series of numbers appear around the frame. These numbers are used to define the location point for leaders. Select a symbol number; the Leader window appears as described in the LEADER option section. Set desired leader parameters. Alternatively, click on the YES button to create the balloon without a leader.
9. Select or indicate a position for the end of the leader.
10. If the position of the leader end is indicated, additional indications can be made to define the path of the leader.
11. Select an element to position the leader end point, or click on the YES button to end the leader creation.

Create roughness symbols

Select: TEXTD2 > CREATE | ROUGHNESS

Prompt: TEXT_POSITION : SEL > ELEM // IND POS

In the ROUGHNESS SYMBOL window, you can select standard symbols and symbols that allow you to add geometric state criteria.

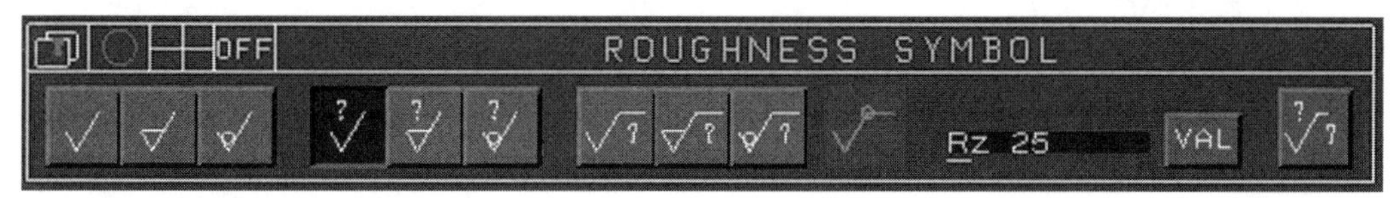

ROUGHNESS SYMBOL window.

1. Indicate a position to define the roughness symbol anchor point. The symbol line created displays with three symbols: 1 and 2 are the end points and 3 is the mid-point. Select the symbol from which you wish the leader to be attached.
2. Select or indicate a position for the leader end point.

✓ **NOTE:** *Upon selecting a symbol from the second or third group of symbols, you can access the VAL button. After clicking on the button, you can select text to be inserted in the symbol from the VALUES window.*

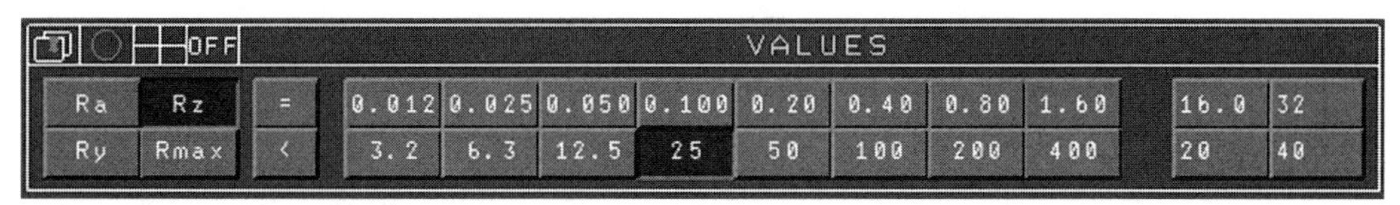

Roughness symbol VALUES window.

It is also possible to define an advanced roughness symbol by selecting the advanced symbol button in the ROUGHNESS SYMBOL window (see left). The button activates the DEFINE SYMBOL dialog from which you can define symbol appearance and text.

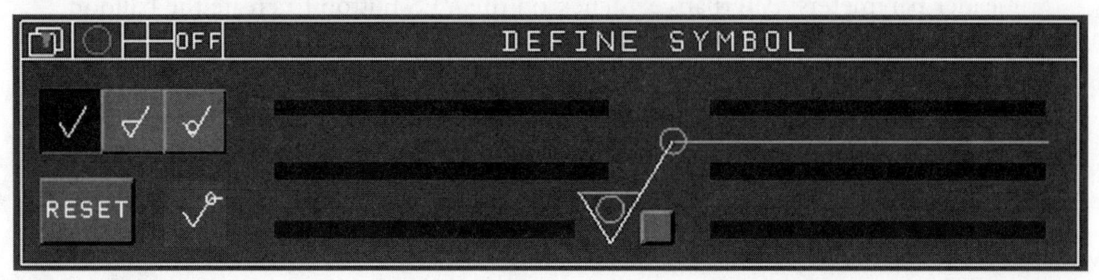

DEFINE SYMBOL window.

MODIFY

Modify text.

Select: TEXTD2 > MODIFY | TEXT or LEADER or SYMBOL or ORIENTAT or LOCATION

Edit text, alter graphic attributes, and update text

Select: TEXTD2 > MODIFY | TEXT | EDIT or GRAPHIC or UPDATE

Edit text

Select: TEXTD2 > MODIFY | TEXT | EDIT

Prompt: TXT_REF : SEL TXT_TXT

1. Select the text to be edited.
2. Key in the desired text to the KEY TEXT window and press <Enter>, or select a piece of existing text to replace the first selected text.

Alter graphic attributes of text

Select: TEXTD2 > MODIFY | TEXT | GRAPHIC

Prompt: MSELW TXTN / DIMN / SYSD // SEL TXT_TXT

1. Select the text to be modified.
2. In the window, modify the values of parameters to be changed. Alternatively, click on the YES button to copy the parameters of another piece of text. Select the text parameters you wish to copy.

Update text affected by change to program, attributes, or identifier

Select: TEXTD2 > MODIFY | TEXT | UPDATE

Prompt: MSELW TXTN / TXTD / DIMN / SYSD

1. Select the text(s) individually or via use of a multiselect option (e.g., **TXT*).
2. Click on the YES button to confirm the update.

Add, delete, or modify leader path

Select: TEXTD2 > MODIFY | LEADER | ADD or DELETE or PATH

Add leader

Select: TEXTD2 > MODIFY | LEADER | ADD

Prompt: SEL TXTN /TXTD

1. Select the leader text to which a leader is to be added. The text selected displays with an underscore and three symbols: 1 and 2 are the end points and 3 is the midpoint.
2. Select the symbol from which you wish the leader to be attached.
3. In the Leader Option windows set the options for the leader creation.
4. Select or indicate a position for the end of the leader.
5. If the position of the leader end is indicated, additional indications can be made to define the path of the leader. Select an element to position the leader end point, or click on the YES button to end the leader creation.

Delete leader

Select: TEXTD2 > MODIFY | LEADER | DELETE

Prompt: SEL TXT_SYM / TXT_LDR ALL_LEADER : SEL TXTN / TXTD

Select the leader to be deleted. Alternatively, select the leader text containing more than one leader to be deleted, and click on the YES button to delete all leaders in the selected text.

Move, add, or delete nodes on leader

Select: TEXTD2 > MODIFY | LEADER | PATH | MOV NODE or ADD NODE or DEL NODE

Move existing node on leader

Select: TEXTD2 > MODIFY | LEADER | PATH | MOV NODE

Prompt: SEL TXT_LDR

1. Select the leader to be modified; the node to be modified will be highlighted.
2. Indicate or select a position for the highlighted node. The leader will be modified to suit the new node position selected or indicated.
3. Click on the YES button to end selection.

Add node to leader

Select: TEXTD2 > MODIFY | LEADER | PATH | ADD NODE

Prompt: SEL TXT_LDR

1. Select the leader to be modified; the node to be modified will be highlighted.
2. Indicate or select a position for the new node. The leader will be modified to suit the new node position selected or indicated.
3. Click on the YES button to end selection.

Delete existing node on leader

Select: TEXTD2 > MODIFY | LEADER | PATH | DEL NODE

Prompt: SEL TXT_LDR

1. Select the leader to be modified; the node to be modified will be highlighted.
2. Click on the YES button to confirm deletion of highlighted node.

Change end symbol

Select: TEXTD2 > MODIFY | SYMBOL

Prompt: SYMB_TXTN_DIMN_SYSD : SEL ELEM

1. Select the symbol to be changed.
2. In the Symbol window, select the desired symbol.

Change text orientation

Select: TEXTD2 > MODIFY | ORIENTAT

Prompt: CHG_ORIENTATION : MSELW TXTN

1. Select the text for which orientation is to be modified.
2. Set the new orientation in the Orientation window.
3. Click on the YES button to apply the new orientation settings.

Locate and anchor text

Select: TEXTD2 > MODIFY | LOCATION

Prompt: CHG_LOCATION : SEL TXTN

Select the text to be relocated. At this point, choose one of the following alternatives.

- In the TEXT POSITION window, you can change the anchor point of the text or its justification. Altering the settings in this window will automatically change the selected text.

TEXT POSITION window.

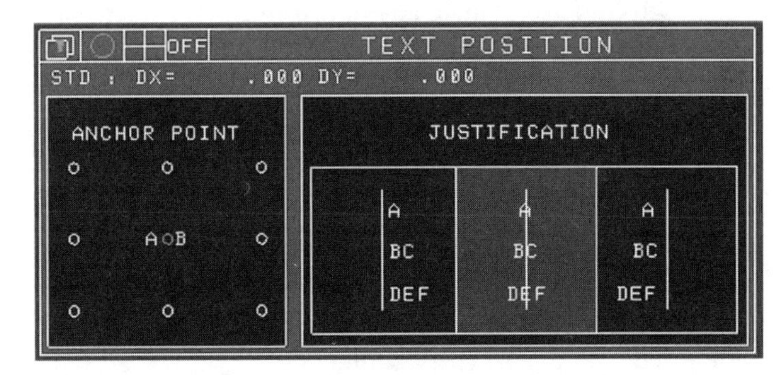

- Key in values to translate the selected text in X,Y format.
- Click on the YES button to accept the default translation values as displayed in the TEXT POSITION window.
- Indicate or select an element to define a new position for the selected text.

ADD

Add new text to existing text.

Select: TEXTD2 > ADD

Prompt: TXT_REF : SEL TXT_TXT // SEL DIM_VAL

1. Select the text to which new text is to be added.
2. From the main window set the parameters for the text to be added (e.g., different height, color). If the new text is to be created on a different line, select the return icon (Button 10, see right).

3. Enter additional text in the KEY TEXT window, and press <Enter>, or select existing text. This step may be repeated to create as many lines of text as desired.
4. Click on the YES button to end the selection.

DELETE

Delete text.

Select: TEXTD2 > DELETE

Prompt: SEL TXT_TXT / TXTN / DIM

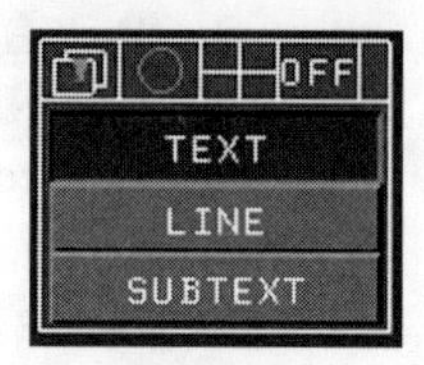

Text Delete window.

Options in the above window are described below.

- TEXT. Delete text created as multiple line text.
- LINE. Delete single line of text from text created as multiple line text.
- SUBTEXT. Delete text within a line of text where the text to be deleted was created by a separate keyed-in entry from the other text in the line.

MANAGE

Manage text. Text here includes text created with the TEXTD2 function, or text and values created with the DIMENS2 function.

Select: TEXTD2 > MANAGE | ISOLATE or VISUALTN or STARTUP

Isolate text(s)

Select: TEXTD2 > MANAGE | ISOLATE

Prompt: MSELW TXTN / DIMN / SYSD / TXTD /DIM

Select the text(s) individually or via use of a multiselect option (e.g., **TXTD*).

Click on the YES button to confirm isolation of selected text(s).

✘ ***WARNING:*** *Once text has been isolated it cannot be reattached to its creation elements.*

Modify text visualization

Text can be shown as boxes or not shown.

Select: TEXTD2 > MANAGE | VISUALTN | SHOW TXT or SHOW BOX or NO SHOW or BLNK LIN

Show text previously modified with VISUALTN options

Select: TEXTD2 > MANAGE | VISUALTN | SHOW TXT

Prompt: MSELW TXTN / DIMN / SYSD / TXTD / DIM

Select text(s) previously modified using TEXTD2 > MODIFY | VISUALTN | SHOW BOX or NO SHOW individually or via use of a multiselect option (e.g., **TXTD*). The text will be restored to normal display mode.

Show text as box

Select: TEXTD2 > MANAGE | VISUALTN | SHOW BOX

Prompt: MSELW TXTN / DIMN / SYSD / TXTD / DIM

Select text(s) to be shown as boxes individually or via use of a multiselect option (e.g., **TXTD*).

Hide text

Select: TEXTD2 > MANAGE | VISUALTN | NO SHOW

Prompt: MSELW TXTN / DIMN / SYSD / TXTD / DIM

Select text(s) to be hidden individually or via use of a multiselect option (e.g., **TXTD*).

Identify blank lines or subtexts

Select: TEXTD2 > MANAGE | VISUALTN | BLNK LIN

Prompt: MSELW TXTN / DIMN / SYSD / TXTD / DIM

Select text(s) to be identified individually or via use of a multiselect option (e.g., **TXTD*).

Reset original status of modal parameters

Select: TEXTD2 > MANAGE | STARTUP

Select this function to reset original modal parameters if certain parameters under DIMENS2 and TEXTD2 were modified during model creation.

A STARTUP LIST window displays containing available startup customization models. Select a model from the list; the model becomes the current customization.

TRANSFOR in DRAW Mode

Chapter 5

The TRANSFOR function in DRAW mode is used to create, apply, and store transformations.

Option menu for TRANSFOR in DRAW mode.

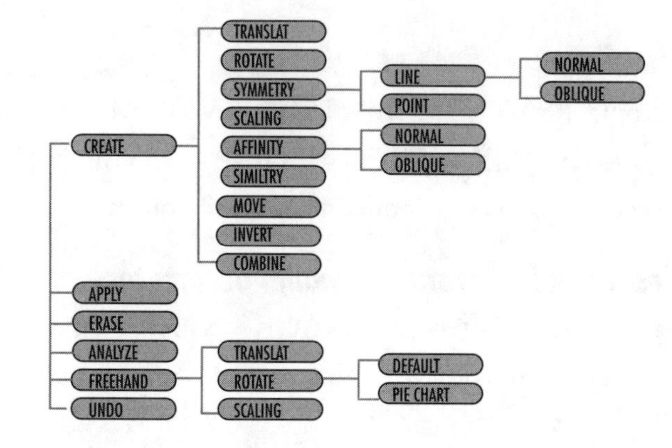

CREATE

Create transformations.

Select: TRANSFOR > CREATE | TRANSLAT or ROTATE or SYMMETRY or SCALING or AFFINITY or SIMILTRY or MOVE or UNSPEC

Create translation

Select: TRANSFOR > CREATE | TRANSLAT

Prompt: START POINT: SEL PT1   SEL CIRD1 / / SEL LND / / KEY DX,DY

1. Define the translation as follows: (a) Select a point or circle (center point will be used) as the start point. (b) Select a point or circle (center will be used) or two lines (intersection point will be used) to define the second point. (c) Select a line.
2. Enter a translation length.
3. Select the arrow to invert as necessary.
4. Enter the first and second points by entering coordinates in the following form: *DX,DY*. The TRANSLATION CREATED message displays.

5. Enter a name for the translation if you wish to store it. The TRANSFORMATION STORED message displays. If you do not wish to store the translation, proceed to the APPLY option.

Create rotation

Select: TRANSFOR > CREATE | ROTATE

Prompt: ROTN CENTER: SEL PTD / / SEL LND1

1. Define the rotation center by selecting a point or two lines (intersection will be used).
2. Define the rotation angle via one of the following alternatives: (a) Select or indicate two points. (b) Select two lines. (c) Enter an angle value. The ROTATION CREATED message displays.
3. Enter a name for the translation if you wish to store it. The TRANSFORMATION STORED message displays. If you do not wish to store the translation, proceed to the APPLY option.

Create symmetry about line

Select: TRANSFOR > CREATE | SYMMETRY | LINE | NORMAL or OBLIQUE

Create normal symmetry about line

Select: TRANSFOR > CREATE | SYMMETRY | LINE | NORMAL

Prompt: SYMMETRY AXIS: SEL LND / / SEL PTD1

1. Define the line of symmetry via one of the following alternatives: (a) Select or indicate two points. (b) Select a line. (c) Select a point and then a line segment (the end point closest to the selection point, meaning that vicinity selection will be used). The TRANSFORMATION CREATED message displays.
2. Enter a name for the translation if you wish to store it. The TRANSFORMATION STORED message displays. If you do not wish to store the translation, proceed to the APPLY option.

Create oblique symmetry about line

Select: TRANSFOR > CREATE | SYMMETRY | LINE | OBLIQUE

Prompt: SYMMETRY AXIS: SELECT LND

1. Define the line of symmetry by selecting a line.
2. Define the direction of oblique symmetry by selecting a second line. The TRANSFORMATION CREATED message displays.

3. Enter a name for the translation if you wish to store it. The TRANSFORMATION STORED message displays. If you do not wish to store the translation, proceed to the APPLY option.

Create symmetry about point

Select: TRANSFOR > CREATE | SYMMETRY | POINT

Prompt: SYMMETRY CENTER: SEL PTD SEL LND / / YES: STORE

1. Select the center of the symmetry point. The TRANSFORMATION CREATED message displays.
2. Enter a name for the translation if you wish to store it. The TRANSFORMATION STORED message displays. If you do not wish to store the translation, proceed to the APPLY option.

Create scaling

Select: TRNSOR > CREATE | SCALING

Prompt: SCALING CENTER: SEL PTD SEL LND1 / / YES: STORE

1. To choose the scaling center point (the point from which the scaling will be calculated), select or indicate a point, or select two lines.
2. Define the scaling ratio via one of the following alternatives: (a) Select two points. The ratio of their distance from the scaling center is the scaling ratio. (b) Select two lines. The ratio of their lengths is the scaling ratio. (c) Select two circles. The ratio of the radii is the scaling ratio. (d) Enter a scaling ratio.
3. The TRANSFORMATION CREATED message displays.
4. Enter a name for the translation if you wish to store it. The TRANSFORMATION STORED message displays. If you do not wish to store the translation, proceed to the APPLY option.

Create affinity (scaling in one direction)

Select: TRANSFOR > CREATE | AFFINITY | NORMAL or OBLIQUE

Create normal affinity

Select: TRANSFOR > CREATE | AFFINITY | NORMAL

Prompt: AFFINITY LINE: SEL LND

1. Select the affinity line. The affinity will be normal to this line.
2. Define the affinity ratio via one of the following alternatives: (a) Select two points. The ratio of their distance from the scaling center is the scaling ratio. (b) Select two

lines. The ratio of their lengths is the scaling ratio. (c) Select two circles. The ratio of the radii is the scaling ratio. (d) Enter a scaling ratio.

3. The TRANSFORMATION CREATED message displays.
4. Enter a name for the translation if you wish to store it. The TRANSFORMATION STORED message displays. If you do not wish to store the translation, proceed to the APPLY option.

Create oblique affinity

Select: TRANSFOR > CREATE | AFFINITY | OBLIQUE

Prompt: AFFINITY LINE: SELECT LND

1. Select the affinity line. The affinity will be normal to this line.
2. Select the affinity direction by selecting a line.
3. Define the affinity ratio via one of the following alternatives: (a) Select two points. The ratio of their distance from the scaling center is the scaling ratio. (b) Select two lines. The ratio of their lengths is the scaling ratio. (c) Select two circles. The ratio of the radii is the scaling ratio. (d) Enter a scaling ratio. The TRANSFORMATION CREATED message displays.
4. Enter a name for the translation if you wish to store it. The TRANSFORMATION STORED message displays. If you do not wish to store the translation, proceed to the APPLY option.

Create similtry

A similtry is defined as the combination of a rotation, translation, and scaling.

Select: TRANSFOR > CREATE | SIMILTRY

Prompt: SEL PTD1 / / SEL LND1

1. Define symmetry by successively selecting four points or two lines. The TRANSFORMATION CREATED message displays.
2. Enter a name for the translation if you wish to store it. The TRANSFORMATION STORED message displays. If you do not wish to store the translation, proceed to the APPLY option.

Create move

A move is defined as the combination of a rotation and translation.

Select: TRANSFOR > CREATE | MOVE

Prompt: SEL AXSD SEL PTD1 / / SEL LND1 / / YES: STORE

1. Define the move by using one of the following alternatives: (a) select two DRAW axes; (b) select four points; or (c) select four lines. The TRANSFORMATION CREATED message displays.
2. Enter a name for the translation if you wish to store it. The TRANSFORMATION STORED message displays. If you do not wish to store the translation, proceed to the APPLY option.

Invert stored transformation

Select: TRANSFOR > CREATE | INVERT

Prompt: KEY (TRA NAME)

1. Upon selection of the INVERT option, the TRANSFORMATIONS DRAW window displays. Select or enter the transformation to be modified.
2. Click on YES to store. Enter a new name. The TRANSFORMATION STORED message displays.

Combine stored transformations

Select: TRANSFOR > CREATE | COMBINE

Prompt: KEY (TRA NAME)

1. Upon selection of the COMBINE option, the TRANSFORMATIONS DRAW window displays. Select or enter the transformation to be modified.
2. Click on YES to store. Enter a new name. The TRANSFORMATION STORED message displays.

APPLY

Apply transformations.

Select: TRANSFOR > APPLY

Prompt: MSELC ELEM ENTER: LIST / / YES: ONCE AGAIN

Upon selection of the APPLY option, the APPLY window displays.

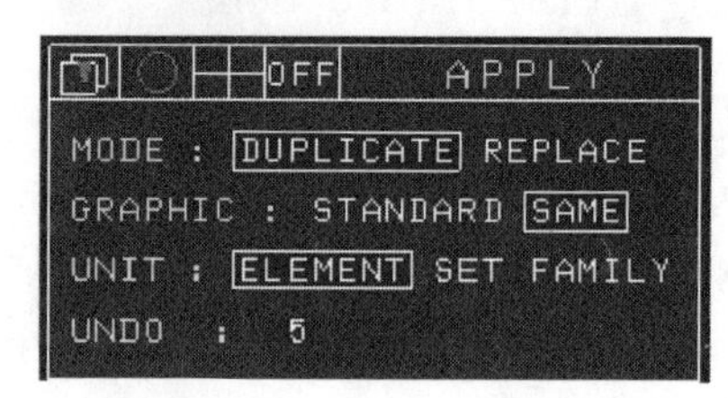

APPLY window.

The window contains the following parameters.

- Duplicate or Replace

- STANDARD or SAME (available under the DUPLICATE option). Determines whether duplicated elements will have the same display characteristics as selected elements, or standard characteristics as defined in the STANDARD function.
- ELEMENT SET or FAMILY. Selection mode.
- NUMBER OF UNDO STEPS.

If desired, press <Enter> to view the list of stored transformations. Alternatively, and if APPLY was selected immediately after CREATE, and depending on parameter settings made in the above window, select elements for transformation individually or by multiselect.

Click on YES to repeat the transformation.

ERASE

Delete stored transformation.

Select: TRANSFOR > ERASE

Prompt: KEY (TRA NAME)

Select (or enter the name of) the transformation to be deleted. Click on YES to confirm deletion.

ANALYZE

Analyze transformation.

Select: TRANSFOR > ANALYZE

Prompt: KEY (TRA NAME)

Upon selection of the ANALYZE option, the ANALYZE window displays.

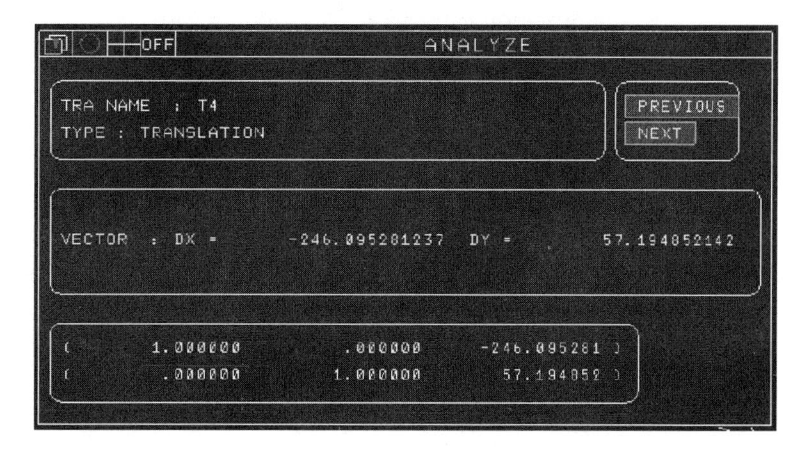

ANALYZE window.

1. Select or enter the name of the transformation to be analyzed.
2. If desired, click on NEXT for analysis of the next transformation in the list. Results display in the alphanumeric window.

FREEHAND

Move elements without defining transformations.

Select: TRANSFOR > FREEHAND | TRANSLAT or ROTATE or SCALING

The APPLY window displays, and is used as described under the APPLY option.

Apply translation without definition

Select: TRANSFOR > FREEHAND | TRANSLAT

Prompt: SEL ELEM

1. Select the desired element for transformation.
2. Select a point. The element will be moved from the point of selection on the element to the selected point.

Apply rotation without definition

Default and pie chart methods of applying the rotation are available.

Select: TRANSFOR > FREEHAND | ROTATE

Prompt: SEL ELEM

Default

1. Select an element (the point of selection is highlighted), and then select a point to serve as the rotation center.
2. Select a point to define the angle and execute the rotation.

Pie chart

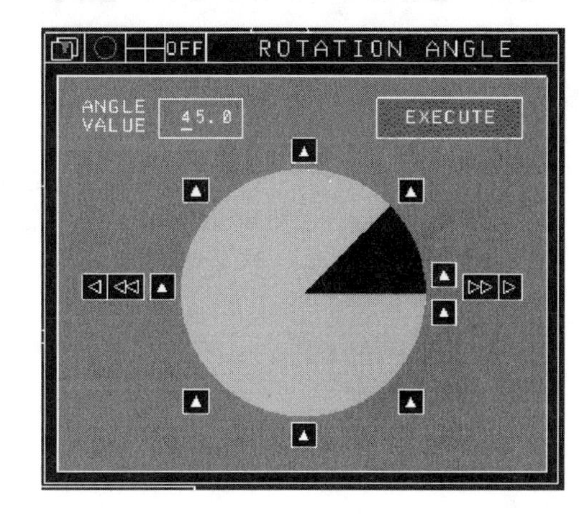

1. Select an element, and then select a point for the rotation center. The ROTATION ANGLE window (PIE CHART) displays.
2. Enter an angle in the angle field in the pie chart. Alternatively, select a single arrowhead to increment by two degrees or a double arrowhead to increment by four degrees.
3. When set as desired, click on YES to execute.

Apply scaling without definition

Select: TRANSFOR > FREEHAND | SCALING

Prompt: SEL ELEM

1. Select an element, and then select the scaling center point (the point from which the scaling is calculated).
2. Select a point to be used as the scaling ratio. Scaling is executed.

UNDO

Undo transformations one by one in reverse order according to settings in APPLY window.

Select: TRANSFOR > UNDO

Prompt: YES: UNDO PREV APPLY

✓ ***NOTE:*** *The first undo operation is performed by clicking on NO before selecting the UNDO option. This is counted as the first undo step.*

Click on YES as many times as desired.

TRANSFOR in 2D SPACE Mode

Chapter 13

The TRANSFOR function in 2D SPACE mode is used to create, apply, and store transformations.

Option menu for TRANSFOR in 2D SPACE mode.

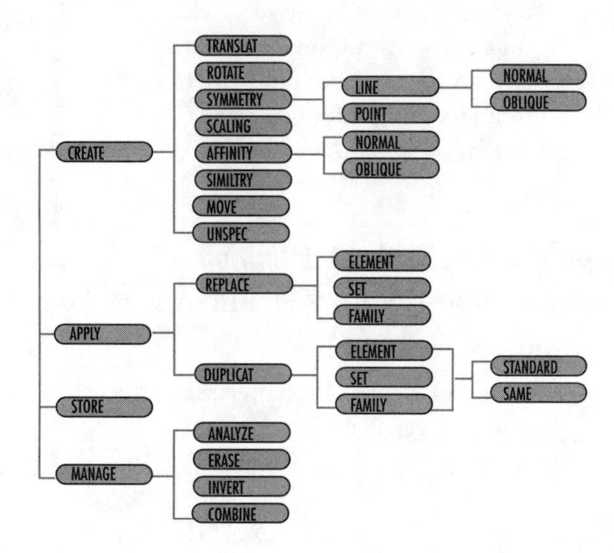

CREATE

Create transformations.

Select: TRANSFOR > CREATE | TRANSLAT or ROTATE or SYMMETRY or SCALING or AFFINITY or SIMILTRY or MOVE or UNSPEC

Create translation

Select: TRANSFOR > CREATE | TRANSLAT

Prompt: KEY TRANSL COMP / / SEL LN/PT/CIR

1. Define the translation as follows: (a) Select a point or circle (center point will be used) as the start point. (b) Select a point or circle (center will be used) or two lines (intersection point will be used) to define the second point. (c) Select a line.
2. Enter a translation length.
3. Select the arrow to invert as necessary.

4. Enter the first and second points by entering coordinates in the following form: *DX,DY*. The TRANSLATION CREATED message displays.
5. Enter a name for the translation if you wish to store it. The TRANSFORMATION STORED message displays. If you do not wish to store the translation, proceed to the APPLY option.

Create rotation

Select: TRANSFOR > CREATE | ROTATE

Prompt: CENTER DEFINITION: SEL PT/LN

1. Define the rotation center by selecting a point or two lines (intersection will be used).
2. Define the rotation angle via one of the following alternatives: (a) Select or indicate two points. (b) Select two lines. (c) Enter an angle value. The ROTATION CREATED message displays.
3. Enter a name for the translation if you wish to store it. The TRANSFORMATION STORED message displays. If you do not wish to store the translation, proceed to the APPLY option.

Create symmetry

Select: TRANSFOR > CREATE | SYMMETRY | LINE or POINT

Create symmetry about line

Select: TRANSFOR > CREATE | SYMMETRY | LINE | NORMAL or OBLIQUE

Create normal symmetry about line

Select: TRANSFOR > CREATE | SYMMETRY | LINE | NORMAL

Prompt: SYMMETRY AXIS DEFINITION: SEL LN

1. Define the line of symmetry via one of the following alternatives: (a) Select or indicate two points. (b) Select a line. (c) Select a point and then a line segment (the end point closest to the selection point, meaning that vicinity selection will be used). The TRANSFORMATION CREATED message displays.
2. Enter a name for the translation if you wish to store it. The TRANSFORMATION STORED message displays. If you do not wish to store the translation, proceed to the APPLY option.

Create oblique symmetry about line

Select: TRANSFOR > CREATE | SYMMETRY | LINE | OBLIQUE

Prompt: SYMMETRY AXIS DEFINITION: SEL LN

1. Define the line of symmetry by selecting a line.
2. Define the direction of oblique symmetry by selecting a second line. The TRANSFORMATION CREATED message displays.
3. Enter a name for the translation if you wish to store it. The TRANSFORMATION STORED message displays. If you do not wish to store the translation, proceed to the APPLY option.

Create symmetry about point

Select: TRANSFOR > CREATE | SYMMETRY | POINT

Prompt: SEL PT/1ST LN

1. Select the center of symmetry point. The TRANSFORMATION CREATED message displays.
2. Enter a name for the translation if you wish to store it. The TRANSFORMATION STORED message displays. If you do not wish to store the translation, proceed to the APPLY option.

Create scaling

Select: TRANSFOR > CREATE | SCALING

Prompt: FIXED POINT DEFINITION: SEL PT/1RST LN

1. To choose the scaling center point (the point from which the scaling will be calculated), select or indicate a point, or select two lines.
2. Define the scaling ratio via one of the following alternatives: (a) Select two points. The ratio of their distance from the scaling center is the scaling ratio. (b) Select two lines. The ratio of their lengths is the scaling ratio. (c) Select two circles. The ratio of the radii is the scaling ratio. (d) Enter a scaling ratio. The TRANSFORMATION CREATED message displays.
3. Enter a name for the translation if you wish to store it. The TRANSFORMATION STORED message displays. If you do not wish to store the translation, proceed to the APPLY option.

Create affinity (scaling in one direction)

Select: TRANSFOR > CREATE | AFFINITY | NORMAL or OBLIQUE

Create normal affinity

Select: TRANSFOR > CREATE | AFFINITY | NORMAL

Prompt: AFFINITY AXIS DEFINITION: SEL LN

1. Select the affinity line. The affinity will be normal to this line.
2. Define the affinity ratio via one of the following alternatives: (a) Select two points. The ratio of their distance from the scaling center is the scaling ratio. (b) Select two lines. The ratio of their lengths is the scaling ratio. (c) Select two circles. The ratio of the radii is the scaling ratio. (d) Enter a scaling ratio. The TRANSFORMATION CREATED message displays.
3. Enter a name for the translation if you wish to store it. The TRANSFORMATION STORED message displays. If you do not wish to store the translation, proceed to the APPLY option.

Create oblique affinity

Select: TRANSFOR > CREATE | AFFINITY | OBLIQUE

Prompt: AFFINITY ASIS DEFINTION: SEL LN

1. Select the affinity line. The affinity will be normal to this line.
2. Select the affinity direction by selecting a line.
3. Define the affinity ratio via one of the following alternatives: (a) Select two points. The ratio of their distance from the scaling center is the scaling ratio. (b) Select two lines. The ratio of their lengths is the scaling ratio. (c) Select two circles. The ratio of the radii is the scaling ratio. (d) Enter a scaling ratio. The TRANSFORMATION CREATED message displays.
4. Enter a name for the translation if you wish to store it. The TRANSFORMATION STORED message displays. If you do not wish to store the translation, proceed to the APPLY option.

Create similtry

A similtry is defined as the combination of a rotation, translation, and scaling.

Select: TRANSFOR > CREATE | SIMILTRY

Prompt: SEL 1SR PT/LN

1. Define the symmetry by successively selecting four points, or two lines. The TRANSFORMATION CREATED message displays.
2. Enter a name for the translation if you wish to store it. The TRANSFORMATION STORED message displays. If you do not wish to store the translation, proceed to the APPLY option.

Create move

A move is defined as the combination of a rotation and translation.

Select: TRANSFOR > CREATE | MOVE

Prompt: SEL 1ST PT/LN/AXS

1. Define the move by successively selecting four points (PT1, PT2, PT3, and PT4), or select four lines (LN1, LN2, LN3, and LN4). The TRANSFORMATION CREATED message displays.
2. Enter a name for the translation if you wish to store it. The TRANSFORMATION STORED message displays. If you do not wish to store the translation, proceed to the APPLY option.

Create transformation by specifying transformation axis system

Select: TRANSFOR > CREATE | UNSPEC

Prompt: KEY TRANSFORMED ORIGIN

1. Enter coordinates of the transformed axis origin.
2. Successively enter components of transformed vectors of the new axis. The TRANSFORMATION CREATED message displays.
3. Enter a name for the translation if you wish to store it. The TRANSFORMATION STORED message displays. If you do not wish to store the translation, proceed to the APPLY option.

APPLY

Apply transformation.

Select: TRANSFOR > APPLY | REPLACE or DUPLICAT

Apply transformation replacement

Select: TRANSFOR > APPLY | REPLACE | ELEMENT or SET or FAMILY

Apply replacement transformation of element

Select: TRANSFOR > APPLY | REPLACE | ELEMENT

Prompt: CUR MULTI-SEL SEL ELEM / / KEY TRA / / ENTER: LIST

Select an element. The element will be transformed.

Apply replacement transformation of set

Select: TRANSFOR > APPLY | REPLACE | SET

Prompt: SEL SET / / KEY TRA / / ENTER: LIST

Select an element in the set to be transformed. Click on YES to confirm.

Apply replacement transformation of family

Select: TRANSFOR > APPLY | REPLACE | FAMILY

Prompt: SEL ELEM / / KEY TRA / / ENTER: LIST

Select an element in the family to be transformed. Click on YES to confirm

Apply transformation duplication

Select: TRANSFOR > APPLY | DUPLICAT | ELEMENT or SET or FAMILY

Apply duplicated transformation of elements

Select: TRANSFOR > APPLY | DUPLICAT | ELEMENT | STANDARD or SAME

Prompt: CUR MULTI-SEL SEL ELEM / / KEY TRA / / ENTER: LIST

Upon using the STANDARD suboption, duplicated elements have display attributes as defined in the STANDARD function. Select the SAME suboption if you wish for duplicated elements to have the same display attributes as the selected elements.

Select elements individually, or use a multiselect option. Click on YES to end, and click on YES again to confirm. The elements are duplicated.

Apply duplicated transformation of set

Select: TRANSFOR > APPLY | DUPLICAT | SET | STANDARD or SAME

Prompt: SEL SET // KET TRA // ENTER: LIST

Use the STANDARD suboption if you wish for duplicated elements to have display attributes as defined in the STANDARD function. Select the SAME suboption if you wish for duplicated elements to have the same display attributes as the selected elements.

Select an element in the set to be duplicated. The set is duplicated.

Apply duplicated transformation of family

Select: TRANSFOR > APPLY | DUPLICAT | FAMILY | STANDARD or SAME

Prompt: SEL ELEM // KET TRA // ENTER: LIST

Use the STANDARD suboption if you wish for duplicated elements to have display attributes as defined in the STANDARD function. Select the SAME suboption if you wish for duplicated elements to have the same display attributes as the selected elements.

Select an element in the family to be duplicated. The family is duplicated.

STORE

Store current transformation. This option is available immediately following creation.

Select: TRANSFOR > STORE

Prompt: KEY TRA IDENT // YES: AUTO IDENT

Enter a name for the current transformation, or click on YES to accept an automatic identifier.

MANAGE

Manage transformations.

Select: TRANSFOR > MANAGE | ANALYZE or ERASE or INVERT or COMBINE

Analyze transformation

Select: TRANSFOR > MANAGE | ANALYZE

Prompt: ENTER: LIST // KEY TRA YES: CUR TRA

Enter the name of the transformation to be analyzed. Alternatively, select the transformation from the list, or click on YES to analyze the current transformation. The alphanumeric window displays the transformation characteristics, and a graphical representation of the transformation definition displays.

Delete transformation

Only stored transformations can be deleted.

Select: TRANSFOR > MANAGE | ERASE

Prompt: SEL TRANSFORMATION

Select transformation to be deleted. Click on YES to confirm.

Create inverted transformation

Only stored transformations can be inverted.

Select: TRANSFOR > MANAGE | INVERT

Prompt: SEL TRANSFORMATION

Select the transformation to be inverted. If desired, enter a new name for the inverted transformation.

Combine transformations

Only stored transformations can be combined.

Select: TRANSFOR > MANAGE | COMBINE

Prompt: SEL TRANSFORMATION

1. Select first transformation.
2. Select second transformation. A combined transformation is created.
3. If desired, enter a name for the combined transformation.

TRANSFOR in 3D SPACE Mode

Chapter 13

The TRANSFOR function in 3D SPACE mode is used to create, apply, and store transformations.

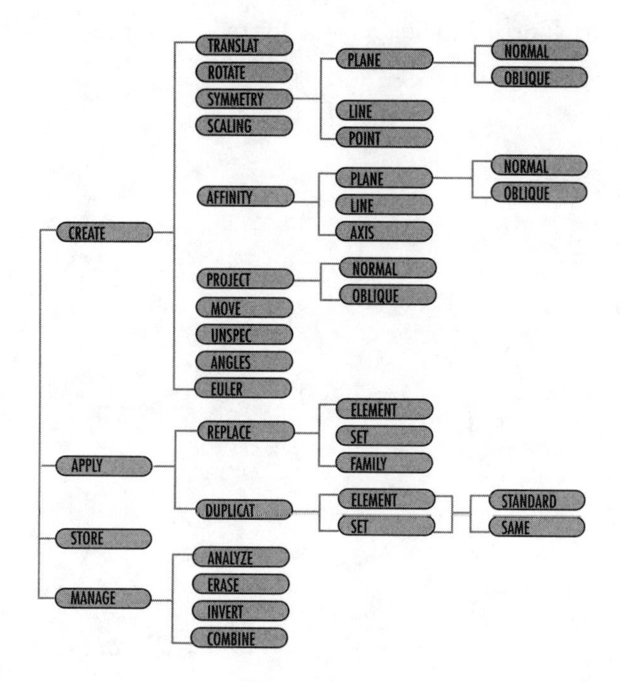

Option menu for TRANSFOR in 3D SPACE mode.

CREATE

Create transformations.

Select: TRANSFOR > CREATE | TRANSLAT or ROTATE or SYMMETRY or SCALING or AFFINITY or PROJECT or MOVE or PROJECT or MOVE or UNSPEC or ANGLES or EULER

Create translation

Select: TRANSFOR > CREATE | TRANSLAT

Prompt: KEY TRANSL COMP / / SEL PT1/LN/PLN

1. Define the translation as follows: (a) Enter components of the transformation vector. (b) Select two points, and then select a line.

2. Enter a translation length.
3. If desired, select arrow to invert.
4. Enter first and second points by inputting coordinates in the *DX,DY* format. The TRANSLATION CREATED message displays.
5. Enter a name for the translation if you wish to store it. The TRANSFORMATION STORED message displays. If you do not wish to store the translation, proceed to the APPLY option.

Create rotation

Select: TRANSFOR > CREATE | ROTATE

Prompt: AXIS DEFINTION SEL LN

1. Define the rotation axis by selecting a line.
2. Define the rotation angle via one of the following alternatives: (a) Select two points (angle between vectors of the points and their projection onto the rotation axis). (b) Select a point and a line (angle between the line and the vector joining the point and its projection onto the rotation axis). (c) Select a point and a plane (angle between a line normal to the plane and a vector joining the point and its projection onto the rotation axis). (d) Select a line and a plane (angle between the line and a line normal to the plane). (e) Select two planes (angle between the lines normal to the planes). Highlighted vectors are displayed describing the defined rotation.
3. If desired, select vectors to invert.
4. The ROTATION CREATED message displays. Enter a name for the translation if you wish to store it. The TRANSFORMATION STORED message displays. If you do not wish to store the translation, proceed to the APPLY option.

Create symmetry

Select: TRANSFOR > CREATE | SYMMETRY | PLANE or LINE or POINT

Create normal symmetry about plane

Select: TRANSFOR > CREATE | SYMMETRY | PLANE | NORMAL

Prompt: SEL PT/LN/PLN

1. Define the plane of symmetry as follows: (a) Successively select three points. (b) Successively selecting two planar lines. (c) Select a point, and then a line. (d) Select a plane. The TRANSFORMATION CREATED message displays.

2. Enter a name for the translation if you wish to store it. The TRANSFORMATION STORED message displays. If you do not wish to store the translation, proceed to the APPLY option.

Create oblique symmetry about plane

Select: TRANSFOR > CREATE | SYMMETRY | PLANE | OBLIQUE

Prompt: SEL PT/LN/PLN

1. Define the plane of symmetry as follows: (a) Successively select three points. (b) Successively select two planar lines. (c) Select a point, and then a line. (d) Select a plane. The TRANSFORMATION CREATED message displays.
2. To define the direction of symmetry, select a line. Enter components to define a vector. The TRANSFORMATION CREATED message displays.
3. Enter a name for the translation if you wish to store it. The TRANSFORMATION STORED message displays. If you do not wish to store the translation, proceed to the APPLY option.

Create symmetry about line

Select: TRANSFOR > CREATE | SYMMETRY | LINE | NORMAL

Prompt: SEL PT/LN/PLN

1. Define the line of symmetry as follows: (a) Select or indicate two points. (b) Select a line. (c) Select two planes. The TRANSFORMATION CREATED message displays.
2. Enter a name for the translation if you wish to store it. The TRANSFORMATION STORED message displays. If you do not wish to store the translation, proceed to the APPLY option.

Create symmetry about point

Select: TRANSFOR > CREATE | SYMMETRY | POINT

Prompt: SEL PT/LN/PLN

1. To define the point of symmetry, select two points or a line and a plane (intersection point will be used). The TRANSFORMATION CREATED message displays.
2. Enter a name for the translation if you wish to store it. The TRANSFORMATION STORED message displays. If you do not wish to store the translation, proceed to the APPLY option.

Create scaling

Select: TRANSFOR > CREATE | SCALING

Prompt: SEL CENTER

1. To select the scaling center point (the point from which the scaling will be calculated), select or indicate a point or select two lines.
2. Define the scaling ratio by selecting two points (the ratio of their distance from the scaling center is the scaling ratio), or entering a scaling ratio. The TRANSFORMATION CREATED message displays.
3. Enter a name for the translation if you wish to store it. The TRANSFORMATION STORED message displays. If you do not wish to store the translation, proceed to the APPLY option.

Create affinity (scaling in one direction)

Select: TRANSFOR > CREATE | AFFINITY | PLANE or LINE or AXIS

Create a normal affinity about plane

Select: TRANSFOR > CREATE | AFFINITY | PLANE | NORMAL

Prompt: SEL PT/LN/PLN

1. To define the affinity plane, take one of the following alternatives: (a) successively select three points; (b) select a line and a point; (c) select two planar lines; or (d) select a plane.
2. Define the affinity ratio by selecting two points (the ratio of their distance from the scaling center is the scaling ratio), or entering a scaling ratio. The TRANSFORMATION CREATED message displays.
3. Enter a name for the translation if you wish to store it. The TRANSFORMATION STORED message displays. If you do not wish to store the translation, proceed to the APPLY option.

Create oblique affinity about plane

Select: TRANSFOR > CREATE | AFFINITY | PLANE | OBLIQUE

Prompt: SEL PT/LN/PLN

1. To define the affinity plane, take one of the following alternatives: (a) successively select two points; (b) select a line and a point; (c) select two planar lines; or (d) select a plane.
2. Select the affinity direction by selecting a line, or entering components.
3. To define the affinity ratio select two points (the ratio of their distance from the scaling center is the scaling ratio), or enter a scaling ratio. The TRANSFORMATION CREATED message displays.

4. Enter a name for the translation if you wish to store it. The TRANSFORMATION STORED message displays. If you do not wish to store the translation, proceed to the APPLY option.

Create affinity about line

Select: TRANSFOR > CREATE | AFFINITY | LINE

Prompt: SEL PT/LN/PLN

1. To define the line of affinity, take one of the following alternatives: (a) select two points; (b) select two planes; or (c) select a line.
2. To define the affinity ratio, select two points (the ratio of their distance from the scaling center is the scaling ratio), or enter a scaling ratio. The TRANSFORMATION CREATED message displays.
3. Enter a name for the translation if you wish to store it. The TRANSFORMATION STORED message displays. If you do not wish to store the translation, proceed to the APPLY option.

Create affinity about three perpendicular planes

Select: TRANSFOR > CREATE | AFFINITY | AXIS

Prompt: SEL AXIS

1. Select an axis system, and then enter an affinity ratio. The TRANSFORMATION CREATED message displays.
2. Enter a name for the translation if you wish to store it. The TRANSFORMATION STORED message displays. If you do not wish to store the translation, proceed to the APPLY option.

Create normal projection onto plane

Select: TRANSFOR > CREATE | PROJECT | NORMAL

Prompt: SEL PT/LN/PLN

1. To define the plane, take one of the following alternatives: (a) successively select three points; (b) select a line and a point; (c) select two planar lines; or (d) select a plane.The TRANSFORMATION CREATED message displays.
2. Enter a name for the translation if you wish to store it. The TRANSFORMATION STORED message displays. If you do not wish to store the translation, proceed to the APPLY option.

Create oblique projection onto plane

Select: TRANSFOR > CREATE | PROJECT | OBLIQUE

Prompt: SEL PT/LN/PLN

1. To define the plane, take one of the following alternatives: (a) successively select three points; (b) select a line and a point; (c) select two planar lines; or (d) select a plane.
2. To define the direction of the projection, enter components of a vector, or select a line. The TRANSFORMATION CREATED message displays.
3. Enter a name for the translation if you wish to store it. The TRANSFORMATION STORED message displays. If you do not wish to store the translation, proceed to the APPLY option.

Create move

A move is defined by two 3D axis systems.

Select: TRANSFOR > CREATE | MOVE

Prompt: SEL AXIS1

1. Successively select two 3D axis systems. The TRANSFORMATION CREATED message displays.
2. Enter a name for the translation if you wish to store it. The TRANSFORMATION STORED message displays. If you do not wish to store the translation, proceed to the APPLY option.

Create transformation by specifying transformation axis system

Select: TRANSFOR > CREATE | UNSPEC

Prompt: KEY TRANSFORMED ORIGIN

1. Enter coordinates of the transformed axis origin.
2. Successively enter components of transformed vectors for the new axis. The TRANSFORMATION CREATED message displays.
3. Enter a name for the translation if you wish to store it. The TRANSFORMATION STORED message displays. If you do not wish to store the translation, proceed to the APPLY option.

Define move using three angles

Select: TRANSFOR > CREATE | ANGLES

Prompt: SEL PT

1. Define the move by selecting a point and entering three angle values. The TRANSFORMATION CREATED message displays.

2. Enter a name for the translation if you wish to store it. The TRANSFORMATION STORED message displays. If you do not wish to store the translation, proceed to the APPLY option.

Define move by Euler angles

Select: TRANSFOR > CREATE | EULER

Prompt: SEL PT

1. Define the move by selecting a point and entering three angle values. The TRANSFORMATION CREATED message displays.
2. Enter a name for the translation if you wish to store it. The TRANSFORMATION STORED message displays. If you do not wish to store the translation, proceed to the APPLY option.

APPLY

Apply transformation.

Select: TRANSFOR > APPLY | REPLACE or DUPLICAT

Apply replacement transformations

Select: TRANSFOR > APPLY | REPLACE | ELEMENT or SET or FAMILY

Apply replaced element transformation

Select: TRANSFOR > APPLY | REPLACE | ELEMENT

Prompt: CUR MULTI-SEL SEL ELEM

Select an element. The element will be transformed.

Apply replaced set transformation

Select: TRANSFOR > APPLY | REPLACE | SET

Prompt: SEL SET

Select an element in the set to be transformed. Click on YES to confirm.

Apply replaced family transformation

Select: TRANSFOR > APPLY | REPLACE | FAMILY

Prompt: SEL ELEM

Select an element in the family to be transformed. Click on YES to confirm.

Apply duplicate transformations

Select: TRANSFOR > APPLY | DUPLICAT | ELEMENT or SET or FAMILY | STANDARD or SAME

Apply duplicate element transformation

Select: TRANSFOR > APPLY | DUPLICAT | ELEMENT | STANDARD or SAME

Prompt: SEL ELEM

1. Select STANDARD if you wish duplicated elements to contain display attributes as defined in the STANDARD function. Select SAME in order for duplicated elements to contain the same display attributes as the selected elements.
2. Select elements individually, or use a multiselect option. Click on YES to end, and then click on YES again to confirm. The elements are duplicated.

Apply duplicate set transformation

Select: TRANSFOR > APPLY | DUPLICAT | SET | STANDARD or SAME

Prompt: SEL ELEM

Select STANDARD if you wish duplicated sets to contain display attributes as defined in the STANDARD function. Select SAME in order for duplicated sets to contain the same display attributes as the selected elements.

Select an element in the set to be duplicated. The set is duplicated.

Apply duplicate family transformation

Select: TRANSFOR > APPLY | DUPLICAT | FAMILY | STANDARD or SAME

Prompt: SEL ELEM

Select STANDARD if you wish duplicated family to contain display attributes as defined in the STANDARD function. Select SAME in order for duplicated family to contain the same display attributes as the selected elements.

Select an element in the family to be duplicated. The family is duplicated.

STORE

Store current transformation.

Select: TRANSFOR > STORE

Prompt: KEY TRA IDENT / / YES: AUTO IDENT

This option is available immediately following creation.

Enter a name for the current transformation, or click on YES to accept an automatic identifier.

MANAGE

Manage transformations.

Select: TRANSFOR > MANAGE | ANALYZE or ERASE or CONVERT or COMBINE

Analyze transformation

Select: TRANSFOR > MANAGE | ANALYZE or ERASE or INVERT or COMBINE

Prompt: YES: CUR TRA

Enter the name of the transformation to be analyzed. Alternatively, select the transformation from the list, or click on YES to analyze the current transformation. Transformation characteristics appear in the alphanumeric window, and a graphical representation of the transformation definition displays.

Delete transformation

Only stored transformations can be deleted.

Select: TRANSFOR > MANAGE | ERASE

Prompt: SEL TRANSFORMATION

Select the transformation to be deleted. Click on YES to confirm.

Create inverted transformation

Only stored transformations can be inverted.

Select: TRANSFOR > MANAGE | INVERT

Prompt: SEL TRANSFORMATION

Select the transformation to be inverted. Enter a new name for the inverted transformation as necessary.

Combine transformations

Only stored transformations can be combined.

Select: TRANSFOR > MANAGE | COMBINE

Prompt: SEL TRANSFORMATION

1. Select the first transformation, and then select the second transformation. A combined transformation is created.
2. Enter a name for the combined transformation as necessary.

UTILITY

The UTILITY function allows you access to a set of utility programs for use with CATIA. Utilities can be accessed via the UTILITY function or by double-clicking on the CatUtil icon in the CATIA_V4 window on the desktop.

✓ **NOTE:** *You can quickly return to the desktop by using <Alt> + <Tab>, or by selecting the Minimize button at the top right corner of the screen.*

Upon selecting the UTILITY function, the list of available utilities displays on the screen as seen in the next illustration.

Utility selection

Available utilities

CATAIX	Convert Catia data from one format to another
CATANSOL	Generative part stress analysis
CATBACK	Migrate data from the current Catia version to an earlier ve
CATCLN	Identify and correct errors in Catia models
CATDATA	List, copy, duplicate, delete, rename, save and free up file
CATDRAW	Extract DRAW views from 3D solid models
CATDXF	Convert CATIA models into DXF/DWG files
CATEXP	Export Catia models or libraries from a Catia site
CATFAM	Manage library families
CATFEA	Finite element solver
CATFONT	Manage Catia fixed or proportional fonts
CATIMD	Image synthesis and processing
CATIMP	Import Catia models or libraries to a Catia site
CATITF	Interference computation
CATKI	Kinematics
CATLIB	Manage library files
CATLSH	List models referenced by sheets and vice versa

String :

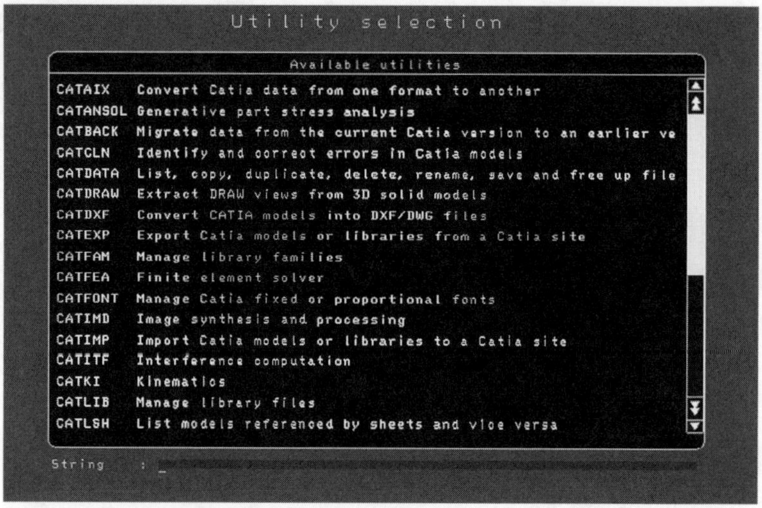

UTILITY selection screen.

CatUtil Icon

Upon selecting the CatUtil icon from the desktop, the list of available utilities displays on the screen as seen in the next illustration.

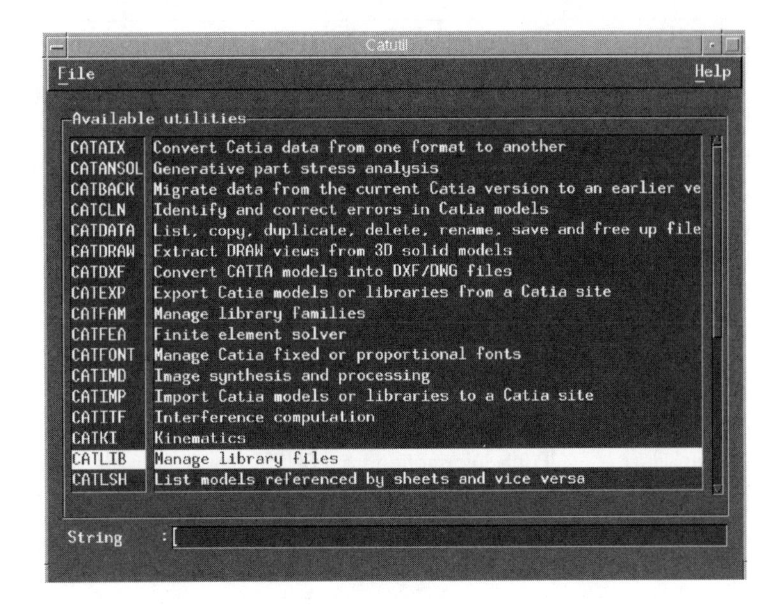

CatUtil selection screen.

Utilities

Available utilities are described in the following table.

Utility	Description
CATAIX	Convert CATIA data produced on IBM S/370 computers in EBCDIC format to ASCII format for IBM RISC System/6000 computers.
CATANSOL	For use with ANSOLID function.
CATBACK	Migrate CATIA data to an earlier version/release.
CATCLN	Check for errors in a specific model. If errors are found, correction measures are proposed and can be applied.
CATDATA	Perform the following functions relating to CATIA files stored on tape or disk: print, copy tape to disk or vice versa, duplicate, rename, delete, and save.
CATDRAW	Update DRAW views similar to AUXVIEW2 I UPDATE.
CATDXF	Convert CATIA models (not including solids, volumes, skins, and other complex entities) to DXF or DWG format for use with other CAD systems.
CATEXP	Extracts CATIA data in preparation for transfer to another site to avoid conflicts with project file at the other site.
CATFAM	Creation and management of CATIA library families.

Utility	Description
CATFEA	For use with CATIA Finite Element Solver.
CATFONT	Create special character fonts.
CATIMD	Generate high quality images as in the SHADES function in batch mode.
CATIMP	Imports previously exported data from another site.
CATITF	Compute interferences between CATIA elements.
CATKI	Used in conjunction with CATIA Kinematics.
CATLIB	Provides additional facilities in CATFAM for the management of CATIA libraries.
CATLSH	Prints information from SHEET files (where filed plots are stored), including list of models referenced by each sheet and the list of sheets in each model.
CATMOD	Perform operations on CATIA models in order to prepare them for transfer to other CAD systems.
CATMESH	Used in conjunction with CATIA Finite Element Analysis products.
CATMSTA	Used in conjunction with CATIA Finite Element Analysis products.
CATNAME	Change CATIA file names.
CATOLER	Analyze 3D CATIA elements, and then modify elements to make them compatible with newly defined tolerances. (Used in conjunction with CATRANS.)
CATPLOT	Works like the /PLOT command. Enables plotting of SHEET, PICTURE, and LIBRARY files as well as buffer plots.
CATPRJ	Print contents of tables in project file.
CATRANS	For transfer of CATIA models between sites that have previously been synchronized.
CATSOE	Analyze and upgrade CATIA version 3 / release 2 solids in a model (works as /M SOLM34).
CATSOL	Create complex volume from solid.
CATSUMR	List information about one or more CATIA models including date and time of creation, execution instructions, messages, associated project file, comments, and model standards.
CATUNIT	Modify units of selected portions or all contents of CATIA model.
DXFCAT	Converts DXF or DWG files into CATIA model.

Appendix A

Element Identifiers Used in CATIA

Appearing below are lists of DRAW element identifiers and SPACE element identifiers. All identifiers can be used for multiselection. When using an identifier, use the asterisk (*) as a prefix (e.g., **PTD*).

DRAW Element Identifiers

Element Identifier	Description
ANND	New dimensions and text
ARWD	Mark up arrow
AXSD	Axis
CIRD	Circle
CND	Schematics connector
CRVD	All types of curves
DIM	All types of dimensions (old and new)
DIMN	New dimension
DITD	Ditto (occurrence of a detail)
ELLD	Ellipses
GDLN	Line grid
GDPT	Point grid
HYPD	Hyperbola
LND	Line
LST	Line string
OCND	Occurrences in schematics

Element Identifier	Description
PARD	Parabola
PTD	Point
SHAP	Shape (pattern hatching)
SPLD	Spline
STRD	Contour
SYMD	Occurrence of a symbol
TXTD	All types of text (old and new)
TXTN	New text
VU	View

SPACE Element Identifiers

Element Identifier	Description
AXS	Axis
CCV	Composite curve
CNP	Tubing connector
CRV	Curve
CST	Constraint
DIT	Ditto (occurrence of a detail)
FAC	Face
FSUR	Functional surface
LN	Line
NET	Network
OCP	Occurrence of piping
PIP	Pipe
PLN	Plane
POL	Polyhedral surface
PT	Point

Element Identifier	Description
SKI	Skin
SOE	Exact solid
SOL	All solids (exact and mock-up)
SOM	Mock-up solid
STR	Structure
SUR	Surface
TXT	All types of text (old and new)
VOL	Volume

✓ **NOTE:** *All of the above keywords can be used when combining multiple selection keywords. For further information, see Appendix B.*

Appendix B

Keywords Used for Multiple Selection

Appearing below are lists of keywords that can be used for multiple selection. When using a keyword, insert an asterisk (*) as a prefix. (For instance, to select all DRAW elements, you would key in **DRW*.)

Single Interaction Keywords Requiring No Additional Information

Keyword	Description
DRW	All DRAW elements
SPC	All SPACE elements

All types of element identifiers (listed in Appendix A) can also be used as single interaction keywords. Examples are **LND* and **LN*.

Single Interaction Keywords Requiring Additional Information

Keyword	Description
COLx	Color, where x = 1 to 125
GRPx	Group, where x = 1 to 3
LAYx	Layer, where x = 0 to 254
LNTx	Line type, where x = 1 to 32
THKx	Line thickness, where x = 1 to 6

The keywords in the above table require a value to enable use as single interactive keywords. For instance, the **COL3* keyword would select all green elements, and the **LNT2* keyword would select all dotted lines.

Keywords Requiring Complementary Interaction

Keyword	Description
COL	Color
GRP	Group

Keyword	Description
LAY	Layer
LIP	Elements of piping logical line
LIS	Elements of schematic logical line
LNT	Line type
OPL	Elements lying on specific plane
SEQ	Sequentially linked monoparametrics
SET	Elements of set
THK	Line thickness
TYP	Type (typical)
VU	Elements of view

Examples of usage for keywords requiring further definition by a complementary interaction follow.

- Key in **COL* and select any element. All elements of that color will be selected.
- Key in **LAY* and select any element. All elements in that layer will be selected.
- Key in **OPL* and select any element in SPACE mode. All elements lying on the same plane as the selected element will be selected.
- Key in **TYP* and select an element. All elements of that type will be selected.

Multiple Interaction Keywords

Keyword	Description
ITRP	Elements wholly inside trap
OTRP	Elements wholly outside trap
PRF	Profile
SEL	Individual selection of several elements
TRP	Elements partially or wholly inside trap
TRP?	Elements partially or wholly inside trap, but excluding elements defined by an additional keyword
XTRP	Elements partially or wholly outside a trap

Keyword	Description
$PAC	Parent faces of VOL or SKI type element

Keywords in the previous table must be further defined by complementary interactions. Examples follow.

- After keying in **TRP*, indicate a series of points to define the required trap. When the trap is complete, click on the YES button to complete the trap definition. Click on YES again to confirm the trap selection.
- After keying in **SEL*, you will have to select a set of elements to include in the selection. When the selection is complete, click on the YES button to end the selection; click on YES again to confirm the selection.

Combining Multiple Selection Keywords

Any keyword listed in Appendices A or B can be combined in order to select several groups of elements in a single operation. Examples of combining keywords are listed below.

- Select all LN (SPACE lines) type elements that are green.
- Select all PTD (DRAW points) type elements and all LND (DRAW line) type elements.
- Select all DRW (DRAW) elements except CRV (curve) type elements.

To combine the keywords, you can use the three types of separators shown below.

- + (plus symbol) — Join the keywords preceding and following the symbol.
- - (minus symbol) — Subtract the keyword following the symbol with the keyword preceding it.
- & (ampersand) — Intersects keywords preceding and following symbol.

Examples of combined multiple selections follow.

- *LN+*COL3 selects all green SPACE lines.
- *PTD+*LND selects all DRAW points and all DRAW lines.
- *DRW-*CRV selects all DRAW elements except the CRV elements.
- *LN&*COL2 selects all red SPACE lines.

You can combine up to 10 keywords in a character string. Examples of combined multiple selections follow.

- *CRV+*LND&LAY10-COL2 selects all curves and lines lying on layer 10 except those that are red.
- *SOL+*VOL&LAY8 selects all solids and volumes that are on layer 8.
- *SOL-*COL3 selects all solids except those that are green.
- *SOL-*COL3+*LN selects all solids except green ones, while also selecting all lines.
- *SPC-*SOL selects all SPACE elements except solids.
- *SOL+*VOL selects all solids and volumes.
- *PT+*LN+*CRV+*PLN selects all SPACE points, lines, curves, and planes.

When using combined multiple selections, the following rules and general observations are pertinent.

- A maximum of 10 keywords can be combined in a character string.
- A separator must not precede the first keyword.
- The string of keywords is analyzed from left to right.
- If an incorrect string of keywords is entered, the following message will be displayed: *IDENTIFER DOES NOT EXIST.*
- Combining keywords requiring graphical interaction after they have been entered is not possible. (An example is **TRP*.)
- The most commonly used combinations of keywords are available for use direct from the SELECT pull-down menu. For further information, see Appendix D.

Appendix C

CATIA General Commands

Appearing below is a list of the most popular general commands. Key in a forward slash before the general command. For example, to clean a model, key in */CLN*, and follow the resulting prompts.

General commands are available regardless of the function you are using. Some of the commands can be used only in specific modes, such as SPACE 3D or 2D, or DRAW. Many general commands are also available in pull-down menus or via fixed menu buttons.

General Command Descriptions

Command	Description
ANACRVT	Analysis of element curvature parameters.
ANADEL	Delete an analysis applied to an element.
ANADRAFT	Check to determine whether an element can be correctly drafted.
ANND34	Upgrade V3R2 dimensions to new dimensions.
CLN	Clean model.
CLOSE	Close model.
COLCHG	Change color of an element.
COLCOP	Copy the color of an element.
COLSTD	Reset the color of an element to the color set in STANDARD.
COMMENT	Model comment display and edit (old style dialog).
COMMENT2	Model comment display and edit (motif style dialog).
CONFIG	Display code level information.
DIMCOORD	Add a box with point coordinates and leader to point (DRAW mode).
DOC	Search available documentation.
EXIT	Close the current session.
GRAB	Generate an image from the screen display using dialog window.
GRABP	Generate an image direct from screen display.

Command	Description
HELP	Start help mode window.
INFO	Display information about current model.
MDLALL	Apply ALL filter.
MDLFIL	Change current layer filter.
MONAXE	Change axis system (AXS button).
MONBRG	Buffer regenerate (BR button).
MONELE	Display element identifiers (ID button).
MONGS	Set graphic standards (ST button).
MONKEY	Define keyboard layouts (KEY button).
MONLAY	Change current layer (Lxxx button).
MONLST	Change between display modes, wireframe, hidden line, or shaded (Set current display mode button).
MONNP	Modify the pick/no pick attributes of elements (NP button).
MONNS	Modify the show/no show attributes of elements (NS button).
MONROT	Rotate the model.
MONS3D	Change between the SPACE 2D and 3D modes (2D/3D button).
MONSCR	Retrieve a stored screen (SC button).
MONSET	Change current set (SET = button).
MONSDR	Change between SPACE and DRAW modes (SP/DR button).
MONV2D	Change current DRAW view (VU button).
MONWSP	Change current workspace (WSP button).
MONWIN	Retrieve a stored window (WI button).
MONZOM	Change image scale (ZM button).
NEW	Create a new model.
NEWS	Review the latest enhancements.
OPEN	Open a model.
PLOT	Plot the current model or models as viewed on the screen.

Command	Description
RECOVERY	Switch on recovery mode for working session.
REFRAME	Image reframe.
REMOTE	Remote access to mainframe files.
SAVE	Save model.
SAVEAS	Save model with a new name.
SAVESES	Save session.
SAVESESA	Save session with a new name.
SETUP	Modify graphics options, cursor type, fast transforms, and preselect highlighting.
SOLPATT	Create pattern of common features when creating solids.
SOLPAR	Generate basic geometric entities from solid primitives.
SOLSIZE	Display data size of a solid.
STCOLD	Set DRAW mode color standard.
STCOLS	Set SPACE mode color standard.
STRETCH	Modify 2D profile with stretching functions.
SWAP	Swap active model.

✓ **NOTE 1:** *Availability of some of the above commands is dependent on your CATIA setup.*

✓ **NOTE 2:** *When using general commands, you may not need to enter the entire command name. In many instances, you need only enter sufficient characters for CATIA to recognize the command (e.g., /STR for /STRETCH).*

Appendix D

CATIA Pull-down Menus

The pull-down menu bar at the top of the CATIA screen provides access to a series of tools. Appearing below are lists of tools or options available in pull-down menus accompanied by usage descriptions. Keyboard access alternatives appear in parentheses.

File

The File pull-down menu contains tools for the management of models, files, and sessions.

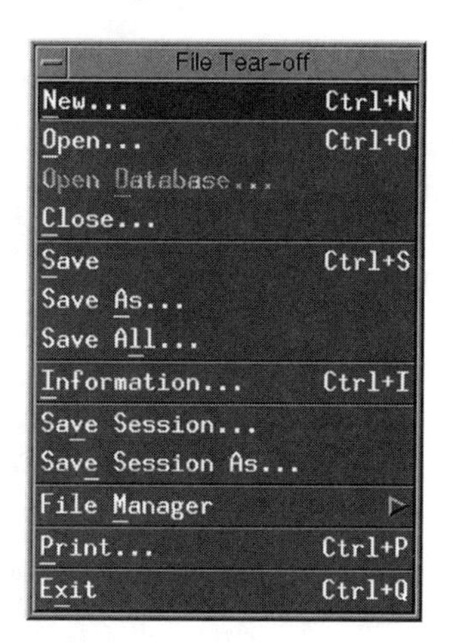

File pull-down menu.

File Pull-down Menu Options/Tools

Option/Tool	Description
New	Create a new model or session (<Ctrl>+<N>).
Open	Open an existing model or session (<Ctrl>+<O>).
Open Database	Open a database (if available).
Close	Close a model.
Save	Save the current model with current name (<Ctrl>+<S>).
Save As	Save the current model with option to specify a new name.
Save All	Save all models in current multimodel environment or session.

Option/Tool	Description
Information	Display information about current model or session (<Ctrl>+<I>).
Save Session	Save session using the Open window.
Save Session As	Save session with a new name using the Open window.
File Manager	Copy, move, rename, or delete models.
Print	Plot the current model(s) window or screen as in PLOT function (<Ctrl>+<P>).
Exit	Exit CATIA (<Ctrl>+<Q>).

Select

The Select pull-down menu provides a list of all element types used, which can also be accessed from the keyboard. (Some options access additional lists, such as SPACE elements.) For a complete listing of the definitions for element abbreviations used in CATIA, see Appendix B.

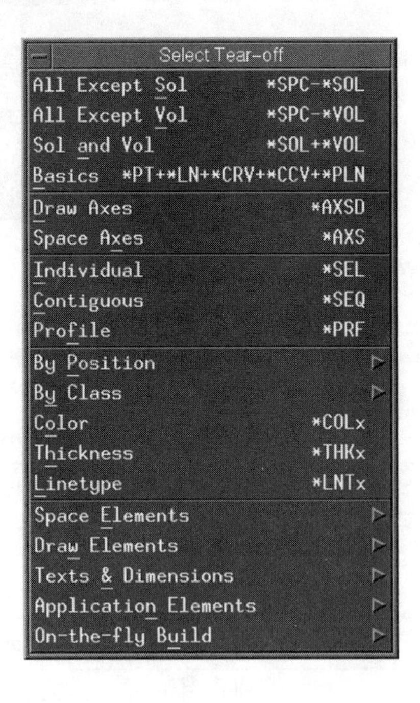

Select pull-down menu.

View

The View pull-down menu contains tools for managing model viewing, and provides fast access to these tools which are also available in various other CATIA functions.

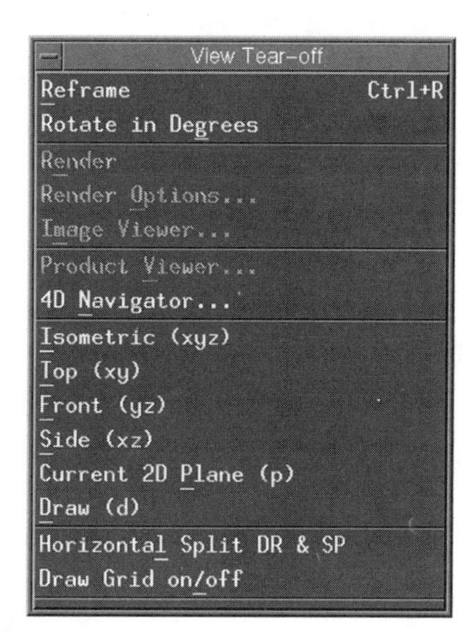

View pull-down menu.

View Pull-down Menu Options/Tools

Option/Tool	Description
Reframe	Reframe the model to its extents (<Ctrl>+<r>).
Rotate	Rotate in degrees; available only in Version 4 R1.8.
Render	Access realistic rendering options and image viewer. Available only to licensees of CATIA Realistic Rendering.
Product Viewer	View the graphical version of an assembly imported from a database. Available only to licensees of CATIA Data Management Access.
4D Navigator	View by navigation through model.
Isometric (xyz)	View model on XYZ.
Top (xy)	View model from above on XY plane.
Front (yz)	View model from front on YZ plane.

Option/Tool	Description
Side (xz)	View model from side on XZ plane.
Current 2D Plane (p)	Move view on current 2D plane parallel to screen.
Draw (d)	Change to DRAW window.
Horizontal Split DR and SP	Split screen horizontally to incorporate SPACE and DRAW windows. Available only in Version 4 R1.8.
Draw Grid on/off	Enable creation of a point grid in DRAW mode. Available only in Version 4 R1.8.
***Note**: A list of windows stored in the model will also be displayed at the bottom of the View pull-down menu.*	

Filter

The Filter pull-down menu enables you to apply filters to a model.

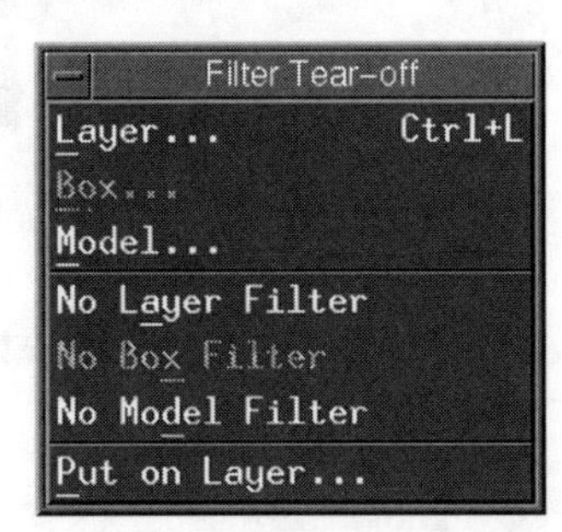

Filter pull-down menu.

Filter Pull-down Menu Options/Tools

Option/Tool	Description
Layer	Apply a layer filter (<Ctrl>+<L>).
Box	Requires third party Box product.
Model	Apply a filter displaying only the current model.
No Layer Filter	Delete the previously applied filter to the current layer.
No Box Filter	Requires third party Box product.
No Model Filter	Delete the previously applied filter to the model.

Option/Tool	Description
Put on Layer	Transfer an element to a different layer. Available only in V 4 R1.8.

Options

The Options pull-down menu is used to modify palette positions.

Options Pull-down Menu Tools

Tool	Description
Full Screen Layout	View maximum possible work area size without palettes or current function displayed.
Palette Placement	Change placement of palettes and function menus.

Options pull-down menu.

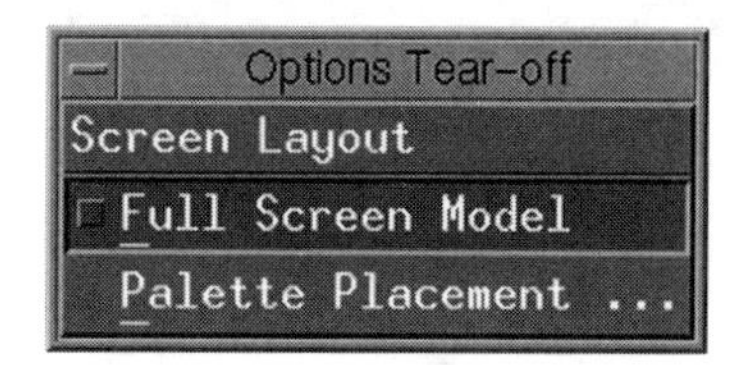

Tools

The Tools pull-down menu provides rapid access to many commands that are also available from the keyboard and within other CATIA functions.

Tools Pull-down Menu Options

Option	Description
Change Color	Access color selection as in GRAPHIC function.
Copy from/to	Copy files from/to.
Reset to standard	Reset to standard colors.
Set SP Element Color	Set SPACE mode element colors.
Set DR Element Color	Set DRAW mode element colors.
Sketcher	Activate the Dynamic Sketcher (<Ctrl>+<K>).

Option	Description
Assembly	Create and manage multilevel assembly of parts.
Grid for Feature	Create or edit feature patterns (for use with part design V 4 R1.8).
Geometry	Create draw points, space points, draw lines, or space lines.
Visualization	Modify visualization of elements or refresh view.
Analyze	Analyze elements, clearances, and clashes.
Management	Write an object into a library, analyze space lines, set up configuration, mount files.
IUA Commands	Access IUA commands.
Conciliation	Activate Conciliation feature for use with concurrent engineering.
Engraving	Access Engraving/Embossing product if available.
Screen Grab	Enable TIFF images to be captured from the screen.
Insert/Edit Hyperlinks	Link external multimedia objects to CATIA elements.
Update	Update exact solids.

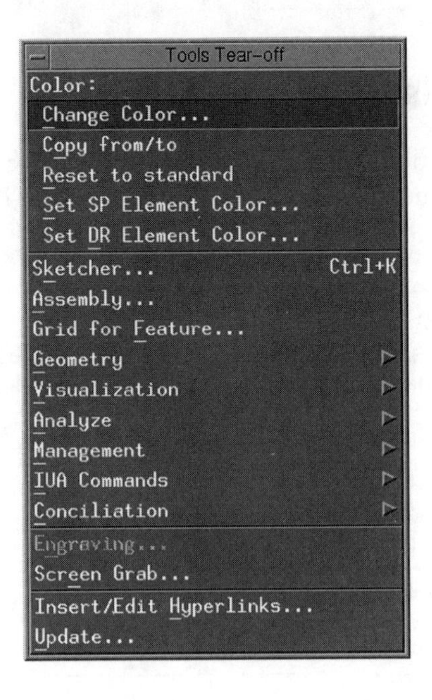

Tools pull-down menu.

Window

The Window pull-down menu lists models contained in the current session and provides tools for managing overlaid models.

Window pull-down menu.

Window Pull-down Menu Options/Tools

Option/Tool	Description
Highlight Active Model	Highlight active model in red.
Show Active Only	Show active model only.
Show/Hide Models	Show/hide specific models.
Swap by Graphic Selection	Change a passive model to active status by selecting it.
Keep Current Layout	Change active model but maintain current screen layout if required.
Keep Current Axis	Keep current axis (or not) on swapping to passive.

Help

The Help pull-down menu provides access to various online help facilities.

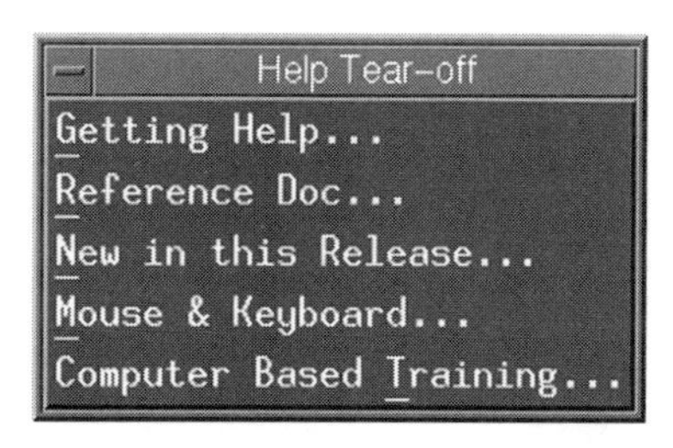

Help pull-down menu.

Help Pull-down Menu Options/Tools

Option/Tool	Description
Getting Help	Activate online help.
Reference Doc	Activate only CATIA reference documentation in Frameviewer.
News in this Release	Display CATIA Solutions Product Enhancements Overview.
Mouse & Keyboard	Display a summary of mouse and keyboard information.
Computer Based Training	Access computer based online training information.

Appendix E

Engineering Symbols

When annotating drawings, having a range of engineering symbols available is useful. Unfortunately, engineering symbols are not available on many UNIX keyboards. However, you can obtain such symbols by holding down <Ctrl>+<Alt> followed by another key.

Key Sequences for Available Engineering Symbols

Key in <Ctrl>+<Alt> followed by	q	w	e	r	t	y	u	i
Resultant symbol	∉	∅	°	Ω	≤	≥	±	μ

Appendix F

Fixed Menu and Dialog Area

The fixed menu (permanent function zone) is always available during CATIA sessions. Communication between software and user occurs in the dialog area. Both are located at the bottom of the CATIA screen. In certain instances certain buttons on the fixed menu will be unavailable (shown in dimmed mode), such as SC and WI while the user is working with the IMAGE function.

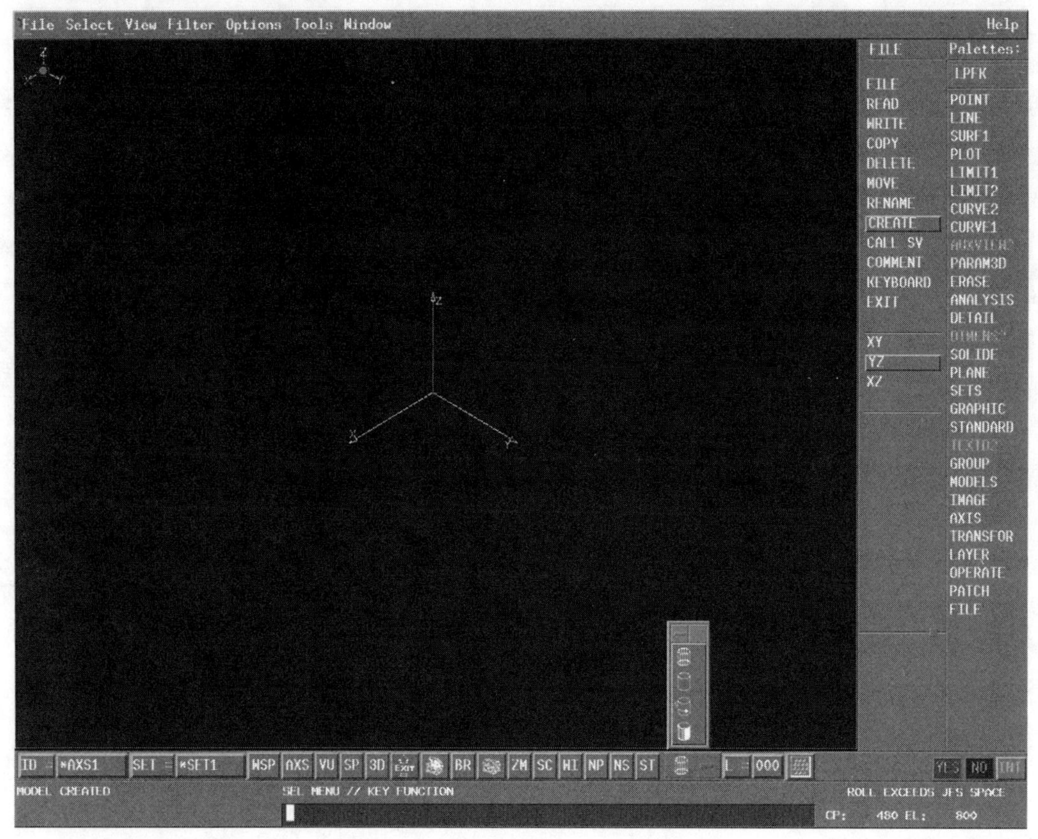

CATIA fixed menu (permanent function zone) and dialog area.

Fixed menu buttons are described below.

ID= *name

Highlights and permits ID change of selected element. Allows scrolling through available elements using YES and NO buttons.

SET= *name

Highlights and permits ID change of selected SET. Allows scrolling through available SETS using YES and NO buttons.

WSP

Provides access to detail workspaces by selection from workspace or list after entering a blank field.

AXS

Highlights the current axis system. Allows swapping to alternative axis by scrolling using YES and NO buttons or by selection in workspace.

VU

Highlights current view. Allows switching to alternative views by scrolling using YES and NO buttons, or selection from list after entering a blank field.

DR/SP

Switches between SPACE and DRAW modes.

3D

Switches between 2D and 3D space modes by selecting planes or elements defining a plane. To return to 3D mode, click on YES button. Available only when SP (SPACE mode) is selected.

EXIT

Exits from the current selection on the fixed menu.

LOCAL SAVE

Temporarily saves model in tmp (temporary) file

BR

Regenerates the screen buffer and refreshes screen visualization if elements have become distorted or have disappeared.

Restore Locally Saved Model

Restores a model which has been saved using the temporary save button.

ZM

Zooms and centers model on the screen by indicating diagonal points, keying in a zoom factor, or clicking YES for zoom factor of 1.

SC

Swaps between screens defined using IMAGE function.

WI

Swaps between windows defined using IMAGE function.

NP

Pick or no pick (making elements nonselectable or selectable) in exactly the same way as ERASE > PICK or NO PICK.

NS

Show or no show (hiding or revealing elements) in exactly the same way as using ERASE > SHOW or NO SHOW.

ST

Access to the DRAW or SPACE ELEMENT GRAPHIC STANDARD SCREEN (depending on the mode selected). Set the current display mode on selection of one of four available icons defined as seen below.

NHR (no hidden line removal) or wireframe mode. All lines are visible.

HLR (static hidden line removal). Hidden lines are removed but BR (buffer regeneration button) is required after manipulation.

HRD (dynamic hidden line removal). Hidden lines are removed with continuous or dynamic buffer regeneration.

SHD (shaded dynamic). Solids appear shaded with continuous or dynamic buffer regeneration.

L=

Sets current layer by selecting an element on the required layer, scrolling using YES and NO buttons, or entering layer number (0 - 255).

Keyboard

Access to the Palette Creation screen.

Appendix G

Display and Manipulation Window

The Display and Manipulation window provides a set of tools with which to alter the onscreen display and manipulate models. Appearing below are ways in which the window can be accessed.

- Press the <F4> function key.
- On a three-button mouse, press and hold button 3 and then press button 1. Release both buttons.
- On a four-button mouse, press button 4.

To clear the window from the screen, take one of the following options.

- Pressing the <F4> function key.
- On a three-button mouse, press and hold button 3 and then press button 1. Release both buttons.
- Select the off option in the window.

The display and manipulation tools are available at any time during a CATIA session. You do not have to close the current function. The only exception is when you are using the IMAGE function. Access the display and manipulation tools by selecting IMAGE > LOCAL TR.

When the BR, WI, SC, RT, or ZM fixed menu bar options are selected, the message LOCKED will appear at the bottom of the 2D and 3D transformation window, and the 2D and 3D transformations will be inactive.

Four menus are available when the window is activated. The following menus can be selected from the tool bar at the top of the window.

- STD. Provides a variety of display and manipulation tools.
- 2D. Enables 2D transformations.
- 3D. Enables 3D transformations.
- COL. Manages color palette.

STD Option Menu

When the STD menu is displayed, the following options are available. All STD options provide ON/OFF selections.

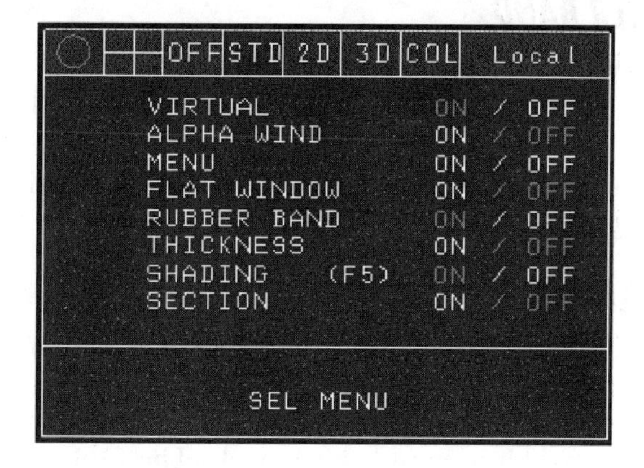

STD menu options.

VIRTUAL

If the size of the model displayed on the screen exceeds the screen limits for a specific zoom factor, response time and performance will decline as the model size increases. With the ON option selected, you can work on a part of the model 256 times greater than the area displayed on the screen. The consequence is poor performance. Upon selecting the OFF option, you can work on only the part of the model displayed on the screen.

ALPHA WIND

With this option, you can display the alphanumeric window (ON) or clear it (OFF). To clear the alphanumeric window from the screen you can either set ALPHA WIND to OFF or select the OFF option in the alphanumeric window.

The alphanumeric window can also be activated by using the following keyboard combinations.

- <Alt>+<+>. Display the alphanumeric window.
- <Alt>+<->. Clear the alphanumeric window.

MENU

The MENU option is inactive on Motif CATIA. On a 5080 CATIA setup this option allows you to switch the side menu off or on. When using Motif CATIA, the alternative to switching the menu off or on is to use the Full Screen Model command in the Options pull-down menu.

FLAT WINDOW

When this option is set to ON a transformable 2D space is generated, and the 3D transformations are inactive. The message LOCKED will appear at the bottom of the 3D menu window. When set to OFF, a 3D space is generated, and the 3D transformations are available. The OFF setting improves model response time and performance.

RUBBER BAND

The RUBBER BAND option, used in the PLOT and LINE functions, allows you to continuously display the position of the line to be created while moving the cursor before confirming creation. The line will then be fixed between the point and the cursor.

The option is enabled when set to ON, and disabled in the OFF setting. Setting the RUBBER BAND option to OFF will increase performance.

THICKNESS

The THICKNESS option permits element thickness set in the STANDARD function to be shown on the display. When the option is set to ON, all elements are shown at normal display thickness. With the option OFF, lines, curves, and so forth are displayed without thickness (width is equal to one pixel). The OFF setting improves model display performance.

SHADING

The SHADING option is used to degrade HRD or SHD modes. When set to ON, HRD or SHD mode remains active. Under the OFF setting, HRD or SHD mode is inactive, which allows you to more easily select elements hidden behind other parts of the model. Only the edges of solids are displayed. This option can be toggled using the <F5> function key.

SECTION

This option activates the solid sectioning mode. Set the option ON to activate the dynamic sectioning mode, thereby switching the BOX option of the 3D menu window to BOX NEAR PLANE ONLY mode. Setting the option to OFF deactivates the dynamic sectioning mode. When the dynamic sectioning mode is not required, setting the option to OFF is recommended in order to improve display performance.

The dynamic sectioning mode allows you to move a sectioning plane backwards and forwards in space. The plane can be located on any point you pick and will allow you to see and select elements from inside a model.

2D Option Menu

The following illustration shows the options available when the 2D menu is displayed.

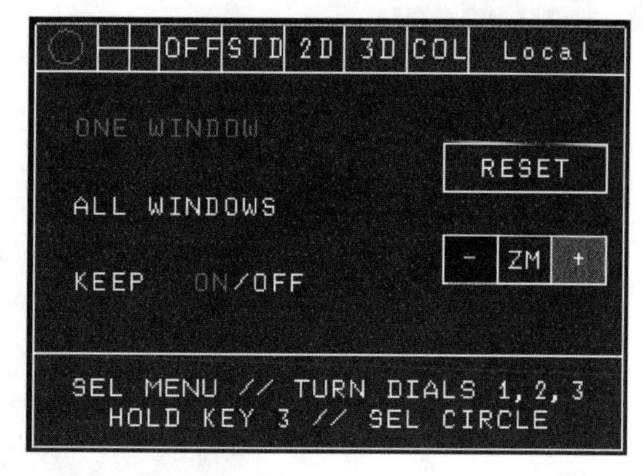

2D menu options.

The window is used to manage 2D transformations. Available 2D transformations include horizontal and vertical translation (dials 1 and 2), and zooming in and out (dial 3).

ONE WINDOW

This option activates 2D transformations in the active window only. With more than one window in the current screen, as defined in the IMAGE function, you can activate the required window by selecting the axis system in the corner of the window. A circle is highlighted on the axis system of the active window axis.

ALL WINDOWS

Activates 2D transformations in all windows belonging to the current screen as defined in the IMAGE function.

KEEP

When this option is set to ON before exiting, 2D transformation capabilities are retained. Set to OFF before exiting has the same effect as exiting the 2D TRANSFORMATION menu, and the highlighted circle disappears. This option is normally set to ON.

RESET

Selecting this option after 2D transformations will restore the window(s) to respective states prior to application of 2D transformations, as long as the BR (buffer regenerate) button has not been selected.

ZOOM

Provides an additional means of zooming the window(s). Select the + (plus) icon to zoom out, and the - (minus) icon to zoom out.

3D Option Menu

The following illustration shows available options when the 3D menu displays.

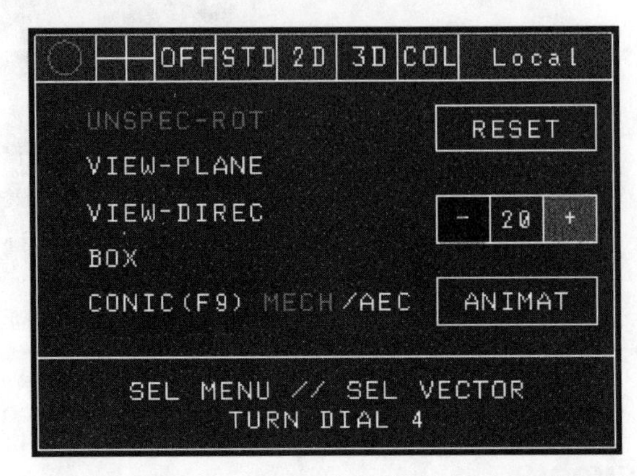

3D menu options.

This window option is used to manage 3D transformations. Three-dimensional transformations are always active, with the following exceptions.

- When using IMAGE function.
- When the FLAT WINDOW option of the STD menu is set to ON.
- When working with an UNSPEC window.
- When working with a CONIC window if the terminal does not allow dynamic conic windows.

UNSPEC-ROT

This option defines the axis of rotation; the model can be rotated using dial 4 or the virtual space ball facility. The current axis of rotation will be shown as a red dashed line with a cross at the center point.

VIEW-PLANE

This option allows viewing of the model projected on to a plane. The plane is defined by selecting lines or curves that lie on the required viewing plane; the plane of projection will be parallel to the screen.

VIEW-DIREC

This option allows viewing of the model along any direction. The direction is defined by selecting a line or curve. If a curve or circle is selected, the viewing direction is defined as a line tangent to the curve and passing through the point at which you selected the curve. The viewing direction will be normal to the screen.

BOX

This option allows you to display a part of the model located between two defined planes parallel to the screen. The model will be clipped by the defined plane boundaries, allowing you to see a cross section of the SPACE model.

✓ **NOTE:** *Setting the SECTION option of the STD menu to ON switches the BOX option to BOX NEAR PLANE ONLY mode to activate dynamic solid sectioning mode.*

CONIC

This option allows you to activate special display modes when working with conic views. CONIC can be set to either of the following options.

MECH

The viewing point is understood as the eye of the observer located outside the model. Viewing the model is similar to moving your head up and down or from side to side, and then looking at the part of the model in front of you. This mode is particularly suited to viewing mechanical parts.

AEC

The viewing point is understood as the eye of the observer located inside the model. Viewing the model is similar to moving your head around inside the model. This mode is particularly suited to AEC work involving large structures such as buildings.

RESET

Selecting this option after 3D transformations will restore the window(s) to their status prior to application of the 3D transformations, as long as the BR (buffer regenerate) button has not been selected.

ANIMAT

Select this option to perform continuously animated rotation. The SPACE model is rotated at the current rotation axis at a speed set in the next option.

Define Angle of Rotation, Animation Speed, and Virtual Spaceball Sensitivity

Define the following parameters using the box in the 3D menu option.

- Angle of rotation used by UNSPEC-ROT
- Animation speed used by ANIMAT
- Virtual spaceball sensitivity

The number in the box defines the value of the increment for any of the abovementioned uses. To increase the value, selecting the plus (+) key, and to decrease the value, select the minus (-) key.

COL Option Menu

The following illustration shows available options when the COL menu displays.

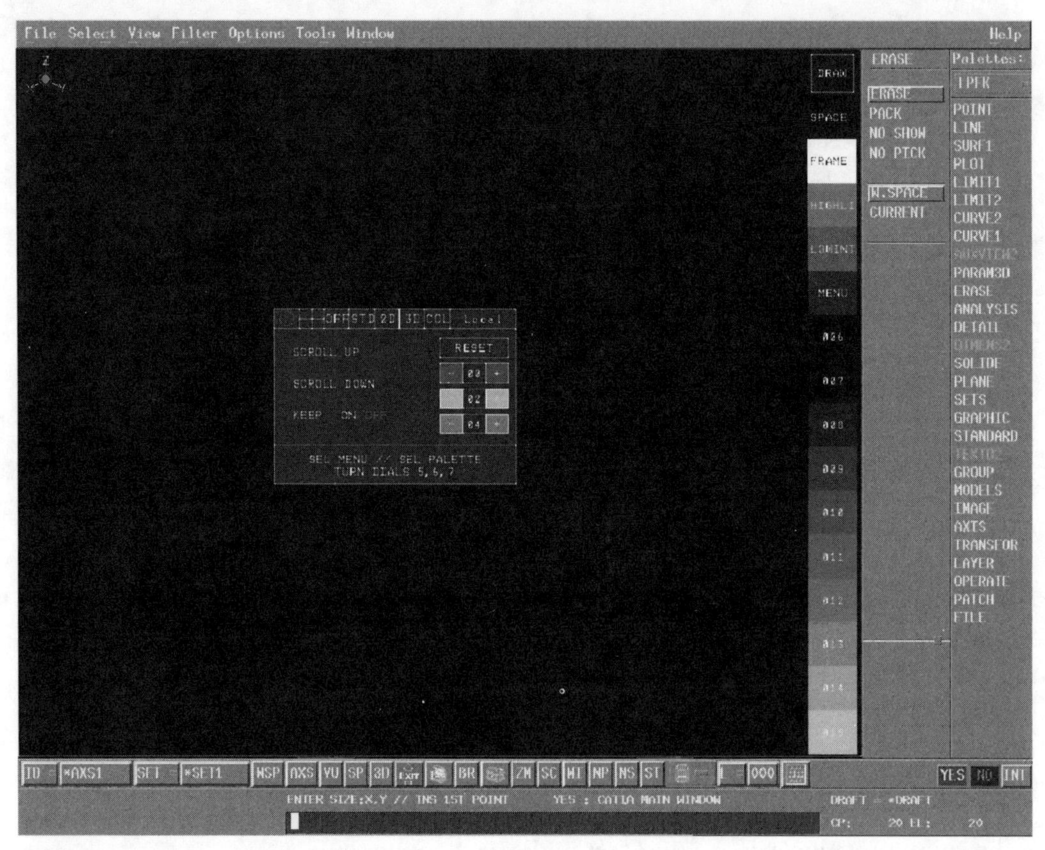

COL menu options and color palette.

With the COL menu you can modify colors available in the palette. Once colors are set, they can be stored in a model using STANDARD > COLOR | TABLE | SAVE. When you select the COL menu option the color palette displays vertically alongside the palette menu area.

- SCROLL UP. Scroll up the palette to access all palette colors.
- SCROLL DOWN. Scroll down the palette to access all palette colors.

- KEEP. When this option is set to ON before exiting the display and manipulation window, the current palette displayed will be retained.
- RESET. Restore the color table saved in the model. If no color table has been stored in the model, the default CATIA color table is restored. If the RESET option is selected, changes to the current color table are lost.

The color palette consists of a list of rectangles filled with colors identified by text or a number as follows.

- DRAW. Background color for DRAW windows.
- SPACE. Background color for SPACE windows.
- FRAME. Color of CATIA frame horizontal and vertical lines, function and item text, and messages.
- HIGHL. Color for highlighting selectable elements.
- LOWINT. Color for dimmed elements.
- MENU. Background color of the display and manipulation tool panel, and of the current function and item.
- 6 to 125. User colors.

The current selected color is indicated by a box around the color rectangle.

Modify Color Palette

All colors in the palette can be modified. First, scroll up or down the plaette and select the the color to be modified. To modify the color's red, green, and blue components, alter the three color tables in the display window. These colors can be altered by selecting the + (plus) or - (minus) icons, or by using dials 5, 6, and 7.

Once colors are changed, they can be stored in your model using STANDARD > COLOR | TABLE | SAVE.

- [illegible]
- [illegible]
- [illegible]
- [illegible]
- [illegible]
- [illegible] Menu and [illegible]

[illegible]

Index

A

access available details using LIBRARY, DETAIL in 3D SPACE mode 104
add back plane to view, AUXVIEW2 37
add breakout, AUXVIEW2 39
add callout to view containing clipping frame, AUXVIEW2 38
add callout to view, AUXVIEW2 34
add clipping frame, AUXVIEW2 38
add leader, TEXTD2 in DRAW mode 477
add models to work area, MODELS in DRAW and SPACE modes 298
add new text to existing text, TEXTD2 in DRAW mode 479
add node to leader, TEXTD2 in DRAW mode 478
add vertex to shape, SHAPE in DRAW mode 374
ADD, TEXTD2 in DRAW mode 479
alter graphic attributes of text, TEXTD2 in DRAW mode 476
analyze and create solid primitive parameters, SOLIDE in SPACE mode 396
analyze and create solid primitive parameters, SOLIDM in SPACE mode 420
analyze axis system, AXIS in 2D SPACE mode 54
analyze axis system, AXIS in 3D SPACE mode 59
analyze children of parametric element, PARAM3D in SPACE mode 308
analyze constraint status of parametric rule, PARAM3D in SPACE mode 308
analyze curvature with ANALYSIS in 3D SPACE mode 12
analyze curve tangents with ANALYSIS in 3D SPACE mode 12
analyze details or workspaces, DETAIL in 3D SPACE mode 108
analyze details or workspaces, DETAIL in DRAW mode 99
analyze details or workspaces, SYMBOL in DRAW mode 449
analyze dimensions, AUXVIEW2 43
analyze inertial characteristics of group of elements (solid assembly), ANALYSIS in 3D SPACE mode 13
analyze inertial characteristics of single element, ANALYSIS in 3D SPACE mode 13
analyze layers in use in current workspace, LAYER in DRAW and SPACE modes 212
analyze logical characteristics of elements, ANALYSIS in 2D SPACE mode 7
analyze logical links of DRAW elements 4
analyze logical links, DRAW elements 1
analyze mechanical data, SOLIDE in SPACE mode 396
analyze modeling data, SOLIDE in SPACE mode 396
analyze models (active and passive), MODELS in DRAW and SPACE modes 294
analyze numerical characteristics of 3D SPACE elements 9
analyze numerical characteristics, ANALYSIS in 2D SPACE mode 5
analyze numerical values of DRAW elements 1
analyze parents of parametric element, PARAM3D in SPACE mode 308
analyze pattern, PATTERN in DRAW mode 317
analyze relationship path dependent on element, PARAM3D in SPACE mode 308
analyze relative position of DRAW elements 2
analyze relative position of group of solids, SOLIDE in SPACE mode 397
analyze relative position of group of solids, SOLIDM in SPACE mode 421
analyze relative position of two solids, SOLIDE in SPACE mode 397
analyze relative position of two solids, SOLIDM in SPACE mode 421
analyze relative values of DRAW elements 1
analyze solid mechanical data, SOLIDM in SPACE mode 420
analyze solid modeling data, SOLIDM in SPACE mode 420
analyze status of filter applied to ditto, LAYER in DRAW and SPACE modes 210
analyze status of filter applied to screen, LAYER in DRAW and SPACE modes 210
analyze status of filter applied to view, LAYER in DRAW and SPACE modes 210
analyze tenon and mortise alignment, SOLIDM in SPACE mode 422
analyze transformation, TRANSFOR in 2D SPACE mode 496
analyze transformation, TRANSFOR in 3D SPACE mode 506
analyze transformations, TRANSFOR in DRAW mode 487
ANALYZE, AXIS in 2D SPACE mode 54
ANALYZE, AXIS in 3D SPACE mode 59
ANALYZE, GRAPHIC in DRAW mode 169
ANALYZE, GRAPHIC in SPACE mode 176–177
ANALYZE, PATTERN in DRAW mode 317
ANALYZE, TRANSFOR in DRAW mode 487–488
ANCHOR POINT in AUXVIEW2 46
ANGLE, DIMENS2 in DRAW mode 120–121

ANGLE, PLANE in SPACE mode 321
ANNOTAT, DRW STD 142–148
apply active work area, MODELS in DRAW and SPACE modes 297
apply and modify display attributes of edges, SOLIDM in SPACE mode 428
apply and modify display attributes of elements, SOLIDM in SPACE mode 428
apply display attributes to elements using other elements, GRAPHIC in SPACE mode 173
apply display attributes to elements, GRAPHIC in DRAW mode 167
apply display parameters of element to other elements, GRAPHIC in SPACE mode 174
apply duplicate element transformation, TRANSFOR in 3D SPACE mode 505
apply duplicate family transformation, TRANSFOR in 3D SPACE mode 505
apply duplicate set transformation, TRANSFOR in 3D SPACE mode 505
apply duplicate transformation of elements, TRANSFOR in 2D SPACE mode 495
apply duplicate transformation of family, TRANSFOR in 2D SPACE mode 495
apply duplicated transformation of set, TRANSFOR in 2D SPACE mode 495
apply layer filter created by combined ditto filters, LAYER in DRAW and SPACE modes 208
apply layer filter created by combined filters, LAYER in DRAW and SPACE modes 207
apply layer filter created by view filters combined in DRAW mode, LAYER in DRAW and SPACE modes 208
apply layer filter to ditto, LAYER in DRAW and SPACE modes 207
apply layer filter to screen display, LAYER in DRAW and SPACE modes 207
apply layer filter to view, LAYER in DRAW and SPACE modes 207
apply multi-area pattern, PATTERN in DRAW mode 314, 315
apply pattern, PATTERN in DRAW mode 313, 315
apply replaced element transformation, TRANSFOR in 3D SPACE mode 504
apply replaced family transformation, TRANSFOR in 3D SPACE mode 504
apply replaced set transformation, TRANSFOR in 3D SPACE mode 504
apply replacement transformation of element, TRANSFOR in 2D SPACE mode 494
apply replacement transformation of family, TRANSFOR in 2D SPACE mode 495
apply replacement transformation of set, TRANSFOR in 2D SPACE mode 494
apply rotation without definition, TRANSFOR in DRAW mode 488, 489
apply scaling, SOLIDE in SPACE mode 400
apply standard display attributes to elements, GRAPHIC in SPACE mode 173
apply transformations to parameterized solids or profiles, PARAM3D in SPACE mode 305
apply transformations to set with parametric data, PARAM3D in SPACE mode 305
apply transformations, TRANSFOR in DRAW mode 486
apply translation without definition, TRANSFOR in DRAW mode 488
APPLY, TRANSFOR in DRAW mode 486–487
APPR CRV, LIMIT1 in 2D SPACE mode 230
APPR CRV, LIMIT1 in 3D SPACE mode 237
APPROXIM, CURVE2 in 2D SPACE mode 81
APPROXIM, CURVE2 in 3D SPACE mode 88–89
APPROXIM, CURVE2 in DRAW mode 71
approximate composite curve by general curve, LIMIT1 in 2D SPACE mode 230
approximate composite curve by general curve, LIMIT1 in 3D SPACE mode 237
approximate curve by another curve at specific degree and tolerance value, CURVE2 in 2D SPACE mode 81
approximate curve by another, CURVE2 in 3D SPACE mode 88
approximate curve by another, CURVE2 in DRAW mode 71
ARROW LENGTH in AUXVIEW2 46
ARROW TYPES in AUXVIEW2 45
assign parameter to relationships or parameters, PARAM3D in SPACE mode 304
AUXVIEW Object window icons 25
AUXVIEW2 menu structure 24

B

break link between detail and library, DETAIL in 3D SPACE mode 109
break link between externally stored detail and library, DETAIL in DRAW mode 101
break link between externally stored symbol and respective library, SYMBOL in DRAW mode 451
BREAK, LIMIT1 in 2D SPACE mode 229
BREAK, LIMIT1 in 3D SPACE mode 235–236
BREAK, LIMIT1 in DRAW mode 222
BREAKOUT, MODELS in DRAW and SPACE modes 297

C

CALL SV, FILE in DRAW and SPACE modes 159–160
cancel comment with COMMANDS window, FILE in DRAW and SPACE modes 162
CatUtil icon, UTILITY 508
center window axis, IMAGE in DRAW and SPACE modes 190

change B-REP type of solid, SOLIDE in SPACE mode 401
change compact dittos to standard dittos, DETAIL in 3D SPACE mode 109
change compact dittos to standard dittos, DETAIL in DRAW mode 100
change current axis system, AXIS in 2D SPACE mode 53
change current axis system, AXIS in 3D SPACE mode 57
change current axis system, AXIS in DRAW mode 49
change current set, SETS in DRAW and SPACE modes 369
change current standard, DRW STD 147
change current view, AUXVIEW in DRAW mode 16
change current workspace, DETAIL in 3D SPACE mode 110
change current workspace, DETAIL in DRAW mode 101
change dress up element visualization, AUXVIEW in DRAW mode 21
change end symbol, TEXTD2 in DRAW mode 478
change frame element visibility, AUXVIEW in DRAW mode 18
change frame scale, AUXVIEW2 36
change graphical representation of clipping frame, AUXVIEW2 38
change graphical representation of elements, AUXVIEW2 30
change graphical representation of section definition, AUXVIEW2 35, 42
change identifier of view, AUXVIEW in DRAW mode 20
change identity of axes, AXIS in DRAW mode 51
change name of model, FILE in DRAW and SPACE modes 159
change normal orientation of selected plane, PLANE in SPACE mode 321
change orientation of text, TEXTD2 in DRAW mode 478
change parameters in DEFAULT panel, AUXVIEW2 33
change size parameter, DRW STD 148
change standard dittos to compact dittos, DETAIL in 3D SPACE mode 109
change standard dittos to compact dittos, DETAIL in DRAW mode 100
change view scale, AUXVIEW2 33
change view to section view, AUXVIEW2 40
CHANGE, AUXVIEW in DRAW mode 16
CHANGE, AXIS in 2D SPACE mode 53
CHANGE, AXIS in 3D SPACE mode 57
CHANGE, AXIS in DRAW mode 49
CHANGE, DETAIL in 3D SPACE mode 110
CHANGE, DETAIL in DRAW mode 101–102
CHANGE, SETS in DRAW and SPACE modes 369
check models for updating, MODELS in DRAW and SPACE modes 295
CIRCLE, CURVE2 in 3D SPACE mode 91
CIRCLE, CURVE2 in DRAW mode 63–66
close open shape, SHAPE in DRAW mode 376
combine multiple selection keywords 516–517
combine stored transformations, TRANSFOR in DRAW mode 486
combine transformations, TRANSFOR in 2D SPACE mode 497
combine transformations, TRANSFOR in 3D SPACE mode 506
COMBINE, COMBIVU in DRAW mode 61–62
commands, general, list of 518–520
COMMENT, FILE in DRAW and SPACE modes 160–162
CONCATEN, LIMIT1 in 2D SPACE mode 230
CONCATEN, LIMIT1 in 3D SPACE mode 236
CONCATEN, LIMIT1 in DRAW mode 222
concatenate element ends into single element, LIMIT1 in 2D SPACE mode 230
concatenate element ends into single element, LIMIT1 in 3D SPACE mode 236
concatenate element ends into single element, LIMIT1 in DRAW mode 222
CONIC, CURVE2 in DRAW mode 67–69
connect two curves by two fourth degree arcs, CURVE2 in 2D SPACE mode 82
connect two curves in space by planar fourth degree arc, CURVE2 in 2D SPACE mode 81
connect two curves in space, CURVE2 in 2D SPACE mode 81
connect two curves in space, CURVE2 in 3D SPACE mode 86
CONNECT, CURVE2 in 3D SPACE mode 86–88
CONNECT, CURVE2 in DRAW mode 71–72
control symbol size parameters and current standards, DRW STD 146
control upgrade of old text and dimensions, DRW STD 150
copy elements into current set, SETS in DRAW and SPACE modes 370
copy elements, MODELS in DRAW and SPACE modes 296
copy families, MODELS in DRAW and SPACE modes 296
copy geometric element from symbol occurrence and superimpose on corresponding element, SYMBOL in DRAW mode 452
copy model to defined receiving file, FILE in DRAW and SPACE modes 157
copy model without defining receiving file, FILE in DRAW and SPACE modes 158
copy or replace selected symbol occurrence or copy with symmetry around X or Y axis, SYMBOL in DRAW mode 447
copy plot sheet from sheet file to another, PLOT in DRAW and SPACE modes 333
copy plot windows from one plot sheet to another, PLOT in DRAW and SPACE modes 336
copy sets, MODELS in DRAW and SPACE modes 297
copy text with LINES COMMANDS window, FILE in

DRAW and SPACE modes 161
COPY, DETAIL in 3D SPACE mode 104–105
COPY, DETAIL in DRAW mode 96
COPY, FILE in DRAW and SPACE modes 157–158
COPY, SETS in DRAW and SPACE modes 370
COPY, SYMBOL in DRAW mode 446
create affinity about line, TRANSFOR in 3D SPACE mode 502
create affinity about three perpendicular planes, TRANSFOR in 3D SPACE mode 502
create and store features with parameterized geometry in library, SOLIDE in SPACE mode 388
create angled line segment through point, LINE in 2D SPACE mode 269
create angled line segment through point, LINE in DRAW mode 248
create angular dimension between lines, DIMENS2 in DRAW mode 121
create approximate solid (SOLM) from exact solid, SOLIDE in SPACE mode 406
create axis in 3D SPACE mode 56
create axis limit point, POINT in 2D SPACE mode 356
create axis limit points, POINT in DRAW mode 344
create balloons, TEXTD2 in DRAW mode 461
create blank model, FILE in DRAW and SPACE modes 159
create breaks in DRAW elements, LIMIT1 in 2D SPACE mode 229
create breaks in DRAW elements, LIMIT1 in 3D SPACE mode 235
create breaks in DRAW elements, LIMIT1 in DRAW mode 222
create center lines through point or circle using AXIS, MARK UP in DRAW mode 288
create chain of lines, LINE in 2D SPACE mode 262
create chain of lines, LINE in 3D SPACE mode 278
create chain of lines, LINE in DRAW mode 241
create chain set of equidistant parallel line segments, LINE in 2D SPACE mode 264
create chain set of equidistant parallel line segments, LINE in DRAW mode 243
create chain set of equidistant unlimited parallel lines, LINE in 2D SPACE mode 264
create chain set of equidistant unlimited parallel lines, LINE in DRAW mode 244
create chain set of points, POINT in 2D SPACE mode 355
create chain set of points, POINT in 3D SPACE mode 363
create chain set of points, POINT in DRAW mode 343
create chamfer by specifying angle and one leg length, LIMIT1 in 2D SPACE mode 227
create chamfer by specifying angle and one leg length, LIMIT1 in 3D SPACE mode 233
create chamfer by specifying angle and one leg length, LIMIT1 in DRAW mode 219
create chamfer by specifying two leg lengths, LIMIT1 in 2D SPACE mode 227
create chamfer by specifying two leg lengths, LIMIT1 in 3D SPACE mode 234
create chamfer by specifying two leg lengths, LIMIT1 in DRAW mode 220
create chamfered edges, SOLIDE in SPACE mode 394
create circle by diametrically opposed points, CURVE2 in 2D SPACE mode 74
create circle limit points, POINT in 2D SPACE mode 356
create circle limit points, POINT in DRAW mode 344
create circle tangent to other elements with or without defined center, CURVE2 in 2D SPACE mode 76
create circle tangent to two other circles with defined center, CURVE2 in 2D SPACE mode 76
create circle tangent to two other circles, CURVE2 in DRAW mode 66
create circle tangent to two other elements with center on line, CURVE2 in 2D SPACE mode 75
create circle using center and diameter, CURVE2 in 2D SPACE mode 74
create circle using center and diameter, CURVE2 in DRAW mode 64
create circle using center and radius, CURVE2 in 2D SPACE mode 74
create circle using center and radius, CURVE2 in DRAW mode 63
create circle using diametrically opposed points, CURVE2 in DRAW mode 64
create circle using threee points, CURVE2 in DRAW mode 64
create circle using two points and radius, CURVE2 in DRAW mode 64
create circle, CURVE2 in 3D SPACE mode 91
create circles using three points, CURVE2 in 2D SPACE mode 74
create circles using two points and radius, CURVE2 in 2D SPACE mode 74
create circular arc joining two elements while relimiting one element, LIMIT1 in 2D SPACE mode 226
create circular arc joining two elements while relimiting single element, LIMIT1 in DRAW mode 218
create circular arc joining two elements with first element limited, LIMIT1 in 3D SPACE mode 233
create circular arc joining two elements without relimiting, LIMIT1 in 2D SPACE mode 226
create circular arc joining two elements without relimiting, LIMIT1 in 3D SPACE mode 233
create circular arc joining two elements without relimiting, LIMIT1 in DRAW mode 219
create circular arc through point with defined center, CURVE2 in 2D SPACE mode 75
create circular arc using three points, CURVE2 in 2D SPACE mode 75

create circular arc using three points, CURVE2 in DRAW mode 64
create circular arc using two points and radius, CURVE2 in 2D SPACE mode 75
create circular arc using two points and radius, CURVE2 in DRAW mode 64
create circular arc with defined center and through two points, CURVE2 in 2D SPACE mode 75
create circular arc with defined center, CURVE2 in DRAW mode 65
create closed polygonal domains, SHAPE in DRAW mode 371
create cone, SOLIDE in SPACE mode 386
create cone, SOLIDM in SPACE mode 413
create conic arc defined by nonparallel tangents, CURVE2 in DRAW mode 68
create conic arc defined by parallel tangents, CURVE2 in DRAW mode 68
create conic arc defined by points and specific condition, CURVE2 in DRAW mode 68
create conic arc defined by two end points, intersection of tangents of points, and specific condition, CURVE2 in 2D SPACE mode 78
create conic arc defined by two end points, tangents of points, and specific condition, CURVE2 in 2D SPACE mode 79
create conic arc using 5 PTS, CURVE2 in DRAW mode 69
create conic arc using end points, CURVE2 in DRAW mode 69
create conic arc, CURVE2 in 2D SPACE mode 79
create conic using five-point method, CURVE2 in DRAW mode 67
create conic using three points, CURVE2 in DRAW mode 67
create conic using three-point method, CURVE2 in DRAW mode 67
create conical dimension of two lines, DIMENS2 in DRAW mode 123
create connecting arcs between pairs of parallel curves, LIMIT1 in 3D SPACE mode 235
create connecting arcs between two parallel lines, LIMIT1 in 2D SPACE mode 227
create connecting arcs between two parallel lines, LIMIT1 in 3D SPACE mode 234
create connecting arcs where connection length provides tangency points, LIMIT1 in 2D SPACE mode 228
create connecting curve, CURVE2 in 3D SPACE mode 86, 87
create connecting curves between pairs of parallel curves, LIMIT1 in 2D SPACE mode 229
create connecting curves between pairs of parallel curves, LIMIT1 in DRAW mode 221
create connection arcs between pairs of parallel lines, LIMIT1 in 3D SPACE mode 235
create constraint at midpoint of element, POINT in 2D SPACE mode 359
create constraint at midpoint of element, POINT in 3D SPACE mode 367
create contour limit points, POINT in DRAW mode 345
create copies of details in element form, DETAIL in DRAW mode 96
create copies of details, DETAIL in 3D SPACE mode 104
create copies of symbols in element form, SYMBOL in DRAW mode 446
create copy of existing work area, MODELS in DRAW and SPACE modes 298
create cross sections of infinite cylinders, COMBIVU in DRAW mode 61
create cuboid, SOLIDE in SPACE mode 385
create cuboid, SOLIDM in SPACE mode 411
create curve between two curves, CURVE2 in DRAW mode 71
create curve diametrical dimension through center point, DIMENS2 in DRAW mode 122
create curve limit points, POINT in 2D SPACE mode 356
create curve limit points, POINT in DRAW mode 344
create curve through point on plane normal to another, CURVE2 in 3D SPACE mode 92
create curve through specified constraint points, CURVE2 in 3D SPACE mode 88
create curves through specified constraint points, CURVE2 in 2D SPACE mode 79
create curves with curvature continuity at arc limits, CURVE2 in 2D SPACE mode 83
create curves with curvature continuity at arc limits, CURVE2 in 3D SPACE mode 89
create curves with tangency continuity at arc limits, CURVE2 in 2D SPACE mode 83
create curves with tangency continuity at arc limits, CURVE2 in 3D SPACE mode 89
create cylinder diametrical dimension, DIMENS2 in DRAW mode 122
create cylinder, SOLIDE in SPACE mode 380
create cylinder, SOLIDM in SPACE mode 411
create datum feature, TEXTD2 in DRAW mode 461, 472
create datum target, TEXTD2 in DRAW mode 461, 473
create definition of descriptions, DRW STD 145
create detail, DETAIL in 3D SPACE mode 107
create dimensions, DIMENS2 in DRAW mode 112, 117
create distance dimension between circles, DIMENS2 in DRAW mode 119
create distance dimension between line and circle, DIMENS2 in DRAW mode 119
create distance dimension between line and point, DIMENS2 in DRAW mode 119
create distance dimension between lines, DIMENS2 in DRAW mode 118
create distance dimension between point and circle,

DIMENS2 in DRAW mode 119
create distance dimension between points, DIMENS2 in DRAW mode 119
create ditto of detail in model, DETAIL in 3D SPACE mode 104
create dittos in grid, DETAIL in DRAW mode 95
create dittos using library details, DETAIL in DRAW mode 96
create DRAW mode views 14
create duplicate of defining view, AUXVIEW in DRAW mode 15
create edge dimension of circle, DIMENS2 in DRAW mode 124
create element constraint on curve arc end points, POINT in 2D SPACE mode 357
create element constraint on curve arc end points, POINT in 3D SPACE mode 365
create ellipse limit points, POINT in 2D SPACE mode 356
create ellipse limit points, POINT in DRAW mode 344
create ellipse using center and vertex and passing through point, CURVE2 in 2D SPACE mode 77
create ellipse using center and vertex, CURVE2 in DRAW mode 67
create ellipse with known axis, CURVE2 in 2D SPACE mode 77
create ellipse with known axis, CURVE2 in DRAW mode 67
create ellipse with known center, CURVE2 in 2D SPACE mode 77
create ellipse with known center, CURVE2 in DRAW mode 66
create empty work area, MODELS in DRAW and SPACE modes 297
create equidistant parallel line segments, LINE in 2D SPACE mode 263
create equidistant parallel line segments, LINE in DRAW mode 243
create equidistant parallel unlimited line, LINE in 2D SPACE mode 264
create equidistant parallel unlimited line, LINE in DRAW mode 244
create equidistant planes along line or curve, PLANE in SPACE mode 324
create equidistant planes between two points along line or curve, PLANE in SPACE mode 324
create equidistant points between two points by specified number of points, POINT in 2D SPACE mode 358, 359
create equidistant points between two points by specified number of points, POINT in 3D SPACE mode 366, 367
create equidistant points between two points separated by specified curvilinear length, POINT in 2D SPACE mode 358, 359
create equidistant points between two points separated by specified curvilinear length, POINT in 3D SPACE mode 366, 367
create equidistant points from specified origin, POINT in 2D SPACE mode 357, 358
create equidistant points from specified origin, POINT in 3D SPACE mode 365, 367
create equidistant points on circle using start point and angular dimensions, POINT in DRAW mode 345
create equidistant points on circle using start point and curvilinear dimensions, POINT in DRAW mode 346
create equidistant points on circle using start/end points and curvilinear dimension, POINT in DRAW mode 346
create equidistant points on circle using start/end points and number of points, POINT in DRAW mode 346
create equidistant points on line with specified distance, POINT in DRAW mode 345
create equidistant points on line with specified number of points, POINT in DRAW mode 345
create equidistant points parallel to line, POINT in 2D SPACE mode 357
create equidistant points parallel to line, POINT in 3D SPACE mode 366
create exact form of cut solid, SPC->DR2 in DRAW mode 431
create fillet between three faces by removing third selected face, SOLIDE in SPACE mode 393
create fillet radius between adjacent faces, SOLIDE in SPACE mode 393
create filter by combining filters using Boolean operators, LAYER in DRAW and SPACE modes 208
create fitted leader text, TEXTD2 in DRAW mode 470
create fitted text, TEXTD2 in DRAW mode 469
create geometric tolerances, TEXTD2 in DRAW mode 461, 470
create geometry and faces by intersecting solids with planes, SOLIDE in SPACE mode 405
create geometry and faces by intersecting solids with planes, SOLIDM in SPACE mode 429
create geometry by projecting solids onto plane, SOLIDE in SPACE mode 405
create geometry by projecting solids onto plane, SOLIDM in SPACE mode 429
create helix of revolution, CURVE2 in 3D SPACE mode 91
create horizontal line and vertical line through points, LINE in DRAW mode 242
create horizontal line and vertical line through two points, LINE in 2D SPACE mode 262
create horizontal line segment at specific distance from origin, LINE in 2D SPACE mode 265
create horizontal line segment at specific distance from origin, LINE in DRAW mode 245
create horizontal line segment tangent to curve, LINE in DRAW mode 245
create horizontal line segment through point, LINE in

DRAW mode 245
create horizontal or vertical line segment tangent to curve, LINE in 2D SPACE mode 265
create horizontal or vertical line segment through point, LINE in 2D SPACE mode 265
create horizontal or vertical line segment, LINE in 2D SPACE mode 262
create horizontal or vertical line segment, LINE in DRAW mode 242
create hyperbola limit points, POINT in 2D SPACE mode 356
create hyperbola limit points, POINT in DRAW mode 344
create infinite point grid, POINT in DRAW mode 349
create inverted transformation, TRANSFOR in 2D SPACE mode 496
create inverted transformation, TRANSFOR in 3D SPACE mode 506
create layer filter from layer filters stored in model, LAYER in DRAW and SPACE modes 208
create layer table in project file, LAYER in DRAW and SPACE modes 212
create leader text, TEXTD2 in DRAW mode 467
create length dimension of curve, DIMENS2 in DRAW mode 120
create length dimension of line, DIMENS2 in DRAW mode 120
create limit planes on multiarc curve, PLANE in SPACE mode 325
create limit planes on single arc curve or line segment, PLANE in SPACE mode 325
create limit point of axis, POINT in 3D SPACE mode 364
create limit point of constraint, POINT in 3D SPACE mode 365
create limit point of exact solid, POINT in 3D SPACE mode 365
create limit point of face, POINT in 3D SPACE mode 364
create limit point of plane, POINT in 3D SPACE mode 364
create limit point of polyhedral surface, POINT in 3D SPACE mode 365
create limit point of solid mock up, POINT in 3D SPACE mode, POINT in 3D SPACE mode 365
create limit point text, POINT in DRAW mode 345
create limit points of circle, POINT in 3D SPACE mode 364
create limit points of curve, POINT in 3D SPACE mode 364
create limit points of ellipse, POINT in 3D SPACE mode 364
create limit points of hyperbola, POINT in 3D SPACE mode 364
create limit points of line segment, POINT in 3D SPACE mode 364
create limit points of parabola, POINT in 3D SPACE mode 364
create limited witness lines in current view, COMBIVU in DRAW mode 61
create line between two point, LINE in 3D SPACE mode 278
create line between two points, LINE in 2D SPACE mode 261
create line between two points, LINE in DRAW mode 241
create line normal to plane of current window, LINE in 3D SPACE mode 285
create line segment equidistant between two parallel lines, LINE in 2D SPACE mode 264
create line segment limit points, POINT in 2D SPACE mode 356
create line segment limit points, POINT in DRAW mode 344
create line segment normal to circle and curve, LINE in 2D SPACE mode 267
create line segment normal to circle and curve, LINE in DRAW mode 246
create line segment normal to curve and through point, LINE in 2D SPACE mode 266
create line segment normal to curve and through point, LINE in DRAW mode 246
create line segment normal to face and through point, LINE in 3D SPACE mode 280
create line segment normal to line and curve, LINE in 2D SPACE mode 267
create line segment normal to line and curve, LINE in DRAW mode 246
create line segment normal to line and through point, LINE in 2D SPACE mode 266
create line segment normal to line and through point, LINE in 3D SPACE mode 281
create line segment normal to line and through point, LINE in DRAW mode 246
create line segment normal to line, parallel to plane, and through point, LINE in 3D SPACE mode 281
create line segment normal to plane and through point, LINE in 3D SPACE mode 280
create line segment normal to surfce and through point, LINE in 3D SPACE mode 280
create line segment normal to two lines and through point, LINE in 3D SPACE mode 280
create line segment on normal common to two lines, LINE in 3D SPACE mode 281
create line segment parallel to another and through point, LINE in 3D SPACE mode 279
create line segment parallel to intersection of two planes, LINE in 3D SPACE mode 279
create line segment parallel to line and tangent to curve, LINE in 2D SPACE mode 264
create line segment parallel to line and tangent to curve, LINE in DRAW mode 243
create line segment parallel to line and through point, LINE

in 2D SPACE mode 263
create line segment parallel to line and through point, LINE in DRAW mode 243
create line segment parallel to plane, normal to line, and through point, LINE in 3D SPACE mode 279
create line segment projected onto plane, LINE in 3D SPACE mode 282
create line segment tangent to curve and through point on curve, LINE in 3D SPACE mode 285
create line segment tangent to curve and through point, LINE in 2D SPACE mode 271
create line segment tangent to curve and through point, LINE in DRAW mode 251
create line segment tangent to curve end point, LINE in 3D SPACE mode 285
create line segment tangent to curve with angle from reference line, LINE in DRAW mode 251
create line segment tangent to curve with specific angle from reference line, LINE in 2D SPACE mode 271
create line segment tangent to two curves, LINE in 2D SPACE mode 271
create line segment tangent to two curves, LINE in DRAW mode 251
create line segment through point forming angle between projections onto two planes, LINE in 3D SPACE mode 283
create line segment through point with specific components, LINE in 3D SPACE mode 284
create line segment using Cartesian equation, LINE in 2D SPACE mode 270
create line segment using Cartesian equation, LINE in DRAW mode 250
create line segment with specific components and through point, LINE in 2D SPACE mode 270
create line segment with specific components and through point, LINE in DRAW mode 250
create lines normal to current 2D plane through point, LINE in 2D SPACE mode 274
create lines on intersection of two planes, LINE in 3D SPACE mode 282
create lines superimposed on edges of solid or polyhedral surface, LINE in 3D SPACE mode 286
create mean line from two or more selected points, LINE in 2D SPACE mode 272
create mean line from two or more selected points, LINE in 3D SPACE mode 286
create mean line from two or more selected points, LINE in DRAW mode 252
create mean plane of at least three points, PLANE in SPACE mode 321
create median segment of line segment, LINE in DRAW mode 247
create median segment of two points or line segment, LINE in 2D SPACE mode 268
create median segment of two points, LINE in DRAW mode 247
create middle plane between two points, PLANE in SPACE mode 324
create middle plane on curve, PLANE in SPACE mode 324
create middle plane on line, PLANE in SPACE mode 324
create midpoint of element, POINT in 2D SPACE mode 358
create midpoint of element, POINT in 3D SPACE mode 366
create midpoint on circle, POINT in DRAW mode 346
create move, TRANSFOR in 2D SPACE mode 493
create move, TRANSFOR in 3D SPACE mode 503
create move, TRANSFOR in DRAW mode 485
create multiple balloons, TEXTD2 in DRAW mode 474
create multiple dittos of existing detail, DETAIL in DRAW mode 94
create multiple lines of simple text, TEXTD2 in DRAW mode 466
create new axis system in current view, AXIS in DRAW mode 49
create new axis system, AXIS in 2D SPACE mode 52
create new axis system, AXIS in 3D SPACE mode 56
create new detail, DETAIL in DRAW mode 98
create new standards for use with DIMENS2 and TEXTD2, DRW STD 152
create new view with same background as defining view, AUXVIEW in DRAW mode 15
create new view, AUXVIEW in DRAW mode 14
create normal affinity about plane, TRANSFOR in 3D SPACE mode 501
create normal affinity, TRANSFOR in 2D SPACE mode 492
create normal affinity, TRANSFOR in DRAW mode 484
create normal infinite point grid, POINT in DRAW mode 349
create normal projection onto plane, TRANSFOR in 3D SPACE mode 502
create normal symmetry about line, TRANSFOR in 2D SPACE mode 491
create normal symmetry about line, TRANSFOR in DRAW mode 483
create normal symmetry about plane, TRANSFOR in 3D SPACE mode 499
create oblique affinity about plane, TRANSFOR in 3D SPACE mode 501
create oblique affinity, TRANSFOR in 2D SPACE mode 493
create oblique affinity, TRANSFOR in DRAW mode 485
create oblique projection onto plane, TRANSFOR in 3D SPACE mode 502
create oblique symmetry about line, TRANSFOR in 2D SPACE mode 491
create oblique symmetry about line, TRANSFOR in

DRAW mode 483
create oblique symmetry about plane, TRANSFOR in 3D SPACE mode 500
create offset solid, SOLIDM in SPACE mode 415
create offset solid, SOLIDE in SPACE mode 388
create open polygonal contours, SHAPE in DRAW mode 373
create or delete feature with BRANCH, SOLIDE in SPACE mode 389
create orthogonal point grid parallel to current axis system, POINT in DRAW mode 347
create orthographic view with AUXVIEW2 22, 25
create parabola limit points, POINT in 2D SPACE mode 356
create parabola limit points, POINT in DRAW mode 344
create parallel curve offset by linear distance, CURVE2 in 2D SPACE mode 80
create parallel curve symmetrical about another curve, CURVE2 in 2D SPACE mode 80
create parallel curves offset by linear distance, CURVE2 in DRAW mode 71
create parallel line segments, LINE in 2D SPACE mode 263
create parallel line segments, LINE in DRAW mode 243
create parameterized text to print on plot as comments, PLOT in DRAW and SPACE modes 331
create parameterized texts to print on plot, PLOT in DRAW and SPACE modes 337
create pipe, SOLIDE in SPACE mode 387
create pipe, SOLIDM in SPACE mode 414
create pitch circle center line using AXIS, MARK UP in DRAW mode 289
create planar curve parallel to another, CURVE2 in 3D SPACE mode 85
create plane by defining respective equation, PLANE in SPACE mode 319
create plane normal to current screen window, through point, and forming angle with line, PLANE in SPACE mode 323
create plane normal to current screen window, through point, and normal to line, PLANE in SPACE mode 323
create plane normal to current screen window, through point and parallel to line, PLANE in SPACE mode 323
create plane normal to current screen window through two points or line, PLANE in SPACE mode 322
create plane parallel to current screen and through point, PLANE in SPACE mode 322
create planes at angle to existing planes passing through line, PLANE in SPACE mode 321
create planes normal to selected elements, PLANE in SPACE mode 320
create planes parallel to plane, PLANE in SPACE mode 319
create plot sheet to store in sheet file, PLOT in DRAW and SPACE modes 332
create point by indication, POINT in 2D SPACE mode 352
create point by indication, POINT in DRAW mode 340
create point by projecting point onto element, POINT in 2D SPACE mode 352
create point by projecting point onto element, POINT in 3D SPACE mode 361
create point by projecting point onto element, POINT in DRAW mode 340
create point midway between two points, POINT in 2D SPACE mode 352
create point midway between two points, POINT in 3D SPACE mode 361
create point midway between two points, POINT in DRAW mode 340
create point on intersection of two elements, POINT in 2D SPACE mode 352
create point on intersection of two elements, POINT in 3D SPACE mode 361
create point on intersection of two elements, POINT in DRAW mode 340
create point via horizontal and vertical projection of point, POINT in 2D SPACE mode 353
create point via horizontal and vertical projection of point, POINT in DRAW mode 341
create point via horizontal projection of point, POINT in 2D SPACE mode 353
create point via horizontal projection of point, POINT in DRAW mode 341
create point via orthogonal projection of point, POINT in 2D SPACE mode 353
create point via orthogonal projection of point, POINT in 3D SPACE mode 361
create point via orthogonal projection of point, POINT in DRAW mode 341
create point via projection of point in specific direction, POINT in 2D SPACE mode 354
create point via projection of point in specific direction, POINT in DRAW mode 342
create point via projection of point in specified direction, POINT in 3D SPACE mode 362
create point via vertical and horizontal projection of point, POINT in 2D SPACE mode 354
create point via vertical and horizontal projection of point, POINT in DRAW mode 342
create point via vertical projection of point, POINT in 2D SPACE mode 353
create point via vertical projection of point, POINT in DRAW mode 341
create points of tangency on curve, POINT in 2D SPACE mode 359
create points of tangency on curve, POINT in 3D SPACE mode 368
create points of tangency on curve, POINT in DRAW mode 346
create points on limits of elements, POINT in 2D SPACE

mode 356
create polyhedral surface, SOLIDM in SPACE mode 416
create prism solid, SOLIDM in SPACE mode 412
create prism, SOLIDE in SPACE mode 379
create pseudo-parallel curves, CURVE2 in 3D SPACE mode 86
create pyramid, SOLIDE in SPACE mode 387
create pyramid, SOLIDM in SPACE mode 414
create quick plot without storage, PLOT in DRAW and SPACE modes 327
create rectangular domain, SHAPE in DRAW mode 372
create representation of screw thread using AXIS, MARK UP in DRAW mode 289
create revolution, SOLIDE in SPACE mode 381
create revolution, SOLIDM in SPACE mode 412
create rotation, TRANSFOR in 2D SPACE mode 491
create rotation, TRANSFOR in 3D SPACE mode 499
create roughness symbols, TEXTD2 in DRAW mode 461, 475
create scaling, TRANSFOR in 2D SPACE mode 492
create scaling, TRANSFOR in 3D SPACE mode 500
create scaling, TRANSFOR in DRAW mode 484
create segment bisector between two angled lines, LINE in 2D SPACE mode 268
create segment bisector or set of equiangular segment lines, LINE in DRAW mode 248
create set of equiangular line segments through point, LINE in 2D SPACE mode 269
create set of equiangular line segments through point, LINE in DRAW mode 248
create set of equiangular lines between two angled lines, LINE in 2D SPACE mode 268
create set of equiangular lines through point, LINE in 2D SPACE mode 269
create set of equiangular lines through point, LINE in DRAW mode 249
create set of equiangular segment lines between two angled lines, LINE in 2D SPACE mode 268
create set of repeated points, POINT in 2D SPACE mode 355
create set of repeated points, POINT in 3D SPACE mode 363
create set of repeated points, POINT in DRAW mode 343
create sets, SETS in DRAW and SPACE modes 369
create several points, POINT in 2D SPACE mode 352
create several points, POINT in 3D SPACE mode 361
create several points, POINT in DRAW mode 340
create shape parallel to another, SHAPE in DRAW mode 373
create shapes of cut polyhedral form of solid, SPC->DR2 in DRAW mode 431
create similtry, TRANSFOR in 2D SPACE mode 493
create similtry, TRANSFOR in DRAW mode 485
create simple text, TEXTD2 in DRAW mode 465
create single balloon, TEXTD2 in DRAW mode 474
create single connecting curve, LIMIT1 in 3D SPACE mode 234
create single connecting curve, LIMIT1 in DRAW mode 220
create single ditto of existing detail, DETAIL in DRAW mode 94
create single line of simple text associated with selected element, TEXTD2 in DRAW mode 466
create single line of simple text, TEXTD2 in DRAW mode 465
create single point, POINT in 2D SPACE mode 355
create single point, POINT in 3D SPACE mode 362
create single point, POINT in DRAW mode 343
create solid by closing surface or nonplanar face, SOLIDE in SPACE mode 388
create solid by closing surface or nonplanar face, SOLIDM in SPACE mode 416
create solid by projecting nonplanar faces onto plane, SOLIDM in SPACE mode 416
create solid by projecting surfaces onto plane, SOLIDM in SPACE mode 416
create solid by projecting surfaces or nonplanar faces onto plane, SOLIDE in SPACE mode 388
create solid from volume, SOLIDE in SPACE mode 388
create solid from volume, SOLIDM in SPACE mode 415
create SPACE mode geometry using DRAW mode elements, DRW>SPC in SPACE mode 140
create spaced dittos of existing detail, DETAIL in DRAW mode 95
create sphere, SOLIDE in SPACE mode 386
create sphere, SOLIDM in SPACE mode 413
create square domain, SHAPE in DRAW mode 372
create standard parallel curve, CURVE2 in DRAW mode 71
create sweep, SOLIDE in SPACE mode 381
create symmetry about line, TRANSFOR in 3D SPACE mode 500
create symmetry about point, TRANSFOR in 2D SPACE mode 492
create symmetry about point, TRANSFOR in 3D SPACE mode 500
create symmetry about point, TRANSFOR in DRAW mode 484
create symmetry around plane, LINE in 3D SPACE mode 287
create symmetry around plane, LINE in DRAW mode 254
create tangent circle defined by center, CURVE2 in DRAW mode 65
create tangent circle with or without defined center, CURVE2 in DRAW mode 65
create temporary elements in model, ANALYSIS in 2D SPACE mode 6
create text associated with SPACE elements, TEXT in

SPACE mode 457
create text, TEXTD2 in DRAW mode 461
create through planes, PLANE in SPACE mode 318
create torus, SOLIDE in SPACE mode 386
create torus, SOLIDM in SPACE mode 414
create transformation by specifying transformation axis system, TRANSFOR in 2D SPACE mode 494
create transformation by specifying transformation axis system, TRANSFOR in 3D SPACE mode 503
create translation, TRANSFOR in 2D SPACE mode 490
create translation, TRANSFOR in 3D SPACE mode 498
create translation, TRANSFOR in DRAW mode 482
create two connecting curves, LIMIT1 in DRAW mode 221
create unlimited angled line through point, LINE in 2D SPACE mode 269
create unlimited angled line through point, LINE in DRAW mode 249
create unlimited bisector between two angled lines, LINE in 2D SPACE mode 268
create unlimited bisector or set of equiangular lines between two angled lines, LINE in DRAW mode 248
create unlimited horizontal line at specific distance from origin, LINE in 2D SPACE mode 266
create unlimited horizontal line at specific distance from origin, LINE in DRAW mode 246
create unlimited horizontal line passing through point, LINE in DRAW mode 245
create unlimited horizontal line tangent to curve, LINE in DRAW mode 245
create unlimited horizontal or vertical line tangent to curve, LINE in 2D SPACE mode 266
create unlimited horizontal or vertical line through point, LINE in 2D SPACE mode 266
create unlimited horizontal or vertical line, LINE in 2D SPACE mode 262
create unlimited horizontal or vertical line, LINE in DRAW mode 242
create unlimited line equidistant between parallel lines, LINE in DRAW mode 244
create unlimited line equidistant between two parallel lines, LINE in 2D SPACE mode 265
create unlimited line normal to another and through point, LINE in 3D SPACE mode 282
create unlimited line normal to circle and curve, LINE in 2D SPACE mode 267
create unlimited line normal to circle and curve, LINE in DRAW mode 247
create unlimited line normal to curve and through point, LINE in 2D SPACE mode 267
create unlimited line normal to curve and through point, LINE in DRAW mode 247
create unlimited line normal to face and through point, LINE in 3D SPACE mode 281
create unlimited line normal to line and curve, LINE in 2D SPACE mode 267
create unlimited line normal to line and curve, LINE in DRAW mode 247
create unlimited line normal to line and through point, LINE in 2D SPACE mode 267
create unlimited line normal to line and through point, LINE in DRAW mode 247
create unlimited line normal to line, parallel to plane, and through point, LINE in 3D SPACE mode 282
create unlimited line normal to plane and through point, LINE in 3D SPACE mode 281
create unlimited line normal to surface and through point, LINE in 3D SPACE mode 281
create unlimited line normal to two lines and through point, LINE in 3D SPACE mode 281
create unlimited line on normal common to two lines, LINE in 3D SPACE mode 282
create unlimited line parallel to another and through point, LINE in 3D SPACE mode 279
create unlimited line parallel to intersection of two planes and through point, LINE in 3D SPACE mode 280
create unlimited line parallel to line and tangent to curve, LINE in 2D SPACE mode 265
create unlimited line parallel to line and tangent to curve, LINE in DRAW mode 244
create unlimited line parallel to line and through point, LINE in 2D SPACE mode 264
create unlimited line parallel to line and through point, LINE in DRAW mode 244
create unlimited line parallel to plane, normal to line, and through point, LINE in 3D SPACE mode 280
create unlimited line projected onto plane, LINE in 3D SPACE mode 283
create unlimited line tangent to curve and through point on curve, LINE in 3D SPACE mode 286
create unlimited line tangent to curve and through point, LINE in 2D SPACE mode 272
create unlimited line tangent to curve and through point, LINE in DRAW mode 252
create unlimited line tangent to curve end point, LINE in 3D SPACE mode 286
create unlimited line tangent to curve with angle from reference line, LINE in DRAW mode 252
create unlimited line tangent to curve with specific angle from reference line, LINE in 2D SPACE mode 272
create unlimited line tangent to two curves, LINE in 2D SPACE mode 272
create unlimited line tangent to two curves, LINE in DRAW mode 252
create unlimited line through point forming angle between projections onto two planes, LINE in 3D SPACE mode 284
create unlimited line through point with specified

components, LINE in 3D SPACE mode 284
create unlimited line using Cartesian equation, LINE in 2D SPACE mode 270
create unlimited line using Cartesian equation, LINE in DRAW mode 250
create unlimited line with specific components and through point, LINE in 2D SPACE mode 270
create unlimited line with specific components and through point, LINE in DRAW mode 250
create unlimited median of line segment, LINE in DRAW mode 247
create unlimited median of two points or line segment, LINE in 2D SPACE mode 268
create unlimited median of two points, LINE in DRAW mode 247
create unlimited vertical line at specific distance from origin, LINE in 2D SPACE mode 266
create unlimited vertical line at specific distance from origin, LINE in DRAW mode 246
create unlimited vertical line tangent to curve, LINE in DRAW mode 245
create unlimited vertical line through point, LINE in DRAW mode 245
create unlimited witness lines in current view, COMBIVU in DRAW mode 60
create unspecified closed shapes, SHAPE in DRAW mode 373
create unspecified line grid, LINE in DRAW mode 256
create unspecified orthogonal line grid, LINE in DRAW mode 256
create unspecified orthogonal or unspecified point grid, POINT in DRAW mode 348
create unspecified point grid, POINT in DRAW mode 350
create vertical line and horizontal line through points, LINE in DRAW mode 242
create vertical line and horizontal line through two points, LINE in 2D SPACE mode 262
create vertical line segment at specific distance from origin, LINE in 2D SPACE mode 265
create vertical line segment at specific distance from origin, LINE in DRAW mode 245
create vertical line segment tangent to curve, LINE in DRAW mode 245
create vertical line segment through point, LINE in DRAW mode 245
create view from existing DRAW elements, AUXVIEW in DRAW mode 15
create view from existing SPACE elements, AUXVIEW in DRAW mode 15
create view with new background, AUXVIEW in DRAW mode 14
create window in stored plot, PLOT in DRAW and SPACE modes 335
create witness lines, COMBIVU in DRAW mode 60
CREATE, AUXVIEW in DRAW mode 14–16
CREATE, AXIS in 2D SPACE mode 52–53
CREATE, AXIS in 3D SPACE mode 56–57
CREATE, AXIS in DRAW mode 49
CREATE, DETAIL in 3D SPACE mode 107
CREATE, DETAIL in DRAW mode 98
CREATE, DIMENS2 in DRAW mode 117–118
CREATE, DRW>SPC in SPACE mode 140
CREATE, FILE in DRAW and SPACE modes 159
CREATE, SETS in DRAW and SPACE modes 369
CREATE, TEXT in SPACE mode 457–458
CREATE, TEXTD2 in DRAW mode 461–??
CURVE, ANALYSIS in 2D SPACE mode 8
CURVE, ANALYSIS in 3D SPACE mode 12
CUSTOM, DRW STD 148–152
cut solid in section or cut view following UN_CUT, AUXVIEW2 32
cut SPACE mode surface elements and project into DRAW mode views, SPC->DRW in DRAW mode 434
cut SPACE mode wireframe elements and project into DRAW mode views, SPC->DRW in DRAW mode 434
CUT, SPC->DRW in DRAW mode 434
CVT CONT, CURVE2 in 2D SPACE mode 83–84
CVT CONT, CURVE2 in 3D SPACE mode 89

D

DATABASE option, PLOT in DRAW and SPACE modes 338
DEFAULT, AUXVIEW2 43–48
define and create auxiliary view, AUXVIEW2 28
define and create copy view, AUXVIEW2 29
define and create detail view, AUXVIEW2 28
define and create isometric view, AUXVIEW2 28
define and create primary view, AUXVIEW2 26
define and create principal view, AUXVIEW2 26
define and create section cut view, AUXVIEW2 28
define and create section view, AUXVIEW2 27
define and manage palettes and LPFK keyboards, FILE in DRAW and SPACE modes 162
define conic projection window, IMAGE in DRAW and SPACE modes 192
define constraints of profile used for solid contours and surfaces, PARAM3D in SPACE mode 301
define current model file, FILE in DRAW and SPACE modes 156
define cylindrical projection window, IMAGE in DRAW and SPACE modes 192
define general pattern visualization, PATTERN in DRAW mode 316
define general standards, STANDARD in DRAW and SPACE modes 436
define geometric standards, STANDARD in DRAW and SPACE modes 440

define graphic attributes of DRAW elements, STANDARD in DRAW and SPACE modes 439
define implicit relationships automatically, PARAM3D in SPACE mode 310
define line segment limitations by algebraic values, LINE in DRAW mode 240
define line segment limitations with types, LINE in DRAW mode 240
define line segments with AUTO LIM, LINE in DRAW mode 241
define line segments with ONE LIM, LINE in DRAW mode 239
define line segments with SYM LIM, LINE in DRAW mode 240
define logical characteristics of elements, ANALYSIS in 2D SPACE mode 7
define move by Euler angles, TRANSFOR in 3D SPACE mode 504
define move using three angles, TRANSFOR in 3D SPACE mode 503
define new symbol, SYMBOL in DRAW mode 448
define or modify color of layer, STANDARD in DRAW and SPACE modes 439
define or modify color of set, STANDARD in DRAW and SPACE modes 439
define parameter to be included in algebraic expression, PARAM3D in SPACE mode 304
define parameters used to create and display text created with TEXT function, TEXT in DRAW mode 454
define profile for parameterization, PARM3D in SPACE mode 302
define relative position of two elements, ANALYSIS in 2D SPACE mode 7
define screen, IMAGE in DRAW and SPACE modes 194
define sheet format, PLOT in DRAW and SPACE modes 329, 337
define standard settings for special elements, STANDARD in DRAW and SPACE modes 437
define step and number of lines in first quadrant only, LINE in DRAW mode 255
define step and number of lines in thrid quadrant only, LINE in DRAW mode 255
define step and number of lines, LINE in DRAW mode 254
define step and two limiting points, LINE in DRAW mode 255
definc symmetry of ditto or copy, DETAIL in 3D SPACE mode 106
define user CATIA model file, IUA 198
define user local modules file, IUA 198
define user panel file, IUA 197
define user procedure file, IUA 197
define valuated relationships and references, PARAM3D in SPACE mode 309
define window to view unspecified object, IMAGE in DRAW and SPACE modes 192
DEFINE, SYMBOL in DRAW mode 448
delete all dittos of used detail, DETAIL in DRAW mode 99
delete all occurrences of symbol and original definition, SYMBOL in DRAW mode 449
delete all occurrences of symbol but retain original definition, SYMBOL in DRAW mode 449
delete all parameterized geometry, PARAM3D in SPACE mode 309
delete back plane in view, AUXVIEW2 37
delete branch from history of solid, SOLIDE in SPACE mode 402
delete branch from history of solid, SOLIDM in SPACE mode 426
delete breakout created via SPACE window, AUXVIEW2 40
delete breakout defined by profile and planes, AUXVIEW2 40
delete callout in view containing clipping frame, AUXVIEW2 38
delete callout in view, AUXVIEW2 34
delete clipping frame, AUXVIEW2 38
delete dimension, PARAM3D in SPACE mode 307
delete dittos of used detail, DETAIL in 3D SPACE mode 107
delete elements from model, ERASE in DRAW and SPACE modes 154
delete elements, ERASE in DRAW and SPACE modes 154
delete existing node on leader, TEXTD2 in DRAW mode 478
delete filters, LAYER in DRAW and SPACE modes 209
delete infinite point grid, POINT in DRAW mode 350
delete leader, TEXTD2 in DRAW mode 477
delete line with LINES COMMANDS window, FILE in DRAW and SPACE modes 161
delete model from model file, FILE in DRAW and SPACE modes 158
delete object from library, LIBRARY in DRAW and SPACE modes 215
delete parameter, PARAM3D in SPACE mode 305
delete parameters of primitive or wireframe, PARAM3D in SPACE mode 309
delete parameters of profile, PARAM3D in SPACE mode 309
delete plot windows, PLOT in DRAW and SPACE modes 329
delete relationship, PARAM3D in SPACE mode 309
delete screen, IMAGE in DRAW and SPACE modes 196
delete section callout, AUXVIEW2 41
delete section, AUXVIEW2 40
delete sets, SETS in DRAW and SPACE modes 369
delete sheet from sheet file, PLOT in DRAW and SPACE modes 333
delete stored keyboard, FILE in DRAW and SPACE modes

164
delete stored transformations, TRANSFOR in DRAW mode 487
delete text in view, AUXVIEW2 35
delete text, TEXTD2 in DRAW mode 479
delete transformation, TRANSFOR in 2D SPACE mode 496
delete transformation, TRANSFOR in 3D SPACE mode 506
delete unused detail after visualization, DETAIL in 3D SPACE mode 107
delete unused detail after visualization, DETAIL in DRAW mode 99
delete unused detail directly, DETAIL in 3D SPACE mode 107
delete unused detail directly, DETAIL in DRAW mode 98
delete unused symbol, SYMBOL in DRAW mode 449
delete used detail and all associated dittos, DETAIL in DRAW mode 99
delete used detail and associated dittos, DETAIL in 3D SPACE mode 107
delete vertex from shape, SHAPE in DRAW mode 374
delete view, AUXVIEW2 29
delete views, AUXVIEW in DRAW mode 20
delete windows from stored sheet, PLOT in DRAW and SPACE modes 336
delete windows, IMAGE in DRAW and SPACE modes 193
delete work area, MODELS in DRAW and SPACE modes 298
DELETE, AUXVIEW in DRAW mode 20
DELETE, DETAIL in 3D SPACE mode 107–108
DELETE, FILE in DRAW and SPACE modes 158
DELETE, LIBRARY in DRAW and SPACE modes 215
DELETE, SETS in DRAW and SPACE modes 369–370
DELETE, TEXTD2 in DRAW mode 479–480
DEPTH, CURVE2 in 3D SPACE mode 90
deselect elements, ERASE in DRAW and SPACE modes 154
DETAIL, for view callout lines in AUXVIEW2 45
DIMENS2 menu structure 113–117
display attributes of elements in alphanumeric window, GRAPHIC in DRAW mode 169
display data with INFORMATION window, FILE in DRAW and SPACE modes 161
display element attributes in alphanumeric window, GRAPHIC in SPACE mode 176
display elements on specific layer, LAYER in DRAW and SPACE modes 211
display graphic attributes of elements, GRAPHIC in SPACE mode 173
display line as rubber band, LINE in DRAW mode 239
display list of utilities, UTILITY 508
display model from current model file, FILE in DRAW and SPACE modes 157
display symmetry around line, LINE in 2D SPACE mode 274
DISTANCE, DIMENS2 in DRAW mode 118–120
DITTO, DETAIL in 3D SPACE mode 103–104
DRAW element identifiers, list of 511–512
DRAW ELT, STANDARD in DRAW and SPACE modes 439
drop models from work area, MODELS in DRAW and SPACE modes 298
DTAILING, AUXVIEW in DRAW mode 21
duplicate branch of solid by rotation, SOLIDE in SPACE mode 403
duplicate branch of solid by rotation, SOLIDM in SPACE mode 427
duplicate branch of solid by symmetry, SOLIDE in SPACE mode 404
duplicate branch of solid by symmetry, SOLIDM in SPACE mode 427
duplicate branch of solid by translation, SOLIDE in SPACE mode 403
duplicate branch of solid using stored translation, SOLIDE in SPACE mode 404
duplicate branch of solid, SOLIDM in SPACE mode 426
duplicate branch of sollid using stored translation, SOLIDM in SPACE mode 427
duplicate solid primitive, SOLIDE in SPACE mode 406

E

EDGE, LINE in 2D SPACE mode 274
EDGE, LINE in 3D SPACE mode 285
edit parameter, PARAM3D in SPACE mode 304
edit SPACE text, TEXT in SPACE mode 459
edit text, TEXTD2 in DRAW mode 476
EDIT, TEXT in SPACE mode 459
eliminate solid from view, AUXVIEW2 31
ELLIPSE, CURVE2 in DRAW mode 66–67
Enclose or trap group of elements, GROUP in DRAW mode 179
Enclose or trap group of elements, GROUP in SPACE mode 183
end session, FILE in DRAW and SPACE modes 164
EQUATION, PLANE in SPACE mode 319
ERASE, ERASE in DRAW and SPACE modes 154
ERASE, TEXT in SPACE mode 458–459
ERASE, TRANSFOR in DRAW mode 487
exclude elements from selected group, GROUP in DRAW mode 179
exclude elements from selected group, GROUP in SPACE mode 183
execute IUA procedure, IUA 198
EXIT, FILE in DRAW and SPACE modes 164
explode ditto in current set, DETAIL in 3D SPACE mode 106

explode ditto in current workspace, DETAIL in 3D SPACE mode 106
explode ditto in model, DETAIL in 3D SPACE mode 106
EXPLODE, DETAIL in 3D SPACE mode 106–107
EXPLODE, DETAIL in DRAW mode 98
EXPLODE, SYMBOL in DRAW mode 448
extend two elements simultaneously to intersection point, LIMIT1 in 2D SPACE mode 224
extend two elements simultaneously to intersection point, LIMIT1 in 3D SPACE mode 231
extend two elements simultaneously to intersection point, LIMIT1 in DRAW mode 217
extract edge geometry from solid, SOLIDE in SPACE mode 406
extract faces from solids, SOLIDE in SPACE mode 407
extract primitive subelements from solid, SOLIDE in SPACE mode 406
extract primitive subelements from solid, SOLIDM in SPACE mode 429
extract skin from primitive subelement, SOLIDE in SPACE mode 406
extract skins from primitives forming solid edge, SOLIDE in SPACE mode 406
extract volumes from solids, SOLIDE in SPACE mode 405
EXTRACT, SYMBOL in DRAW mode 452
EXTRAPOL, LIMIT1 in 2D SPACE mode 230
EXTRAPOL, LIMIT1 in 3D SPACE mode 236–237
EXTRAPOL, LIMIT1 in DRAW mode 223
extrapolate a line or curve, LIMIT1 in 3D SPACE mode 236
extrapolate line or curve, LIMIT1 in 2D SPACE mode 230
extrapolate line or curve, LIMIT1 in DRAW mode 223

F

FAMILY, LIBRARY in DRAW and SPACE modes 214
FEATURE, SOLIDE in SPACE mode 388–389
FILE in DRAW and SPACE modes 156
FILE, LIBRARY in DRAW and SPACE modes 213
fillet edge, SOLIDE in SPACE mode 392
fix axis, AXIS in 2D SPACE mode 54
fix axis, AXIS in 3D SPACE mode 58
fix axis, AXIS in DRAW mode 50
FIXED, AXIS in 2D SPACE mode 54
FIXED, AXIS in 3D SPACE mode 58
FIXED, AXIS in DRAW mode 50
free constraints at point on spline, CURVE2 in DRAW mode 70

G

generate elements using curves and elements, COMBIVU in DRAW mode 62
generate elements using planes and elements, COMBIVU in DRAW mode 61

H

HELIX, CURVE2 in 3D SPACE mode 91–92
HIDDEN LINES in AUXVIEW2 46
hide elements permanently, ERASE in DRAW and SPACE modes 155
hide elements, ERASE in DRAW and SPACE modes 154
hide geometric element from symbol occurrence, SYMBOL in DRAW mode 452
hide or make frames visible, AUXVIEW2 36
hide text, TEXTD2 in DRAW mode 481
highlight identifiers for element by type using LOGICAL | CHILDREN 8
highlight identifiers for element by type using LOGICAL | PARENTS 8
highlight identifiers in family of element by type using LOGICAL | FAMILY 8
highlight or make elements blink, GRAPHIC in DRAW mode 167
highlight or make elements blink, GRAPHIC in SPACE mode 172
horizontal definition, LINE in 2D SPACE mode 259

I

identify blank lines or subtexts, TEXTD2 in DRAW mode 481
import previously published passive solids into model, SOLIDE in SPACE mode 389
impose curvature, CURVE2 in DRAW mode 70
impose points through which spline passes, CURVE2 in DRAW mode 69
impose symmetry on models, MODELS in DRAW and SPACE modes 296
impose tangent, CURVE2 in DRAW mode 69
inactivate solid features, SOLIDE in SPACE mode 395
include elements in selected group, GROUP in DRAW mode 178
include elements in selected group, GROUP in SPACE mode 182
insert gaps in callout lines, AUXVIEW2 34, 41
insert line with LINES COMMANDS window, FILE in DRAW and SPACE modes 161
insert operation into history of solid, SOLIDE in SPACE mode 402
insert operation into history of solid, SOLIDM in SPACE mode 425
INTENSITY in AUXVIEW2 46
INTERSEC, LINE in 3D SPACE mode 282
intersect two closed shapes, SHAPE in DRAW mode 375
invert axes of axis system, AXIS in DRAW mode 50
invert axes, AXIS in 2D SPACE mode 54
invert axes, AXIS in 3D SPACE mode 58
invert curves, CURVE2 in 2D SPACE mode 84
invert curves, CURVE2 in DRAW mode 72

invert elements CCV, CRV, and LN, CURVE2 in 3D SPACE mode 92
invert stored transformation, TRANSFOR in DRAW mode 486
invert view, AUXVIEW2 34
INVERT, AXIS in 2D SPACE mode 53–54
INVERT, AXIS in 3D SPACE mode 58
INVERT, AXIS in DRAW mode 50
INVERT, CURVE2 in 2D SPACE mode 84
INVERT, CURVE2 in 3D SPACE mode 92
INVERT, CURVE2 in DRAW mode 72
isolate parameter from relationship, PARAM3D in SPACE mode 304
isolate text, TEXTD2 in DRAW mode 480
isolate view to prevent revision, AUXVIEW2 32

J

join lines by two curves, LIMIT1 in 3D SPACE mode 235
join lines with two curves tangent to each line, LIMIT1 in 2D SPACE mode 228

K

KEEP, KEEP in DRAW and SPACE modes 204–205
KEYBOARD, FILE in DRAW and SPACE modes 162–164
keywords requiring complementary interaction, list of 514–515
keywords, combining multiple selection 516–517
keywords, multiple interaction, list of 515–516
keywords, single interaction requiring additional information, list of 514
keywords, single interaction requiring no additional information, list of 514

L

lay out window, PLOT in DRAW and SPACE modes 337
lay out windows, PLOT in DRAW and SPACE modes 329
LENGTH, DIMENS2 in DRAW mode 120
limit or break shape using line, SHAPE in DRAW mode 376
limit two DRAW elements with circular arc, LIMIT1 in 2D SPACE mode 225
limit two DRAW elements with circular arc, LIMIT1 in DRAW mode 218
limit two space elements with circular arc, LIMIT1 in 3D SPACE mode 232
line limitation using AUTO LIM, LINE in 2D SPACE mode 261
line limitation using AUTO LIM, LINE in 3D SPACE mode 278
line limitation using ONE LIM, LINE in 2D SPACE mode 259
line limitation using ONE LIM, LINE in 3D SPACE mode 276
line limitation using SYM LIM, LINE in 2D SPACE mode 261
line limitation using SYM LIM, LINE in 3D SPACE mode 277
line limitation using TWO LIM, LINE in 2D SPACE mode 260
line limitation using TWO LIM, LINE in 3D SPACE mode 277
line limitation, LINE in 2D SPACE mode 259
line limitation, LINE in 3D SPACE mode 276
LINES, COMBIVU in DRAW mode 60–61
LINETYPE, STANDARD in DRAW and SPACE modes 442
link a set with current set, SETS in DRAW and SPACE modes 370
LINK, PARAM3D in SPACE mode 303
LINK, SETS in DRAW and SPACE modes 370–??
list models contained in work area, MODELS in DRAW and SPACE modes 299
list plot windows for information only, PLOT in DRAW and SPACE modes 329
locally modify graphic characteristics of pattern, PATTERN in DRAW mode 316
locate and anchor text, TEXTD2 in DRAW mode 478
lock (free) view for modification, AUXVIEW2 32
lock availability of descriptions, DRW STD 145
LOGICAL | FAMILY 12
LOGICAL | PARENTS 12
LOGICAL, ANALYSIS in 2D SPACE mode ??–8
LOGICAL, ANALYSIS in 3D SPACE mode 11–12
analyze logical characteristics of elements
LOGICAL | CHILDREN 12
LOGICAL, ANALYSIS in DRAW mode 4

M

make connecting curve pass through specified point, CURVE2 in 2D SPACE mode 82, 83
make connecting curve through specified point, CURVE2 in 3D SPACE mode 87, 88
make elements unselectable, ERASE in DRAW and SPACE modes 155
make temporary elements permanent, ANALYSIS in 2D SPACE mode 7
manage and create patterns used with PATTERN, DRW STD 153
manage display of infinite grid, POINT in DRAW mode 350
manage parameterized texts, PLOT in DRAW and SPACE modes 329
manage plotting scale and format, PLOT in DRAW and SPACE modes 332
MANAGE, DETAIL in 3D SPACE mode 108–109
manipulate layers independently of filters, LAYER in DRAW and SPACE modes 211

MEAN, LINE in 2D SPACE mode 272
MEAN, LINE in 3D SPACE mode 286
MEAN, LINE in DRAW mode 252
MEAN, PLANE in SPACE mode 321–322
menu, File pull-down options/tools, list of 521–522
menu, Select pull-down 522
menu, View pull-down options/tools, list of 523–524
menus, pull-down 521–528
merge contents of two plot sheets, PLOT in DRAW and SPACE modes 334
merge passive model with active model, KEEP in DRAW and SPACE modes 205
MERGE, KEEP in DRAW and SPACE modes 205
MIN GAP, for fillet representation in AUXVIEW2 47
MOD GEN, GRAPHIC in SPACE mode 171–173
MOD SPEC, GRAPHIC in SPACE mode 173
MOD VISU, GRAPHIC in SPACE mode 173–175
MODEL, STANDARD in DRAW and SPACE modes 440–442
modify axis in 3D SPACE mode 56
modify back plane in view, AUXVIEW2 37
modify circle, CURVE2 in 3D SPACE mode 91
modify circle, CURVE2 in DRAW mode 66
modify color of elments, GRAPHIC in SPACE mode 172
modify color of primitive, SOLIDE in SPACE mode 405
modify color of primitive, SOLIDM in SPACE mode 428
modify connecting curve, CURVE2 in 3D SPACE mode 86, 87
modify defining plane in view, AUXVIEW2 34
modify detail comment, DETAIL in 3D SPACE mode 109
modify dimensions using DIMENS2 112
modify discretization, GRAPHIC in SPACE mode 174
modify display attributes of elements with other elements, GRAPHIC in DRAW mode 167
modify display attributes of elements with other elements, GRAPHIC in SPACE mode 172
modify display attributes of point type elements, GRAPHIC in DRAW mode 166
modify display attributes of point type elements, GRAPHIC in SPACE mode 171
modify display attributes of vector type elements, GRAPHIC in DRAW mode 166
modify display attributes of vector type elements, GRAPHIC in SPACE mode 171
modify display characteristics of dimensions, PARAM3D in SPACE mode 307
modify display mode type, GRAPHIC in SPACE mode 174
modify DRAW mode views 14
modify element color, GRAPHIC in DRAW mode 167
modify end point of line segment or stretch line segment, LINE in 2D SPACE mode 274
modify end point of line segment or stretch line segment, LINE in 3D SPACE mode 287
modify end point of line segment or stretch line segment, LINE in DRAW mode 254
modify hidden line parameters, STANDARD in DRAW and SPACE modes 438
modify hidden line removal, GRAPHIC in SPACE mode 174
modify identifier of solid subelement, SOLIDM in SPACE mode 428
modify keywords of library object, LIBRARY in DRAW and SPACE modes 215
modify layer table in project file, LAYER in DRAW and SPACE modes 212
modify line geometry, LINE in 2D SPACE mode 273
modify or create view frame, AUXVIEW2 36
modify orthographic view with AUXVIEW2 22, 25
modify parameter values, PARAM3D in SPACE mode 303
modify parameters of solid primitive, SOLIDE in SPACE mode 401
modify parameters of solid primitive, SOLIDM in SPACE mode 424
modify parameters of SPACE text, TEXT in SPACE mode 458
modify position of circle center, CURVE2 in 2D SPACE mode 77
modify position of circle center, CURVE2 in DRAW mode 66
modify position of shape vertices, SHAPE in DRAW mode 374
modify primitives with contours, SOLIDE in SPACE mode 400
modify primitives with contours, SOLIDM in SPACE mode 425
modify quick plot window, PLOT in DRAW and SPACE modes 327
modify radius of circle, CURVE2 in 2D SPACE mode 76
modify radius of circle, CURVE2 in DRAW mode 66
modify scale and center of view with clipping frame, AUXVIEW in DRAW mode 18
modify scale of ditto or copy, DETAIL in DRAW mode 97
modify section, AUXVIEW2 41
modify size of clipping frame, AUXVIEW in DRAW mode 17
modify solid thickness, SOLIDE in SPACE mode 391
modify status of filter applied to current view, LAYER in DRAW and SPACE modes 209
modify status of filter applied to ditto or macroprimitive, LAYER in DRAW and SPACE modes 209
modify status of filter, LAYER in DRAW and SPACE modes 209
modify status of general applied filter, LAYER in DRAW and SPACE modes 209
modify symbol scale, SYMBOL in DRAW mode 447
modify symbol symmetry, SYMBOL in DRAW mode 447
modify symmetry of ditto or copy, DETAIL in DRAW mode 97

modify text in view, AUXVIEW2 35
modify thickness of vector type elements, GRAPHIC in DRAW mode 166
modify thickness of vector type elments, GRAPHIC in SPACE mode 171
modify transparency of view for SPACE elements, AUXVIEW in DRAW mode 20
modify transparency, GRAPHIC in SPACE mode 174
modify view scale, AUXVIEW in DRAW mode 18
modify visualization of SPACE text, TEXT in SPACE mode 460
modify window display in screen plane, IMAGE in DRAW and SPACE modes 189
modify windows in stored plot, PLOT in DRAW and SPACE modes 335
MODIFY, AUXVIEW in DRAW mode 16–20
MODIFY, DETAIL in 3D SPACE mode 105–106
MODIFY, GRAPHIC in DRAW mode 166–168
MODIFY, LIBRARY in DRAW and SPACE modes 215
MODIFY, LINE in DRAW mode 252–254
MODIFY, PARAM3D in SPACE mode 303
MODIFY, TEXT in SPACE mode 458
move callout in current view, AUXVIEW2 34
move current point, CURVE2 in DRAW mode 70
move dimensions between views, AUXVIEW2 42
move display in window, IMAGE in DRAW and SPACE modes 190
move existing node on leader, TEXTD2 in DRAW mode 477
move model from one file to another, FILE in DRAW and SPACE modes 158
move models, MODELS in DRAW and SPACE modes 296
move parameters from LIST window into DEFINITION window, DRW STD 150
move parameters within DEFINITION window, DRW STD 150
move plot sheet from one sheet file to another, PLOT in DRAW and SPACE modes 333
move SPACE text, TEXT in SPACE mode 459
move text with LINES COMMANDS window, FILE in DRAW and SPACE modes 161
move view clipping frame, AUXVIEW in DRAW mode 16
MOVE, FILE in DRAW and SPACE modes 158
MOVE, TEXT in SPACE mode 459

N

NO PICK/PICK, ERASE in DRAW and SPACE modes 155
NO SHOW//SHOW, ERASE in DRAW and SPACE modes 155
NORMAL, PLANE in SPACE mode 320–321
NUMERIC, ANALYSIS in 2D SPACE mode 5–7
NUMERIC, ANALYSIS in 3D SPACE mode 9–11
  analyze elements 9
  create temporary elements in model 11
NUMERIC, ANALYSIS in DRAW mode 1–2

OFFSET, SHAPE in DRAW mode 373–374
ORIENTN, PLANE in SPACE mode 321
orthogonally project edges of solids onto one or more DRAW views, SPC->DR2 in DRAW mode 432
orthogonally project one or more complex faces onto DRAW views, SPC->DR2 in DRAW mode 432
orthogonally project one or more complex volumes onto DRAW views, SPC->DR2 in DRAW mode 432
orthogonally project one or more exact solids onto DRAW views, SPC->DR2 in DRAW mode 432

P

pack model, ERASE in DRAW and SPACE modes 154
PACK, ERASE in DRAW and SPACE modes 154–155
PARALLEL, CURVE2 in 3D SPACE mode 85–86
PARALLEL, CURVE2 in DRAW mode 70–71
PARALLEL, PLANE in SPACE mode 319–320
parameterize solid primitives or wireframe elements, PARAM3D in SPACE mode 302
PART EDITOR 407–409
Part Editor window, SOLIDE in SPACE mode 407
PART EDITOR, SOLIDE in SPACE mode 407–409
PATTERN, DRW STD 153
perform assemble operation, SOLIDM in SPACE mode 419
perform draft operation on solid, SOLIDE in SPACE mode 392
perform intersection operation, SOLIDM in SPACE mode 417
perform intersection operations, SOLIDE in SPACE mode 390
perform local transformations, IMAGE in DRAW and SPACE modes 196
perform rotation on solid geometry, SOLIDM in SPACE mode 423
perform rotation on solid geometry, SOLLIDE in SPACE mode 399
perform sewing operation, SOLIDE in SPACE mode 395
perform shell operation, SOLIDE in SPACE mode 394
perform sort out operation, SOLIDE in SPACE mode 395
perform sort out operation, SOLIDM in SPACE mode 419
perform split operation, SOLIDE in SPACE mode 394
perform split operation, SOLIDM in SPACE mode 418
perform stored transformation on solid geometry, SOLIDE in SPACE mode 399
perform stored transformations on solid geometry, SOLIDM in SPACE mode 424
perform subtraction operation, SOLIDM in SPACE mode 418
perform subtraction operation, SOLIDE in SPACE mode

391
perform symmetry on solid geometry, SOLIDE in SPACE mode 399
perform symmetry on solid geometry, SOLIDM in SPACE mode 424
perform transformation on solid geometry, SOLIDE in SPACE mode 398
perform union operation, SOLIDM in SPACE mode 417
perform union operation, SOLIDE in SPACE mode 390
place ditto of detail in model, DETAIL in 3D SPACE mode 104
place elements in alternative workspaces, DETAIL in 3D SPACE mode 103
place library symbols, SYMBOL in DRAW mode 445
place multiple symbols in model, SYMBOL in DRAW mode 444
place single symbol in model, SYMBOL in DRAW mode 444
place spaced symbols in model, SYMBOL in DRAW mode 444
place symbols in grid, SYMBOL in DRAW mode 445
POL EDGE, LINE in 3D SPACE mode 286
preview plot, PLOT in DRAW and SPACE modes 329, 337
preview setup from PRINT option, PLOT in DRAW and SPACE modes 332
PRIMITIV, PARAM3D in SPACE mode 302
print plot, PLOT in DRAW and SPACE modes 331
PRL WINW, PLANE in SPACE mode 322
profile plots, PLOT in DRAW and SPACE modes 332
project SPACE mode wireframe and surface elements into DRAW mode views, SPC->DRW in DRAW mode 434
PROJECT, SPC->DRW in DRAW mode 434–435
PROJECTION MODE in AUXVIEW2 46
PROJECTION, for view callout lines in AUXVIEW2 44
provide tangency points of connecting curves with connection length, LIMIT1 in DRAW mode 221
PTS CST, CURVE2 in 2D SPACE mode 79–80
PTS CST, CURVE2 in 3D SPACE mode 88

R

read object from library family into current model, LIBRARY in DRAW and SPACE modes 214
read sheet from sheet file, PLOT in DRAW and SPACE modes 333
READ, FILE in DRAW and SPACE modes 157
READ, LIBRARY in DRAW and SPACE modes 214
recall screen, IMAGE in DRAW and SPACE modes 195
recall windows, IMAGE in DRAW and SPACE modes 193
redisplay hidden elements, ERASE in DRAW and SPACE modes 155
reframe window, IMAGE in DRAW and SPACE modes 191
RELATIVE, ANALYSIS in 2D SPACE mode ?7
RELATIVE, ANALYSIS in 3D SPACE mode 11
RELATIVE, ANALYSIS in DRAW mode 2–4
relimit one end of DRAW element, LIMIT1 in 2D SPACE mode 225
relimit one end of DRAW element, LIMIT1 in 3D SPACE mode 232
relimit one end of DRAW element, LIMIT1 in DRAW mode 218
remove all default parameters from DEFINITION window into LIST window, DRW STD 149
remove default parameters from DEFINITION window to LIST window, DRW STD 149
remove hidden lines, SOLIDE in SPACE mode 404
remove unoccupied space from model, ERASE in DRAW and SPACE modes 154
rename available descriptions, DRW STD 146
rename detail (or modify detail comment), DETAIL in DRAW mode 101
rename detail, DETAIL in 3D SPACE mode 109
rename filters, LAYER in DRAW and SPACE modes 209
rename identifiers according to string, IDENTIFY in DRAW and SPACE modes 187
rename identifiers by type, IDENTIFY in DRAW and SPACE modes 187
rename identifiers selected from list, IDENTIFY in DRAW and SPACE modes 186
rename model and unlink from duplicates using BREAKOUT, MODELS in DRAW and SPACE modes 297
rename screen, IMAGE in DRAW and SPACE modes 196
rename selected identifiers, IDENTIFY in DRAW and SPACE modes 186
rename sheet in sheet file, PLOT in DRAW and and SPACE modes 334
rename symbols (or modify symbol comment), SYMBOL in DRAW mode 451
rename windows, IMAGE in DRAW and SPACE modes 193
RENAME, AUXVIEW in DRAW mode 20
RENAME, AXIS in DRAW mode 51
RENAME, FILE in DRAW and SPACE modes 159
renumber identifiers, IDENTIFY in DRAW and SPACE modes 188
repeat line with LINES COMMANDS window, FILE in DRAW and SPACE modes 161
replace all references to pattern with different pattern, PATTERN in DRAW mode 315
replace branch in history of solid, SOLIDE in SPACE mode 402
replace branch in history of solid, SOLIDM in SPACE mode 426
replace ditto or copy with symmetry around X or Y axis, DETAIL in DRAW mode 97

replace ditto with another ditto, DETAIL in DRAW mode 100
replace ditto with another, DETAIL in 3D SPACE mode 108
replace reference with parametric relationship, PARAM3D in SPACE mode 303
replace relationship with reference element, PARAM3D in SPACE mode 303
replace symbol with another symbol, SYMBOL in DRAW mode 450
REPLACE, PATTERN in DRAW mode 315
reposition complete dimension, PARAM3D in SPACE mode 307
reposition dimension value, PARAM3D in SPACE mode 307
reset all display parameters to default settings, GRAPHIC in SPACE mode 175
reset filters applied in any workspace, LAYER in DRAW and SPACE modes 210
reset or empty selected group, GROUP in DRAW mode 181
reset or empty selected group, GROUP in SPACE mode 185
reset original status of modal parameters, TEXTD2 in DRAW mode 481
restore color table, STANDARD in DRAW and SPACE modes 440
restore default limits of DRAW elements, LIMIT1 in 2D SPACE mode 225
restore default limits of DRAW elements, LIMIT1 in 3D SPACE mode 232
restore default limits of DRAW elements, LIMIT1 in DRAW mode 218
restore deleted section callout, AUXVIEW2 41
restore model saved with SV button, FILE in DRAW and SPACE modes 159
restore previously deleted dimension, PARAM3D in SPACE mode 306
restore solid following UN_USE, AUXVIEW2 31
restore solid to state of last update, SOLIDM in SPACE mode 430
RESTORE, SOLIDM in SPACE mode 430
retain selected items in model, KEEP in DRAW and SPACE modes 204
retrieve identifiers from deleted elements, IDENTIFY in DRAW and SPACE modes 188
retrieve keyboard, FILE in DRAW and SPACE modes 163
rotate ditto or copy, DETAIL in 3D SPACE mode 105
rotate ditto or copy, DETAIL in DRAW mode 97
rotate image horizontally, IMAGE in DRAW and SPACE modes 191
rotate image vertically, IMAGE in DRAW and SPACE modes 191
rotate image, unspecified rotation, IMAGE in DRAW and SPACE modes 191
rotate line, LINE in 2D SPACE mode 273
rotate line, LINE in DRAW mode 253
rotate models, MODELS in DRAW and SPACE modes 295
rotate plane, IMAGE in DRAW and SPACE modes 191
rotate symbol, SYMBOL in DRAW mode 447
rotate view, AUXVIEW in DRAW mode 19
rotate view, AUXVIEW2 33
rubber banding, LINE in 2D SPACE mode 259
rubber banding, LINE in 3D SPACE mode 276

S

SAME, for DRAW generated graphical attributes in AUXVIEW2 47
save color table, STANDARD in DRAW and SPACE modes 440
scale ditto or copy, DETAIL in 3D SPACE mode 105
SECTION CUT, for view callout lines in AUXVIEW2 45
SECTION, COMBIVU in DRAW mode 61
SECTION, for view callout lines in AUXVIEW2 45
select application elements to keep, KEEP in DRAW and SPACE modes 204
select application sets to keep, KEEP in DRAW and SPACE modes 204
select applications, KEEP in DRAW and SPACE modes 204
select details to keep, KEEP in DRAW and SPACE modes 203
select details to merge, MERGE in DRAW and SPACE modes 293
select elements inside or overlapping trap boundary, GROUP in DRAW mode 179
select elements inside or overlapping trap boundary, GROUP in SPACE mode 183
select elements inside trap boundary, GROUP in DRAW mode 179
select elements inside trap boundary, GROUP in SPACE mode 183
select elements outside or overlapping trap boundary, GROUP in DRAW mode 179
select elements outside or overlapping trap boundary, GROUP in SPACE mode 183
select elements outside trap boundary, GROUP in DRAW mode 179
select elements outside trap boundary, GROUP in SPACE mode 184
select elements to keep, KEEP in DRAW and SPACE modes 203
select elements to merge, MERGE in DRAW and SPACE modes 292
select family containing objects to be handled, LIBRARY in DRAW and SPACE modes 214
select library containing objects to be handled, LIBRARY

in DRAW and SPACE modes 213
select line or curve element to define end point, POINT in 3D SPACE mode 351, 360
select line or curve element to define end point, POINT in DRAW mode 339
select pattern type, PATTERN in DRAW mode 313, 314
select points used in creating lines in LINE, LINE in DRAW mode 238
select receiving workspace, DETAIL in 3D SPACE mode 110
select sets to keep, KEEP in DRAW and SPACE modes 203
select sets to merge, MERGE in DRAW and SPACE modes 293
select SPACE text or element to which text is associated, TEXT in SPACE mode 459
select views to merge, MERGE in DRAW and SPACE modes 292
send elements of receiving workspace, DETAIL in 3D SPACE mode 110
separate screen, IMAGE in DRAW and SPACE modes 194
set available unspecified line types, STANDARD in DRAW and SPACE modes 442
set display standards, SOLIDM in SPACE mode 429
set total workspace or current set and view, GROUP in DRAW mode 180
set total workspace or current set and view, GROUP in SPACE mode 184
SHOW or NO SHOW dimensions, PARAM3D in SPACE mode 306
show text as box, TEXTD2 in DRAW mode 480
show text modified with VISUALTN options, TEXTD2 in DRAW mode 480
SHOW, TEXT in SPACE mode 460
smooth curves with ANALYSIS in 3D SPACE mode 12
SPACE element identifiers, list of 512–513
specify group, GROUP in DRAW mode 178
specify group, GROUP in SPACE mode 182
SPINE, CURVE2 in 3D SPACE mode 92
SPLINE, CURVE2 in DRAW mode 69–70
STANDARD, for DRAW generated graphical attributes in AUXVIEW2 47
STANDARD, TEXT in DRAW mode 454–456
store current transformation, TRANSFOR in 2D SPACE mode 496
store current transformation, TRANSFOR in 3D SPACE mode 505
store keyboard, FILE in DRAW and SPACE modes 164
store screen, IMAGE in DRAW and SPACE modes 195
store windows, IMAGE in DRAW and SPACE modes 193
STORE, TRANSFOR in 2D SPACE mode 496
STORE, TRANSFOR in 3D SPACE mode 505–506
subtract closed shape from another, SHAPE in DRAW mode 375
swap axes of axis system, AXIS in DRAW mode 50
swap axes, AXIS in 2D SPACE mode 53
swap two axes, AXIS in 3D SPACE mode 58
SWAP, AXIS in 2D SPACE mode 53
SWAP, AXIS in 3D SPACE mode 58
SWAP, AXIS in DRAW mode 50
switch layers, IDENTIFY in DRAW and SPACE modes 187

T

T.NODE, TEXT in DRAW mode 453–454
TANGENT, POINT in 2D SPACE mode 359
TANGENT, POINT in 3D SPACE mode 368
TANGENT, POINT in DRAW mode 346–347
TEXTURE in AUXVIEW2 46
TGT CONT, CURVE2 in 2D SPACE mode 83
TGT CONT, CURVE2 in 3D SPACE mode 89
THICK, for fillet representation in AUXVIEW2 47
THROUGH, PLANE in SPACE mode 318–319
transfer all DRAW elements, GROUP in DRAW mode 181
transfer all SPACE elements, GROUP in SPACE mode 185
transfer elements between views, AUXVIEW in DRAW mode 21
transfer elements from layer of selected element, GROUP in DRAW mode 180
transfer elements from layer of selected element, GROUP in SPACE mode 184
transfer elements from layer to layer, LAYER in DRAW and SPACE modes 211
transfer elements from one workspace to another, DETAIL in DRAW mode 102
transfer elements in group containing selected element, GROUP in DRAW mode 180
transfer elements in group containing selected element, GROUP in SPACE mode 184
transfer elements into current set, SETS in DRAW and SPACE modes 370
transfer elements of family of selected element, GROUP in DRAW mode 180
transfer elements of family of selected element, GROUP in SPACE mode 184
transfer elements of same view as selected element, GROUP in DRAW mode 181
transfer elements of selected element set, GROUP in DRAW mode 181
transfer elements of selected element set, GROUP in SPACE mode 185
transfer elements of specified type, GROUP in DRAW mode 180
transfer elements of specified type, GROUP in SPACE mode 184
transfer elements with same attributes as selected element, GROUP in DRAW mode 180
transfer elements with same attributes as selected element,

GROUP in SPACE mode 185
transfer elements with specified identifier, GROUP in DRAW mode 180
transfer elements with specified identifier, GROUP in SPACE mode 184
transfer individual elements, GROUP in DRAW mode 180
transfer individual elements, GROUP in SPACE mode 184
transfer patterns into show or no show mode, PATTERN in DRAW mode 316
transfer relationships among elements, PARAM3D in SPACE mode 302
TRANSFER, AUXVIEW in DRAW mode 21
TRANSFER, DETAIL in 3D SPACE mode 110–111
TRANSFER, DETAIL in DRAW mode 102
TRANSFER, SETS in DRAW and SPACE modes 370
transform compact symbol occurrences into standard symbol occurrences, SYMBOL in DRAW mode 451
transform ditto into copy, DETAIL in DRAW mode 98
transform solid geometry, SOLIDM in SPACE mode 422
transform standard symbol occurrences into compact symbol occurrences, SYMBOL in DRAW mode 451
transform symbol occurrence into copy, SYMBOL in DRAW mode 448
transform to wireframe via rotation, CURVE2 in 3D SPACE mode 90
transform to wireframe via translation, CURVE2 in 3D SPACE mode 90
transform to wireframe, using stored transformation, CURVE2 in 3D SPACE mode 90
translate ditto or copy, DETAIL in 3D SPACE mode 105
translate ditto or copy, DETAIL in DRAW mode 96
translate line segment, LINE in 2D SPACE mode 273
translate line segment, LINE in 3D SPACE mode 287
translate line segment, LINE in DRAW mode 253
translate models, MODELS in DRAW and SPACE modes 295
translate screen, IMAGE in DRAW and SPACE modes 194
translate symbol, SYMBOL in DRAW mode 446
translate unlimited line, LINE in 2D SPACE mode 273
translate unlimited line, LINE in 3D SPACE mode 287
translate unlimited line, LINE in DRAW mode 253
translate view and clipping frame, AUXVIEW in DRAW mode 19
translate view center but not clipping frame, AUXVIEW in DRAW mode 19
translate view to new position, AUXVIEW2 29
trim two elements simultaneously to intersection point, LIMIT1 in 2D SPACE mode 224
trim two elements simultaneously to intersection point, LIMIT1 in 3D SPACE mode 231
trim two elements simultaneously to intersection point, LIMIT1 in DRAW mode 217

U

uncut solid in section or cut view, AUXVIEW2 31
undo blink (highlight) of elements, GRAPHIC in DRAW mode 167
Undo blink (highlight) of elements, GRAPHIC in SPACE mode 172
undo transformations, TRANSFOR in DRAW mode 489
UNDO, TRANSFOR in DRAW mode 489
unfix axis, AXIS in 2D SPACE mode 54
unfix axis, AXIS in 3D SPACE mode 58
unfix axis, AXIS in DRAW mode 50
UNFIXED, AXIS in 2D SPACE mode 54
UNFIXED, AXIS in 3D SPACE mode 58
UNFIXED, AXIS in DRAW mode 50
union two closed shapes, SHAPE in DRAW mode 375
UNLINK, PARAM3D in SPACE mode 303
unlock availability of descriptions, DRW STD 146
UPD ALL, AUXVIEW2 48
update all views with single command, AUXVIEW2 48
update back panel, AUXVIEW2 37
update boundary representation (B-REP) of solid, SOLIDE in SPACE mode 407
update element identifiers after merge, IDENTIFY in DRAW and SPACE modes 187
update external (library) detail vis-a-vis current library, DETAIL in DRAW mode 99
Update external (library) detail, DETAIL in 3D SPACE mode 108
update external (library) symbol vis-a-vis current library, SYMBOL in DRAW mode 450
update individual view frame, AUXVIEW2 36
update individual view, AUXVIEW2 26, 33
update library detail to show changes to object in model, LIBRARY in DRAW and SPACE modes 215
update pattern when geometry modified, PATTERN in DRAW mode 317
update solid boundary representation (B-REP), SOLIDM in SPACE mode 430
update text affected by change to program, attributes, or identifier, TEXTD2 in DRAW mode 477
update view for breakout manipulation, AUXVIEW2 39
update view for clipping frame manipulation, AUXVIEW2 37
update view for moving and analyzing dimensions, AUXVIEW2 42
update view for section manipulation, AUXVIEW2 40
update view for text manipulation, AUXVIEW2 35
UPDATE, IDENTIFY in DRAW and SPACE modes 187
UPDATE, LIBRARY in DRAW and SPACE modes 215–216
UPDATE, PATTERN in DRAW mode 317
UPDATE, SOLIDE in SPACE mode 407
UPDATE, SOLIDM in SPACE mode 430

use diagonal point and number of lines in grid, LINE in DRAW mode 254
use IUA commands transparently, IUA 199
USE, AUXVIEW2 25–43
utilities, list of 509–510

V

VANISH, SYMBOL in DRAW mode 452
verify elements displayed in blink mode, GRAPHIC in SPACE mode 176
verify elements displayed in blink mode, GRAPHIC in DRAW mode 169
verify elements displayed in steady mode, GRAPHIC in DRAW mode 169
verify elements displayed in steady mode, GRAPHIC in SPACE mode 176
verify elements with same color, GRAPHIC in DRAW mode 168
verify elements with same color, GRAPHIC in SPACE mode 175
verify elements with same display attributes, GRAPHIC in DRAW mode 169
verify elements with same display attributes, GRAPHIC in SPACE mode 176
verify elements with same linetype display attributes, GRAPHIC in DRAW mode 168
verify elements with same linetype display attributes, GRAPHIC in SPACE mode 175
verify elements with same point type display attributes, GRAPHIC in DRAW mode 168
verify elements with same point type display attributes, GRAPHIC in SPACE mode 175
verify elements with same thickness, GRAPHIC in DRAW mode 168
verify elements with same thickness, GRAPHIC in SPACE mode 176
verify elements with standard display attributes, GRAPHIC in DRAW mode 169
verify elements with standard display attributes, GRAPHIC in SPACE mode 176
verify nature of dittos, DETAIL in 3D SPACE mode 109
verify nature of dittos, DETAIL in DRAW mode 100
verify nature of symbols, SYMBOL in DRAW mode 450
VERIFY, GRAPHIC in DRAW mode 168–169
VERIFY, GRAPHIC in SPACE mode 175–176
vertical definition, LINE in 2D SPACE mode 259
vicinity selection, LINE in 2D SPACE mode 258
vicinity selection, LINE in 3D SPACE mode 275
VICINITY SELECTION, POINT in 3D SPACE mode 351, 360
VICINITY SELECTION, POINT in DRAW mode 339
view line number in comment with LINE NUMBER window, FILE in DRAW and SPACE modes 162
view model before changing name, FILE in DRAW and SPACE modes 159
view model before copying, FILE in DRAW and SPACE modes 157
view model before deletion, FILE in DRAW and SPACE modes 158
VISU STD, SOLIDM in SPACE mode 429
visualize available descriptions, DRW STD 142

W

write objects from model into library, LIBRARY in DRAW and SPACE modes 214
write or save model to current model file, FILE in DRAW and SPACE modes 157
WRITE, FILE in DRAW and SPACE modes 157
WRITE, LIBRARY in DRAW and SPACE modes 214–215

Z

zoom screen frame, IMAGE in DRAW and SPACE modes 193
zoom screen, IMAGE in DRAW and SPACE modes 194